Graduate Texts in Mathematics   13

Graduate Texts in Mathematics

continued after index

Frank W. Anderson   Kent R. Fuller

Rings and Categories of Modules

Second Edition

Springer-Verlag
New York Berlin Heidelberg London Paris
Tokyo Hong Kong Barcelona Budapest

Frank W. Anderson
University of Oregon
Department of Mathematics
Eugene, OR 97403
USA

Kent R. Fuller
University of Iowa
Department of Mathematics
Iowa City, IA 52242
USA

Mathematics Subject Classifications (1991): 13-01, 16-01

Library of Congress Cataloging-in-Publication Data
Anderson, Frank W. (Frank Wylie), 1928–
     Rings and categories of modules / Frank W. Anderson, Kent R.
Fuller.—2nd ed.
        p.   cm.—(Graduate texts in mathematics; 13)
     Includes bibliographical references and index.
     ISBN 0-387-97845-3
     1. Modules (Algebra)   2. Rings (Algebra)   3. Categories
(Mathematics)   I. Fuller, Kent R.   II. Title.   III. Series.
     QA247.A55   1992
512′.4—dc20   92-10019

Printed on acid-free paper.

Production managed by Bill Imbornoni; manufacturing supervised by Vincent Scelta.
Printed and bound by R.R. Donnelley & Sons, Harrisonburg, VA.
Printed in the United States of America.

9 8 7 6 5 4 3 2 1

ISBN 0-387-97845-3 Springer-Verlag New York Berlin Heidelberg
ISBN 3-540-97845-3 Springer-Verlag Berlin Heidelberg New York

Preface

This book is intended to provide a reasonably self-contained account of a major portion of the general theory of rings and modules suitable as a text for introductory and more advanced graduate courses. We assume the familiarity with rings usually acquired in standard undergraduate algebra courses. Our general approach is categorical rather than arithmetical. The continuing theme of the text is the study of the relationship between the one-sided ideal structure that a ring may possess and the behavior of its categories of modules.

Following a brief outline of set-theoretic and categorical foundations, the text begins with the basic definitions and properties of rings, modules and homomorphisms and ranges through comprehensive treatments of direct sums, finiteness conditions, the Wedderburn-Artin Theorem, the Jacobson radical, the hom and tensor functions, Morita equivalence and duality, decomposition theory of injective and projective modules, and semiperfect and perfect rings. In this second edition we have included a chapter containing many of the classical results on artinian rings that have helped to form the foundation for much of the contemporary research on the representation theory of artinian rings and finite dimensional algebras. Both to illustrate the text and to extend it we have included a substantial number of exercises covering a wide spectrum of difficulty. There are, of course, many important areas of ring and module theory that the text does not touch upon. For example, we have made no attempt to cover such subjects as homology, rings of quotients, or commutative ring theory.

This book has evolved from our lectures and research over the past several years. We are deeply indebted to many of our students and colleagues for their ideas and encouragement during its preparation. We extend our sincere thanks to them and to the several people who have helped with the preparation of the manuscripts for the first two editions, and/or pointed out errors in the first.

Finally, we apologize to the many authors whose works we have used but not specifically cited. Virtually all of the results in this book have appeared in some form elsewhere in the literature, and they can be found either in the books and articles that are listed in our bibliography, or in those listed in the collective bibliographies of our citations.

Eugene, OR
Iowa City, IA

Frank W. Anderson
Kent R. Fuller
January 1992

Contents

Rings and Categories of Modules

§0. Preliminaries

In this section is assembled a summary of various bits of notation, terminology, and background information. Of course, we reserve the right to use variations in our notation and terminology that we believe to be self-explanatory without the need of any further comment.

A word about categories. We shall deal only with very special concrete categories and our use of categorical algebra will be really just terminological —at a very elementary level. Here we provide the basic terminology that we shall use and a bit more. We emphasize though that our actual use of it will develop gradually and, we hope, naturally. There is, therefore, no need to try to master it at the beginning.

0.1. Functions. Usually, but not always, we will write functions "on the left". That is, if f is a function from A to B, and if $a \in A$, we write $f(a)$ for the value of f at a. Notation like $f : A \to B$ denotes a function from A to B. The elementwise action of a function $f : A \to B$ is described by

$$f : a \mapsto f(a) \qquad (a \in A).$$

Thus, if $A' \subseteq A$, the *restriction* $(f \mid A')$ of f to A' is defined by

$$(f \mid A') : a' \mapsto f(a') \qquad (a' \in A').$$

Given $f : A \to B$, $A' \subseteq A$, and $B' \subseteq B$, we write

$$f(A') = \{f(a) \mid a \in A'\} \qquad \text{and} \qquad f^\leftarrow(B') = \{a \in A \mid f(a) \in B'\}.$$

For the *composite* or *product* of two functions $f : A \to B$ and $g : B \to C$ we write $g \circ f$, or when no ambiguity is threatened, just gf; thus, $g \circ f : A \to C$ is defined by $g \circ f : a \mapsto g(f(a))$ for all $a \in A$. The resulting operation on functions is associative wherever it is defined. The *identity function* from A to itself is denoted by 1_A. The set of all functions from A to B is denoted by B^A or by $Map(A, B)$:

$$B^A = Map(A, B) = \{f \mid f : A \to B\}.$$

So A^A is a monoid (= semigroup with identity) under the operation of composition.

A diagram of sets and functions *commutes* or is *commutative* in case travel around it is independent of path. For example, the first diagram commutes iff $f = hg$. If the second is commutative,

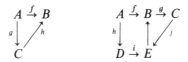

then in particular, travel from A to E is independent of path, whence $jgf = ih$.

A function $f : A \to B$ is *injective* (*surjective*) or is an *injection* (*surjection*) in case it has a *left (right) inverse* $f':B \to A$; that is, in case $f'f = 1_A$ ($ff' = 1_B$) for some $f':B \to A$. So (see (0.2)) $f:A \to B$ is injective (surjective) iff it is one-to-one (onto B). A function $f:A \to B$ is *bijective* or a *bijection* in case it is both injective and surjective; that is, iff there exists a (necessarily unique) inverse $f^{-1}:B \to A$ with $ff^{-1} = 1_B$ and $f^{-1}f = 1_A$.

If $A \subseteq B$, then the function $i = i_{A \subseteq B}:A \to B$ defined by $i = (1_B \mid A):a \mapsto a$ for all $a \in A$ is called the *inclusion map* of A in B. Note that if $A \subseteq B$ and $A \subseteq C$, and if $B \neq C$, then $i_{A \subseteq B} \neq i_{A \subseteq C}$. Of course $1_A = i_{A \subseteq A}$.

With every pair $(0, 1)$ there is a *Kronecker delta*; that is, a function $\delta:(\alpha, \beta) \mapsto \delta_{\alpha\beta}$ on the class of all ordered pairs defined by

$$\delta_{\alpha\beta} = \begin{cases} 1 & \text{if } \alpha = \beta \\ 0 & \text{if } \alpha \neq \beta. \end{cases}$$

Whenever we use a Kronecker delta, the context will make clear our choice of the pair $(0, 1)$.

0.2. The Axiom of Choice. Let A be a set, let $\mathcal{S}$ be a collection of non-empty subsets of B, and let σ be a function from A to $\mathcal{S}$. Then the *Axiom of Choice* states that there is a function $g:A \to B$ such that

$$g(a) \in \sigma(a) \qquad (a \in A).$$

Suppose now that $f:B \to A$ is onto A; that is, $f(B) = A$. Then for each $a \in A$, there is a non-empty subset $\sigma(a) = f^{\leftarrow}(\{a\}) \subseteq B$. Applying the Axiom of Choice to A, the function $\sigma:a \mapsto \sigma(a)$, and the collection $\mathcal{S}$ of subsets of B produces a right inverse g for f, so as claimed in (0.1), f is surjective.

Let $\sim$ be an equivalence relation on a set A. A subset R of A is a (*complete*) *irredundant set of representatives* of the relation $\sim$ in case for each $a \in A$ there is a unique $\sigma(a) \in R$ such that $a \sim \sigma(a)$. The Axiom of Choice guarantees the existence of such a set of representatives for each equivalence relation.

0.3. Cartesian Products. A function $\sigma:A \to X$ will sometimes be called an *indexed set* (*in X indexed by A*) or an *A-tuple* (in X) and will be written as

$$\sigma = (x_\alpha)_{\alpha \in A}$$

where $x_\alpha = \sigma(\alpha)$. If $A = \{1, ..., n\}$, then we also use the standard variation $(x_\alpha)_{\alpha \in A} = (x_1, ..., x_n)$. Let $(X_\alpha)_{\alpha \in A}$ be an indexed set of non-empty subsets of a set X. Then the (*cartesian*) *product* of $(X_\alpha)_{\alpha \in A}$ is

$$\mathsf{X}_A X_\alpha = \{\sigma:A \to X \mid \sigma(\alpha) \in X_\alpha \quad (\alpha \in A)\}.$$

That is, $\mathsf{X}_A X_\alpha$ is just the set of all A-tuples $(x_\alpha)_{\alpha \in A}$ such that $x_\alpha \in X_\alpha$ ($\alpha \in A$). By the Axiom of Choice $\mathsf{X}_A X_\alpha$ is non-empty. If $A = \{1, ..., n\}$, then we allow the notational variation

$$\mathsf{X}_A X_\alpha = X_1 \times ... \times X_n.$$

Note that if $X = X_\alpha$ ($\alpha \in A$), then the cartesian product $\mathsf{X}_A X_\alpha$ is simply X^A, the set of all functions from A to X. For each $\alpha \in A$ the α-*projection*

$\pi_\alpha : \mathsf{X}_A X_\alpha \to X_\alpha$ is defined via

$$\pi_\alpha : \sigma \mapsto \sigma(a) \qquad (\sigma \in \mathsf{X}_A X_\alpha).$$

In A-tuple notation, $\pi_\alpha((x_\beta)_{\beta \in A}) = x_\alpha$. An easy application of the Axiom of Choice shows that each π_α is surjective. Observe that if σ and σ' are in this cartesian product, then $\sigma = \sigma'$ iff $\pi_\alpha \sigma = \pi_\alpha \sigma'$ for all $\alpha \in A$. This fact establishes the uniqueness assertion in the following result. This result, whose easy proof we omit, is used in making certain definitions *coordinatewise*.

0.4. *Let* $(X_\alpha)_{\alpha \in A}$ *be an indexed set of non-empty sets, let Y be a set, and for each $\alpha \in A$, let* $f_\alpha : Y \to X_\alpha$. *Then there is a unique* $f : Y \to \mathsf{X}_A X_\alpha$ *such that* $\pi_\alpha f = f_\alpha$ *for each* $\alpha \in A$.

0.5. Posets and Lattices. A relation $\leq$ on a set P is a *partial order* on P in case it is reflexive ($a \leq a$), transitive ($a \leq b$ and $b \leq c \Rightarrow a \leq c$), and anti-symmetric ($a \leq b$ and $b \leq a \Rightarrow a = b$). A pair $(P, \leq)$ consisting of a set and a partial order on the set is called a *partially ordered set* or a *poset*. If the partial order is a *total order* ($a \leq b$ or $b \leq a$ for every pair a, b), then the poset is a *chain*. If $(P, \leq)$ is a poset and if $P' \subseteq P$, then $(P', \leq')$ is a subposet in case $\leq'$ is the restriction of $\leq$ to P'; of course, this requires that $(P', \leq')$ be a poset. Henceforth, we will usually identify a poset $(P, \leq)$ with its underlying set P.

Let P be a poset and let $A \subseteq P$. An element $e \in A$ is a *greatest* (*least*) element of A in case $a \leq e$ ($e \leq a$) for all $a \in A$. Not every subset of a poset has a greatest or a least element, but clearly if one does exist, it is unique. (See Example (2) below.) An element $b \in P$ is an *upper bound* (*lower bound*) for A in case $a \leq b$ ($b \leq a$) for all $a \in A$. So a greatest (least) element, if it exists, is an upper (lower) bound for A. If the set of upper bounds of A has a least element, it is called the *least upper bound* (lub), *join*, or *supremum* (sup) of A; if the set of lower bounds has a greatest element, it is called the *greatest lower bound* (glb), *meet*, or *infimum* (inf) of A. A *lattice* (*complete lattice*) is a poset P in which every pair (every subset) of P has both a least upper bound and a greatest lower bound in P.

Examples. (1) Let X be a set. The *power set* of X is the set $\mathscr{P}(X)$ of all subsets of X. Then $\mathscr{P}(X)$ is certainly a poset under the partial order of set inclusion. This poset is a complete lattice for if $\mathscr{A}$ is a subset of $\mathscr{P}(X)$, then its join in $\mathscr{P}(X)$ is its union $\cup \mathscr{A}$ and its meet in $\mathscr{P}(X)$ is its intersection $\cap \mathscr{A}$.

(2) Let X be a set and let $\mathscr{F}(X)$ be the set of all finite subsets of X. Then $\mathscr{F}(X)$ is a poset under set inclusion, and it is a lattice for if $A, B \in \mathscr{F}(X)$, then $A \cup B$ and $A \cap B$ are their join and meet. Since these are also join and meet of A, B in $\mathscr{P}(X)$, it follows that $\mathscr{F}(X)$ is a sublattice of $\mathscr{P}(X)$. But note that if X is infinite, $\mathscr{F}(X)$ is not complete.

(3) Let X be the closed unit interval on the real line. Then the set $\mathscr{J}(X)$ of all closed intervals in X is certainly a subposet of $\mathscr{P}(X)$. Also the inter-section ($=$ meet in $\mathscr{P}(X)$) of any subset of $\mathscr{J}(X)$ is again in $\mathscr{J}(X)$. The convex closure of the union of any subset $\mathscr{A}$ of $\mathscr{J}(X)$ is in $\mathscr{J}(X)$ and is clearly the join of $\mathscr{A}$ in $\mathscr{J}(X)$. So $\mathscr{J}(X)$ is a complete lattice. But $\mathscr{J}(X)$ is not a sublattice

of $\mathscr{P}(X)$ precisely because the join in $\mathscr{J}(X)$ of some pairs of elements of $\mathscr{J}(X)$ is not their join ($=$ union) in $\mathscr{P}(X)$.

(4) Let X be a two-dimensional real vector space and let $\mathscr{S}(X)$ be the set of all subspaces. Then $\mathscr{S}(X)$ is a subposet of $\mathscr{P}(X)$, and the intersection of any subset of $\mathscr{S}(X)$ is again in $\mathscr{S}(X)$. The join in $\mathscr{S}(X)$ of any subset $\mathscr{A}$ of $\mathscr{S}(X)$ is the subspace spanned by the union $\cup \mathscr{A}$ (not necessarily $\cup \mathscr{A}$ itself). So $\mathscr{S}(X)$ is a complete lattice but it is not a sublattice of $\mathscr{P}(X)$.

Let P be a lattice. Then each pair $a, b \in P$ has both a join and a meet in P; let us denote these by $a \vee b$ and $a \wedge b$, respectively. Then the maps $\vee$ and $\wedge$ from $P \times P$ to P defined by

$$(a, b) \mapsto a \vee b \qquad \text{and} \qquad (a, b) \mapsto a \wedge b$$

are binary operations on P. It is easy to see that both $(P, \vee)$ and $(P, \wedge)$ are commutative semigroups with

$$a \vee a = a = a \wedge a \qquad (a \in P).$$

The lattice is said to be *modular* in case it satisfies the *modularity condition*: For all $a, b, c \in P$

$$a \geq b \text{ implies } a \wedge (b \vee c) = b \vee (a \wedge c).$$

Most lattices we encounter will be modular (but note (3) above). The lattice is *distributive* in case it satisfies the stronger property: For all $a, b, c \in P$

$$a \wedge (b \vee c) = (a \wedge b) \vee (a \wedge c).$$

Examples (1) and (2) above are distributive, but (4) is not.

0.6. *A partially ordered set P is a complete lattice if P has a join (i.e., P contains a greatest element) and every non-empty subset of P has a meet in P.*

Proof. It will suffice to prove that if $B \subseteq P$, then B has a join in P. Let $e \in P$ be the greatest element of P. Then $e \geq x$ for all $x \in P$. In particular, the set of upper bounds of B is non-empty, so it has a meet. Clearly this meet of the upper bounds of B is an upper bound of B and hence the join of B. $\quad\square$

0.7. Lattice Homomorphisms. Let P and P' be posets. A map $f: P \to P'$ is *order preserving (order reversing)* in case whenever $a \leq b$ in P, then $f(a) \leq f(b)$ $(f(b) \leq f(a))$ in P'. If P and P' are lattices, then f is a *lattice homomorphism (lattice antihomomorphism)* in case whenever $a, b \in P$,

$$f(a \vee b) = f(a) \vee f(b) \qquad (f(a \vee b) = f(a) \wedge f(b))$$
$$f(a \wedge b) = f(a) \wedge f(b) \qquad (f(a \wedge b) = f(a) \vee f(b)).$$

It is easy to see (using $a \leq b \Leftrightarrow a = a \wedge b$) that a lattice-homomorphism is order preserving. The converse, however, is false (try the inclusion map $\mathscr{J}(X) \to \mathscr{P}(X)$ in example (3) of (0.5)). A bijective lattice (anti-) homomorphism is a *lattice (anti-)isomorphism*. It is a simple exercise to prove the following useful test:

0.8. *Let P and P' be lattices, and let $f:P \to P'$ be bijective with inverse $f^{-1}:P' \to P$. Then f is a lattice isomorphism if and only if both f and f^{-1} are order preserving.*

0.9. The Maximal Principle. Let P be a poset. An element $m \in P$ is *maximal (minimal)* in P in case $x \in P$ and $x \geq m$ $(x \leq m)$ implies $x = m$. Clearly, a greatest (least) element in P, if it exists, is maximal (minimal) in P; on the other hand, a poset may have many maximal (minimal) elements and no greatest (least) element.

A poset P is *inductive* in case every subchain of P has an upper bound in P; that is, for every subset C of P that is totally ordered by the partial ordering of P, there is an element of P greater than or equal to every element of C. The *Maximal Principle* (frequently called Zorn's Lemma) is an equivalent form of the Axiom of Choice (see Stoll [63] for the details). It states:

Every non-empty inductive poset has at least one maximal element.

0.10. Cardinal Numbers. Two sets A and B are *cardinally equivalent* or have the *same cardinal* in case there is a bijection from A to B (and hence one from B to A). Since this clearly defines an equivalence relation, the class of all sets (see (0.11)) can be partitioned into its classes of cardinally equivalent sets. These classes are the *cardinal numbers*. The class of a set A is denoted by $card\ A$:

$$card\ A = \{B \mid \text{there is a bijection } A \to B\}.$$

Given two sets A and B we write

$$card\ A \leq card\ B$$

in case there is an injection from A to B (or, equivalently, a surjection from B to A). Clearly this is independent of the representatives A and B. Given sets A and B there is always an injection from one to the other. The Cantor–Schröder–Bernstein Theorem states that

If $card\ A \leq card\ B$ and $card\ B \leq card\ A$, then $card\ A = card\ B$.

Thus the relation $\leq$ is a total order on the class of cardinal numbers.

Let $\mathbb{N} = \{1, 2, ...\}$ be the natural numbers. Its cardinality is often denoted by $card\ \mathbb{N} = \aleph_0$. A set A is *finite* if $card\ A < card\ \mathbb{N}$. Of course, $card(\{1, ..., n\}) = n$ and $card\ \varnothing = 0$. If $card\ A \leq card\ \mathbb{N}$, then A is *countable*. If $card\ A \geq card\ \mathbb{N}$, then A is *infinite*.

The operations of cardinal arithmetic are given by

$$card\ A + card\ B = card((A \times \{1\}) \cup (B \times \{2\}))$$

$$card\ A \cdot card\ B = card(A \times B)$$

$$(card\ A)^{(card\ B)} = card(A^B)$$

If A and B are finite sets these operations agree with ordinary addition, multiplication and exponentiation. Moreover, they satisfy:

(1) *If A is infinite then, card A + card B = max{card A, card B}.*

(2) *If A is infinite and B $\neq \varnothing$, then*

$$card\ A \cdot card\ B = \max\{card\ A, card\ B\}.$$

(3) *For all sets A, B, and C,*

$$((card\ A)^{(card\ B)})^{(card\ C)} = (card\ A)^{(card\ B) \cdot (card\ C)}.$$

(4) *If card B ≥ 2, then $(card\ B)^{(card\ A)} > card\ A$.*

It is easy to establish the existence of a bijection between the power set $\mathscr{P}(A)$ and the set of functions from A to $\{1, 2\}$. Thus $card(\mathscr{P}(A)) = 2^{(card\ A)} > card\ A$. However, the set of finite subsets of any infinite set A has the same cardinality as A. For further details see Stoll [63].

0.11. Categories. The term "class", like that of "set", will be undefined. Every set is a class, and there is a class containing all sets. Note that if A is a set and $\mathscr{C}$ is a class, then an indexed class $(A_C)_{C \in \mathscr{C}}$ in $\mathscr{P}(A)$ has a union and an intersection in A. Let $\mathscr{C}$ be a class for each pair $A, B \in \mathscr{C}$, let $mor_C(A, B)$ be a set; write the elements of $mor_C(A, B)$ as "arrows" $f: A \to B$ for which A is called the *domain* and B the *codomain*. Finally, suppose that for each triple $A, B, C \in \mathscr{C}$ there is a function

$$\circ : mor_C(B, C) \times mor_C(A, B) \to mor_C(A, C).$$

We denote the arrow assigned to a pair

$$g: B \to C \qquad f: A \to B$$

by the arrow $gf: A \to C$. The system $\mathsf{C} = (\mathscr{C}, mor_C, \circ)$ consisting of the class $\mathscr{C}$, the map $mor_C: (A, B) \mapsto mor_C(A, B)$, and the rule $\circ$ is a *category* in case:

(C.1) For every triple $h: C \to D, g: B \to C, f: A \to B$,

$$h \circ (g \circ f) = (h \circ g) \circ f.$$

(C.2) For each $A \in \mathscr{C}$, there is a unique $1_A \in mor_C(A, A)$ such that if $f: A \to B$ and $g: C \to A$, then

$$f \circ 1_A = f \qquad \text{and} \qquad 1_A \circ g = g.$$

If C is a category, then the elements of the class $\mathscr{C}$ are called the *objects* of the category, the "arrows" $f: A \to B$ are called the *morphisms*, the partial map $\circ$ is called the *composition*, and the morphisms 1_A are called the *identities* of the category. A morphism $f: A \to B$ in C is called an *isomorphism* in case there is a (necessarily unique) morphism $f^{-1}: B \to A$ in C such that $f^{-1} \circ f = 1_A$ and $f \circ f^{-1} = 1_B$.

For our purpose the most interesting categories are certain "concrete" categories. Let $\mathsf{C} = (\mathscr{C}, mor_C, \circ)$ be a category. Then C is *concrete* in case there is a function u from $\mathscr{C}$ to the class of sets such that for each $A, B \in \mathscr{C}$

$$mor_C(A, B) \subseteq Map(u(A), u(B)),$$

$$1_A = 1_{u(A)},$$

and such that ∘ is the usual composition of functions. Here an isomorphism $f: A \to B$ is a bijection $f: u(A) \to u(B)$.

Examples. (1) Let $\mathscr{S}$ be the class of all sets; for each $A, B \in \mathscr{S}$, let $mor_s(A, B) = Map(A, B)$, and for each $A, B, C \in \mathscr{S}$, let $\circ: mor_s(B, C) \times mor_s(A, B) \to mor_s(A, C)$ be the composition of functions. Then $S = (\mathscr{S}, mor_s, \circ)$ is a concrete category where $u(A) = A$ for each $A \in \mathscr{S}$. Call S *the category of sets.*

(2) Let $\mathscr{G}$ be the class of all groups, let $mor_G(G, H)$ be the set of all group homomorphisms from G to H, and again let ∘ be the usual composition of functions. Then $G = (\mathscr{G}, mor_G, \circ)$ is a concrete category, the *category of groups*, where $u(G)$ is the underlying set of G.

(3) The *category of* V *real vector* spaces is the category $(\mathscr{V}, mor_V, \circ)$ where $\mathscr{V}$ is the class of real vector spaces, $mor_V(U, V)$ is the set of linear transformations from U to V, and ∘ is the usual composition. This category is concrete where $u(V)$ is the underlying set of V.

(4) Let $\mathscr{P}$ be the class of all posets, $mor_P(P, Q)$ the set of all monotone maps (order preserving and order reversing ones), and ∘ the usual composition. Then $(\mathscr{P}, mor_P, \circ)$ is not a category, for ∘ is not as required—the composite of two monotone functions need not be monotone.

If $C = (\mathscr{C}, mor_C, \circ)$ is a concrete category, then the set $u(A)$ is called the *underlying* set of $A \in \mathscr{C}$.

A category $D = (\mathscr{D}, mor_D, \circ)$ is a *subcategory* of $C = (\mathscr{C}, mor_C, \circ)$ provided $\mathscr{D} \subseteq \mathscr{C}$, $mor_D(A, B) \subseteq mor_C(A, B)$ for each pair $A, B \in \mathscr{D}$, ∘ in D is the restriction of ∘ in C. If in addition $mor_D(A, B) = mor_C(A, B)$ for each $A, B \in \mathscr{D}$, then D is a *full* subcategory of C.

It is clear that the class of abelian groups is the class of objects of a full subcategory of the category of groups, and that this category has a full subcategory whose objects are the finite abelian groups. It is a common practice in algebra to identify an object in a category with its underlying set. Thus for example, we usually identify a group $(G, \circ)$, consisting of a set G and an operation ∘, with its underlying set G. Note, however, that the category of groups is not a subcategory of the category of sets, quite simply because for groups $(G, \circ)$, $(H, \circ)$ in $\mathscr{G}$

$$mor_G((G, \circ), (H, \circ)) \subseteq Map(G, H)$$

and

$$mor_G((G, \circ), (H, \circ)) \nsubseteq Map((G, \circ), (H, \circ)).$$

0.12. Functors. A functor is a thing that can be viewed as a "homomorphism of categories". Let $C = (\mathscr{C}, mor_C, \circ)$ and $D = (\mathscr{D}, map_D, \circ)$ be two categories. A pair of functions $F = (F', F'')$ is a *covariant functor* from C to D in case F' is a function from $\mathscr{C}$ to $\mathscr{D}$, F'' is a function from the morphisms of C to those of D such that for all $A, B, C \in \mathscr{C}$ and all $f: A \to B$ and $g: B \to C$ in C,

(F.1) $F''(f): F'(A) \to F'(B)$ in D;

(F.2) $F''(g \circ f) = F''(g) \circ F''(f)$;

(F.3) $F''(1_A) = 1_{F'(A)}$.

Thus, a covariant functor sends objects to objects, maps to maps, identities to identities, and "preserves commuting triangles":

A *contravariant functor* is a pair $F = (F', F'')$ satisfying instead of (F.1) and (F.2) their duals

(F.1)* $F''(f): F'(B) \to F'(A)$ in D;

(F.2)* $F''(g \circ f) = F''(f) \circ F''(g)$;

(F.3) $F''(1_A) = 1_{F'(A)}$.

So a contravariant functor is "arrow reversing".

Examples. (1) Given a category $C = (\mathscr{C}, mor_C, \circ)$, there is the *identity functor* $1_C = (1_C', 1_C'')$ from C to C defined by $1_C'(A) = A$ and $1_C''(f) = f$.

(2) Let $C = (\mathscr{C}, mor_C, \circ)$ be a concrete category. For each $A \in \mathscr{C}$, let $F'(A) = u(A)$ be the underlying set of A. For each morphism f of C, let $F''(f) = f$. Then clearly $F = (F', F'')$ is a covariant functor from C to the category of sets. It is called a *forgetful functor* (because it "forgets" all the "structure" on the objects of C). It should be evident there are "partially forgetful functors" of various kinds—for example, the covariant functor from the category of real vector spaces to the category of abelian groups that "forgets" the scalar multiplication.

(3) Let $(G, +)$ be an abelian group. If A is a set, then $(G^A, +)$ is an abelian group where for $\sigma, \tau \in G^A$, the sum $\sigma + \tau \in G^A$ is defined by $(\sigma + \tau): a \mapsto \sigma(a) + \tau(a)$. (Note that $(G^A, +)$ is simply the cartesian product of A copies of G with coordinatewise addition.) Define $F'(A) = (G^A, +)$. If A, B are sets, and if $f: A \to B$, then define $F''(f): G^B \to G^A$ by

$$F''(f)(\sigma) = \sigma \circ f \qquad (\sigma \in G^B).$$

Then $F''(f)$ is a group homomorphism, and $F = (F', F'')$ is a contravariant functor from the category of non-empty sets to the category of abelian groups. All kinds of contravariant functors can be built in this way. For example, if $(G, +, \circ)$ were a real vector space, then G^A can be made into a vector space with coordinatewise operations, and a contravariant functor into the real vector spaces results.

Given a functor $F = (F', F'')$, then rather than bother with all the primes, we shall usually write $F(A)$ and $F(f)$ instead of $F'(A)$ and $F''(f)$. The relatively minor formal objection is that a morphism f of the category may also be an object of the category whence $F'(f)$ and $F''(f)$ may both make sense yet be different.

0.13. Natural Transformations. A natural transformation is a thing that compares two functors between the same categories. Let C and D be categories. Let F and G be functors from C to D, say both covariant. Let $\eta = (\eta_A)_{A \in C}$ be an indexed class of morphisms in D indexed by $\mathscr{C}$ such that for each $A \in \mathscr{C}$,

$$\eta_A \in mor_\mathsf{D}(F(A), G(A)).$$

Then η is a *natural transformation* from F to G in case for each pair, $A, B \in \mathscr{C}$, and each $f \in mor_\mathsf{C}(A, B)$ the diagram

$$\begin{array}{ccc} F(A) & \xrightarrow{F(f)} & F(B) \\ \eta_A \downarrow & & \downarrow \eta_B \\ G(A) & \xrightarrow{G(f)} & G(B) \end{array}$$

commutes; that is $\eta_B \circ F(f) = G(f) \circ \eta_A$. If each η_A is an isomorphism, then η is called a *natural isomorphism*. (If both F and G were contravariant, the only change would be to reverse the arrows $F(f)$ and $G(f)$.) The crucial property of functors is that "they preserve commuting triangles"; then a natural transformation η achieves a "translation of commuting triangles"

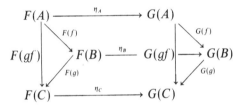

In fact notice that any commutative diagram Δ in C when operated on elementwise by F and G produces a pair of commutative diagrams $F(\Delta)$ and $G(\Delta)$ in D (because F and G are functors). Then a natural transformation η from F to G "translates" commutatively $F(\Delta)$ onto $G(\Delta)$. Because of the technical clumsiness in defining many interesting functors at this stage, we shall postpone giving examples until such time as we have an abundance of functors (see §20).

Some Special Notation

$\mathbb{N}_0 = \{0, 1, 2, \ldots\}$, the non negative integers;
$\mathbb{N} \ = \{1, 2, \ldots\}$, the positive integers;
$\mathbb{P} \ = \{p \in \mathbb{N} \mid p \text{ is prime}\}$;
$\mathbb{Z} \ = $ the set of integers;
$\mathbb{Z}_n = \{0, 1, \ldots, n - 1\}$;
$\mathbb{Q} \ = $ the set of rational numbers;
$\mathbb{R} \ = $ the set of real numbers;
$\mathbb{C} \ = $ the set of complex numbers;
$\emptyset \ = $ the empty set.

Chapter 1

Rings, Modules and Homomorphisms

The subject of our study is ring theory. In this chapter we introduce the fundamental tools of this study. Section 1 reviews the basic facts about rings, subrings, ideals, and ring homomorphisms. It also introduces some of the notation and the examples that will be needed later.

Rings admit a valuable and natural representation theory, analogous to the permutation representation theory for groups. As we shall see, each ring admits a vast horde of representations as an endomorphism ring of an abelian group. Each of these representations is called a *module*. A substantial amount of information about a ring can be learned from a study of the class of modules it admits. Modules actually serve as a generalization of both vector spaces and abelian groups, and their basic behavior is quite similar to that of the more special systems. In Sections 2 and 3 we introduce modules and their homomorphisms. In Section 4 we see that these form various natural and important categories, and we begin our study of categories of modules.

§1. Review of Rings and their Homomorphisms

Rings and Subrings

By a ring we shall always mean an associative ring with identity. Formally, then, a *ring* is a system $(R, +, \cdot, 0, 1)$ consisting of a set R, two binary operations, addition $(+)$ and multiplication $(\cdot)$, and two elements $0 \neq 1$ of R such that $(R, +, 0)$ is an abelian group, $(R, \cdot, 1)$ is a monoid (i.e., a semigroup with identity 1) and multiplication is both left and right distributive over addition. A ring whose multiplicative structure is commutative is called a *commutative ring*. We assume that the reader is versed in the elementary arithmetic of rings and we shall therefore use that arithmetic without further mention. We shall also invoke the time-honored convention of identifying a ring with its underlying set whenever there is no real risk of confusion. Of course, when we are dealing with more than one ring we may modify our notation to eliminate ambiguity. Thus, for example, if R and S are two rings, we may distinguish their identities by such self-explanatory notation as 1_R and 1_S.

Often in practice, particularly in some areas of analysis, one encounters "rings without identity". Nevertheless the severity of our requirement of an identity is more imaginary than real. Indeed a ring without identity can be embedded naturally in a ring with identity (see Exercise (1.1)). Thus our requirement involves no substantive restrictions, but it does allow considerable streamlining of the theory.

In this section and its exercises we treat very briefly several of the more basic concepts and examples that serve as tools for our study of rings.

Let R be a ring. Then an element $a \in R$ is said to be:

(1) *cancellable on the left* (or *left cancellable*) in case for all $x, y \in R$

$$ax = ay \quad \text{implies} \quad x = y;$$

(2) *a left zero divisor* in case there is an element $b \neq 0$ in R with $ab = 0$;

(3) *invertible on the left* (or *left invertible*) in case there is an element $a' \in R$, called a *left inverse* for a, such that $a'a = 1$.

The meanings of the right and two-sided ($=$ left and right) versions, such as right cancellable and cancellable, should be clear. (See Exercises for some of the arithmetic properties of these special elements.)

These arithmetical concepts provide the means for an important classification of rings. A ring R is an *integral domain* in case each of its non-zero elements is cancellable (or equivalently, it has no non-zero divisors of zero). Note that integral domains need not be commutative. A *division ring* is a ring each of whose non-zero elements is invertible (see Exercise (1.2)); thus a division ring is an integral domain. A commutative division ring is a *field*.

We reserve the term "subring" for what is sometimes called a "unital subring". Thus, if R and S are rings, we say that S is a *subring* of R and that R is an *overring* of S, and write $S \leq R$ in case additively S is a subgroup of R and multiplicatively S is a submonoid of R; so in particular, for S to be a subring of R, it must contain the identity 1 of R.

Observe that every subring of an integral domain is again an integral domain, but that an overring of an integral domain need not be one. For example, the ring of all continuous functions from $\mathbb{R}$ to $\mathbb{R}$ is not an integral domain, but the constant functions form a subring that is a field. Also observe that the ring of integers $\mathbb{Z}$ (an integral domain) has a natural embedding as a subring of the rational numbers $\mathbb{Q}$ (a field). In general, every commutative integral domain has a natural overfield, called its *field of fractions* (or *quotient field*), which is constructed in the same way that $\mathbb{Q}$ is constructed from $\mathbb{Z}$.

Ring Homomorphisms

Consistent with our requirement of identities for rings we shall require that ring homomorphisms preserve these identities. Thus, if R and S are rings, a function $\phi : R \to S$ is a *(ring) homomorphism* in case ϕ is simultaneously an additive group homomorphism and a multiplicative monoid homomorphism. That is, the function ϕ is a ring homomorphism if and only if for all $a, b \in R$

$$\phi(a + b) = \phi(a) + \phi(b); \quad \phi(ab) = \phi(a)\phi(b); \quad \phi(1_R) = 1_S.$$

The composition of two ring homomorphisms (where defined as a function) is again a ring homomorphism and the identity map $1_R : R \to R$ is a ring homomorphism. (The ambiguity of the notation 1_R is, in practice, not at all disturbing. In fact, if we think of the elements of R as "multiplications" $R \to R$, then the ambiguity vanishes.) Thus, the collection of rings and ring homomorphisms with the usual composition is a concrete category (0.11).

A ring homomorphism $\phi : R \to S$ that is bijective (as a function) is called a *(ring) isomorphism*. If ϕ is such an isomorphism, then as a function from R to S it has an inverse; i.e., there exists a (necessarily unique) function $\psi : S \to R$ such that

$$\psi \circ \phi = 1_R \qquad \text{and} \qquad \phi \circ \psi = 1_S.$$

Indeed this ψ must be a ring isomorphism; for first if $s, s' \in S$, then

$$\phi(\psi(ss')) = 1_S(ss') = ss' = 1_S(s)1_S(s')$$
$$= \phi(\psi(s))\phi(\psi(s')) = \phi(\psi(s)\psi(s')),$$

and so, since ϕ is injective $\psi(ss') = \psi(s)\psi(s')$. Similarly, one checks that ψ is an additive homomorphism and that it preserves 1. Thus,

1.1. Proposition. *Let R and S be rings and let $\phi : R \to S$ be a ring homomorphism. Then ϕ is an isomorphism if and only if there exist functions $\psi, \psi' : S \to R$ such that*

$$\psi \circ \phi = 1_R \qquad \text{and} \qquad \phi \circ \psi' = 1_S.$$

Moreover, if the latter condition holds, then $\psi = \psi'$ is a ring isomoprhism. $\square$

If R and S are rings, then we say they are *isomorphic*, and we write

$$R \cong S,$$

in case there is a (ring) isomorphism $\phi : R \to S$. Since the identity map on a ring is clearly an isomorphism from the ring to itself, we have as an easy application of (1.1) that the relation of "being isomorphic" satisfies the usual equivalence properties.

Of course, the behavior of the subrings of one ring is virtually the same as that of the subrings of any isomorphic ring. For ring homomorphisms we have the following easily proved result:

1.2. Proposition. *Let R and S be rings and let $\phi : R \to S$ be a ring homomorphism. Then for each subring R' of R, its image $\phi(R')$ under ϕ is a subring of S and*

$$(\phi \,|\, R') : R' \to \phi(R')$$

is a surjective ring homomorphism. On the other hand, for each subring S' of S, its preimage $\phi^{\leftarrow}(S')$ is a subring of R, and

$$\phi(\phi^{\leftarrow}(S')) \leq S'.$$ $\square$

Ideals and Factor Rings

Like structure preserving maps in general, ring homomorphisms are effectively determined by congruence relations. For rings these are characterized by ideals. Specifically, a subset I of a ring R is a (*two-sided*) *ideal* of R in case it is an additive subgroup such that for all $x \in I$ and all $a, b \in R$

$$axb \in I.$$

Note that the two subsets $\{0\}$ and R are both ideals of R; these are called the *trivial* ideals of R. Any ideal of R other than R itself is called a *proper* ideal. The ideal $\{0\}$, which we frequently denote simply by 0, is called the *zero* ideal. Observe that if $a \in R$, then $a = a \cdot 1 \cdot 1$, so it is immediate that an ideal I is all of R if and only if $1 \in I$. Moreover, if $a \in R$ is left invertible, say $a'a = 1$, then $1 = a'a1$, so R is the only ideal that contains a left invertible (or a right invertible) element.

The ring R is *simple* in case 0 and R are the only ideals of R. Thus, every division ring is a simple ring. On the other hand, every commutative simple ring is a field, but in general, simple rings need not be division rings and division rings need not be commutative. (See Exercises (1.6), (1.7).) Using just these few elementary concepts we have already identified and compared (modulo a few exercises) several very important classes of rings. There is one further fact about these concepts that it not so easy. Not every division ring is a field, but Wedderburn proved in 1905 that every finite division ring is a field. From this remarkable result it follows (see Exercise (1.2)) that every finite integral domain is a field. We shall not include a proof since it is arithmetic and would lead us too far astray. (See Jacobson [64].)

The collection of all ideals of a ring R is a complete lattice partially ordered by set inclusion. The proof of this will follow trivially from (2.5); see also (1.9) and (2.13). In any event this lattice we shall call the *ideal lattice* of R.

Given a ring homomorphism $\phi : R \to S$, the *image Im* ϕ and the *kernel Ker* ϕ of ϕ are defined by

$$Im\,\phi = \{\phi(x) \,|\, x \in R\} \qquad Ker\,\phi = \{x \in R \,|\, \phi(x) = 0\}.$$

Then by (1.2) $Im\,\phi$ is a subring of S, and $Ker\,\phi$ is easily seen to be a proper ideal of R. The kernel characterizes the equivalence relation induced on R by ϕ via

$$\phi(a) = \phi(b) \qquad \text{iff} \qquad a - b \in Ker\,\phi.$$

Thus, every ring homomorphism gives rise to a proper ideal, its kernel, which describes the classes of the homomorphism. Before we proceed, there is one (now trivial) fact that we should record:

1.3. Proposition. *Let R and S be rings and let $\phi : R \to S$ be a ring homomorphism. Then*

(1) ϕ *is onto S if and only if $Im\,\phi = S$;*

(2) ϕ *is an injection if and only if $Ker\,\phi = 0$.* $\qquad\qquad\square$

Now we can prove a fundamental result, one that is a ring theoretic version of part of The Factor Theorem. (See (3.6).)

1.4. Theorem *Let R, S, and S' be rings, let $\phi : R \to S$ and $\phi' : R \to S'$ be ring* *homomorphisms with ϕ' surjective, and let $K = \mathrm{Ker}\,\phi$ and $K' = \mathrm{Ker}\,\phi'$. If $K' \subseteq K$, then there is a unique ring homomorphism $\psi : S' \to S$ such that $\psi \circ \phi' = \phi$. Moreover, ψ is injective if and only if $K = K'$.*

Proof. Assume that $K' \subseteq K$, and let $x', y' \in S'$. Since ϕ' is surjective, there exist $x, y \in R$ such that $\phi'(x) = x'$ and $\phi'(y) = y'$. Now if $x' = y'$, then

$$\phi'(x - y) = \phi'(x) - \phi'(y) = x' - y' = 0,$$

whence $x - y \in K' \leq K$, and so $\phi(x) = \phi(y)$. In other words, there is a function $\psi : S' \to S$ such that $\psi(\phi'(x)) = \phi(x)$ for all $x \in R$. It is easy to check that ψ is a homomorphism. For example, with x, x', y, and y' as above

$$\psi(x' + y') = \psi(\phi'(x) + \phi'(y)) = \psi\phi'(x + y)$$
$$= \phi(x + y) = \phi(x) + \phi(y)$$
$$= \psi(\phi'(x)) + \psi(\phi'(y)) = \psi(x') + \psi(y').$$

That ψ is unique with $\psi \circ \phi' = \phi$ follows from the fact that $\mathrm{Im}\,\phi' = S'$. Finally, ψ is injective if and only if $\mathrm{Ker}\,\psi = 0$ (1.3), but clearly $\mathrm{Ker}\,\psi = \phi'(K)$, and $\phi'(K) = 0$ if and only if $K \subseteq K'$.   □

Suppose next that I is a proper ideal of a ring R. Then I determines a both additive and multiplicative congruence relation on R defined by

$$a \equiv b(\mathrm{mod}\,I) \qquad \text{in case} \qquad a - b \in I.$$

The congruence class of any element $a \in R$ is its coset

$$a + I = \{a + x \mid x \in I\}$$

and the factor set R/I of these cosets of I is a ring with operations

$$(a + I) + (b + I) = (a + b) + I, \quad (a + I)(b + I) = (ab) + I,$$

and having additive and multiplicative identities

$$0 + I \qquad \text{and} \qquad 1 + I,$$

respectively. We call the ring R/I the *factor ring* (of R) *modulo* I. Moreover, the natural map

$$n_I : R \to R/I \quad \text{via} \quad n_I : a \mapsto a + I \qquad (a \in R)$$

is a surjective ring homomorphism with $\mathrm{Ker}\,n_I = I$. With this terminology we now have what is perhaps the single most important application of Theorem (1.4).

1.5. Corollary. *Let R and S be rings and let $\phi : R \to S$ be a surjective ring homomorphism with kernel*

$$K = Ker\, \phi.$$

Then there is a unique isomorphism $\psi : R/K \to S$ with $\psi \circ n_K = \phi$.

Another immediate consequence of (1.3) and (1.4) is that a ring R is simple if and only if every ring homomorphism $\phi : R \to S$ is injective. We shall postpone further review of the ideal structure of a ring until we have developed enough additional information to treat it as a part of module theory.

Some Special Rings

We conclude this section with several odds and ends of examples, notation, and special constructions that we shall need subsequently.

1.6 The notation $\mathbb{Z}$, $\mathbb{Q}$, $\mathbb{R}$, and $\mathbb{C}$ for the sets of integers, rational numbers, real numbers, and complex numbers will also be used to denote these sets with their usual ring structures. Of course, as rings they are all commutative integral domains, and $\mathbb{Q}$, $\mathbb{R}$, and $\mathbb{C}$ are fields. As an abelian group $\mathbb{Z}$ is cyclic, so every subgroup is cyclic. Thus every ideal of the ring $\mathbb{Z}$ is *principal* (see (2.13)); i.e., is of the form $\mathbb{Z}n = \{an \mid a \in \mathbb{Z}\}$ for some unique $n \geq 0$. For each $n > 1$ and each $a \in \mathbb{Z}$, denote by $[a]_n$ the least positive remainder of a divided by n; that is, $[a]_n$ is the unique element of

$$\mathbb{Z}_n = \{0, 1, \ldots, n-1\}$$

in the coset $a + \mathbb{Z}n$. Now $\mathbb{Z}_n$ is a ring under the usual operations of residues modulo n, and it is easy to check that $r_n : a \mapsto [a]_n$ is also a surjective ring homomorphism $\mathbb{Z} \to \mathbb{Z}_n$ with kernel $\mathbb{Z}n$. So (1.5), $\mathbb{Z}_n \cong \mathbb{Z}/\mathbb{Z}n$ as rings.

1.7. Polynomial Rings. We shall relegate the definitions and general treatment of polynomial rings to the exercises. (See Exercises (1.16)–(1.18).) Here we wish to point out that if R is a ring, then we write

$$R[X_1, \ldots, X_n]$$

for the ring of all polynomials over R in the commuting indeterminants $X_1, \ldots, X_n$. Note that R is not a subring of $R[X_1, \ldots, X_n]$ but that it is isomorphic, under the obvious map, to the subring of "constant polynomials". Thus we shall feel free to identify R with its natural image in $R[X_1, \ldots, X_n]$. Note also that as a notational consequence of this identification

$$R[X_1][X_2] \ldots [X_n] = R[X_1, X_2, \ldots, X_n].$$

1.8. Products and Function Rings. Let $(R_\alpha)_{\alpha \in A}$ be a non-empty indexed set of rings, and let

$$R = \times_A R_\alpha$$

be the cartesian product of this indexed set of sets. Then the ring structures on the factors R_α induce a ring structure, defined "coordinatewise" on the product R. That is, with respect to the operations

$$(r + s)(\alpha) = r(\alpha) + s(\alpha), \quad (rs)(\alpha) = r(\alpha)s(\alpha) \qquad (\alpha \in A),$$

for all $r, s \in R$, R is a ring with additive and multiplicative identities 0 and 1 defined by

$$0(\alpha) = 0_\alpha \quad \text{and} \quad 1(\alpha) = 1_\alpha \qquad (\alpha \in A).$$

Using the "A-tuple" notation (0.3), the operations are given by

$$(r_\alpha)_{\alpha \in A} + (s_\alpha)_{\alpha \in A} = (r_\alpha + s_\alpha)_{\alpha \in A}, \qquad (r_\alpha)_{\alpha \in A}(s_\alpha)_{\alpha \in A} = (r_\alpha s_\alpha)_{\alpha \in A}$$

and the identities by

$$(0_\alpha)_{\alpha \in A} \qquad \text{and} \qquad (1_\alpha)_{\alpha \in A}.$$

The resulting ring R is called the (cartesian) product of the rings $(R_\alpha)_{\alpha \in A}$, and is denoted

$$R = \Pi_A R_\alpha.$$

Let R be the product of the rings $(R_\alpha)_{\alpha \in A}$. Then the canonical projections $\pi_\alpha : R \to R_\alpha$ $(\alpha \in A)$ are surjective ring homomorphisms. The canonical injections $\iota_\alpha : R_\alpha \to R$ $(\alpha \in A)$ defined coordinatewise (see (0.4)) by

$$\pi_\beta \iota_\alpha = \delta_{\alpha\beta} 1_{R_\alpha} \qquad (\beta \in A)$$

preserve both operations and are injections, but if A has at least two elements, then the ι_α are not ring homomorphisms.

A special case of a product ring is a function ring. That is, if A is a non-empty set and if R is a ring, then the set

$$R^A = \{ f \mid f : A \to R \}$$

of all functions from A to R becomes a ring with "pointwise" operations

$$(f + g)(\alpha) = f(\alpha) + g(\alpha), \quad (fg)(\alpha) = f(\alpha)g(\alpha)$$

and with identities the "constant functions"

$$0(\alpha) = 0, \qquad 1(\alpha) = 1$$

for all $\alpha \in A$. Now define a function $A \to \{R\}$ by $\alpha \mapsto R_\alpha = R$. Thus $(R_\alpha)_{\alpha \in A}$ is an indexed class of "A copies of R". Then it is easy to check that R^A is precisely the same as the product of $(R_\alpha)_{\alpha \in A}$. Therefore we shall denote this ring by

$$R^A = \Pi_A R.$$

1.9. Let R be a ring and let $A \subseteq R$. Then the set $\mathscr{A}$ of all subrings of R that contain A is not empty for $R \in \mathscr{A}$. Moreover, it is easy to check that the intersection $\bigcap \mathscr{A}$ is a subring of R; it is called the subring of R *generated* by A. Thus, the subring of R generated by A is the unique smallest subring of R that contains A. Different subsets of R may generate the same subring. Indeed in any ring $\varnothing$, $\{0\}$, and $\{1\}$ all generate the same subring of R. This subring can be characterized as the image of the unique ring homomorphism $\chi : \mathbb{Z} \to R$ and so is isomorphic to some factor ring of $\mathbb{Z}$. (See Exercise (1.11).)

1.10. The Center of a Ring. Let R be a ring. Then its *center* is

$$Cen\,R = \{r \in R \,|\, rx = xr \quad (x \in R)\}.$$

It is easy to check that $Cen\,R$ is a subring of R. Of course, $Cen\,R$ is commutative and R is commutative if and only if R is equal to its center. But it is not true in general that $Cen\,R$ is a maximal commutative subring. We may say that an element $r \in R$ is *central* in case $r \in Cen\,R$. Note that if $A \subseteq Cen\,R$, then the subring of R generated by A is also in the center of R.

1.11. Algebras. Let R be a ring, K a commutative ring, and $\phi : K \to Cen\,R$ a ring homomorphism. The resulting system (R, K, ϕ) is called a K-*algebra.* In practice we tend to suppress the ϕ and we speak of R as a K-algebra or as an algebra over K. Thus by (1.5) R is a K-algebra (with respect to some ϕ) if and only if there is an ideal I of K with K/I isomorphic to a subring of $Cen\,R$. Therefore, since (see Exercise (1.11)) there is a unique ring homomorphism $\chi : \mathbb{Z} \to Cen\,R$, the ring R is (in one and only one way) a $\mathbb{Z}$-algebra.

Classically this concept has its greatest importance when K is a field and the homomorphism ϕ is necessarily injective. In this case the entire concept of a K-algebra R is equivalent to the requirements that, in addition to being a ring, R be a K-vector space satisfying

$$\alpha(ab) = a(\alpha b) = (\alpha a)b$$

for all $\alpha \in K$, and all $a, b \in R$.

If R and R' are K-algebras, via ϕ and ϕ', respectively, then a ring homomorphism $\sigma : R \to R'$ is a K-*algebra homomorphism* in case for each $\alpha \in K, a \in R$,

$$\sigma(\phi(\alpha)a) = \phi'(\alpha)\sigma(a).$$

It is easy to check that the class of K-algebras together with all K-algebra homomorphisms and the usual composition is a concrete category (0.11).

1.12. The Opposite Ring of a Ring. Let R be a ring. From this we construct a new ring R^{op}, called the *opposite ring of R.* Both the underlying set and the additive structure of R^{op} are just those of R. But the multiplication on R^{op}, which for the present we shall denote by $(r,s) \mapsto r * s$, is defined by

$$r * s = sr.$$

It is easy to check that R^{op} is a ring with these operations and that the identities of R^{op} are those of R. Clearly, $Cen\,R = Cen\,R^{op}$ and R is commutative if and only if $R = R^{op}$.

Suppose that R and S are rings. Then a function $\phi : R \to S$ is a *ring anti-homomorphism* in case ϕ is an abelian group homomorphism, $\varphi(1) = 1$, and

$$\phi(ab) = \phi(b)\phi(a).$$

Thus, the function $\phi : R \to S$ is a ring anti-homomorphism if and only if the same function $\phi : R^{op} \to S$ is a ring homomorphism.

1.13. Matrix Rings. Particularly for constructing examples it is often a great convenience to have a generalization of the familiar rings of $n \times n$ matrices over a field. Indeed we want rings of matrices of infinite dimension. Clearly the usual multiplication will not generalize without some adjustment quite simply because in the infinite case "row dot column" can result in infinite sums. Fortunately the adjustment is natural, so we shall permit ourselves a relaxed treatment omitting some of the rather dull details and formalities. Let R be a ring and let Γ and Λ be non-empty sets. Then a $\Gamma \times \Lambda$-*matrix over* R is simply a function $A : \Gamma \times \Lambda \to R$. Let A be a $\Gamma \times \Lambda$-matrix over R. For each $(\alpha, \beta) \in \Gamma \times \Lambda$ let $A(\alpha, \beta) = a_{\alpha\beta} \in R$; then we call $a_{\alpha\beta}$ the (α, β) *entry* in A and we write

$$A = [\![a_{\alpha\beta}]\!]_{\Gamma \times \Lambda}.$$

When there is no likelihood of confusion about the sets Γ and Λ we may simply write $A = [\![a_{\alpha\beta}]\!]$. If $\Gamma' \subseteq \Gamma$ and $\Lambda' \subseteq \Lambda$ are non-empty subsets, then the restriction of A to $\Gamma' \times \Lambda'$ is a *submatrix* of A and may be denoted $[\![a_{\alpha\beta}]\!]_{\Gamma' \times \Lambda'}$.

Let $\alpha \in \Gamma$ and $\beta \in \Lambda$. Then $[\![a_{\alpha\beta}]\!]_{\{\alpha\} \times \Lambda}$ and $[\![a_{\alpha\beta}]\!]_{\Gamma \times \{\beta\}}$ are called the α *row* of A and the β *column* of A, respectively. The matrix A is said to be *row finite* (*column finite*) in case each row (column) of A has at most finitely many non-zero entries. In practice we shall be interested mainly in matrices that are row finite or column finite. The collection of all $\Gamma \times \Lambda$-matrices over the ring R will be denoted by

$$\mathbb{M}_{\Gamma \times \Lambda}(R)$$

and the subsets of row finite and column finite matrices by

$$\mathbb{RFM}_{\Gamma \times \Lambda}(R) \qquad \text{and} \qquad \mathbb{CFM}_{\Gamma \times \Lambda}(R),$$

respectively. If $\Gamma = \Lambda$, then we write simply $\mathbb{M}_\Gamma(R)$, $\mathbb{RFM}_\Gamma(R)$ and $\mathbb{CFM}_\Gamma(R)$, and we call the entries Γ-*square* or $\Gamma \times \Gamma$-*square* matrices. The *diagonal* of a $\Gamma \times \Gamma$-square matrix $A = [\![a_{\alpha\beta}]\!]$ is the indexed set $(a_{\alpha\alpha})_{\alpha \in \Gamma}$.

Of course, $\mathbb{M}_{\Gamma \times \Lambda}(R)$ is simply $R^{\Gamma \times \Lambda}$ and so (see 1.8) it has a natural group structure; in particular, it is an abelian group with "pointwise" addition. In matrix notation, let

$$A = [\![a_{\alpha\beta}]\!], \quad B = [\![b_{\alpha\beta}]\!]$$

be elements of $\mathbb{M}_{\Gamma \times \Lambda}(R)$; then this pointwise or *matrix addition* is given somewhat imprecisely by

$$[\![a_{\alpha\beta}]\!] + [\![b_{\alpha\beta}]\!] = [\![a_{\alpha\beta} + b_{\alpha\beta}]\!].$$

The identity of this group structure on $\mathbb{M}_{\Gamma \times \Lambda}(R)$ is the *zero matrix* $0 = [\![0_{\alpha\beta}]\!]$, and the inverse (negative) of A is $-A = [\![-a_{\alpha\beta}]\!]$.

Now suppose that Γ, Λ, and Ω are non-empty sets, and that

$$A = [\![a_{\alpha\beta}]\!] \in \mathbb{M}_{\Gamma \times \Lambda}(R), \qquad B = [\![b_{\beta\gamma}]\!] \in \mathbb{M}_{\Lambda \times \Omega}(R).$$

For each $\alpha \in \Gamma$ and $\gamma \in \Omega$ consider the formal series $\Sigma_{\beta \in \Lambda} a_{\alpha\beta} b_{\beta\gamma}$. If either A is row finite or B is column finite, then this series has at most finitely many non-zero terms which sum to a unique element $c_{\alpha\gamma} \in R$, and the $\Gamma \times \Omega$-matrix

$$AB = [\![\Sigma_{\beta \in \Lambda} a_{\alpha\beta} b_{\beta\gamma}]\!]_{\Gamma \times \Omega}$$

is called the *(matrix) product* of A and B (in that order). Note that if both A and B are column finite (row finite) then AB is column finite (row finite). It is easy (but tedious) to show that wherever this product is defined, it is associative and that it distributes over addition on both the right and left. Now let I_Γ be the $\Gamma \times \Gamma$-square matrix over R

$$I_\Gamma = [\![\delta_{\alpha\beta}]\!]$$

where $\delta_{\alpha\beta}$ denotes the Kronecker delta over R (0.1). Then clearly, I_Γ is both row finite and column finite.

We call I_Γ the $\Gamma \times \Gamma$-*identity* matrix over R. Now the point is that on both of the sets $\mathbb{RFM}_\Gamma(R)$ and $\mathbb{CFM}_\Gamma(R)$ the matrix product defines a binary operation, which we call *matrix multiplication*.

1.14. Proposition. *Let R be a ring and let Γ be a non-empty set. Then with pointwise addition and matrix multiplication*

$$\mathbb{RFM}_\Gamma(R) \qquad and \qquad \mathbb{CFM}_\Gamma(R) \qquad\qquad \square$$

are rings.

In the case where $\Gamma = \{1, \ldots, m\}$ and $\Lambda = \{1, \ldots, n\}$ are finite, all matrices are both row finite and column finite, and we write simply

$$\mathbb{M}_{m \times n}(R) \qquad \mathbb{M}_n(R)$$

for the sets of these $m \times n$-*matrices over R* and the set of $n \times n$-square matrices over R, respectively. From (1.14) $\mathbb{M}_n(R)$ is a ring, called the *ring of $n \times n$-matrices over R*, with respect to matrix addition and matrix multiplication. Also, as usual, we shall adopt the familiar rectangular array notation for an $m \times n$-matrix $[\![a_{ij}]\!]$ over R.

The ideal structure of the matrix rings $\mathbb{M}_n(R)$ is quite easy. It can be shown (see Exercise (1.8)) that a subset K of $\mathbb{M}_n(R)$ is an ideal in $\mathbb{M}_n(R)$ if and only if there is an ideal I of R with $K = \{[\![a_{ij}]\!] \in \mathbb{M}_n(R) \,|\, a_{ij} \in I\}$.

With a slight perversion of notation this says that

$$I \mapsto \mathbb{M}_n(I)$$

defines an isomorphism between the ideal lattices of R and of $\mathbb{M}_n(R)$. In particular, $\mathbb{M}_n(R)$ is a simple ring if and only if R is a simple ring.

On the other hand, if Γ is infinite, then the ideal structure of $\mathbb{CFM}_\Gamma(R)$, say, is not quite so clear. However, in the case of greatest interest, where R is a field, the ideal lattice of $\mathbb{CFM}_\Gamma(R)$ is a chain of more than two elements. (See Exercise (14.13).)

Among the many interesting subrings of $\mathbb{CFM}_\Gamma(R)$ there is one that we shall refer to frequently. If Γ is linearly ordered by $\leq$, a $\Gamma \times \Gamma$-square matrix $A = [\![a_{\alpha\beta}]\!]$ is *upper triangular* (*lower triangular*) in case for all $\alpha, \beta \in \Gamma$

$$\alpha > \beta \quad \text{implies} \quad a_{\alpha\beta} = 0 \qquad (\alpha < \beta \quad \text{implies} \quad a_{\alpha\beta} = 0).$$

Of course every scalar matrix is both upper and lower triangular. Moreover, it is easy to see that the set of upper triangular matrices in $\mathbb{CFM}_\Gamma(R)$ forms a subring of $\mathbb{CFM}_\Gamma(R)$ and the set of upper triangular matrices in $\mathbb{RFM}_\Gamma(R)$ forms a subring of $\mathbb{RFM}_\Gamma(R)$. Of course, parallel statements hold for the sets of lower triangular matrices.

1.15. Endomorphism Rings. We look next at a class of examples that motivates much of our subsequent work. Thus, let A be an abelian group written additively. By an *endomorphism* of A we mean of course just a group homomorphism $f: A \to A$; in other words, if we write our functions on the left,

$$f(a + b) = f(a) + f(b) \qquad (a, b \in A).$$

It is easy to check that the set E of all such endomorphisms of A forms an abelian group with respect to the addition $(f, g) \mapsto f + g$ defined by

$$(f + g)(a) = f(a) + g(a) \qquad (a \in A).$$

Of course the identity and the inverse ($=$ negative) are given by

$$0(a) = 0 \quad \text{and} \quad (-f)(a) = -f(a).$$

Now on E it also happens that composition of functions is an associative operation that distributes over the additive operation on E. So if $A \neq 0$ (i.e., if E has at least two elements), then E is actually a ring whose identity is the identity map $1_A: A \to A$. But note that if $f, g \in E$, then in general, the product fg in E depends on whether we consider these as functions operating on the left or on the right:

$$(fg)(a) = f(g(a)); \qquad (a)(fg) = ((a)f)g.$$

In other words, there arise naturally for every (non-zero) abelian group A two endomorphism rings, a *ring of left endomorphisms* and a *ring of right endomorphisms*, denoted

$$End^l(A) \qquad \text{and} \qquad End^r(A),$$

respectively. The fates being what they are, we shall have need for both of these rings. When we have an $f \in End^l(A)$, we are considering it as a "left" endomorphism and shall denote its values $f(a)$. On the other hand, if we have $f \in End^r(A)$, then we are considering f as a "right" endomorphism and shall denote its values by $(a)f$. Of course,

$$End^l(A) = (End^r(A))^{op}.$$

It turns out that such endomorphism rings (pick a side) play a role in ring theory entirely analogous to that played by the symmetric groups in the theory of groups. In fact, there is a perfect analogue of "Cayley's Theorem" to the effect that every ring is isomorphic to a subring of an endomorphism ring of an abelian group. (See Exercise (1.10).)

1.16. Idempotents. Let R be a ring. An element $e \in R$ is an *idempotent* in case $e^2 = e$. A ring always has at least two idempotents, namely 0 and 1. An idempotent e of R is a *central idempotent* in case it is in the center of R. As we shall see, the arithmetic of idempotents plays a fundamental role in the study of rings. For the most part, however, the details of this arithmetic are quite straightforward and will be relegated to the exercises. As one small example, we note here that if $e \in R$ is an idempotent, then so is $1 - e$, for

$$(1 - e)^2 = 1 - e - e + e^2 = 1 - e - e + e = 1 - e.$$

Also it is easy to check that if e is central, then so is $1 - e$.

Each non-zero idempotent e of a ring R determines a second ring, namely

$$eRe = \{exe \mid x \in R\},$$

with addition and multiplication that of R restricted to eRe, and with identities $0 = e0e$ and $e = e1e$. If $e \neq 1$, then the ring eRe is not a subring of R and if e is not central, eRe need not be a homomorphic image of R. Of course, if e is a central idempotent, then the map

$$\tau_e : x \mapsto exe \qquad (x \in R)$$

is a surjective ring homomorphism R onto eRe with kernel $(1 - e)R(1 - e)$.

There is one easy but important class of examples of this last phenomenon. Thus, let R be the cartesian product $R = \Pi_A R_\alpha$ of rings $(R_\alpha)_{\alpha \in A}$. Let $\alpha \in A$. Then there is an element $e_\alpha \in R$ defined coordinatewise (0.4) by

$$\pi_\beta(e_\alpha) = \delta_{\alpha\beta} 1_\beta.$$

That is, $e_\alpha = \iota_\alpha(1_\alpha)$ is the identity of R_α at the αth coordinate and 0 elsewhere. Now it is easy to see that e_α is a central idempotent of R and that the ring $e_\alpha R e_\alpha$ is isomorphic to R_α via

$$(\pi_\alpha \mid e_\alpha R e_\alpha) : e_\alpha R e_\alpha \to R_\alpha.$$

Moreover, in the case where A is finite, the existence of a ring isomorphism $R \cong \Pi_A R_\alpha$ can be determined by means of the behavior of the central idempotents of R. (See §7.)

As another important example of idempotents, let R be a ring, let $n > 0$ be an integer and consider the matrix ring $\mathbb{M}_n(R)$. Let $1 \leq m \leq n$ and let $e = [\![a_{ij}]\!]$ be the matrix defined via

$$a_{ij} = \begin{cases} 1 & \text{if } i = j \leq m \\ 0 & \text{otherwise.} \end{cases}$$

Then it is easy to check that e is a non-zero idempotent and that as rings

$$e\mathbb{M}_n(R)e \cong \mathbb{M}_m(R).$$

Incidentally, we can draw the same conclusions provided just that e has exactly m non-zero entries, each of which is a 1 on the diagonal. It is important to realize that these examples by no means describe all of the idempotents of the matrix ring $\mathbb{M}_n(R)$.

1.17. Nilpotent Elements. The antithesis of the idempotents of a ring are its nilpotent elements. An element x of a ring R is *nilpotent* in case there is a natural number n such that

$$x^n = 0;$$

the least such n is called the *nilpotency index* of the element. Clearly 0 is the only element of a ring that is simultaneously idempotent and nilpotent.

If R is a ring and if $n > 1$, then the matrix ring $\mathbb{M}_n(R)$ is fairly rich in nilpotent elements. Indeed, every strictly upper triangular matrix (i.e., upper-triangular with 0 diagonal) and every strictly lower-triangular matrix in $\mathbb{M}_n(R)$ is nilpotent with nilpotency index at most n.

There is one "zero-like" property of nilpotent elements that is of some importance. Indeed if x is nilpotent, then $1 - x$ is invertible. For if $x^n = 0$, then

$$(1 - x)(1 + x + \ldots + x^{n-1}) = 1 \quad \text{and} \quad (1 + x + \ldots + x^{n-1})(1 - x) = 1.$$

The elementwise concept of nilpotence can be extended. Thus, a subset A of a ring R is *nilpotent* in case there is an integer $n > 0$ such that

$$x_1 x_2 \ldots x_n = 0$$

for every sequence $x_1, x_2, \ldots, x_n$ in A. Also, a subset A of a ring is *nil* in case each of its elements is nilpotent. Thus, every nilpotent subset of R is certainly nil; but there are nil subsets of rings that are not nilpotent. (See Exercise (1.14).)

As we shall see, the analysis of a ring and its arithmetic is very dependent on the behavior of its idempotents and its nilpotent elements. One seeks to learn the idempotents of a ring to a large extent because locally they behave like the identity; indeed a non-zero idempotent $e \in R$ *is* the identity of the induced ring eRe. In a sense the nilpotent elements are relatively weak and the extent to which they permeate the ring provides a measure of the arithmetic strength of the ring. For example (see Exercise (15.14)) a commutative ring with no non-zero nilpotent elements can be embedded in a cartesian product of fields. On the other hand, rings having substantial amounts of nilpotence often suffer from some very weird pathologies.

1. Exercises

1. Let $(R, +, \cdot, 0)$ be a system satisfying all the requirements for a ring except the existence of a multiplicative identity. Prove that there is a ring

$(\bar{R}, +, \cdot, 0, 1)$ in which $(R, +, \cdot, 0)$ is an ideal. [Hint: On $R \times \mathbb{Z}$ define addition and multiplication by $(r, n) + (s, m) = (r + s, n + m)$ and $(r, n)(s, m) = (rs + mr + ns, nm)$.]

2. (1) Prove that a ring in which each non-zero element is left cancellable (left invertible) is an integral domain (division ring).

   (2) Prove that every finite integral domain is a division ring.

3. Let $a \in R$, a ring. Prove that if a has more than one left inverse, then it has infinitely many. [Hint: Set $A = \{a' \in R \mid a'a = 1\}$. Then $A \neq \varnothing$. Fix $a_0 \in A$. Observe that $a' \mapsto aa' - 1 + a_0$ defines an injection from A to a proper subset of itself.]

4. Show that the matrix $[\![\delta_{i\,2j}]\!] \in \mathbb{CFM}_\mathbb{N}(\mathbb{R})$ is left invertible but not even right cancellable.

5. Let R and S be rings and $\phi : R \to S$ a surjective ring homomorphism. Prove that if $a \in R$ is invertible, central, idempotent, or nilpotent, respectively, then so is $\phi(a)$ in S. How about converses?

6. Let H be the subset of $\mathbb{M}_2(\mathbb{C})$, the 2×2 matrices over the complex field, of all elements of the form

$$q = \begin{bmatrix} a + ib & c + id \\ -c + id & a - ib \end{bmatrix}$$

with $a, b, c, d \in \mathbb{R}$. Show that H is a subring of $\mathbb{M}_2(\mathbb{C})$. Consider the elements

$$\mathbf{1} = \begin{bmatrix} 1 & 0 \\ 0 & 1 \end{bmatrix} \quad \mathbf{i} = \begin{bmatrix} i & 0 \\ 0 & -i \end{bmatrix} \quad \mathbf{j} = \begin{bmatrix} 0 & 1 \\ -1 & 0 \end{bmatrix} \quad \mathbf{k} = \begin{bmatrix} 0 & i \\ i & 0 \end{bmatrix}$$

in H. Thus, the above "typical" element q of H is $q = a\mathbf{1} + b\mathbf{i} + c\mathbf{j} + d\mathbf{k}$. Show that if $q \neq 0$, then it is invertible. Deduce that H is a non-commutative division ring. It is called the ring of *quaternions*.

7. Let K be a field. Prove that:

   (1) $\mathbb{M}_n(K)$ is a simple ring.

   (2) In the ring $\mathbb{CFM}_\mathbb{N}(K)$, the set T of matrices that have only a finite number of rows with non-zero entries is a non-trivial ideal. [Hint: If $a_{ij} = 0$ whenever $i > n$ and $[\![c_{ij}]\!] = [\![b_{ij}]\!] \cdot [\![a_{ij}]\!]$ then $c_{ij} = 0$ whenever $i > \max\{k \mid b_{kj} \neq 0 \text{ and } 1 \leq j \leq n\}$.]

   (3) $\mathbb{CFM}_\mathbb{N}(K)$ has exactly one non-trivial ideal.

8. Let R be a ring and let $n > 1$ be a natural number. For each ideal I of R set

$$\mathbb{M}_n(I) = \{[\![a_{ij}]\!] \in \mathbb{M}_n(R) \mid a_{ij} \in I \ (i, j = 1, \ldots, n\}.$$

   (1) Prove that $I \mapsto \mathbb{M}_n(I)$ defines an isomorphism from the lattice of ideals of R onto the lattice of ideals of $\mathbb{M}_n(R)$. This generalizes the first part of Exercise (1.7): The ring of $n \times n$ matrices over a simple ring is a simple ring. [Hint: If $\mathbb{I}$ is an ideal of $\mathbb{M}_n(R)$, then the collection of all entries from elements of $\mathbb{I}$ forms an ideal I of R.]

   (2) Prove that if I is an ideal of R, then $\mathbb{M}_n(R)/\mathbb{M}_n(I) \cong \mathbb{M}_n(R/I)$.

9. Let R be a ring and $A \subseteq R$. The R-centralizer of A is $Cen_R(A) = \{x \in R \mid ax = xa \ (a \in A)\}$. Thus $Cen\ R = Cen\ _R(R)$. Prove that:
   (1) $Cen_R(A)$ is a subring of R.
   (2) A is a maximal commutative subring of R iff $A = Cen_R(A)$.
   (3) If $x \in Cen_R(A)$ is invertible in R, then its inverse is in $Cen_R(A)$.
   (4) Infer that the center of a simple ring is a field. [Hint: If $x \in Cen\ R$, then $\{rx \mid r \in R\}$ is an ideal of R.]

10. Denote the underlying additive group of the ring R by R^+. For each $r \in R$ define two functions $\lambda_r, \rho_r : R \to R$ by

$$\lambda_r : x \mapsto rx \qquad \text{and} \qquad \rho_r : x \mapsto xr.$$

   Write each λ_r as a left operator and each ρ_r as a right operator.
   (1) Prove that $\lambda : r \mapsto \lambda_r$ defines an injective ring homomorphism into $End^l(R^+)$ and that $\rho : r \mapsto \rho_r$ defines an injective ring homomorphism into $End^r(R^+)$. Thus a ring R is isomorphic to a ring of left endomorphisms of an abelian group as well as to a ring of right endomorphisms of an abelian group.
   (2) Prove that if R^+ is cyclic, then R is commutative and both λ and ρ are isomorphisms.

11. Let R be a ring. Prove that there is a unique ring homomorphism $\chi : \mathbb{Z} \to R$. The kernel of χ is of the form $\mathbb{Z}n$ for some unique $n \geq 0$ (1.6); this n is the *characteristic* of R.

12. (1) Let R be a ring and $A \subseteq R$. Suppose R is generated by A (i.e., R is the only subring of R that contains A (1.9)). Prove that if $\phi : R \to S$ is a ring homomorphism, then $Im\ \phi$ is the subring of S generated by $\phi(A)$.
   (2) Let $\chi : \mathbb{Z} \to R$ (see Exercise (1.11)). Deduce that $Im\ \chi$ is the subring of R generated by $\{1\}$. Infer that if R is an integral domain, its characteristic is either 0 or a prime.

13. A ring is a *Boolean ring* in case each of its elements is idempotent. Prove that:
   (1) Every Boolean ring R is commutative and $a = -a$ for all $a \in R$. [Hint: Square $(a + a)$ and $(a - b)$.]
   (2) Every subring and every factor ring of a Boolean ring is a Boolean ring.
   (3) Every simple Boolean ring is isomorphic to $\mathbb{Z}_2$.
   (4) If A is a set and if R is a Boolean ring, then R^A is a Boolean ring.

14. Let $p \in \mathbb{P}$ be a prime. Prove that for each natural number n the ideals of $\mathbb{Z}_{p^n}$ form a chain and that each proper ideal is nilpotent. Then show that the product

$$R = \underset{n > 1}{\text{X}}\ \mathbb{Z}_{p^n}$$

   has a nil ideal that is not nilpotent. [Hint: For each $n > 1$ let I_n be a proper ideal of $\mathbb{Z}_{p^n}$. Let I be the set of all $\sigma \in R$ such that $\sigma(n) \in I_n$ and is not zero for at most finitely many n.]

15. Let G be a non-empty set and let R be a ring. A function $f : G \to R$ is *zero almost always* in case its *support* $S(f) = \{x \in G \mid f(x) \neq 0\}$ is finite. The

set $R^{(G)}$ of all functions $G \to R$ that are zero almost always is clearly a subgroup of the additive group of $R^{(G)}$ under the addition $(f, g) \mapsto f + g$ where $(f + g)(x) = f(x) + g(x)$ for all $x \in G$, for $S(f + g) \subseteq S(f) \cup S(g)$. Now suppose that G is a semigroup (written multiplicatively) with identity e. For each pair $f, g \in R^{(G)}$ define

$$(fg)(x) = \Sigma_{yz = x} f(y)g(z) \qquad (x \in G).$$

(1) Prove that with respect to this addition and multiplication $R^{(G)}$ is a ring with identity the function $\xi(e) : x \mapsto \delta_{ex}$ in $R^{(G)}$. This ring is called the *semigroup ring* (or *group ring* if G is a group) *of G over R*. It is denoted RG.
(2) For each $r \in R$ and each $x \in G$ define $\sigma(r)$ and $\xi(x)$ in RG by

$$\sigma(r)(x) = \delta_{ex} r \qquad \text{and} \qquad \xi(x)(y) = \delta_{xy}.$$

Prove that $\sigma : r \mapsto \sigma(r)$ defines an injective ring homomorphism $R \to RG$ and that $\xi : x \mapsto \xi(x)$ defines an injective monoid homomorphism $G \to RG$ into the multiplicative semigroup of RG.
(3) Prove that for each non-zero $f \in R^{(G)}$ there is a unique sequence $r_1, \ldots, r_n$ of non-zero elements of R and distinct $x_1, \ldots, x_n \in G$ such that $f = \sigma(r_1)\xi(x_1) + \ldots + \sigma(r_n)\xi(x_n)$. For this reason it is a common practice to write f simply in the form $r_1 x_1 + \ldots + r_n x_n$. Observe that in this notation, the canonical image of $r \in R$ (under σ) in RG is re, the identity of RG is $1e$, and (with the obvious simplification that may be possible on the right)

$$(s_1 y_1 + \ldots + s_m y_m)(r_1 x_1 + \ldots + r_n x_n) = \Sigma_{i,j=1,1}^{m,n} s_i r_j y_i x_j.$$

(4) Let S be a ring and suppose that there is a ring homomorphism $\phi : R \to S$ and a monoid homomorphism $\theta : G \to S$ such that for each $r \in R$ and $x \in G$, $\phi(r)\theta(x) = \theta(x)\phi(r)$. Prove that there is a unique ring homomorphism $\psi : RG \to S$ such that $\psi \circ \sigma = \phi$ and $\psi \circ \xi = \theta$.

16. Using the concept of a semigroup ring, polynomial rings can be treated without recourse to the artificial invention of an indeterminant. The non-negative integers $\mathbb{N}_0 = \{0, 1, 2, \ldots\}$ form a commutative monoid under addition. Let R be a ring. Adopting the notation of Exercise (1.15.2) (with $G = \mathbb{N}_0$ and $e = 0$), let $X = \xi(1) \in R\mathbb{N}_0$. We call the ring $R\mathbb{N}_0$ the *ring of polynomials in one indeterminant* (i.e., X) *over R*, and we normally denote it by $R[X]$. The elements of $R[X]$ are called *polynomials in X over R*.
(1) Prove that in $R[X]$, if $n \in \mathbb{N}_0$, then $\xi(n) = X^n$. [Remember that $\mathbb{N}_0$ is an additive semigroup.] Infer that for each non-zero polynomial $f \in R[X]$ there is a unique n and a unique sequence $r_0, r_1, \ldots, r_n$ in R with $r_n \neq 0$ and $f = r_0 X^0 + r_1 X + \ldots + r_n X^n$. We call this n the *degree* of f (the zero polynomial is assigned degree $-\infty$) and write $\deg f = n$, call $r_0, r_1, \ldots, r_n$ the *coefficients* of f and call r_n the *leading coefficient* of f.
(2) Prove that if R is commutative, then so is $R[X]$.
17. Let S be a ring and let R be a subring of S. Let $x \in S$ such that $rx = xr$ for all $r \in R$. Prove that there is a unique ring homomorphism $\psi : R[X] \to S$

such that

$$\psi : r_0 X^0 + r_1 X + \ldots + r_n X^n \mapsto r_0 + r_1 x + \ldots + r_n x^n$$

for each $r_0, r_1, \ldots, r_n \in R$. [See Exercise (1.15.4). Prove then that the image of ψ is the subring of S generated by $R \cup \{x\}$.]

18. Let $\mathbb{N}_0^n$ be the n-fold cartesian product of $\mathbb{N}_0$. Under coordinatewise addition $\mathbb{N}_0^n$ is a commutative monoid. Let R be a ring. Prove that the semigroup ring $R\mathbb{N}_0^n$ is isomorphic to the (iterated) polynomial ring $R[X_1][X_2]\ldots[X_n]$. We usually denote this ring by $R[X_1, X_2, \ldots, X_n]$.

§2. Modules and Submodules

Let R be a ring. Then a pair (M, λ) is a *left R-module* in case M is an abelian group (which we shall write additively) and λ is a map from R to the set of left endomorphisms of M such that if M is not zero,

$$\lambda : R \to End^l(M)$$

is a ring homomorphism. This means simply that for each $a \in R$, there is a mapping $\lambda(a) : M \to M$ such that for all $a, b \in R$ and all $x, y \in M$

$$\lambda(a)(x + y) = \lambda(a)(x) + \lambda(a)(y), \qquad \lambda(ab)(x) = \lambda(a)(\lambda(b)(x)),$$

$$\lambda(a + b)(x) = \lambda(a)(x) + \lambda(b)(x), \qquad \lambda(1)(x) = x.$$

In practice we usually are able to suppress the λ and the excess parentheses. Writing just ax for $\lambda(a)(x)$ we may think of λ as defining a "left scalar multiplication" $R \times M \to M$ via $(a, x) \mapsto ax$ satisfying for all $a, b \in R$ and $x, y \in M$ the axioms for a "left R-vector space":

$$a(x + y) = ax + ay, \qquad (ab)x = a(bx),$$

$$(a + b)x = ax + bx, \qquad 1x = x.$$

At the same time we shall usually say simply that M, rather than (M, λ), is the left R-module. This allows some potential ambiguity, for a given abelian group may admit more than one left R-module structure. In only a few instances will this ambiguity be significant, and in these we shall be able to eliminate the ambiguity with special notation.

By a *right R-module* we mean an abelian group M and a ring homomorphism ρ of R into the right endomorphism ring of M. Shorn of unnecessary notation this means that there is a "right scalar multiplication"

$$(x, a) \mapsto xa \qquad (x \in M, a \in R)$$

from $M \times R$ to M satisfying for all $a, b \in R$ and $x, y \in M$

$$(x + y)a = xa + ya, \qquad x(ab) = (xa)b,$$

$$x(a + b) = xa + xb, \qquad x1 = x.$$

Thus, it is intuitively obvious (but see Exercise (2.1)) that the right R-modules are essentially the same as the left R^{op}-modules. So in particular, if R is commutative, we may be allowed to view the two concepts as identical.

2.1. Examples. (1) If D is a division ring, then a left D-module is simply our old friend a left D-vector space. In most elementary courses, we encounter only vector spaces over fields and hence are not concerned with sides. But for non-commutative division rings D, a left D-vector space is not the same as a right D-vector space.

(2) If V is a vector space of dimension n over a field K, then the ring $R = \mathbb{M}_n(K)$ of $n \times n$-matrices over K operates as K-linear transformations, and hence as abelian group endomorphisms, on V. Here in particular we have considerable choice. If we view R as operating from the left on column vectors, then V acquires the structure of a left R-module. If we let R operate on the right on row vectors, then V has the structure of a right R-module. But there is more. In either case the way that R operates on V is determined by the choice of basis; each such choice giving a different module structure to V. Of course all of the left structures obtained this way are in some sense "isomorphic" as are the various right structures. Still it must be recognized that strictly speaking these structures are different.

(3) For a ring R there is a unique ring homomorphism from $\mathbb{Z}$ to R (see Exercise (1.11)). So for every abelian group M there is a unique $\mathbb{Z}$-module structure on M. This is simply the structure given by the usual "multiple function"

$$(n, x) \mapsto nx.$$

(4) In Exercise (1.10) we found homomorphisms λ and ρ of the ring R into the left and right endomorphism rings, respectively, of the additive group of R. Thus each ring R induces a left R-module structure on its additive group and a right R-module structure on its additive group via the scalar multiplications

$$(a, x) \mapsto ax, \quad \text{and} \quad (x, a) \mapsto xa,$$

where ax and xa denote products in the ring R. These modules induced on the additive group of a ring R will be called the *regular left* and *regular right modules of R*, respectively.

(5) This last example admits an important extension. Let R and S be rings and let $\phi : R \to S$ be a ring homomorphism. Then ϕ induces both a left and a right R-module structure on the additive group of S. Indeed, the scalar multiplication, for the left R-module S, is given by

$$(r, s) \mapsto \phi(r)s \qquad (r \in R, s \in S)$$

where the product $\phi(r)s$ is computed in the ring S. The right R-module structure on S is defined similarly. Clearly this is an extension of the familiar business of viewing a field S as a vector space over each of its subfields R.

(6) There is one particularly important way of constructing new modules from old ones. The general theory will be discussed in §6. For now, however,

suppose $M_1, \ldots, M_n$ is a sequence of left R-modules. Then the cartesian product $M_1 \times \ldots \times M_n$ admits a natural R-module structure. That is, writing the elements of this product as n-tuples $(x_1, \ldots, x_n)$, the module operations are defined by the formulas

$$(x_1, \ldots, x_n) + (y_1, \ldots, y_n) = (x_1 + y_1, \ldots, x_n + y_n)$$

$$r(x_1, \ldots, x_n) = (rx_1, \ldots, rx_n).$$

This module, which we continue to denote by $M_1 \times \ldots \times M_n$ is called the *cartesian product (module)* of $M_1, \ldots, M_n$.

Except for a few exercises we shall not treat much of the elementary arithmetic of modules. Indeed, with few exceptions this elementary arithmetic differs only superficially from that of vector spaces. Perhaps the most dramatic difference is that with general modules we can expect $ax = 0$ even though neither a nor x is zero. The interested reader can find the general material in several standard texts.

The concept of a bimodule arises most naturally in the context of endomorphism rings of modules (see §4). Nevertheless, bimodules are simple enough to introduce directly. Thus, let R and S be two rings. An abelian group M is a *left R- right S-bimodule* in case M is both a left R-module and a right S-module for which the two scalar multiplications jointly satisfy

$$r(xs) = (rx)s \qquad (r \in R, \ s \in S, \ x \in M).$$

There are other styles of bimodules depending on the sides on which R and S operate. The crucial identity in the definition of, say, a *left R- left S-bimodule* is then

$$r(sx) = s(rx) \qquad (r \in R, \ s \in S, \ x \in M).$$

There is a very concise and suggestive notational device for describing the various flavors of modules. The following partial dictionary should suffice to explain this device:

$$_R M \text{ means } M \text{ is a left } R\text{-module}$$
$$M_R \text{ means } M \text{ is a right } R\text{-module}$$
$$_R M_S \text{ means } M \text{ is a left } R\text{- right } S\text{-bimodule}$$
$$_{R-S} M \text{ means } M \text{ is a left } R\text{- left } S\text{-bimodule}.$$

The bimodule $_{R-S}M$ is in essence the same object as the bimodule $_R M_{S^{op}}$ (see Exercise (2.1)). Thus we shall generally deal with left-right bimodules, and simply refer to $_R M_S$ as an (R, S)-bimodule. Note also that the $\mathbb{Z}$-module structure that the abelian group M admits (2.1.3) makes $_R M$ into a bimodule $_R M_{\mathbb{Z}}$.

Linear Combinations and Submodules

Let M be a left R-module. Then an abelian subgroup N of M is a *(left R-) submodule* of M in case N is stable under the endomorphisms of M induced by R. In other words, N is a submodule of M if and only if it is a subgroup of M "closed" under scalar multiplication by R. In particular, a submodule N of M is a left R-module on its own right. It is possible for a subgroup N of a left R-module M to be an R-module in terms of some representation $R \to End^l(N)$ without being an R-submodule of M. (See Exercise (2.2).)

If $X \subseteq M$ and $A \subseteq R$, then any element of M of the form

$$a_1 x_1 + \ldots + a_n x_n = \Sigma_{i=1}^n a_i x_i$$

with $x_1, \ldots, x_n \in X$ and $a_1, \ldots, a_n \in A$ is a *linear combination* of X with *coefficients* in A, or simply an *A-linear combination* of X. We shall denote the set of all such A-linear combinations of X by AX.

2.2. Proposition. *Let M be a left R-module and let X be a non-empty subset of M. Then RX is an R-submodule of M.*

Proof. The R-linear combinations of X are clearly closed under the group operation of M, and the identity

$$a(r_1 x_1 + \ldots + r_n x_n) = (ar_1)x_1 + \ldots + (ar_n)x_n$$

finishes the job. □

The subset $\{0\}$ of a module M is clearly a submodule of M. We call it the *zero submodule* and usually denote it by 0 alone. To avoid a special case later, we agree

$$R\varnothing = 0;$$

that is, 0 is the unique R-linear combination of $\varnothing$. The following, which is an easy exercise, characterizes submodules as those non-empty subsets "closed" under all R-linear combinations.

2.3. Proposition. *Let M be a left R-module and let N be a non-empty subset of M. Then the following are equivalent:*
   (a) *N is a submodule of M;*
   (b) *$RN = N$;*
   (c) *For all $a, b \in R$ and all $x, y \in N$*

$$ax + by \in N.$$ □

Of course for each of the various types of modules there is a corresponding notion of submodule, and there are results analogous to (2.2) and (2.3). For example, given $_R M_S$ a subset N is an (R, S)-submodule (strictly speaking, a left R-, right S-submodule) of M iff N is simultaneously an R-submodule and an S-submodule. Also, in this setting, for example, an (R, S)-linear combination of $X \subseteq M$ is simply an element of the form

$$r_1 x_1 s_1 + \ldots + r_n x_n s_n$$

with $r_i \in R$, $s_i \in S$, and $x_i \in X$ $(i = 1, \ldots, n)$. The set of all of these is abbreviated RXS. Then a non-empty subset N of M is an (R, S)-submodule if and only if $N = RNS$. This in turn is equivalent to containing those (R, S)-linear combinations of the form $rxs + r'x's'$ with $x, x' \in N$.

Like the subgroups of a group or the subspaces of a vector space, the set of submodules of a module M forms a complete modular lattice with respect to the partial order of set inclusion. Thus suppose that M is a module. If N is a submodule of M, we denote this fact by

$$N \leq M.$$

To avoid occasional ambiguity about the ring of scalars, we may also use such self-explanatory variations as

$$_R N \leq {_R M} \qquad \text{or} \qquad _R P_S \leq {_R Q_S}.$$

Let M be a left R-module and let $L \leq M$ and $N \leq M$ be submodules. Then it is clear from (2.3) that

$$L \leq N \qquad \text{iff} \qquad L \subseteq N.$$

In particular, the set $\mathscr{S}(M)$ of all submodules of M is partially ordered by $\leq$ (which on $\mathscr{S}(M)$ coincides with set inclusion). The submodules 0 and M of M are the unique smallest and largest elements of $\mathscr{S}(M)$. Moreover, if $\mathscr{A}$ is any non-empty subset of $\mathscr{S}(M)$, then it is an immediate consequence of (2.3) that

$$\cap \mathscr{A} \in \mathscr{S}(M).$$

Since clearly $\cap \mathscr{A}$ must be the greatest lower bound of $\mathscr{A}$ in $\mathscr{S}(M)$, we infer that $\mathscr{S}(M)$ is a complete lattice (see (0.6)). Although the partial order for the lattice $\mathscr{S}(M)$ is set inclusion and the greatest lower bound is intersection, the least upper bound of $\mathscr{A} \subseteq \mathscr{S}(M)$ is not generally its union. Indeed the union of two submodules is rarely a submodule. (See Example 4 of (0.5).) To characterize the join in $\mathscr{S}(M)$ we introduce some special, but entirely standard, notation; if $M_1, \ldots, M_n$ are non-empty subsets of M, we set

$$M_1 + \ldots + M_n = \{x_1 + \ldots + x_n \mid x_i \in M_i \ (i = 1, \ldots, n)\}.$$

Another easy consequence of (2.3) is

2.4. Lemma. *If M is a left R-module and if $M_1, \ldots, M_n$ are submodules of M, then $M_1 + \ldots + M_n$ is also a submodule of M. In fact, $M_1 + \ldots + M_n$ is the set of all R-linear combinations of $M_1 \cup \ldots \cup M_n$.* $\qquad\square$

If $\mathscr{X}$ is an arbitrary collection of subsets of M, then there is no reasonable concept of "sum" of $\mathscr{X}$. However, motivated by this last lemma we do define the sum $\Sigma \mathscr{A}$ of a family $\mathscr{A}$ of submodules of M to be the set of all R-linear combinations of $\cup \mathscr{A}$. It is easy to see that if $\mathscr{A} = \{M_\alpha \mid \alpha \in A\}$, then

$$\Sigma \mathscr{A} = \Sigma_A M_\alpha = \cup \{M_{\alpha_1} + \ldots + M_{\alpha_n} \mid a_1, \ldots, \alpha_n \in A \ (n = 1, 2, \ldots)\},$$

i.e., each element of $\Sigma_A M_\alpha$ can be written as a finite sum

$$\Sigma_{i=1}^n x_{\alpha_i} \qquad (x_{\alpha_i} \in M_{\alpha_i}, \alpha_i \in A).$$

By (2.2) then, the sum $\Sigma \mathscr{A}$ of a set $\mathscr{A}$ of submodules is again a submodule, and moreover, if N is any submodule of M containing all the submodules in $\mathscr{A}$, then by (2.3) it must contain the sum $\Sigma \mathscr{A}$. Therefore it is this sum $\Sigma \mathscr{A}$ that is the least upper bound of $\mathscr{A}$ in $\mathscr{S}(M)$.

Let H, K, and L be submodules of M. Then it is easy to check that

$$H \cap (K + L) \geq (H \cap K) + (H \cap L).$$

In general this inequality may be strict, i.e., $\mathscr{S}(M)$ need not be a distributive lattice. However, if $H \geq K$ and $h \in H, k \in K, l \in L$ with $h = k + l$ then, since $k \in K = H \cap K$, $\mathscr{S}(M)$ does satisfy the modularity condition (see (0.5)).

In summary we have

2.5. Proposition. *If M is a left R-module, then the set $\mathscr{S}(M)$ of submodules of M is a complete modular lattice with respect to $\leq$. In this lattice, if $\mathscr{A}$ is a non-empty set, then its join and meet are given by*

$$\Sigma \mathscr{A} \qquad and \qquad \cap \mathscr{A},$$

respectively. In particular, if K and L are submodules of M, then

$$K + L \qquad and \qquad K \cap L$$

are their join and meet, respectively; and if H is another submodule of M, then

$$K \leq H \qquad implies \qquad H \cap (K + L) = K + (H \cap L). \qquad \square$$

These submodule lattices $\mathscr{S}(M)$ provide a great deal of information about the nature of the modules, and hence about the scalar ring. In many instances we are able to obtain very explicit information about the ring from knowledge of these lattices. Conversely, for certain rings the behavior of these lattices is quite civilized; a familiar example is offered by modules (= vector spaces) over fields. In general, however, modules can be very unpredictable; just some of the less extreme pathology they display will be considered in the exercises.

Given a module M and a subset $X \subseteq M$, the set $\mathscr{A}$ of all submodules of M that contain X contains M and so is non-empty. Its intersection $\cap \mathscr{A}$ is again a submodule of M and it is, in fact, the unique smallest submodule of M that contains X. We call it the submodule of M *spanned* by X.

2.6. Proposition. *If M is a left R-module and if X is a subset of M, then the submodule of M spanned by X is just RX, the set of all R-linear combinations of X.*

Proof. By (2.2), RX is a submodule of M and since $1x = x$ for all $x \in M$, we certainly have $X \subseteq RX$. Finally, by (2.3), any submodule that contains X must contain the linear combinations RX. $\qquad \square$

If $(M_\alpha)_{\alpha \in A}$ are submodules of M, then $\Sigma_A M_\alpha$ is the submodule *spanned* by $(M_\alpha)_{\alpha \in A}$. Thus if

$$M = \Sigma_A M_\alpha,$$

then we say that the submodules $(M_\alpha)_{\alpha \in A}$ *span* M. If X is a subset of ${}_R M$ such that

$$RX = M,$$

then X is said to *span* M, and X is called a *spanning set for* M. A module with a finite spanning set is said to be *finitely spanned*. A module with a single element spanning set is a *cyclic* module. Thus a cyclic left module is one of the form $M = R\{x\}$ where x is some element of M; and we write

$$M = Rx = \{rx \mid r \in R\}.$$

Of course, the regular modules ${}_R R$ and R_R are cyclic. Now it is clear that every module is spanned by the set of its cyclic submodules.

2.7. Proposition. *If X is a spanning set for ${}_R M$, then*

$$M = \Sigma_{x \in X} Rx. \qquad \square$$

A module M is *simple* in case $M \neq 0$ and it has no non-trivial submodules. Not only is such a module cyclic, but clearly a non-zero module is simple iff it is spanned by each of its non-zero elements. Somewhat like the primes in arithmetic the simple modules are basic building blocks in the theory of modules. Indeed note that an abelian group is simple iff it is isomorphic to $\mathbb{Z}_p$ for some prime $p \in \mathbb{P}$.

Clearly the module itself is the greatest element in its lattice of submodules; hence in the terminology of posets it is a maximal (indeed the only maximal) submodule of itself. But it is the next level, the maximal proper submodules, that is of real interest. Dually, the zero submodule is of little consequence, but the minimal non-zero submodules of a module are very important. As a result one rather weird bit of terminology has evolved. That is,

maximal submodule means *maximal proper submodule*

minimal submodule means *minimal non-zero submodule*.

For example, M is simple (hence non-zero) iff M is a minimal and 0 is a maximal submodule! The question of existence of minimal or maximal submodules is critical and not trivial. Note for example that the abelian group $\mathbb{Z}$ has no minimal subgroup ($= \mathbb{Z}$-submodule). (See also Exercise (2.8).) On the other hand there is at least one very important class of modules, each with maximal submodules.

2.8 Theorem. *Let M be a non-zero left R-module with a finite spanning set. Then every proper submodule of M is contained in a maximal submodule. In particular, M has a maximal submodule.*

Proof. Let K be a proper submodule of M. Then there is a finite sequence $x_1, \ldots, x_n \in M$ such that

$$M = K + Rx_1 + \ldots + Rx_n.$$

So certainly among all such sequences there is one of minimal length (presumably there are several such sequences), and so we may assume that $x_1, \ldots, x_n$ has minimal length. Then

$$L = K + Rx_2 + \ldots + Rx_n$$

is a proper submodule of M (otherwise the too short sequence $x_2, \ldots, x_n$ would do for $x_1, x_2, \ldots, x_n$). Let $\mathscr{P}$ be the set of all proper submodules of M that contain L. Clearly, $\mathscr{P}$ is a non-empty subposet of the lattice of submodules of M for $L \in \mathscr{P}$. Now a submodule N that contains L is in $\mathscr{P}$ iff $x_1 \notin N$. We apply the Maximal Principle (0.9) to $\mathscr{P}$. Suppose $\mathscr{C}$ is a non-empty chain in the poset $\mathscr{P}$. Set $V = \cup \mathscr{C}$. We claim that V is a submodule of M. For if $a, b \in R$ and $x, y \in V$, then for some $N_x, N_y \in \mathscr{C}$, $x \in N_x$ and $y \in N_y$. Since $\mathscr{C}$ is a chain, we may assume $N_x \leq N_y$. So $x, y \in N_y$ and (2.3) $ax + by \in N_y \subseteq V$. Thus (2.3) V is a submodule of M as claimed. But clearly since x_1 is in no element of $\mathscr{C}$, $x_1 \notin V$. We have shown then that every non-empty chain in $\mathscr{P}$ has an upper bound in $\mathscr{P}$, namely its union, so by the Maximal Principle $\mathscr{P}$ has a maximal element, say N. Because N is maximal in $\mathscr{P}$ any strictly larger submodule of M is not in $\mathscr{P}$, and so contains x_1. But then any such module must contain $N + Rx_1 \geq L + Rx_1 = M$. Thus N is a maximal (proper) submodule of M containing K. For the final statement of the Theorem let $K = 0$. □

There is in one case a significant difference between left modules and bimodules. If $_R M_S$ is an R-S-bimodule, and if $x \in M$, then the cyclic submodule spanned by x is $RxS = (Rx)S$, but this need not be just the elements rxs. (See Exercise (2.3).)

Factor Modules

Just as for vector spaces, there is a factor module of a module with respect to each of its submodules. Let M be a left R-module and let K be a submodule. Then it is easy to see that the set of cosets

$$M/K = \{x + K \mid x \in M\}$$

is a left R-module relative to the addition and scalar multiplication defined via

$$(x + K) + (y + K) = (x + y) + K, \qquad a(x + K) = ax + K.$$

Of course, the additive identity and inverses are given by

$$K = 0 + K \qquad \text{and} \qquad -(x + K) = -x + K.$$

The resulting module M/K is called (the *left R-factor module of*) M modulo K. Entirely similar constructions exist for other types of factor modules. Thus, for instance, if we have $_R M_S$ and an R-S-submodule K, then the factor group M/K is a left R- right S-bimodule via

$$r(x + K)s = rxs + K$$

for all $r \in R$, $x \in M$, $s \in S$.

Let K be a submodule of M. Then it is easy to see that the set

$$\mathscr{S}(M)/K = \{H \in \mathscr{S}(M) \,|\, K \leq H\}$$

is a sublattice of $\mathscr{S}(M)$. Moreover, for each H in this sublattice

$$n_K(H) = H/K$$

is obviously a submodule of the factor module M/K. Since clearly $H \leq H'$ implies $n_K(H) \leq n_K(H')$, we have that n_K defines an order-preserving function from $\mathscr{S}(M)/K$ to $\mathscr{S}(M/K)$. On the other hand, if T is a submodule of M/K, then

$$n_K^{\leftarrow}(T) = \{x \in M \,|\, x + K \in T\}$$

is a submodule of M, and, since $0 + K = k + K \in T$ for all $k \in K$, clearly $K \leq n_K^{\leftarrow}(T)$. We see at once that $n_K n_K^{\leftarrow}(T) = T$ and $n_K^{\leftarrow} n_K(H) \geq H$ for all $T \in \mathscr{S}(M/K)$ and for all $H \in \mathscr{S}(M)/K$. But if $x \in n_K^{\leftarrow} n_K(H)$, then $x + K = a + K$ for some $a \in H$ and so since $K \leq H$, we have $x \in H$. Thus, n_K and $n_K^{\leftarrow}$ define inverse bijections. Finally, since $n_K^{\leftarrow}$ is also order-preserving, we have by (0.8) the important

2.9. Proposition. *Let M be a left R-module and let K be a submodule of M. Then the lattice of submodules of the factor module M/K is lattice isomorphic to the submodules of M that contain K via the inverse maps*

$$n_K : H \mapsto H/K = \{x + K \,|\, x \in H\}$$

$$n_K^{\leftarrow} : T \mapsto n_K^{\leftarrow}(T) = \{x \in M \,|\, x + K \in T\}. \qquad \square$$

Since a module is simple iff its lattice of submodules is a two element chain, we have the

2.10. Corollary. *A factor module M/K is simple if and only if K is a maximal submodule of M.* $\qquad \square$

Change of Rings

There are certain important and natural ways that a module over one ring inherits a module structure over a second. For example, every module over one ring R is, in a completely natural way, a module over all subrings of R. In general, some of these "changes of rings" are induced by ring homomorphisms. Thus suppose that M is a left S-module, that R is a second ring, and that $\phi : R \to S$ is a ring homomorphism. If the structure $_SM$ is obtained from the ring homomorphism $\lambda : S \to End^l(M)$, then

$$\lambda \phi : R \to End^l(M)$$

induces a left R-structure on M. Here the scalar multiplication is given by $(r, x) \mapsto \phi(r)x$. Thus $_SM$ is a left R-module $_RM$ with

$$rx = \phi(r)x \qquad (r \in R, \; x \in M).$$

If S' is a subring of S, then the inclusion map $i_{S'}: S' \to S$, a ring homomorphism, induces an S'-module structure $_{S'}M$ on each $_S M$ with the induced multiplication

$$(s', x) \mapsto s'x = i_{S'}(s')x.$$

So with each S-module $_S M$ and each ring homomorphism $\phi: R \to S$, there are four modules

$$_S M, \quad _R M, \quad _{\phi(R)}M, \quad _{\mathbb{Z}}M.$$

Clearly each submodule of any one of these is a submodule of each subsequent one and since the meet and join of submodules are just intersection and sum, the submodule lattice of any of these is a sublattice of that of the subsequent modules. Notice also that the submodule lattices of $_R M$ and $_{\phi(R)}M$ are the same. These nearly trivial but important facts we state formally in

2.11. Proposition. *Let $\phi: R \to S$ be a ring homomorphism and let M be an abelian group that is simultaneously a left R-module and a left S-module such that for all $r \in R$, $x \in M$, $rm = \phi(r)m$. Then, as lattices*

$$\mathscr{S}(_S M) \leq \mathscr{S}(_R M) = \mathscr{S}(_{\phi(R)}M) \leq \mathscr{S}(_{\mathbb{Z}}M). \qquad \square$$

Of course the inclusions stated in (2.11) in general are not equalities. For example, in a one-dimensional $\mathbb{R}$-vector space there are $\mathbb{Q}$-subspaces that are not $\mathbb{R}$-subspaces, and there are abelian subgroups that are not $\mathbb{Q}$-subspaces.

Suppose now that M is a left R-module via the ring homomorphism $\lambda: R \to End^l(M)$. As usual though, we abbreviate $ax = \lambda(a)(x)$. Then the kernel

$$K = Ker\,\lambda = \{a \in R \,|\, ax = 0 \ (x \in M)\}$$

is a two-sided ideal of R, called the *annihilator of M in R*. If $K = 0$ (i.e., λ is injective), we say that M is a *faithful* left R-module. By (1.4) we have that for every ideal I of R contained in this annihilator, there is a unique ring homomorphism

$$\eta: R/I \to End^l(M)$$

such that $\eta n_I = \lambda$. Thus, for each such ideal I there is induced on M a left R/I module structure called the *natural R/I-structure $_{R/I}M$*. In this case the scalar multiplication is given by

$$(a + I, x) \mapsto (a + I)x = ax \qquad (a \in R, x \in M).$$

Thus the R-structure on M is just that induced as in the previous paragraph by the R/I structure and the surjective ring homomorphism $n_I: R \to R/I$; so by (2.11) and (1.4) we have

2.12. Corollary. *Let M be a left R-module, and let I be an ideal of R contained in the annihilator of M. Then*

(1) *a subgroup of M is an R-submodule iff it is an R/I-submodule. That is, the lattices of R-submodules and R/I-submodules coincide;*
(2) *M is faithful as a left R/I-module iff I is the annihilator of M.* ☐

As an easy and important illustration of some of this, let M be a finite non-zero $\mathbb{Z}$-module (i.e., abelian group). Then its annihilator K in $\mathbb{Z}$ is a proper ideal. Indeed, K is the (principal) ideal $\mathbb{Z}k$ where k is the least positive integer with $kx = 0$ for all $x \in M$. Then M becomes a $\mathbb{Z}/\mathbb{Z}k(\cong \mathbb{Z}_k)$ module with respect to the scalar multiplication

$$(m + \mathbb{Z}k)x = mx.$$

In particular, if $K = \mathbb{Z}p$ with p prime, then $\mathbb{Z}_p$ is a field, M is naturally a $\mathbb{Z}_p$-vector space, and the lattice of subgroups of M is precisely that of the lattice of $\mathbb{Z}_p$-subspaces of the $\mathbb{Z}_p$-vector space $_{\mathbb{Z}_p}M$.

2.13. Rings as Bimodules. As we saw in (2.1.4) the additive group of a ring R is both a left R-module and a right R-module via left and right ring multiplications. Since ring multiplication is associative,

$$a(xb) = (ax)b,$$

it follows that these left and right multiplications give R the structure of a bimodule. Henceforth when we speak of a ring R as a module (left, right, or two-sided) over itself, we mean this regular module structure, and unless we explicitly suspend this agreement,

$$_RR, \qquad R_R, \qquad \text{and} \qquad _RR_R$$

will denote these regular modules.

Clearly, if R is a ring, then a non-empty subset I of R is an ideal iff it is a submodule of the bimodule $_RR_R$. This is equivalent to

$$I = RIR.$$

More generally, the submodules of $_RR$ are called *left ideals* of R and the submodules of R_R are called *right ideals* of R. Thus a non-empty set I of R is a left ideal iff

$$I = RI.$$

A left, right, or two-sided cyclic ideal is usually said to be *principal*.

Since the ideals—left, right, or two-sided—of R are merely the submodules of certain special modules, we have in particular that each of these sets is a complete lattice with respect to set inclusion. And of course these lattices satisfy the general properties of lattices of submodules. The convention about maximal and minimal submodules also is adopted: maximal (left, right, two-sided) ideal means proper and minimal means non-zero. Also for example, if A is a non-empty subset of R, then

$$RA, \qquad AR, \qquad \text{and} \qquad RAR$$

are the left ideal, the right ideal, and the (two-sided) ideal of R spanned by A, respectively. In the same way, for example, if I is an ideal of R, then the lattice of left ideals of R/I is isomorphic to the lattice of left ideals of R that contain I. Therefore, we shall feel free to apply our various results about modules in general to these special ones. However, as we shall see in the next section, when we are considering homomorphisms we shall have to exercise caution about such translations.

Annihilators

Given a left R-module M we should like to read properties of R from M and conversely, properties of M from R. One very valuable tool for obtaining some of this exchange of information is provided by "annihilators". We have mentioned the annihilator in a ring of an R-module M. More generally, let M be a left R-module. Then for each $X \subseteq M$, the (left) annihilator of X in R is

$$l_R(X) = \{r \in R \,|\, rx = 0 \quad (x \in X)\},$$

and, for each $A \subseteq R$, the (right) annihilator of A in M is

$$r_M(A) = \{x \in M \,|\, ax = 0 \quad (a \in A)\}.$$

For singletons $\{x\}$ and $\{a\}$, we usually abbreviate to $l_R(x)$ and $r_M(a)$. When there is no chance for ambiguity, we may omit the subscripts R and M. Also, of course, beginning with a right R-module M, we encounter the right annihilator $r_R(X)$ and the left annihilator $l_M(A)$. There is some other fairly obvious terminology; for example, if $A \subseteq l_R(X)$, we may say that A annihilates X. The value of these annihilators will not be evident until much later, but their basic properties are easy enough to obtain now.

2.14. Proposition. Let $_RM_S$ be a bimodule, let $X \subseteq M$ and let $A \subseteq R$. Then
(1) $l_R(X)$ is a left ideal of R;
(2) $r_M(A)$ is a submodule of M_S.
Moreover, if X is a submodule of $_RM$, then $l_R(X)$ is an ideal of R. If A is a right ideal of R, then $r_M(A)$ is a submodule of $_RM_S$. If R is commutative, then $l_R(X)$ is an ideal and $r_M(A)$ is a submodule of $_RM_S$.

Proof. Consider (2). Let $x, y \in r_M(A)$ and let $s, s' \in S$. Then for each $a \in A$, we have (since $_RM_S$ is a bimodule):

$$a(xs + ys') = a(xs) + a(ys') = (ax)s + (ay)s' = 0.$$

So (2.3.c) since $r_M(A) \neq \varnothing$ (it must contain 0), it is a submodule of M_S. The rest of the proof is equally simple and will be omitted. $\qquad\square$

2.15. Proposition. Let $_RM$ be a left R-module, let X, Y be subsets of M and let A, B be subsets of R. Then
(1) $X \subseteq Y$ implies $l_R(X) \supseteq l_R(Y)$ $\quad$ and $\quad$ $A \subseteq B$ implies $r_M(A) \supseteq r_M(B)$.
(2) $X \subseteq r_M l_R(X)$ $\quad$ and $\quad$ $A \subseteq l_R r_M(A)$.
(3) $l_R(X) = l_R r_M l_R(X)$ $\quad$ and $\quad$ $r_M(A) = r_M l_R r_M(A)$.

Proof. (1) and (2) are really easy. For (3) apply (2) to $l_R(X)$ for $l_R(X) \subseteq l_R r_M l_R(X)$. Then apply (1) to the first assertion of (2) for $l_R(X) \supseteq l_R r_M l_R(X)$.   □

One very significant bit of information has emerged already. Suppose $_R M_S$ is a bimodule. Then

$$r_M : I \mapsto r_M(I) \qquad \text{and} \qquad l_R : K \mapsto l_R(K)$$

are order-reversing maps between the poset of left ideals I of R and the poset of right S-submodules K of M. Of course these maps are not always bijective since the inclusion of (1) and (2) can be strict (consider $M = {}_\mathbb{Z}\mathbb{Z}_\mathbb{Z}$). Also in general they are not lattice anti-homomorphisms (see Exercise (2.15)) but as the next result shows, they come close. Curiously enough there is an important lattice anti-isomorphism (of new lattices) lurking in the wings. (See Exercise (2.16).)

2.16. Proposition. *Let $_R M$ be a left R-module. Let $(K_\alpha)_{\alpha \in A}$ and $(I_\alpha)_{\alpha \in A}$ be subgroups of the additive groups of M and R, respectively. Then*
(1) $l_R(\Sigma_A K_\alpha) = \cap_A l_R(K_\alpha)$ *and* $r_M(\Sigma_A I_\alpha) = \cap_A r_M(I_\alpha)$;
(2) $\Sigma_A l_R(K_\alpha) \subseteq l_R(\cap_A K_\alpha)$ *and* $\Sigma_A r_M(I_\alpha) \subseteq r_M(\cap_A I_\alpha)$.

Proof. (1) Since $K_\beta \leq \Sigma_A K_\alpha$ for each β, apply (2.15.1) to get $l_R(\Sigma_A K_\alpha) \subseteq l_R(K_\alpha)$ for each α. On the other hand, if an element $r \in R$ annihilates every K_α, then it certainly annihilates every sum of elements from these K_α. Thus, $\cap_A l_R(K_\alpha) \subseteq l_R(\Sigma_A K_\alpha)$. A similar argument handles the right annihilators.
(2) Clearly, $\cap_A K_\alpha \subseteq K_\beta$ for each $\beta \in A$. So by (2.15.1), $l_R(K_\beta) \subseteq l_R(\cap_A K_\alpha)$ for each β. Now $l_R(\cap_A K_\alpha)$ is a left ideal of R (2.14.1), so $\Sigma_{\beta \in A} l_R(K_\beta) \subseteq l_R(\cap_A K_\alpha)$. Again a similar argument for right annihilators.   □

2. Exercises

1. Let R be a ring and let R^{op} be its opposite ring. Denote the multiplication operation in R^{op} by $*$, thus, $r * s = sr$ for all $r, s \in R$.
   (1) Let M be a left R-module. Define a function $* : M \times R^{op} \to M$ via $(x, r) \mapsto x * r = rx$. Prove that under this operation M is a right R^{op} module; we shall denote it by M^{op}. [Hint: See (1.15).]
   (2) Let S be a second ring and let M be a left R- left S-bimodule. Prove that with the operation $* : M \times R^{op} \to M$ of part (1), M is a left S- right R^{op}-bimodule.
2. (1) Let M be a non-zero abelian group, $L = End^l(M)$, and $R = End^r(M)$. Then $_L M$ and M_R. Show that $_L M_R$ iff L is commutative. [Hint: Exercise (2.1).]
   (2) Let $_\mathbb{C} V$ be a non-zero complex vector space. The abelian group V becomes a left $\mathbb{C}$-vector space $_\mathbb{C} \bar{V}$ with the scalar multiplication $(\alpha, x) \mapsto \bar{\alpha} x$ ($\bar{\alpha}$ is the conjugate of α). Prove that neither of these $\mathbb{C}$-vector

spaces $_{\mathbb{C}}V$ and $_{\mathbb{C}}\overline{V}$ is a subspace of the other and that these two $\mathbb{C}$-scalar multiplications do not form a $(\mathbb{C}, \mathbb{C})$-bimodule.

3. Let R be a ring and $x \in R$. Prove that the ideal (i.e., (R, R)-submodule of $_RR_R$) of R generated by x is the set of all finite sums $r_1xs_1 + \dots + r_nxs_n$ for $n \in \mathbb{N}$, $r_1, \dots, r_n$, $s_1, \dots, s_n \in R$. Give an example showing that $\{rxs \mid r, s \in R\}$ need not be an ideal of R. [Hint: Consider $\mathbb{M}_2(\mathbb{Q})$.]

4. Let M be a left R-module, let $A, B \subseteq R$, and let $X \subseteq M$. Prove that:
   (1) $AX \le M$ whenever A is a left ideal of R;
   (2) $A(BX) = (AB)X$;
   (3) $A(\Sigma_C X_\gamma) = \Sigma_C AX_\gamma$ whenever each X_γ is a subgroup of $_{\mathbb{Z}}M$ ($\gamma \in C$);
   (4) $(\Sigma_C A_\gamma)X = \Sigma_C A_\gamma X$ whenever each A_γ is a subgroup of $_{\mathbb{Z}}R$.

5. (1) Let I be left ideal and J be a two-sided ideal in R. Prove that if I and J are nil (nilpotent), then so is the left ideal $I + J$. [Hint: Consider $(I + J)/J$ in R/J. (Also note that I nilpotent means $I^n = 0$ for some $n \in \mathbb{N}$.)]
   (2) Prove that if I is a left ideal in R, then I is nilpotent iff IR is nilpotent. Conclude that if I and K are nilpotent left ideals then so is $I + K$.
   (3) Let $\mathscr{I}$ be a set of nilpotent left ideals of R. Prove that $\Sigma\mathscr{I}$ is a nil left ideal.
   (4) Let $\mathscr{J}$ be a set of nil ideals in R. Prove that $\Sigma\mathscr{J}$ is a nil ideal.

6. These are (*Hasse*) *diagrams* of three finite posets

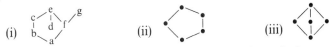

For example, the first poset has seven elements, a through g, a is the meet of $\{b, f\}$ and also of $\{c, f\}$, e is the join of $\{c, d\}$ and of $\{a, d\}$, c is the join of $\{b, c\}$ and the pair $\{d, g\}$ has neither a join nor a meet.
   (1) Prove that if a lattice L has a sublattice with diagram (ii), (respectively (iii)), then L is not modular (distributive). Conversely, prove that a non-modular lattice actually contains a sublattice like (ii). (A non-distributive lattice must also contain a copy of (ii) or (iii). (See Birkhoff [66].)
   (2) Let L be a modular lattice and let $a, b, c \in L$ with $a \le b$. Prove that if $a \vee c = b \vee c$ and $a \wedge c = b \wedge c$, then $a = b$.
   (3) Prove that a lattice L is distributive iff $a \vee (b \wedge c) = (a \vee b) \wedge (a \vee c)$ for all $a, b, c \in L$. (See (0.5).)

7. (1) Sketch the (Hasse) diagrams of the lattices of submodules of the $\mathbb{Z}$-modules $\mathbb{Z}_8$, $\mathbb{Z}_{24}$, $\mathbb{Z}_{15}$, $\mathbb{Z}_{30}$. Using the test claimed in Exercise (2.6) determine whether any of these are distributive.
   (2) Let R be the ring of all 2×2 upper triangular matrices over the field $\mathbb{Z}_2$. Sketch the diagrams of the lattices of submodules of $_RR$, of R_R, and of $_RR_R$.
   (3) Which, if any, of these modules are spanned by their minimal submodules?
   (4) For each module in (1) and (2) determine the intersection of the maximal submodules.

8. Let $p \in \mathbb{P}$ be a positive prime. Then $M = \{a/p^n \in \mathbb{Q} \mid a \in \mathbb{Z} \text{ and } n \in \mathbb{N}\}$ is

an additive subgroup of $\mathbb{Q}$ with subgroup $\mathbb{Z}$. Denote the factor group $M/\mathbb{Z}$ by $\mathbb{Z}_{p^\infty}$. Clearly the elements $a/p^n \in M$ with $0 \le a \le p^n$ form a set of representatives.

(1) Prove that for each $x \in \mathbb{Z}_{p^\infty}$ and each $m \ne 0$ in $\mathbb{Z}$, there is a $y \in \mathbb{Z}_{p^\infty}$ with $x = my$.

(2) Prove that every proper subgroup of $\mathbb{Z}_{p^\infty}$ is cyclic and spanned by $1/p^n$ for some n. Then deduce that the lattice of subgroups of $\mathbb{Z}_{p^\infty}$ is a well ordered chain and that $\mathbb{Z}_{p^\infty}$ has no maximal subgroup.

9. Let M be a non-zero module, let N be a proper submodule, and let $x \in M \backslash N$. Prove that:

(1) M has a submodule K maximal with respect to $N \le K$ and $x \notin K$.

(2) If $M = Rx + N$, then M has a maximal submodule K with $N \le K$ and $x \notin K$.

10. Let I and M be proper ideals of a ring R. Prove that:

(1) M is a maximal ideal iff R/M is a simple ring.

(2) R has a maximal ideal that contains I.

(3) R has at least one maximal ideal.

11. Let R be commutative. A proper ideal P of R is *prime* in case $ab \in P$ implies that $a \in P$ or $b \in P$. Prove that

(1) A proper ideal P is prime iff the factor ring R/P is an integral domain. Thus every maximal ideal is prime.

(2) There exist chains of prime ideals of arbitrary length. [Hint: Consider $\mathbb{Z}[X_1, \dots, X_n]$.]

(3) If there is an $n \in \mathbb{N}$ such that $x^n = x$ for each $x \in R$, then every prime ideal is maximal.

(4) Every prime ideal of R contains a minimal (possibly 0) prime ideal. [Hint: Apply the dual of the Maximal Principle.]

12. A commutative ring is a *local ring* in case it has a unique maximal ideal. (See §15 for the non-commutative generalization.)

(1) Prove that every commutative ring whose ideal lattice is a chain (e.g., $\mathbb{Z}_{p^n}$) is a local ring. Can you give an example of a commutative local ring whose ideal lattice is not a chain? [Hint: Try a factor ring of $\mathbb{Q}[X, Y]$.]

(2) Let $p \in \mathbb{Z}$ be a prime and set

$$\mathbb{Z}_{(p)} = \{a/b \in \mathbb{Q} \mid b \notin \mathbb{Z}p \,(a/b \text{ in lowest terms})\}$$

Prove that $\mathbb{Z}_{(p)}$ is a local ring with maximal ideal $p\mathbb{Z}_{(p)}$.

(3) Prove that the ideal lattice of the local ring $\mathbb{Z}_{(p)}$ is co-well-ordered (i.e., every non-empty set has a greatest element).

13. Generalize Exercise (1.9) by proving that if a ring R, not necessarily commutative, has a unique maximal ideal, then $Cen\,R$ is a local ring. Thus, every such ring R can be viewed as a (central) algebra over a local ring. Show, however, that a ring whose center is local (even a field!) need not have a unique maximal ideal. [Hint: Exercise (2.7.2).]

14. (1) Let I be a left ideal of a ring R. Prove that $I_R(R/I)$ is the unique largest two-sided ideal of R that is contained in I.

(2) On the other hand show that it is possible for I to be a maximal left ideal yet contain no maximal ideal of R. [Hint: Try Exercise (1.7).]

15. Let R be the subring of $\mathbb{Z}[X] \times \mathbb{Z}[Y]$ of all (f, g) with $f(0) = g(0)$ (i.e., all pairs with the same constant term). It is easy to see that R is isomorphic to the factor ring of $\mathbb{Z}[X, Y]$ modulo the principal ideal generated by XY. Let $M = {}_R R$. Show that the inclusions of (2.16.2) can be strict even if we choose the K_α's to be annihilators of ideals and the I_α's to be annihilators of submodules. [Hint: Consider for example the submodule generated by $(X, 0)$.]

16. Let M be a left R-module. Let $\mathscr{L}_R(M) = \{l_R(X) \mid X \subseteq M\}$ and $\mathscr{R}_M(R) = \{r_M(A) \mid A \subseteq R\}$. Observe that $A \in \mathscr{L}_R(M)$ iff $A = l_R r_M(A)$ and $X \in \mathscr{R}_M(R)$ iff $X = r_M l_R(X)$. Now prove that:

(1) Both $\mathscr{L}_R(M)$ and $\mathscr{R}_M(R)$ are closed under arbitrary intersections, whence as posets partially ordered by set inclusion $\mathscr{L}_R(M)$ and $\mathscr{R}_M(R)$ are complete lattices. Moreover, if $\mathscr{A} \subseteq \mathscr{L}_R(M)$ and $\mathscr{X} \subseteq \mathscr{R}_M(R)$, then

$$\inf \mathscr{A} = \cap \mathscr{A}, \qquad \sup \mathscr{A} = l_R(r_M(\Sigma \mathscr{A}))$$

$$\inf \mathscr{X} = \cap \mathscr{X}, \qquad \sup \mathscr{X} = r_M(l_R(\Sigma \mathscr{X})).$$

(2) The maps $r_M : A \mapsto r_M(A)$ and $l_R : X \mapsto l_R(X)$ are lattice anti-isomorphisms between $\mathscr{L}_R(M)$ and $\mathscr{R}_M(R)$.

17. Let V be a left vector space over a division ring D. A subset $X \subseteq V$ is *linearly independent* in case for every finite sequence $x_1, \ldots, x_n$ of distinct elements of X and every $d_1, \ldots, d_n \in D$

$$d_1 x_1 + \ldots + d_n x_n = 0 \quad \text{implies} \quad d_1 = \ldots = d_n = 0.$$

A linearly independent spanning set of V is called a *basis* for V. (Note that $\varnothing$ is a basis for 0.)

(1) Use the Maximal Principle to prove that if Y is a linearly independent subset of V and X is a spanning set, then there exists a subset $X' \subseteq X$ such that $Y \cup X'$ is a basis for V.

(2) Prove that every vector space has a basis, and that every maximal linearly independent subset and every minimal spanning set of a vector space is a basis.

(3) Prove that if $W \leq V$ and X is a basis for V, then there is a subset $X' \subseteq X$ such that $V = W + DX'$ and $W \cap DX' = 0$.

18. If X is a basis for a vector space V, then the *dimension* of V is $\dim V = \operatorname{card} X$. This is independent of the choice of basis. Indeed, prove that if X and Y are bases for V, then $\operatorname{card} X = \operatorname{card} Y$. [Hint: If X is finite, use Exercise (2.17.1) and induction to show that $\operatorname{card} Y \leq \operatorname{card} X$. On the other hand suppose that X is infinite. Then for each $x \in X$ there is a finite subset $F(x) = \{y_1, \ldots, y_n\}$ of Y such that $x \in Dy_1 + \ldots + Dy_n$. Show that $Y = \cup_{x \in X} F(x)$ and hence that $\operatorname{card} Y \leq \operatorname{card}(\mathbb{N} \times X) = \operatorname{card} X$. (See (0.10).)]

§3. Homomorphisms of Modules

If M and N are two left R-modules, then a function $f : M \to N$ is a (*left R-*)
homomorphism in case for all $a, b \in R$ and all $x, y \in M$

$$f(ax + by) = af(x) + bf(y);$$

i.e., in case f is R-*linear*. Note that here "left" has nothing to do with the side
on which we write f. Thus, if we write f on the right,

$$(ax + by)f = a((x)f) + b((y)f).$$

Or if M_R and N_R, then $f : M \to N$ is a right R-homophorism iff

$$f(xa + yb) = f(x)a + f(y)b.$$

The point really is that to be a module homomorphism f must preserve the
defining structure. Thus, if the abelian groups M and N are, say, left R-modules
via ring homomorphisms λ and λ' of R into their left endomorphism rings,
then an abelian group homomorphism $f : M \to N$ is an R-homomorphism iff
for each $a \in R$ the diagram

commutes.

For bimodules we have the obvious variations. Given bimodules $_R M_S$ and
$_R N_S$, an (R,S)-homomorphism from M to N is simply a function $f : M \to N$
that is linear over both R and S; this can be expressed "jointly". Thus,
$f : M \to N$ is an (R,S)-homomorphism iff for all $r, r' \in R$, $s, s' \in S$, and $x, x' \in M$

$$f(rxs + r'x's') = rf(x)s + r'f(x')s'.$$

Since the arithmetic of module homomorphisms is clearly analogous to
that of abelian group homomorphisms and of linear transformations of
vector spaces, we shall not discuss it here. We do note, however, that whenever
two, say, left R-homomorphisms compose as functions, then the resulting
function is again an R-homomorphism. And that for $_R M$ the identity map
$1_M : M \to M$ is an R-homomorphism. Thus the class of all left R-modules and
all R-homomorphisms between them form a concrete category; although it
will be important later, for the present we shall not be concerned with this
fact.

Let M and N be left R-modules and let $f : M \to N$ be a left R-homo-
morphism. Then the *image of f*, *Im f*, and the *kernel of f*, *Ker f*, are defined by

$$Im\, f = \{f(x) \in N \mid x \in M\}, \qquad Ker\, f = \{x \in M \mid f(x) = 0\}.$$

These are readily seen to be submodules of N and M, respectively. The *coimage* of f and the *cokernel* of f are defined by

$$\text{Coim } f = M/\text{Ker } f, \quad \text{Coker } f = N/\text{Im } f.$$

The linearity of an R-homomorphism tells us that its behavior is completely determined by its action on a spanning set. That is,

3.1 Proposition. *Let M and N be left R-modules, let X span M and let $f : M \to N$ be an R-homomorphism. Then $\text{Im } f$ is spanned by $f(X)$. Moreover, if g is also an R-homomorphism from M to N, then*

$$f = g \quad \text{iff} \quad f(x) = g(x) \quad (x \in X).$$

Proof. The first statement follows from (2.7) in view of the fact that
$$\text{Im } f = f(RX) = Rf(X).$$

One implication in the final statement is trivial. For the converse, suppose $f(x) = g(x)$ for all $x \in X$. It is easy to check that

$$K = \{y \in M \mid f(y) = g(y)\}$$

is a submodule of M. But since $X \subseteq K$, we have $M = RX \subseteq K$ by (2.3). Thus $f(y) = g(y)$ for all $y \in M$. $\square$

Epimorphisms and Monomorphisms

A homomorphism $f : M \to N$ is called an *epimorphism* in case it is surjective (i.e., onto N). It is called a *monomorphism* in case it is injective (i.e., one-to-one). From time to time we shall use self-explanatory variations of these terms (e.g., *epic* and *monic*) to simplify our sentence structure.

If M is a left R-module, then every submodule of M is actually the image of some monomorphism. For if K is a submodule of M, then the inclusion map $i_K = i_{K \leq M} : K \to M$ (see (0.1)) is an R-monomorphism, also called the *natural embedding of K in M*, with image K. Every submodule of M is also the kernel of an epimorphism. For let K be a submodule of M. Then the mapping $n_K : M \to M/K$ from M onto the factor module M/K defined by

$$n_K(x) = x + K \in M/K \quad (x \in M)$$

is seen to be an R-epimorphism with kernel K. We call n_K the *natural epimorphism of M onto M/K*.

An R-homomorphism $f : M \to N$ is an $(R\text{-})$ *isomorphism* in case it is a bijection. Two modules M and N are said to be $(R\text{-})$ *isomorphic*, abbreviated

$$M \cong N$$

in case there is an R-isomorphism $f : M \to N$. It is easy to check that this relation (of "being isomorphic") is an equivalence relation.

3.2. Story of "0". Given any pair of left R-modules M and N there is always one R-homomorphism from M to N, namely, the *zero homomorphism* $0:M \to N$ defined via

$$0:x \mapsto 0 \in N \qquad (x \in M)$$

Fortunately, the ambiguity of our multiple use of the symbol "0" for all zero elements, all zero submodules, and now all zero homomorphisms, turns out to be of no real consequence in practice. Indeed, we note that the zero submodule of a module M is the unique single element submodule of M, that one whose single element is zero, and that not only are any two zero submodules isomorphic but there is a unique isomorphism between them, namely the zero homomorphism.

While we are on the subject, there are several other conventions concerning 0. Since between any two zero modules there is a unique isomorphism, we shall feel free to identify all zero modules. Also, given a module M there is a unique homomorphism $M \to 0$, that is necessarily epic, and there is a unique homomorphism $0 \to M$ that is necessarily monic. When we write something such as $M \to 0$ or $0 \to M$, we have in mind these unique module homomorphisms. Now finally, for any module M,

$$n_0:M \to M/O \quad \text{and} \quad n_M:M/M \to 0$$

are isomorphisms, and again we shall usually identify the factor modules $M/0$ and M/M with M and 0, respectively.

We now state various characterizations of epimorphisms and monomorphisms analogous to those for surjections and injections in the category of sets and functions. For homomorphisms we have the advantage of the 0-function, but we no longer can characterize, say, monomorphisms as we did injections by means of a one-sided inverse.

3.3 Proposition. *Let M and N be left R-modules and let $f:M \to N$ be an R-homomorphism. Then the following statements are equivalent:*

(a) *f is an epimorphism onto N;*

(b) *$\text{Im } f = N$;*

(c) *For every ${}_RK$ and every pair $g, h:N \to K$ of R-homomorphisms, $gf = hf$ implies $g = h$.*

(d) *For every ${}_RK$ and every R-homomorphism $g:N \to K$, $gf = 0$ implies $g = 0$.*

Proof. (a) $\Leftrightarrow$ (b) and (a) $\Rightarrow$ (c) are trivial.

(c) $\Rightarrow$ (d). Let $h:N \to K$ be the zero homomorphism. Then $gf = 0$ means $gf = hf$; so assuming (c), we have $g = h = 0$.

(d) $\Rightarrow$ (b). Let $I = \text{Im} f$. Then $n_I:N \to N/I = \text{Coker} f$ clearly satisfies $n_I f = 0$. So assuming (d) this means that $n_I = 0$. But since n_I is onto N/I, we infer $N/I = 0$ whence $I = N$. ☐

3.4. Proposition. *Let M and N be left R-modules and let $f:M \to N$ be an R-homomorphism. Then the following statements are equivalent:*

(a) *f is a monomorphism;*

(b) *$\text{Ker} f = 0$;*

(c) *For every $_RK$ and every pair $g, h : K \to M$ of R-homomorphisms, $fg = fh$ implies $g = h$;*
   (d) *For every $_RK$ and every R-homomorphism $g : K \to M$, $fg = 0$ implies $g = 0$.*

Proof. The implication (d) $\Rightarrow$ (b) is the only one that offers any challenge. But let $K = \operatorname{Ker} f$. Then $i_K : K \to M$ is an R-homophorphism and $fi_K = 0$. So assuming (d) we have $i_K = 0$. But then $K = \operatorname{Im} i_K = 0$. □

3.5. Proposition. *Let M and N be left R-modules and let $f : M \to N$ be an R-homomorphism. Then f is an isomorphism iff there are functions $g, h : N \to M$ such that*
$$fg = 1_N \quad and \quad hf = 1_M.$$
When these last conditions are satisfied, $g = h$ is an isomorphism.

Proof. Of course the implication ($\Leftarrow$) and the uniqueness assertion are easy ($g = 1_M g = hfg = h1_N = h$). For the converse we observe that if f is an isomorphism, and hence a bijection, then there is a function $g : N \to M$ such that $fg = 1_N$ and $gf = 1_M$. (See (0.1).) To complete the proof we need to check that g is R-linear. But since f is, we have
$$f(g(ax + by)) = ax + by = f(ag(x) + bg(y)),$$
and then since f is injective, we have the R-linearity of g. □

When $f : M \to N$ is an isomorphism, the unique R-homomorphism $g : N \to M$ satisfying the condition of (3.5) is the *inverse* of f and is denoted by f^{-1}. (See (0.1).) Note that in (3.3) and (3.4) we did not claim as an equivalent condition the existence of one-sided inverses. As we shall see, this omission was not accidental.

The Factor Theorem

A homomorphism $f : M \to N$ that is the composite of homomorphisms
$$f = gh,$$
is said to *factor through g and h*. The following result essentially says that a homomorphism f factors uniquely through every epimorphism whose kernel is contained in that of f and through every monomorphism whose image contains the image of f.

3.6. The Factor Theorem. *Let $M, M', N,$ and N' be left R-modules and let $f : M \to N$ be an R-homomorphism.*
   (1) *If $g : M \to M'$ is an epimorphism with $\operatorname{Ker} g \subseteq \operatorname{Ker} f$, then there exists a unique homomorphism $h : M' \to N$ such that*
$$f = hg.$$
Moreover, $\operatorname{Ker} h = g(\operatorname{Ker} f)$ and $\operatorname{Im} h = \operatorname{Im} f$, so that h is monic iff $\operatorname{Ker} g = \operatorname{Ker} f$ and h is epic iff f is epic.

(2) *If* $g:N' \to N$ *is a monomorphism with* $Im f \subseteq Im g$, *then there exists a unique homomorphism* $h:M \to N'$ *such that*

$$f = gh.$$

Moreover, $Ker h = Ker f$ *and* $Im h = g^{\leftarrow}(Im f)$, *so that* h *is monic iff* f *is monic and* h *is epic iff* $Im g = Im f$.

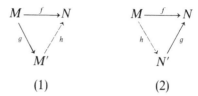

(1)   (2)

Proof. (1) Since $g:M \to M'$ is epic, for each $m' \in M'$ there is at least one $m \in M$ with $g(m) = m'$. If also $l \in M$ with $g(l) = m'$, then clearly $m - l \in Ker g$. But since $Ker g \subseteq Ker f$, we have that $f(m) = f(l)$. Thus, there is a well defined function $h:M' \to N$ such that $f = hg$. To see that h is actually an R-homomorphism, let $x', y' \in M'$ and let $x, y \in M$ with $g(x) = x'$, $g(y) = y'$. Then for each $a, b \in R$, $g(ax + by) = ax' + by'$, so that

$$h(ax' + by') = f(ax + by)$$
$$= af(x) + bf(y) = ah(x') + bh(y').$$

The uniqueness of h with these properties is assured by (3.3.c) since g is an epimorphism. The final assertion is trivial.

(2) For each $m \in M$, $f(m) \in Im f \subseteq Im g$. So since g is monic, there is a unique $n' \in N'$ such that $g(n') = f(m)$. Therefore, there is a function $h:M \to N'$ (viz., $m \mapsto n'$) such that $f = gh$. The rest of the proof is also easy.   □

As consequences of the first part of the factor theorem we have the all-important Noether Isomorphism Theorems.

3.7. Corollary [The Isomorphism Theorems]. *Let* M *and* N *be left* R-*modules.*

(1) *If* $f: M \to N$ *is an epimorphism with* $Ker f = K$, *then there is a unique isomorphism* $\eta:M/K \to N$ *such that* $\eta(m + K) = f(m)$ *for all* $m \in M$.

(2) *If* $K \le L \le M$, *then* $M/L \cong (M/K)/(L/K)$.

(3) *If* $H \le M$ *and* $K \le M$, *then* $(H + K)/K \cong H/(H \cap K)$.

Proof. (1) Let $M' = M/K$ and let g be the natural epimorphism $g = n_K:M \to M/K$ in (3.6.1).

To prove (2) and (3) apply (1) to the epimorphism $f':M/K \to M/L$ via $f'(m + K) = m + L$ and to the epimorphism $f'':H \to (H + K)/K$ via $f''(h) = h + K$, respectively.   □

3.8. Corollary. *Let* M *and* N *be left* R-*modules and let* $f:M \to N$ *be an* R-*epimorphism with kernel* K. *Then*

$$L \mapsto f(L) = \{f(x) \mid x \in L\}$$
$$P \mapsto f^{\leftarrow}(P) = \{x \in M \mid f(x) \in P\}$$

are inverse lattice isomorphisms between the lattices $\mathscr{S}(M)/K$ of submodules of M that contain K and $\mathscr{S}(N)$ of all submodules of N.

Proof. By The First Isomorphism Theorem (3.7.1) we have an isomorphism $\eta : M/K \to N$ such that

commutes. Clearly, η induces a lattice isomorphism between $\mathscr{S}(M/K)$ and $\mathscr{S}(N)$. But by (2.9) n_K induces one between $\mathscr{S}(M)/K$ and $\mathscr{S}(M/K)$. Now it is simply a matter of checking that these isomorphisms compose into the ones claimed. □

It is now easy to characterize (to within isomorphism) the cyclic modules and the (subclass of) simple modules. This is not to be scoffed at, for as we saw in (2.8) every module is spanned by its cyclic submodules and as we shall see in §9, modules spanned by simple ones are very well behaved. Given $_RM$ and $x \in M$ right multiplication by x

$$\rho_x : r \mapsto \rho_x(r) = rx \qquad (r \in R)$$

is a left homomorphism from R onto the cyclic submodule Rx with kernel

$$Ker\, \rho_x = l_R(x) = \{r \in R \mid rx = 0\}$$

the left annihilator (in R) of x. So by (3.7.1) $R/l_R(x) \cong Rx$. On the other hand, if I is a left ideal of R, then R/I is a cyclic left module spanned by $1 + I$ and with $l_R(1 + I) = I$. Thus with an assist from (2.10)

3.9. Corollary. *A left R-module M is cyclic if and only if it is isomorphic to a factor module of $_RR$. If $M = Rx$, then $\rho_x : R \to M$ is an epimorphism with kernel $l_R(x)$, so $M \cong R/l_R(x)$ and M is simple if and only if $l_R(x)$ is a maximal left ideal.* □

As a final application here of The Factor Theorem, we give

3.10. Corollary. *Let M and K be left R-modules and let $j : K \to M$ be an R-monomorphism with Im $j = I$. Then there is a unique isomorphism $v : I \to K$ such that $jv = i_I$.*

Proof. Let $I = M$, $M = N$, $K = N'$ and $i_I = f$, and $j = g$ in (3.6.2). □

3.11. Rings and Other Modules. Again we have been stating our definitions and results for left R-modules. Fortunately, with no difficulty everything translates to other styles of modules. There is, however, one possible source of misunderstanding, namely, that there is a real difference between ring homomorphisms of a ring R and bimodule homomorphisms of $_RR_R$. (See Exercise (3.6).) Let R and S be rings and let $\phi : R \to S$ be a ring homomorphism. Then (see §2) ϕ induces a bimodule structure $_R S_R$ on S via

$$(r, s) \mapsto \phi(r)s \quad \text{and} \quad (s, r) \mapsto s\phi(r).$$

Moreover, ϕ is then an (R,R)-bimodule homomorphism

$$\phi : {}_R R_R \to {}_R S_R$$

and its image is not only an (R,R)-submodule, but also a subring of S. If ϕ is actually onto S, then the left and right R-submodules of S coincide with the same sided ideals of the ring S and we have the ring theoretic version of (3.8), namely, that if $\phi : R \to S$ is a ring homomorphism onto S with kernel K, then ϕ induces lattice isomorphisms between the left, right and two-sided ideal lattices of S and, respectively, the sublattices of left, right and two-sided ideals of R that contain K. (Note, however, that if ϕ is not onto S, then the images of ideals of R, although (R,R)-submodules of S, need not be ideals of S.)

Exactness

A pair of homomorphisms

$$M' \xrightarrow{f} M \xrightarrow{g} M''$$

is said to be *exact at M* in case $Im f = Ker g$. We also say that a single homomorphism $M' \xrightarrow{f} M$ is exact at both M' and M. Finally, a sequence (finite or infinite) of homomorphisms

$$\cdots \xrightarrow{f_{n-1}} M_{n-1} \xrightarrow{f_n} M_n \xrightarrow{f_{n+1}} M_{n+1} \longrightarrow \cdots$$

is *exact* in case it is exact at each M_n; i.e., in case for each successive pair f_n, f_{n+1}

$$Im f_n = Ker f_{n+1}.$$

Immediate from the definition is the following set of special cases.

3.12. Proposition. *Given modules M and N and a homomorphism $f : M \to N$, the sequence*
(1) $0 \to M \xrightarrow{f} N$ *is exact iff f is monic;*
(2) $M \xrightarrow{f} N \to 0$ *is exact iff f is epic;*
(3) $0 \to M \xrightarrow{f} N \to 0$ *is exact iff f is an isomorphism.* □

To some extent the status of the kernel and cokernel of a homomorphism can be summarized as a certain exact sequence.

3.13. Proposition. *If M and N are modules and if $f : M \to N$ is a homomorphism, then*

$$0 \to Ker f \xrightarrow{i} M \xrightarrow{f} N \xrightarrow{n} Coker f \to 0$$

is exact where i is the inclusion map and n is the natural epimorphism $N \to N/Im f$. □

This result, whose proof is trivial, has as special cases the two facts that $f: M \to N$ is monic iff

$$0 \to M \xrightarrow{f} N \xrightarrow{n} \mathrm{Coker}\, f \to 0$$

is exact, whereas it is epic iff

$$0 \to \mathrm{Ker}\, f \xrightarrow{i} M \xrightarrow{f} N \to 0$$

is exact. In general, an exact sequence of the form

$$0 \to K \xrightarrow{f} M \xrightarrow{g} N \to 0$$

is called a *short exact sequence*. By (3.12) in such a sequence f is a mono-morphism and g is an epimorphism. Thus by (3.7.1) and (3.10) there exist unique isomorphisms v and η such that

$$0 \to K \xrightarrow{f} M \xrightarrow{g} N \to 0$$

$$\mathrm{Im}\, f \quad M/\mathrm{Ker}\, g$$

commutes where i is the inclusion map and n is the natural epimorphism. But by exactness $\mathrm{Im}\, f = \mathrm{Ker}\, g$, so v and η are isomorphisms such that

$$0 \to K \xrightarrow{f} M \xrightarrow{g} N \to 0$$

$$v \uparrow \qquad 1_M \uparrow \qquad \uparrow \eta$$

$$0 \to \mathrm{Im}\, f \xrightarrow{i} M \xrightarrow{n} M/\mathrm{Im}\, f \to 0$$

commutes. That is, every short exact sequence is "isomorphic" in this latter sense to one of the form

$$0 \to M' \xrightarrow{i} M \xrightarrow{n} M/M' \to 0$$

where i is an inclusion map of a submodule M' of M and n is the natural epimorphism. A short exact sequence

$$0 \to K \xrightarrow{f} M \xrightarrow{g} N \to 0$$

is also called an *extension of K by N*. (See Exercise (3.13).)

Remark. In order to simplify matters we shall try to omit certain un-necessary symbols. In a given diagram if we fail to specify some homo-morphism, it is generally because there is really only one natural candidate. For example, if we were to write, "consider the short exact sequence $0 \to \mathrm{Ker}\, f \to M \xrightarrow{f} N \to 0$," we clearly intend $\mathrm{Ker}\, f \to M$ to be the identity embedding. On the other hand, it is often helpful to add more than necessary. Examples of this occur in statements such as "consider a monomorphism $0 \to K \xrightarrow{f} M$," or such as "given an epimorphism $M \xrightarrow{g} N \to 0$." For several standard lemmas about commuting diagrams, the usual proofs involve a technique known as "diagram chasing". In our next result, one of these diagram lemmas, we illustrate this technique.

3.14 Lemma. *Suppose that the following diagram of modules and homo-morphisms*

$$A \xrightarrow{\;f\;} B \xrightarrow{\;g\;} C$$
$$\alpha \downarrow \quad \beta \downarrow \quad \gamma \downarrow$$
$$A' \xrightarrow{\;f'\;} B' \xrightarrow{\;g'\;} C'$$

is commutative and has exact rows.

(1) *If α, γ, and f' are monic, then so is β;*

(2) *If α, γ, and g are epic, then so is β;*

(3) *If β is monic, and if α and g are epic, then γ is monic;*

(4) *If β is epic, and if f' and γ are monic, then α is epic.*

Proof. (1) It will suffice to show $Ker\,\beta = 0$. So let $b \in Ker\,\beta$. Since the diagram commutes, $\gamma g(b) = g'\beta(b) = 0$. Since γ is monic, $g(b) = 0$ whence $b \in Ker\,g$. But the top row is exact, so $Ker\,g = Im\,f$. Thus, there is an $a \in A$ such that $b = f(a)$. Now since the diagram commutes $f'\alpha(a) = \beta f(a) = \beta(b) = 0$. Finally, f' and α are monic, so $a = 0$, whence $b = f(a) = 0$.

(4) Let $a' \in A'$. Then since β is epic, there is a $b \in B$ such that $\beta(b) = f'(a')$. Since the diagram commutes and the bottom row is exact $\gamma g(b) = g'\beta(b) = g'f'(a') = 0$. But γ is monic, so $b \in Ker\,g = Im\,f$. Thus, there is an $a \in A$ with $f(a) = b$. So $f'\alpha(a) = \beta f(a) = \beta(b) = f'(a')$. Finally, f' is monic, so $\alpha(a) = a'$. □

3.15. The "Five Lemma". *Suppose that the following diagram of modules and homomorphisms*

$$A \longrightarrow B \longrightarrow C \longrightarrow D \longrightarrow E$$
$$\alpha \downarrow \quad \beta \downarrow \quad \gamma \downarrow \quad \delta \downarrow \quad \varepsilon \downarrow$$
$$A' \longrightarrow B' \longrightarrow C' \longrightarrow D' \longrightarrow E'$$

is commutative and has exact rows.

(1) *If α is epic and β and δ are monic, then γ is monic.*

(2) *If ε is monic and β and δ are epic, then γ is epic.*

(3) *If α, β, δ, and ε are isomorphisms, then so is γ.*

Proof. By diagram chasing.   □

3. Exercises

1. Let M be a left R-module. Prove that the following assertions are equivalent: (a) $M = 0$; (b) For each left R-module N there is a unique R-homomorphism $M \to N$; (c) For each left R-module N there is a unique R-homomorphism $N \to M$.

2. Let M be a left R-module. Prove that the following are equivalent: (a) M is simple; (b) Every non zero homomorphism $M \to N$ is a monomorphism; (c) Every non-zero homomorphism $N \to M$ is an epimorphism.

3. Let $f: M \to N$ be an R-homomorphism. Prove that if f is monic, then $l_R(M) \supseteq l_R(N)$, whereas if f is epic, then $l_R(M) \subseteq l_R(N)$.

4. Let $\mathbf{C}$ be a category and let $f: A \to B$ be a morphism in $\mathbf{C}$. Then f is a *monomorphism* (*epimorphism*) in case it is cancellable on the left (right) in $\mathbf{C}$; i.e., in case for each pair of morphisms $g, h: C \to A(g, h: B \to C)$ in $\mathbf{C}$, if $fg = fh$ (if $gf = hf$), then $g = h$. The morphism f is an *isomorphism* in case it is invertible in $\mathbf{C}$; i.e., in case there is a morphism $g: B \to A$ with $gf = 1_A$ and $fg = 1_B$.
   (1) Prove that if $f: A \to B$ and $f': B \to C$ are both monomorphisms, epimorphisms, isomorphisms, respectively, then $f'f$ is a monomorphism, epimorphism, isomorphism.
   (2) Prove that in the concrete category $\mathbf{R}$ of all rings and ring homomorphisms, the inclusion map $f: \mathbb{Z} \to \mathbb{Q}$ is both a monomorphism and an epimorphism but not an isomorphism. (Also, see Exercise (4.2).)

5. Let R be a ring and let $\alpha: R \to R$ be a ring automorphism. For each left R-module M define a map $*: R \times M \to M$ via $(r, m) \mapsto r * m = \alpha(r)m$. Show that with respect to this operation M is a left R-module; we denote this module by M^α. (See Exercise (2.2).) However, show that in general, M and M^α need not be R-isomorphic. [Hint: Let $R = \mathbb{Q}[X, Y]$, α the automorphism that interchanges X and Y, and $M = R/I$ where I is the ideal generated by X. Then use Exercise (3.3).]

6. Let R and S be rings and let $\phi: R \to S$ be a ring homomorphism, so that S is an (R,R)-bimodule with respect to the scalar multiplications $(r, s) \mapsto \phi(r)s$ and $(s, r) \mapsto s\phi(r)$.
   (1) Prove that if ϕ is surjective, then the (R,R)-submodules of S are precisely the ideals of the ring S, but that if ϕ is not surjective, then the images of the ideals of R need not be ideals of S. [Hint: Let S be a field.]
   (2) Let $\sigma: \mathbb{Q} \to \mathbb{M}_2(\mathbb{Q})$ be defined by

$$\sigma: r \mapsto \begin{bmatrix} r & 0 \\ 0 & r \end{bmatrix}.$$

Then σ is a ring homomorphism, whence via σ, $\mathbb{M}_2(\mathbb{Q})$ is an $(\mathbb{Q},\mathbb{Q})$-bimodule. Show that the mapping $f: \mathbb{Q} \mapsto \mathbb{M}_2(\mathbb{Q})$ defined via

$$f: r \mapsto \begin{bmatrix} r & 0 \\ 0 & -r \end{bmatrix}.$$

is an $(\mathbb{Q},\mathbb{Q})$-homomorphism but not a ring homomorphism.

7. Let $f: M \to N$ be an epimorphism and let $K \leq M$. Prove that
   (1) If $K \cap \operatorname{Ker} f = 0$, then $(f \,|\, K): K \to N$ is a monomorphism.
   (2) If $K + \operatorname{Ker} f = M$, then $(f \,|\, K): K \to N$ is an epimorphism.

8. (1) Prove that if M is a finite cyclic $\mathbb{Z}$-module, then there is a short exact sequence

$$0 \to \mathbb{Z} \xrightarrow{f} \mathbb{Z} \xrightarrow{g} M \to 0.$$

(2) Prove that there exists an exact sequence (over $\mathbb{Z}$)

$$0 \to \mathbb{Z}_2 \to \mathbb{Z}_4 \to \mathbb{Z}_4 \to \mathbb{Z}_2 \to 0.$$

(3) Prove that there exists an exact sequence (over $\mathbb{Z}$)

$$\dots \to \mathbb{Z}_4 \to \mathbb{Z}_4 \to \mathbb{Z}_4 \to \mathbb{Z}_4 \to \dots.$$

9. Let U, V, W be real vector spaces of dimension one, three, and two respectively. Let $\{u\}$ be a basis for U, $\{v_1, v_2, v_3\}$ a basis for V, and $\{w_1, w_2\}$ a basis for W. Finally, let $f:U \to V$ be the $\mathbb{R}$-homomorphism defined by $f(\alpha u) = \alpha v_1 + \alpha v_2$ and $g:V \to W$ be the $\mathbb{R}$-homomorphism defined by $g(\alpha_1 v_1 + \alpha_2 v_2 + \alpha_3 v_3) = \alpha_1 w_1 + \alpha_3 w_2$.

(1) Prove that the sequence $0 \to U \xrightarrow{f} V \xrightarrow{g} W \to 0$ is exact at U and W but not at V.
(2) Prove that there exists $g':V \to W$ with $0 \to U \xrightarrow{f} V \xrightarrow{g'} W \to 0$ exact.
(3) Prove that there exists $f':U \to V$ with $0 \to U \xrightarrow{f'} V \xrightarrow{g} W \to 0$ exact.

10. Let R be a ring and let $M' \xrightarrow{f} M \xrightarrow{g} M''$ be a sequence of R-modules and R-homomorphisms. Prove that this sequence is exact iff there exists a commutative diagram

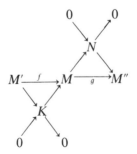

of R-modules and R-homomorphisms in which the "diagonal" sequences are all exact.

11. (1) Prove "The Five Lemma" (3.15).
(2) Suppose that the following diagram of modules and homomorphisms is commutative and has exact rows:

$$0 \to A \xrightarrow{f} B \xrightarrow{g} C \to 0$$
$$\downarrow{\alpha} \quad \downarrow{\beta} \quad \downarrow{\gamma}$$
$$0 \to A' \xrightarrow{f'} B' \xrightarrow{g'} C' \to 0$$

Assume that β is an isomorphism. Prove that α is monic and γ is epic, and that α is epic iff γ is monic.

12. Suppose that the following diagram of modules and homomorphisms is commutative and has exact rows:

$$
\begin{array}{ccc}
0 & 0 & 0 \\
\downarrow & \downarrow & \downarrow \\
0 \to A' \to B' \to C' \to 0 \\
\downarrow & \downarrow & \downarrow \\
0 \to A \to B \to C \to 0 \\
\downarrow & \downarrow & \downarrow \\
0 \to A'' \to B'' \to C'' \to 0 \\
\downarrow & \downarrow & \downarrow \\
0 & 0 & 0
\end{array}
$$

Prove that if the middle column is exact, then the last column is exact iff the first column is exact.

13. Two extensions of K by N

$$0 \to K \xrightarrow{f} M \xrightarrow{g} N \to 0$$

$$0 \to K \xrightarrow{f'} M' \xrightarrow{g'} N \to 0$$

are *equivalent* in case there is a homomorphism $h: M \to M'$ such that

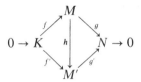

commutes.

(1) Prove that if the above two extensions are equivalent (via h), then h is an isomorphism.

(2) Prove that the relation of "being equivalent" is an equivalence relation on the class of all extensions of K by N.

(3) Given K and N, there is at least one extension of K by N

$$0 \to K \xrightarrow{\iota} K \times N \xrightarrow{\pi} N \to 0$$

where $\iota : k \mapsto (k, 0)$ and $\pi : (k, n) \mapsto n$.

(4) Prove that there are (at least) two inequivalent extensions of $\mathbb{Z}_2$ by $\mathbb{Z}_4$.

14. Let R be a commutative integral domain and let M be a left R-module. The set $T(M) = \{x \in M \mid l_R(x) \neq 0\}$ is a submodule (proof?) called the *torsion* submodule of $_R M$. If $T(M) = M$, then M is *torsion*; if $T(M) = 0$, then M is *torsion free*. Let $_R M$ and $_R N$ and let $f: M \to N$ be an R-homomorphism. Prove that

(1) $T(M/T(M)) = 0$.

(2) If N is torsion free, then $T(M) \leq \operatorname{Ker} f$.

(3) If M is torsion, then $\operatorname{Im} f \leq T(N)$.

15. Let R be a commutative integral domain. A left R-module M is $(R\text{-})$ *divisible* in case $aM = M$ for each $0 \neq a \in R$; i.e., in case $m = ax$ has a solution $x \in M$ for each $m \in M$ and $0 \neq a \in R$. Let Q be the field of fractions of the integral domain R. Prove that
(1) If M is a Q-vector space, then as an R-module it is divisible and torsion free.
(2) If M is a divisible and torsion free R-module via $\lambda : R \to End^l(M)$, then there is a unique ring homomorphism $\theta : Q \to End^l(M)$ with $(\theta \mid R) = \lambda$; in particular θ induces a Q-vector space structure on M.
(3) If $f : M \to N$ is an R-epimorphism and if M is R-divisible, then so is N.

16. Let $p \in \mathbb{P}$. Consider the abelian group $\mathbb{Z}_{p^\infty}$. (See Exercise (2.8).)
(1) Prove that $\mathbb{Z}_{p^\infty}$ is divisible (see Exercise (3.15)). Moreover, show that if $n \in \mathbb{Z}$ is not divisible by p, then $x \mapsto nx$ defines an isomorphism $\mathbb{Z}_{p^\infty} \to \mathbb{Z}_{p^\infty}$.
(2) Prove that an abelian group M is isomorphic to $\mathbb{Z}_{p^\infty}$ iff M is spanned by a countable set $g_1, g_2, \ldots$ of non zero elements satisfying $pg_1 = 0$ and $pg_{n+1} = g_n \ (n = 1, 2, \ldots)$.

17. Let $p \in \mathbb{P}$. Consider the abelian group $\mathbb{Z}_{p^\infty}$ and let $\mathbb{A}(p) = End^l(\mathbb{Z}_{p^\infty})$. The ring $\mathbb{A}(p)$ is called the ring of p-*adic integers*.
(1) Prove that $\mathbb{A}(p)$ has a (unique) subring isomorphic to the ring $\mathbb{Z}_{(p)}$. (See Exercises (2.12), (3.15.2), and 3.16.1).)
(2) Let $g_1, g_2, \ldots$ be a spanning set for $\mathbb{Z}_{p^\infty}$ with $pg_1 = 0$ and $pg_{n+1} = g_n$ $(n = 1, 2, \ldots)$. (See Exercise (3.16.2).) Prove that for each

$$\sigma = (a_n)_{n \in \mathbb{N}} \in \mathbb{Z}_p^{\mathbb{N}},$$

there is an endomorphism $\bar{\sigma} \in \mathbb{A}(p)$ defined by

$$\bar{\sigma}(g_n) = a_1 g_n + a_2 g_{n-1} + \ldots + a_n g_1,$$

and that the map $\sigma \mapsto \bar{\sigma}$ is a bijection from $\mathbb{Z}_p^{\mathbb{N}}$ onto $\mathbb{A}(p)$. Infer that $\mathbb{A}(p)$ is not isomorphic to $\mathbb{Z}_{(p)}$.
(3) Denote the endomorphism $\bar{\sigma}$ (see (2)) as a "power series"

$$\bar{\sigma} = a_1 + a_2 p + a_3 p^2 + \ldots.$$

Show that computations in $\mathbb{A}(p)$ with these power series are the natural extensions of the usual computations with integers represented to the "base" p. In particular, deduce that $\mathbb{A}(p)$ is commutative.

18. An indexed set $(x_\alpha)_{\alpha \in A}$ in a vector space V is *linearly independent* in case $\{x_\alpha \mid \alpha \in A\}$ is linearly independent and $\alpha \neq \beta$ implies $x_\alpha \neq x_\beta$. Thus if $(x_\alpha)_{\alpha \in A}$ is a basis for V, then $dim\ V = card\ A$. (See Exercises (2.17) and (2.18).) Let V and W be left D-vector spaces. Prove
(1) If $(x_\gamma)_{\gamma \in C}$ is independent in V and $(y_\gamma)_{\gamma \in C}$ is an indexed set in W, then there exists a linear transformation $(= D$-homomorphism$)\ f : V \to W$ such that $f(x_\gamma) = y_\gamma \ (\gamma \in C)$.
(2) There exists an epimorphism (monomorphism) $f : V \to W$ iff $dim\ V \geq$

$dim\ W\ (dim\ V \leq dim\ W)$, and if f is epic (monic) then $dim\ V = dim(Ker\ f)$ $+\ dim\ W\ (dim\ W = dim\ V + dim(Coker\ f))$.

(3) A linear transformation $f: V \to W$ is epic (monic) iff f has a right (left) inverse $f': W \to V$.

(4) $V \cong W$ iff $dim\ V = dim\ W$.

§4. Categories of Modules; Endomorphism Rings

Given two modules M and N, say left R-modules, every R-homomorphism from M to N is an element of the set of functions from M to N; in particular, these homomorphisms form a set. The standard notation for this set is
$$Hom_R(M, N).$$
Suppose we have left R-modules $_RM$ and $_RN$. Then for each pair f, g in $Hom_R(M, N)$ define the functions $f + g$ and $(-f)$ from M to N by

$$f + g : x \mapsto f(x) + g(x) \qquad (x \in M).$$

$$(-f) : x \mapsto -f(x) \qquad (x \in M).$$

It is completely elementary, if somewhat tedious, to check that each of these is a left R-homomorphism from M to N. Using the "negative", $(-f)$, and the zero homomorphism, an easy computation gives

4.1. Proposition. *If M and N are left R-modules, then $Hom_R(M, N)$ is an abelian group with respect to the operation of addition $(f, g) \mapsto f + g$ defined by*

$$(f + g)(x) = f(x) + g(x) \qquad (x \in M). \qquad \square$$

Although we shall not need the terminology for several sections, we begin now to acclimate ourselves to thinking of modules categorically. Given a ring R the *category of left R-modules* is the system

$$_R\mathsf{M} = (_R\mathscr{M}, Hom_R, \circ)$$

where $_R\mathscr{M}$ is the class of all left R-modules, $Hom_R : (M, N) \mapsto Hom_R(M, N)$, and $\circ$ is the usual composition of functions. Clearly, this is a concrete category (0.11) whose objects are left R-modules, (M, λ) with underlying set M, and whose morphisms are left R-homomorphisms. Allowing a slight perversion of our notation, we may write things like $M \in {}_R\mathsf{M}$ and $f \in {}_R\mathsf{M}$ to indicate that M is an object in $_R\mathsf{M}$ and f is a morphism in $_R\mathsf{M}$. As we shall see, $_R\mathsf{M}$ has a rich structure. One thing that already distinguishes it from other categories is that for each pair $M, N \in {}_R\mathsf{M}$, the set $Hom_R(M, N)$ has the structure of an abelian group such that the composition of $_R\mathsf{M}$ distributes over this addition.

There are many other module categories that are of interest. Thus there is the concrete category $\mathsf{M}_R = (\mathscr{M}_R, Hom_R, \circ)$ of *right R-modules* and right R-homomorphisms. There is the category $_R\mathsf{M}_S$ of *left R- right S-bimodules* and their homomorphisms. Two other very important ones are the category $_R\mathsf{FM}$ of all *finitely spanned left R-modules* and its sister FM_R of all *finitely* *spanned right R-modules.* These latter are full subcategories of $_R$M and M$_R$, respectively. For more on the subject, including a few examples of functors, see the exercises.

Concerning the homomorphism groups there are two rather simple facts that we should note. Suppose we have two S-modules $_SM$ and $_SN$ and a ring homomorphism $\phi: R \to S$. Then ϕ induces an R-module structure on both M and N and every S-homomorphism $f: M \to N$ is also an R-homomorphism. For

$$f(rx) = f(\phi(r)x) = \phi(r)f(x) = rf(x).$$

(Indeed ϕ induces a functor from $_SM$ to $_RM$; see Exercise (4.15).) Thus as abelian groups

4.2. $Hom_S(M, N) \leq Hom_R(M, N) = Hom_{\phi(R)}(M, N) \leq Hom_\mathbb{Z}(M, N)$

Now suppose that $_RM$ and $_RN$ are R-modules and that I is an ideal of R that annihilates both M and N. Then (see §2), M and N both carry natural R/I module structures, and certainly every R-homomorphism between them is an R/I-homomorphism. Therefore, with one inclusion from (4.2) we have

4.3. $Hom_R(M, N) = Hom_{R/I}(M, N)$.

Again, let $_RM$ and $_RN$ be left R-modules. In general we cannot expect the abelian group $Hom_R(M, N)$ to be an R-module. Of course there is always the temptation prompted by our experience in linear algebra, to define a scalar product $af: M \to N$ by $af: x \mapsto af(x) \, (x \in M)$. This is a perfectly good function, but it may fail to be "R-linear" unless $a \in Cen \, R$. For in general we cannot conclude that $(af)(bx) = af(bx) = abf(x)$ and $b(af)(x) = b(af(x)) = baf(x)$ are equal. However, if $M = {}_RM_S$ is a bimodule, then for each $s \in S$ the function obtained by first multiplying by s and then applying $f \in Hom_R(M, N)$

$$sf: x \mapsto f(xs) \qquad (x \in M)$$

is an R-homomorphism $sf \in Hom_R(M, N)$. Indeed, it is clearly additive and

$$(sf)(rx) = f((rx)s) = f(r(xs)) = r(f(xs)) = r(sf)(x).$$

In other words, $Hom_R(_RM_S, {}_RN)$ is a left S-module with scalar multiplication $(s, f) \mapsto sf$ defined by

$$(sf)(x) = f(xs) \qquad (x \in M).$$

Thus from the right action of S on M we get a left action of S on $Hom_R(M, N)$. On the other hand if $N = {}_RN_T$, then $Hom_R(_RM, {}_RN_T)$ is a right T-module with scalar multiplication $(f, t) \mapsto ft$ defined by

$$(ft)(x) = f(x)t \qquad (x \in M).$$

Here the right action of T on N induced a right action of T on $Hom_R(M, N)$. Note also that if $s \in S$, $t \in T$, then

$$((sf)t)(x) = ((sf)(x))t = f(xs)t = (ft)(xs) = (s(ft))(x).$$

These and other equally easy computations give

4.4. Proposition. *Let M and N be abelian groups and let R, S, and T be rings. Then the module structure*
(1) $_RM_S$, $_RN_T$ *induces a left S- right T-bimodule structure on* $\text{Hom}_R(M, N)$ *via*

$$(sf)(x) = f(xs) \quad and \quad (ft)(x) = f(x)t;$$

(2) $_SM_R$, $_TN_R$ *induces a left T- right S-bimodule structure on* $\text{Hom}_R(M, N)$ *via*

$$(tf)(x) = tf(x) \quad and \quad (fs)(x) = f(sx). \qquad \square$$

This business of transferring the action of a ring on M or N to action on $\text{Hom}_R(M, N)$ has many rather obvious variations. But about all of them there are two things to remember. First, the basic R-action on M and N does not transfer to $\text{Hom}_R(M, N)$. And second, the sides change when transferring from the first variable M, but stay the same when transferring from the second variable N. (Incidentally, this latter "contravariant-covariant" phenomenon will be treated more fully in subsequent sections.) Concerning notation: when, for example, we write $\text{Hom}_R(_RM_S, {}_RN_T)$ we are viewing $\text{Hom}_R(M, N)$ with the bimodule structure of (4.4.1).

The regular bimodule $_RR_R$ gives rise to important applications of (4.4). By (4.4.1) for each $_RM$ we have a left R-module $\text{Hom}_R(_RR_R, {}_RM)$ and a right R-module $\text{Hom}_R(_RM, {}_RR_R)$. The second of these, called the R-dual of M, will receive attention in later sections. The first is just another copy of M:

4.5. Proposition. *Given a left R-module* $_RM$ *there is a left R-isomorphism* $\rho : M \to \text{Hom}_R(R, M)$ *defined by*

$$\rho(x)(a) = ax \quad (x \in M, a \in R).$$

Moreover, if M is a bimodule $_RM_S$ *then* ρ *is an* (R, S) *isomorphism.*

Proof. First we see that $\rho(x)$ is an R-homomorphism from R to M; for

$$\rho(x)(ab + a'b') = (ab + a'b')x = a\rho(x)(b) + a' \rho(x)(b')$$

for all $a, a', b, b' \in R$. Then ρ itself is R-linear, for

$$\rho(ax + by)(c) = c(ax + by) = cax + cby$$
$$= \rho(x)(ca) + \rho(y)(cb) = (a\rho(x) + b\rho(y))(c)$$

by (4.4.1). Now ρ is monic for $\rho(x) = 0$ forces $x = \rho(x)(1) = 0$. And ρ is epic for if $f \in \text{Hom}_R(R, M)$, then $f(a) = af(1) = \rho(f(1))(a)$. The last statement is now easy to check. $\qquad \square$

In §20 we shall see that this isomorphism ρ between M and $\text{Hom}_R(R, M)$ is "natural" in that it defines a natural transformation of functors (see (0.13)). There is a generalization of this that will be quite useful. Let $e \in R$ be a non-zero idempotent. Then eRe is a ring with identity (1.16). Moreover, if $_RM$ is a left R-module, then

$$eM = \{ex \mid x \in M\}$$

is clearly a subgroup of M and acquires a left eRe-module structure via the scalar multiplication

$$(ere, ex) \mapsto erex.$$

Actually, left multiplication by e defines a covariant functor from $_R M$ to $_{eRe} M$ (see Exercise (4.17)). For now, however, we record only the following easy generalization of (4.5) whose proof we leave as an exercise. (See Exercise (4.9).)

4.6. Proposition. *Let* $e \in R, f \in S$ *be non-zero idempotents and let* $_R M_S$ *be a bimodule. Then* $_{eRe} eM_S$ *and* $_R Mf_{fSf}$ *are bimodules, and*

$$\rho : eM \to Hom_R(Re, M) \quad and \quad \lambda : Mf \to Hom_S(fS, M)$$

defined via

$$\rho \, (em)(re) = rem \quad and \quad \lambda(mf)(fr) = mfr$$

are bimodule isomorphisms.   □

As a particularly important special case we have

4.7. Corollary. *If* e *and* f *are idempotents in a ring* R, *then*

$$Hom_R(Re, Rf) \cong \,_{eRe} eRf_{fRf} \cong Hom_R(fR, eR).$$ □

An R-homomorphism of an R-module M to itself is called an $(R\text{-})endo$-$morphism$ of M. An R-isomorphism from M to itself is an $(R\text{-})automorphism$ of M. As we have seen (4.1), the set

$$Hom_R(M, M)$$

of all R-endomorphisms of $_R M$ is an abelian group. Since it is also closed under the usual product ($=$ composition) of maps, if $M \neq 0$, then $Hom_R(M, M)$ is a subring of the ring of endomorphisms of the abelian group M. But just as for groups, there are two such rings and we must distinguish between them. Let

$$End_R^l(M) \qquad and \qquad End_R^r(M)$$

denote the ring of endomorphisms of $_R M$ treated as left operators on M and as right operators on M, respectively. Thus, these are opposite rings of each other. There usually turns out to be a preferred side. Indeed since we shall almost always want to write endomorphisms, when considered as elements in the endomorphism ring, on the side opposite the scalars, we adopt a convention. For a left R-module M we write

$$End(_R M) = End_R^r(M)$$

for the endomorphism ring of M *operating on the right* and for a right R-module N we write

$$End(N_R) = End_R^l(N)$$

for the endomorphism ring of N operating on the left. In other words the endo-morphisms will operate on the side opposite the interior subscript.

Before proceeding with real business at hand we shall note versions of (4.2) and (4.3) for endomorphism rings. Suppose then that $_S M$ is a non-zero left S-module and that $\phi : R \to S$ is a ring homomorphism. Then considering $_R M$ with the R-structure induced by ϕ, it is clear from (4.2) that as rings

4.8. $End(_S M) \leq End(_R M) = End(_{\phi(R)} M) \leq End^r_{\mathbb{Z}}(M)$.

Also, if $_R M$ is non-zero and if I is an ideal of R that annihilates M, then (see (4.3)) we have that

4.9. $End(_R M) = End(_{R/I} M)$.

If M is a non-zero left R-module, then $End(_R M)$, the ring of R-endo-morphisms of M viewed as right operators, is actually a subring of $End^r_{\mathbb{Z}}(M)$. (See (4.8).) This means simply that (M, i) is a right $End(_R M)$-module where $i : End(_R M) \to End^r_{\mathbb{Z}}(M)$ is the inclusion map. That each $f \in End(_R M)$ is an R-endomorphism means that for each such f, each $r \in R$, and each $x \in M$,

$$(rx)f = r(xf).$$

In other words, M is a left R- right-$End(_R M)$-bimodule

$$_R M_{End(_R M)}.$$

This simple fact is really the starting point for the concept of bimodules. For suppose also that S is a ring and that M is a right S-module via a ring homo-morphism $\rho : S \to End^r_{\mathbb{Z}}(M)$. If $s \in S$, then $\rho(s)$ is in the subring $End(_R M)$ iff for all $r \in R$ and $x \in M$,

$$r(x\rho(s)) = (rx)\rho(s),$$

or writing $\rho(s)$ as a scalar product,

$$r(xs) = (rx)s.$$

Therefore, the image of ρ is in the subring $End(_R M)$ iff this identity holds for all $s \in S$, all $r \in R$, and all $x \in M$; that is, iff M is an (R,S)-bimodule. Reversing the roles of R and S will clearly produce a similar conclusion. From this it should be clear how the concept of a bimodule enriches the theory—a bimodule is simply the representation of one ring as a ring of endomorphisms of a module over another ring. Formally summarizing these observations we have

4.10. Proposition. *Let R and S be rings and M an abelian group. If M is a left R-module via $\lambda : R \to End^l(M)$ and a right S-module via $\rho : S \to End^r(M)$, then the following are equivalent:*

(a) $_R M_S$;

(b) $\lambda : R \to End(M_S)$ *is a ring homomorphism;*

(c) $\rho : S \to End(_R M)$ *is a ring homomorphism.* $\square$

Thus for a bimodule $_R M_S$ we have the canonical ring homomorphisms "left and right multiplication"

$$\lambda : R \to End(M_S) \qquad \text{and} \qquad \rho : S \to End(_R M)$$

such that for $r \in R$, $x \in M$ and $s \in S$

$$\lambda(r) : x \mapsto rx \qquad \text{and} \qquad \rho(s) : x \mapsto xs.$$

Here $_R M$ (respectively, M_S) is faithful iff λ (respectively, ρ) is injective. If both λ and ρ are surjective we say that $_R M_S$ is a *balanced bimodule*. In other words, the bimodule $_R M_S$ is balanced in case every S-endomorphism of M is "multiplication by" an element of R and every R-endomorphism of M is "multiplication by" an element of S. If λ and ρ are isomorphisms, then $_R M_S$ is called a *faithfully balanced bimodule*.

There is an example, perhaps familiar, from elementary linear algebra. Let S be a field, let M_S be a non-zero vector space over S, and let $R = End(M_S)$ be the ring of S-linear transformations of M viewed as left operators. Then, it is clear that $_R M_S$ (see (4.10)) and both $_R M$ and M_S faithful. In particular, right multiplication by each scalar $s \in S$ is an endomorphism of $_R M$. But an easy argument (see Exercise (4.4)) shows that every $\sigma \in End(_R M)$ is in fact just such a scalar multiplication. Therefore $_R M_S$ is a faithfully balanced bimodule.

Another important example of a faithfully balanced bimodule is given in

4.11. Proposition. *If R is a ring and if λ and ρ denote left and right multiplication, then*

$$\lambda : R \to End(R_R) \qquad \text{and} \qquad \rho : R \to End(_R R)$$

are ring isomorphisms; i.e., the regular bimodule $_R R_R$ is faithfully balanced.

Proof. That λ and ρ are ring homomorphisms follows from (4.10). That they are bijective follows from (4.5) and its right-hand version.   □

Consider a left R-module M and its endomorphism ring

$$T = End(_R M).$$

By (4.10) there is a bimodule $_R M_T$ where the T-action is induced by the identity homomorphism $T \to End(_R M)$. The endomorphism ring B of M_T, called the *biendomorphism ring of $_R M$*, is abbreviated

$$B = BiEnd(_R M) = End(M_T).$$

The elements of B are called the *biendomorphisms* of $_R M$. Since $_R M_T$ is a bimodule, (4.10) implies that if the module action of R is given by λ, then $\lambda(r) \in BiEnd(_R M)$ for all $r \in R$. That is, λ is a ring homomorphism

$$\lambda : R \to BiEnd(_R M);$$

we call this the *natural homomorphism of R into the biendomorphism ring* of $_R M$. On the other hand, by (4.10), the left R-module M can be made into a bimodule $_B M_T$ with multiplication

$$(b, x) \mapsto b(x) \qquad \text{and} \qquad (x, t) \mapsto xt.$$

It is, in fact, a balanced bimodule.

4.12. Proposition. *If M is a left R-module, then*

$$_{BiEnd(_RM)}M_{End(_RM)}$$

is a faithfully balanced bimodule.

Proof. Let $T = End(_RM)$ and $B = BiEnd(_RM) = End(M_T)$. Then $_BM$ and M_T are automatically faithful; and every T-homomorphism is, by definition, (multiplication by) an element of B. But because of the ring homomorphism $\lambda : R \to B$ with $\lambda(r)x = rx$, we have $End(_BM) \leq End(_RM) = T$ (see (4.8)); and the proposition is proved. $\qquad\square$

There is a parallel theory for right modules. If N_R, then

$$_{End(N_R)}N_{BiEnd(N_R)}$$

is faithfully balanced. Also right multiplication $\rho : R \to BiEnd(N_R)$ is called the natural homomorphism of R into $BiEnd(N_R)$.

4.13. Remark. By (4.12) the natural development that leads from R to $T = End(_RM)$ to $B = BiEnd(_RM)$ stabilizes. That is, T is the "triendo-morphism ring" $T = End(_BM) = BiEnd(M_T)$.

There is an important variation of the concept of a balanced bimodule. We say that a non-zero left R-module $_RM$ is *balanced* in case the derived bimodule

$$_RM_{End(_RM)}$$

is a balanced bimodule. Thus, $_RM$ is balanced iff the natural homomorphism

$$\lambda : R \to BiEnd(_RM)$$

is surjective, and $_RM$ is both balanced and faithful iff λ is an isomorphism. Again there is an obvious corresponding notion for right modules. There is a more than formal difference between balanced bimodules and balanced one-sided modules. On the one hand, we see at once that

4.14. Proposition. *If $_RM_S$ is a (faithfully) balanced bimodule, then $_RM$ and M_S are (faithful and) balanced modules.* $\qquad\square$

On the other hand the converse of (4.14) is false. For if $_QM$ is a 2-dimensional vector space, both $_QM$ and M_Q are balanced but the bimodule $_QM_Q$ is not. We deduce from (4.11) and (4.14) that both $_RR$ and R_R are balanced. From (4.12) and (4.14) we have, in rather imprecise terminology, that every module is balanced both over its endomorphism ring and over its biendomorphism ring.

Let $e \in R$ be a non-zero idempotent. An argument very much like that used for (4.11) allows us to characterize the ring eRe as the endomorphism ring of the principal left ideal Re of R. Again as we shall see in §7 these left ideals and their endomorphism rings are of considerable significance in analyzing R.

4.15. Proposition. *If e is a non-zero idempotent in a ring R, then*

$$\rho : eRe \rightarrow End(_R Re) \qquad and \qquad \lambda : eRe \rightarrow End(eR_R)$$

defined via

$$\rho(ere) : ae \mapsto aere \qquad and \qquad \lambda(ere) : ea \mapsto erea$$

are ring isomorphisms. In particular,

$$BiEnd(_R Re) = End(Re_{eRe}).$$ □

Remark. Looking somewhat far ahead we can perhaps get an idea of the significance of this study of the biendomorphism ring. Given a specific ring R it may be possible to find an especially well-behaved representation $\lambda : R \rightarrow End^l(M)$. By "well-behaved" we might mean any of a variety of things—for example, $_R M$ may be simple, or faithful, or its endomorphism ring T may be a simple ring, an integral domain, etc. In any event, from this good behavior we may be able to deduce the structure of the biendomorphism ring $B = BiEnd(_R M)$. This is not too far-fetched for we have assumed some reasonable behavior for $_R M$, we know how to compute B in terms of $_R M$, and with B we have achieved a certain stability (4.13).

4. Exercises

1. Let $p \in \mathbb{P}$ be a positive prime. Compute each of the following abelian groups:

$$Hom_{\mathbb{Z}}(\mathbb{Q}, \mathbb{Z}), \qquad Hom_{\mathbb{Z}}(\mathbb{Z}_{p^\infty}, \mathbb{Q}), \qquad Hom_{\mathbb{Z}}(\mathbb{Z}_{p^\infty}, \mathbb{Z}).$$

[Hint: Exercises (3.14), (3.15), and (3.16).]

2. Let F be the full subcategory of $_{\mathbb{Z}}M$ whose object class consists of all torsionfree groups (see Exercise (3.14)) and let D be the full subcategory of $_{\mathbb{Z}}M$ whose object class consists of all divisible groups (see Exercise (3.15)).
   (1) Prove that in F the morphism $f : \mathbb{Z} \rightarrow \mathbb{Z}$ defined via $f : x \mapsto 2x$ is an epimorphism even though in the category $_{\mathbb{Z}}M$ it is not an epimorphism. (See Exercise (3.4).)
   (2) Prove that in D the natural epimorphism $f : \mathbb{Q} \rightarrow \mathbb{Q}/\mathbb{Z}$ is a monomorphism even though it is not a monomorphism in $_{\mathbb{Z}}M$. (See Exercise (3.4).)

3. Let R be a ring, let $K = Cen\, R$, and let $_R M$ be a non-zero left R-module. Prove that $End(_R M)$ is a K-algebra (see (1.11)) via $\phi : K \rightarrow End(_R M)$ where $(x)\phi(\alpha) = \alpha x$ for all $\alpha \in K$ and $x \in M$. Moreover, if $_R M$ is faithful, then ϕ is an injection into $Cen(End(_R M))$.

4. Let D be a division ring, let V_D be a non-zero vector space, and let $R = End(V_D)$. Then $_R V_D$ is a bimodule that is both R and D faithful (4.10). Prove that

(1) $_R V_D$ is balanced. [Hint: Let $\sigma \in End(_R V)$, let $x \in V$, and suppose x and $x\sigma$ are D-linearly independent. Then there is an $r \in R$ with $rx = 0$ and $r(x\sigma) = x$!]

(2) If X is a basis for V_D then $R \cong \mathbb{CFM}_X(D)$.

5. Let $_R M_S$ be a faithfully balanced bimodule. Show that there is a ring isomorphism $\phi : Cen\, R \to Cen\, S$ such that $km = m\phi(k)$ for all $m \in M$ and $k \in Cen\, R$.

6. Let $_R M$ be a non-zero module. Prove that

   (1) As sets of functions, $Cen(BiEnd(_R M)) = Cen(End(_R M))$.

   (2) If R is commutative, then $_R M$ is balanced iff every element in $Cen(End(_R M))$ is multiplication by an element of R.

7. Every abelian group M admits a unique $(\mathbb{Z}, \mathbb{Z})$-bimodule structure $_{\mathbb{Z}} M_{\mathbb{Z}}$. Prove that if M is a finitely generated abelian group, then $_{\mathbb{Z}} M_{\mathbb{Z}}$ is balanced iff M is cyclic.

8. Let I be the ideal of the polynomial ring $\mathbb{Q}[X, Y]$ generated by $\{X^2, XY, Y^2\}$ and let $R = \mathbb{Q}[X, Y]/I$. By Proposition 4.11 the regular bimodule $_R R_R$ is faithfully balanced. Prove that $_R R_R$ has submodules and factor modules that are not balanced.

9. (1) Prove Proposition 4.6.

   (2) Prove Proposition 4.15.

10. Compute both the endomorphism ring T and the biendomorphism ring B of each of the modules

    (1) $_{\mathbb{Z}} \mathbb{Q}$.

    (2) $_R Re$ where R is the ring of 3×3 lower triangular matrices over a field K and

$$e = \begin{bmatrix} 1 & 0 & 0 \\ 0 & 0 & 0 \\ 0 & 0 & 0 \end{bmatrix}.$$

11. Let M be a non-zero abelian group with $S = End^l(M)$. Suppose that M is a left R-module via $\lambda : R \to S$. Let R' be the S-centralizer of $\lambda(R)$,

$$R' = Cen_S(\lambda(R)).$$

(See Exercise (1.9).) Then $_{R'} M$. Set $R'' = Cen_S(R')$, etc. The rings R' and R'' are sometimes called the *first* and *second* centralizers of $_R M$, respectively. Prove that

(1) $R' = (End(_R M))^{op}$ and $R'' = BiEnd(_R M)$.

(2) $R''' = R'$.

12. Let I be a two sided ideal of a ring R. Then R/I is both a ring and a left R-module. Let $f : R \to R/I$ be the natural left R-epimorphism with kernel I. Prove that as rings $End(_R(R/I)) \cong R/I$ and that there is a ring homomorphism $\phi : R \to End(_R(R/I))$ such that $f(xr) = f(x)\phi(r)$ for all $x, r \in R$.

13. Generalize the result of Exercise (4.12). That is, let M and N be left R-modules and let $f : M \to N$ be an R-epimorphism. Suppose further that $Ker\, f$ is stable under $End(_R M)$; i.e., $Ker\, f$ is a right $End(_R M)$ sub-

module of M. Prove that there is a ring homomorphism $\phi : End(_RM) \to End(_RN)$ such that $f(m\gamma) = f(m)\phi(\gamma)$ for all $m \in M$, $\gamma \in End(_RM)$. Also show that

$$Ker\ \phi = r_{End(_RM)}(M/Ker\ f).$$

14. Let M and N be left R-modules and let $f : M \to N$ be an R-isomorphism. Prove that

    (1) There is a ring isomorphism $\phi_1 : End(_RM) \to End(_RN)$ such that $f(m\gamma) = f(m)\phi_1(\gamma)$ for all $m \in M$, $\gamma \in End(_RM)$. [Hint: See Exercise (4.13).]

    (2) There is a ring isomorphism $\phi_2 : BiEnd(_RM) \to BiEnd(_RN)$ such that $f(bm) = \phi_2(b)f(m)$ for all $m \in M$, $b \in BiEnd(_RM)$.

    (3) If $_RM$ is balanced, then so is $_RN$.

15. Let R and S be rings and let $\phi : R \to S$ be a ring homomorphism. For each $_SM$ let $T_\phi(M)$ be the R-module (M, γ) where $\gamma(r)(x) = \phi(r)x$ for each $r \in R$, $x \in M$. For each pair $_SM$, $_SN$ and each $f \in Hom_S(M, N)$, let $T_\phi(f) \in Hom_R(T_\phi(M), T_\phi(N))$ be $T_\phi(f) = f$. Prove that

    (1) T_ϕ defines a covariant functor from $_S\mathsf{M}$ to $_R\mathsf{M}$.

    (2) Unless ϕ is surjective, T_ϕ restricted to the full subcategory $_S\mathsf{FM}$ of finitely spanned left S-modules need not be a functor to $_R\mathsf{FM}$, the category of finitely spanned left R-modules.

16. Let I be an ideal of R. For each $_RM$ let $F(M)$ be the left R/I-module M/IM. For each $_RM$, $_RN$ and each $f \in Hom_R(M, N)$ let $F(f) : F(M) \to F(N)$ be defined by $F(f) : x + IM \mapsto f(x) + IN$. Prove that F defines a covariant functor from $_R\mathsf{M}$ to $_{R/I}\mathsf{M}$. Show that F restricted to $_R\mathsf{FM}$ is a functor to $_{R/I}\mathsf{FM}$.

17. Let R be a ring and let $e \in R$ be a non-zero idempotent. For each $_RM$ define $T_e : M \mapsto eM$, where eM is the left eRe-module defined on page 58. For each pair $_RM$, $_RN$ and each left R-homomorphism $f : M \mapsto N$, let $T_e : f \mapsto (f \mid eM)$. Prove that T_e defines a covariant additive functor from $_R\mathsf{M}$ to $_{eRe}\mathsf{M}$.

Chapter 2

Direct Sums and Products

For each ring R we have derived several module categories—among these the category $_R\mathsf{M}$ of left R-modules. This derivation is not entirely reversible for, in general, $_R\mathsf{M}$ does not characterize R. However, as we shall see in Chapter 6 it does come close. Thus, we can expect to uncover substantial information about R by mining $_R\mathsf{M}$. So in this chapter we start to probe more deeply into the structure of the modules themselves. In so far as possible we propose to do this in the context of the category $_R\mathsf{M}$ for in this way at any subsequent stage we shall be able to apply the general machinery of category theory.

We begin with the general decomposition theory of modules. This parallels closely the more special theories for vector spaces and for abelian groups, so many of the fundamental ideas are fairly transparent. In Sections 5 and 6 we develop the general theory of both internal and external decompositions. The substance is reasonably clear, but the necessary formalities are occasionally tedious.

In Section 7 we apply the theory to the regular modules $_R R$, R_R, and $_R R_R$ to obtain some of the fundamental results on the theory of ring decompositions. Finally, in Section 8 we make a natural application to obtain a general treatment of the concepts of generating one class of modules by another and of its less familiar dual of cogenerating one class by another.

§5. Direct Summands

Given two modules M_1 and M_2 we can construct their cartesian product $M_1 \times M_2$. The structure of this product module is then determined "coordinatewise" from that of the factors M_1 and M_2. In this section we shall begin by considering when this process can be reversed. That is, given a module M we shall concern ourselves with when it can be "factored" in some fashion as a type of product of other modules.

Split Homomorphisms

Let M_1 and M_2 be submodules of a module M. Recall that they *span* M in case

$$M_1 + M_2 = M;$$

i.e., in case their supremum in $\mathscr{S}(M)$ is M. At the other extreme, they are *independent* in case

$$M_1 \cap M_2 = 0;$$

i.e., in case their infimum in $\mathscr{S}(M)$ is 0. Now there is a canonical R-homomorphism i from the cartesian product $M_1 \times M_2$ module (2.1.6) to M defined via

$$i : (x_1, x_2) \mapsto x_1 + x_2 \qquad ((x_1, x_2) \in M_1 \times M_2)$$

with image and kernel

$$Im\, i = M_1 + M_2 \qquad \text{and} \qquad Ker\, i = \{(x, -x) \,|\, x \in M_1 \cap M_2\}.$$

So i is epic iff M_1 and M_2 span M, and monic iff M_1 and M_2 are independent. If this canonical homomorphism i is an isomorphism (i.e., if M_1 and M_2 are independent and span M), then M is the (*internal*) *direct sum* of its submodules M_1 and M_2, and we write

$$M = M_1 \oplus M_2.$$

Thus $M = M_1 \oplus M_2$ iff for each $x \in M$ there exist unique elements $x_1 \in M_1$ and $x_2 \in M_2$ such that

$$x = x_1 + x_2.$$

Not every submodule of a module M need appear in such a direct factorization of M. Those that do, however, are of considerable interest. A submodule M_1 of M is a *direct summand* of M in case there is a submodule M_2 of M with $M = M_1 \oplus M_2$; such an M_2 is also a direct summand, and M_1 and M_2 are *complementary direct summands* or *direct complements* of each other. Of course, even in vector spaces direct summands need not have unique complements.

The following fundamental result shows how one encounters direct sums and direct summands in the study of homomorphisms with "one-sided inverses".

5.1. Lemma. *Let $f : M \to N$ and $f' : N \to M$ be homomorphisms such that*

$$ff' = 1_N.$$

Then f is an epimorphism, f' is a monomorphism and

$$M = Ker\, f \oplus Im\, f'.$$

Proof. Clearly (see (3.3) and (3.4)), f is epic, f' is monic. If $x = f'(y) \in Ker\, f \cap Im\, f'$, then $0 = f(x) = ff'(y) = y$ and $x = f'(y) = 0$. If $x \in M$, then $f(x - f'f(x)) = f(x) - f(x) = 0$, and $x = (x - f'f(x)) + f'f(x) \in Ker\, f + Im\, f'$. $\qquad\square$

If $f : M \to N$ and $f' : N \to M$ are homomorphisms with $ff' = 1_N$, we say that f is a *split epimorphism*, and we write

$$M \xrightarrow{\ f\ } \oplus \to N \to 0;$$

and we say that f' is a *split monomorphism*, and we write

$$0 \to N \xrightarrow{f'} \oplus \to M.$$

A short exact sequence (see §3)

$$0 \to M_1 \xrightarrow{f} M \xrightarrow{g} M_2 \to 0$$

is *split* or is *split exact* in case f is a split monomorphism and g is a split epimorphism. As we see next, if an exact sequence is split at either end, then it is split at both ends.

5.2. Proposition. *The following statements about a short exact sequence*

$$0 \to M_1 \xrightarrow{f} M \xrightarrow{g} M_2 \to 0$$

in $_R M$ are equivalent:

(a) *The sequence is split;*
(b) *The monomorphism $f: M_1 \to M$ is split;*
(c) *The epimorphism $g: M \to M_2$ is split;*
(d) *$\operatorname{Im} f = \operatorname{Ker} g$ is a direct summand of M;*
(e) *Every homomorphism $h: M_1 \to N$ factors through f;*
(f) *Every homomorphism $h: N \to M_2$ factors through g.*

Proof. (a) $\Rightarrow$ (b) and (a) $\Rightarrow$ (c) are trivial, and (b) $\Rightarrow$ (d) and (c) $\Rightarrow$ (d) are by (5.1). Since (b) and (c) together give (a), it will suffice to prove (d) $\Rightarrow$ (e) $\Rightarrow$ (b) and (d) $\Rightarrow$ (f) $\Rightarrow$ (c).

(d) $\Rightarrow$ (e). Suppose $M = \operatorname{Im} f \oplus K$ and $h: M_1 \to N$. Since f is monic, for each $m \in M$ there is a unique $m_1 \in M_1$ and $k \in K$ with $m = f(m_1) + k$. Define $\bar{h}: M \to N$ by

$$\bar{h}: m = f(m_1) + k \mapsto h(m_1).$$

Then clearly $\bar{h}$ is an R-homomorphism with $\bar{h}f = h$.

(d) $\Rightarrow$ (f). Suppose $M = \operatorname{Ker} g \oplus K$ and $h: N \to M_2$. Since $K \cap \operatorname{Ker} g = 0$ and $g(M) = g(K)$, we see that $(g \mid K): K \to M_2$ is an isomorphism. Let $g': M_2 \to K$ be its inverse. Then $\bar{h} = g'h: N \to M$ is an R-homomorphism with $g\bar{h} = h$.

(e) $\Rightarrow$ (b) and (f) $\Rightarrow$ (c). Let $h = 1_N$ where in the first case $N = M_1$ and in the other $N = M_2$.   □

Let M_1 and M_2 be two modules. Then with their product module $M_1 \times M_2$ are associated the *natural injections* and *projections*

$$\iota_j: M_j \to M_1 \times M_2 \qquad \text{and} \qquad \pi_j: M_1 \times M_2 \to M_j$$

$(j = 1, 2)$, defined by

$$\iota_1(x_1) = (x_1, 0), \qquad\qquad \iota_2(x_2) = (0, x_2),$$

and

$$\pi_1(x_1, x_2) = x_1, \qquad\qquad \pi_2(x_1, x_2) = x_2.$$

These are clearly R-homomorphisms for which

$$0 \to M_1 \xrightarrow{\iota_1} M_1 \times M_2 \xrightarrow{\pi_2} M_2 \to 0$$

$$0 \to M_2 \xrightarrow{\iota_2} M_1 \times M_2 \xrightarrow{\pi_1} M_1 \to 0$$

are exact. Moreover, since

$$\pi_1 \iota_1 = 1_{M_1} \qquad \text{and} \qquad \pi_2 \iota_2 = 1_{M_2},$$

these sequences are split exact.

Observe also that

$$\pi_i \iota_j = \delta_{ij} 1_{M_i} \qquad \text{and} \qquad \iota_1 \pi_1 + \iota_2 \pi_2 = 1_{M_1 \times M_2}.$$

We now prove that, as we might expect, these sequences are the prototypes of all split exact sequences.

5.3. Proposition. *For a sequence of R-homomorphisms*

$$0 \to M_1 \xrightarrow{f_1} M \xrightarrow{g_2} M_2 \to 0$$

the following statements are equivalent:

(a) *The sequence is split exact;*

(b) *There exists a sequence of R-homomorphisms*

$$0 \to M_2 \xrightarrow{f_2} M \xrightarrow{g_1} M_1 \to 0$$

(necessarily split exact) such that for $i, j \in \{1, 2\}$,

$$g_i f_j = \delta_{ij} 1_{M_i} \qquad \text{and} \qquad f_1 g_1 + f_2 g_2 = 1_M;$$

(c) *There exists an isomorphism $h: M_1 \times M_2 \to M$ such that the following diagram commutes:*

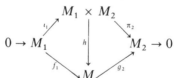

Proof. (a) $\Rightarrow$ (c). To prove this implication define $h: M_1 \times M_2 \to M$ by $h(x_1, x_2) = f_1(x_1) + f_2(x_2)$ where $f_2: M_2 \to M$ satisfies $g_2 f_2 = 1_{M_2}$. Then the diagram commutes, and h is an isomorphism by The Five Lemma (3.15).

(c) $\Rightarrow$ (b). Given an isomorphism h making the diagram commute, define $f_2 = h\iota_2$ and $g_1 = \pi_1 h^{-1}$. Then

$$g_i f_j = \pi_i h^{-1} h\iota_j = \pi_i \iota_j = \delta_{ij} 1_{M_i}$$

and

$$f_1 g_1 + f_2 g_2 = h\iota_1 \pi_1 h^{-1} + h\iota_2 \pi_2 h^{-1}$$

$$= h(\iota_1 \pi_1 + \iota_2 \pi_2) h^{-1} = hh^{-1} = 1_M.$$

(b) $\Rightarrow$ (a). Assume (b). Then $f_1 g_1 + f_2 g_2 = 1_M$, so $M = \operatorname{Im} f_1 + \operatorname{Im} f_2$.

But $g_2 f_1 = 0$ implies $Im f_1 \subseteq Ker\, g_2$, and $g_2 f_2 = 1_{M_2}$ implies (see (5.1)) $M = Ker\, g_2 \oplus Im f_2$. So by modularity

$$Ker\, g_2 = Im f_1 + (Ker\, g_2 \cap Im f_2) = Im f_1.$$

Thus the sequence

$$0 \to M_1 \xrightarrow{f_1} M \xrightarrow{g_3} M_2 \to 0$$

is exact at M; and since $g_1 f_1 = 1_{M_1}$ and $g_2 f_2 = 1_{M_2}$, it is split exact.   □

Projections

Let K be a direct summand of M with complementary direct summand K', so $M = K \oplus K'$. Then

$$p_K : k + k' \mapsto k \qquad (k \in K, k' \in K')$$

defines an epimorphism

$$p_K : M \to K$$

called *the projection of M on K along K'*.

5.4. Proposition. *If $M = K \oplus K'$, then the projection of M on K along K' is the unique epimorphism*

$$M \xrightarrow{p_K} K \to 0$$

satisfying

$$(p_K | K) = 1_K \qquad and \qquad Ker\, p_K = K'.$$

Proof. That p_K does satisfy these conditions is an immediate consequence of its definition. If $g : M \to K$ is such that $(g | K) = 1_K$ and $Ker\, g = K'$, then for all $k \in K$, $k' \in K'$, $g(k + k') = g(k) + g(k') = k = p_K(k + k')$.   □

Again let K be a direct summand of M with complementary direct summand K',

$$M = K \oplus K'.$$

Then K' is a direct summand of M with complementary direct summand K. Moreover, if p_K is the projection of M on K along K', then the projection $p_{K'}$ of M on K' along K can be characterized by

$$p_{K'} : m \mapsto m - p_K(m) \qquad (m \in M).$$

Now if $i_K : K \to M$ and $i_{K'} : K' \to M$ are the inclusion maps, then by (5.4) and (5.2),

$$0 \to K' \xrightarrow{i_{K'}} M \xrightarrow{p_K} K \to 0$$

$$0 \to K \xrightarrow{i_K} M \xrightarrow{p_{K'}} K' \to 0$$

are split exact. Also it is clear that, with the obvious notational changes, these maps satisfy the identities of (5.3).

In general, a direct summand of a module has many complementary direct summands; the projections provide a useful characterization of these.

5.5. Proposition. *Let* $M = K \oplus K'$, *let* p_K *be the projection of* M *on* K *along* K', *and let* L *be a submodule of* M. *Then*

$$M = L \oplus K'$$

if and only if

$$(p_K | L) : L \to K$$

is an isomorphism.

Proof. Let $L \le M = K \oplus K'$. Then $Ker(p_K | L) = L \cap Ker\, p_K = L \cap K'$ so that $(p_K | L)$ is monic iff $L \cap K' = 0$. On the other hand, since $(p_K | K) = 1_K$ and $Ker\, p_K = K'$,

$$p_K(L) = p_K(L + K') = p_K((L + K') \cap (K + K'))$$
$$= p_K(((L + K') \cap K) + K') = p_K((L + K') \cap K)$$
$$= (L + K') \cap K$$

so that $p_K(L) = K$ iff $K \subseteq L + K'$ iff $L + K' = M$.   □

Idempotent Endomorphisms

Suppose that $_RM = K \oplus K'$ and p_K is the projection of M on K along K'. Define $e_K \in End(_RM)$ by

$$e_K : x \mapsto p_K(x) \qquad (x \in M).$$

Then since $(p_K | K) = 1_K$, e_K is an idempotent endomorphism of M,

$$e_K = e_K^2 \in End(_RM),$$

and (note that e_K is a right operator on M)

$$K = Me_K.$$

Thus each direct summand of M is the image of an idempotent endomorphism of M. As we see from the following lemma, the converse is also true.

5.6. Lemma. *Let* e *be an idempotent in* $End(_RM)$. *Then* $1 - e$ *is an idempotent in* $End(_RM)$ *such that*

$$Ker\, e = \{x \in M \mid x = x(1 - e)\} = Im(1 - e),$$
$$Im\, e = \{x \in M \mid x = xe\} = Ker(1 - e)$$

and $M = Me \oplus M(1 - e)$.

Proof. In (1.16) we saw that $1 - e$ is an idempotent. Since $e^2 = e$, $(1 - e)^2 = (1 - e)$, and $e(1 - e) = (1 - e)e = 0$, we have at once the inclusions

$$Im\, e \subseteq \{x \in M \mid x = xe\} \subseteq Ker(1 - e),$$

$$Im(1 - e) \subseteq \{x \in M \mid x = x(1 - e)\} \subseteq Ker\, e.$$

But since $x = xe + x(1 - e)$ for all $x \in M$, these are not strict and $M = Me + M(1 - e)$. Finally, $Me \cap M(1 - e) = 0$, for if $xe = y(1 - e)$, then $xe = xe^2 = (y(1 - e))e = 0$. $\qquad\qquad\square$

5.7. Proposition. *If* $_RM = K \oplus K'$, *then there is a unique idempotent* $e_K \in End(_RM)$ *such that*

$$K = Me_K \qquad and \qquad K' = M(1 - e_K).$$

Proof. The proposition follows from (5.4) and (5.6) which combine to tell us that if $e \in End(_RM)$ is idempotent, then $x \mapsto xe$ is the projection of M on Me along $M(1 - e)$. $\qquad\qquad\square$

5.8. Corollary. *A submodule* $K \leq M$ *is a direct summand of* M *if and only if* $K = Im\, e$ *for some idempotent endomorphism* e *of* M. $\qquad\qquad\square$

It should be noted that a direct summand K of M can be the image of several different idempotent endomorphisms (see Exercise (5.13)), but for each decomposition $M = K \oplus K'$ the associated pair of idempotents (5.7) is unique. The idempotents of a direct summand K provide a tool for computing the endomorphism ring of K.

5.9. Proposition. *Let* e *be an idempotent in* $End(_RM)$. *Then there is a ring isomorphism*

$$\phi : e\, End(_RM)e \to End(_RMe),$$

such that for all $s \in End(_RM)$ *and all* $x \in M$

$$\phi(ese) : xe \mapsto xese.$$

Proof. It is a routine matter to check that there is an injective ring homomorphism ϕ from $e\, End(_RM)e$ into $End(_RMe)$ satisfying the required condition. Now $e : Me \to M$ is a split monomorphism (5.6). So (5.2) if $g \in End(_RMe)$ (i.e., $g : Me \to Me \leq M$), then g factors through e

Thus for each $g \in End(_RMe)$ there is a $\bar{g} \in End(_RM)$ such that for all $xe \in Me$

$$xe\phi(e\bar{g}e) = xe\bar{g}e = xege = xeg.$$

Thus ϕ is an isomorphism. $\qquad\qquad\square$

It is clear that every non-zero module M has at least two direct summands, namely, 0 and M. A non-zero module M is *indecomposable* if 0 and M are its only direct summands. Such indecomposable modules will play a central role in our work.

A pair of idempotents e_1 and e_2 in a ring R are said to be *orthogonal* if

$$e_1 e_2 = 0 = e_2 e_1.$$

An idempotent $e \in R$ is called a *primitive idempotent* in case $e \neq 0$ and for every pair e_1, e_2 of orthogonal idempotents

$$e = e_1 + e_2 \qquad \text{implies} \qquad e_1 = 0 \quad \text{or} \quad e_2 = 0.$$

If $e = e^2 \in R$, then e and $1 - e$ are orthogonal idempotents such that $1 = e + (1 - e)$.

Thus, applying (5.8) and (5.6) we have

5.10. Proposition. *Let M be a non-zero module. Then the following are equivalent:*

(a) *M is indecomposable.*

(b) *0 and 1 are the only idempotents in $\mathrm{End}(M)$.*

(c) *1 is a primitive idempotent in $\mathrm{End}(M)$.*   □

If e is a non-zero idempotent in a ring R, then e is primitive if and only if the identity e of the ring eRe is a primitive idempotent. Indeed, if $e = e_1 + e_2$ where e_1 and e_2 are orthogonal idempotents in R, then $e_1 = e e_1 e \in eRe$ and $e_2 = e e_2 e \in eRe$. Hence the preceding two propositions yield

5.11. Corollary. *Let e be a non-zero idempotent endomorphism of a left module M. Then the direct summand Me of M is indecomposable if and only if e is a primitive idempotent in $\mathrm{End}(M)$.*   □

Essential and Superfluous Submodules

A submodule K or M is a direct summand of M iff there is a submodule K' of M with

$$K \cap K' = 0 \qquad \text{and} \qquad K + K' = M,$$

that is, iff K is complemented in the lattice of submodules of M. For any submodule K of M we can always find a submodule satisfying, with K, one or the other of these conditions. Indeed,

$$K \cap 0 = 0 \qquad \text{and} \qquad K + M = M.$$

Those submodules for which one of these is the "best" turn out to be of great significance in our subsequent work. Specifically, a submodule K of M is *essential* (or *large*) in M, abbreviated

$$K \trianglelefteq M,$$

in case for every submodule $L \leq M$,

$$K \cap L = 0 \qquad \text{implies} \qquad L = 0.$$

Dually, a submodule K of M is *superfluous* (or *small*) in M, abbreviated

$$K \ll M,$$

in case for every submodule $L \leq M$

$$K + L = M \qquad \text{implies} \qquad L = M.$$

The three concepts, direct summand, essential submodule, and super-fluous submodule, are reminiscent of the topological concepts of connected component, dense, and nowhere dense. In a sense an essential submodule of M dominates the lattice of submodules in that it is independent of no non-zero submodule, whereas a superfluous submodule is quite ineffective in that it contributes nothing to spanning M. Note, however, that a submodule can be both essential and superfluous; indeed this is true of every non-trivial submodule of $\mathbb{Z}_{p^\infty}$.

A monomorphism $f : K \to M$ is said to be *essential* in case $Im f \trianglelefteq M$. An epimorphism $g : M \to N$ is *superfluous* in case $Ker\, g \ll M$. As we shall see below (particularly (5.13) and (5.15)) these two concepts are dual in the category $_R\mathsf{M}$. That is, any statement about an essential monomorphism stated in terms of $_R\mathsf{M}$ is true iff the statement obtained by reversing the arrows is true about superfluous epimorphisms. Thus perhaps it would be best to state our results in "arrowese" where their duals are natural. However, in practice we shall more often be concerned with the behavior of the essential and superfluous submodules of M in the lattice $\mathscr{S}(M)$. Therefore most of the results are in lattice theoretic terms. See the exercises for the categorical formulation. (Particularly, Exercises (5.14)–(5.16).)

5.12. Proposition. *For a submodule K of M the following statements are equivalent:*

(a) $K \trianglelefteq M$.
(b) *The inclusion map $i_K : K \to M$ is an essential monomorphism.*
(c) *For every module N and for each $h \in Hom(M, N)$,*

$$(Ker\, h) \cap K = 0 \qquad \text{implies} \qquad Ker\, h = 0.$$

Proof. (a) $\Leftrightarrow$ (b) and (a) $\Rightarrow$ (c) are both clear.
(c) $\Rightarrow$ (a). Suppose that $L \leq M$ and $K \cap L = 0$. Let $n_L : M \to M/L$ be the natural epimorphism. Then clearly $(Ker\, n_L) \cap K = 0$. So assuming (c), $L = Ker\, n_L = 0$. $\qquad\square$

Suppose f is a monomorphism and h is a homomorphism such that $f \circ h$ is monic. Then clearly h is also a monomorphism. On the other hand,

5.13. Corollary. *A monomorphism $f : L \to M$ is essential if and only if, for all homomorphisms (equivalently, epimorphisms) h, if hf is monic, then h is monic.*

Proof. Let $K = Im f$. Then by (3.10) there is an isomorphism $v : K \to L$ such that $fv = i_K$. Thus it follows that hf is monic iff hi_K is monic. But the latter condition holds iff $(Ker\, h) \cap K = 0$. For the parenthetical version note that h is an epimorphism onto $Im\, h$. $\qquad\square$

The proofs of the following duals to Proposition (5.12) and Corollary (5.13) are left as exercises.

5.14. Proposition. *For a submodule K of M the following statements are equivalent:*

(a) $K \ll M$.

(b) *The natural map $p_K : M \to M/K$ is a superfluous epimorphism.*

(c) *For every module N and for every $h \in \operatorname{Hom}(N, M)$*

$$(\operatorname{Im} h) + K = M \qquad implies \qquad \operatorname{Im} h = M. \qquad \square$$

5.15. Corollary. *An epimorphism $g : M \to N$ is superfluous if and only if for all homomorphisms (equivalently, monomorphisms) h, if gh is epic, then h is epic.* $\qquad \square$

The essential submodules of M form an important sublattice of the lattice of all submodules of M; specifically,

5.16. Proposition. *Let M be a module with submodules $K \leq N \leq M$ and $H \leq M$. Then*

(1) $K \trianglelefteq M$ iff $K \trianglelefteq N$ and $N \trianglelefteq M$;

(2) $H \cap K \trianglelefteq M$ iff $H \trianglelefteq M$ and $K \trianglelefteq M$.

Proof. (1) Let $K \trianglelefteq M$ and suppose $0 \neq L \leq M$, then $L \cap K \neq 0$. In particular this is true if $L \leq N$, so $K \trianglelefteq N$. But also $K \leq N$ so $L \cap N \neq 0$ whence $N \trianglelefteq M$.

Conversely, if $K \trianglelefteq N$ and $N \trianglelefteq M$ and $L \leq M$, then $L \cap K = 0$ implies $L \cap N = 0$ implies $L = 0$.

(2) One implication follows at once from (1). For the other, suppose $H \trianglelefteq M$ and $K \trianglelefteq M$. If $L \leq M$ with $L \cap H \cap K = 0$, then $L \cap H = 0$ because $K \trianglelefteq M$. Whence $L = 0$ because $H \trianglelefteq M$. $\qquad \square$

This result has a natural dual for superfluous submodules. In categorical terminology its statement is quite obvious. The lattice theoretic version, whose proof we omit, goes as follows.

5.17. Proposition. *Let M be a module with submodules $K \leq N \leq M$ and $H \leq M$. Then*

(1) $N \ll M$ iff $K \ll M$ and $N/K \ll M/K$;

(2) $H + K \ll M$ iff $H \ll M$ and $K \ll M$. $\qquad \square$

The following lemma concerning superfluous submodules also has a dual. However, we shall relegate it to the exercises where we can give a proper formulation.

5.18. Lemma. *If $K \ll M$ and $f : M \to N$ is a homomorphism then $f(K) \ll N$. In particular, if $K \ll M \leq N$ then $K \ll N$.*

Proof. Let $L \leq N$ and assume $L + f(K) = N$. Then $f^{\leftarrow}(L) + K = M$. Since $K \ll M$, this implies $K \leq M = f^{\leftarrow}(L)$, so $f(K) \leq L$, and $L = N$. $\qquad \square$

Our next lemma gives an extraordinarily useful test for essential inclusions.

5.19. Lemma. *A submodule $K \leq M$ is essential in M if and only if for each $0 \neq x \in M$ there exists an $r \in R$ such that $0 \neq rx \in K$.*

Proof. ($\Rightarrow$) If $K \trianglelefteq M$ and $0 \neq x \in M$, then $Rx \cap K \neq 0$.

($\Leftarrow$) If the condition holds and $0 \neq x \in L \leq M$, then there is an $r \in R$ such that $0 \neq rx \in K \cap L$. $\qquad\qquad\qquad\qquad\qquad\qquad\qquad\qquad\qquad\qquad\qquad\square$

Using these two lemmas we have

5.20. Proposition. *Suppose that $K_1 \leq M_1 \leq M$, $K_2 \leq M_2 \leq M$, and $M = M_1 \oplus M_2$; then*

(1) $K_1 \oplus K_2 \ll M_1 \oplus M_2$ *iff* $K_1 \ll M_1$ *and* $K_2 \ll M_2$;

(2) $K_1 \oplus K_2 \trianglelefteq M_1 \oplus M_2$ *iff* $K_1 \trianglelefteq M_1$ *and* $K_2 \trianglelefteq M_2$.

Proof. (1) Let $p_i : M \to M_i$ denote the projection of M on M_i along M_j ($i \neq j$). Then $K_i = p_i(K_i)$; so necessity follows from (5.18).

Conversely, if $K_i \ll M_i \leq M$ ($i = 1, 2$), then by (5.18) and (5.17.2) $K_1 \oplus K_2 = K_1 + K_2 \ll M$.

(2) Suppose, say, K_1 is not essential in M_1, i.e., $K_1 \cap L_1 = 0$ for some $0 \neq L_1 \leq M_1$. Then the necessity is proved by observing that

$$(K_1 + K_2) \cap L_1 = 0;$$

for if $k_1 \in K_1$, $k_2 \in K_2$ and $l_1 \in L_1$ with $k_1 + k_2 = l_1$, then

$$k_2 = l_1 - k_1 \in M_1 \cap M_2 = 0.$$

For the sufficiency suppose that $K_i \trianglelefteq M_i$ and $0 \neq x_i \in M_i$ ($i = 1, 2$), then by (5.19) there is an $r_1 \in R$ such that $0 \neq r_1 x_1 \in K_1$. If $r_1 x_2 \in K_2$ then, by independence, $0 \neq r_1 x_1 + r_1 x_2 \in K_1 \oplus K_2$. If $r_1 x_2 \notin K_1$ then again by (5.19) there is an $r_2 \in R$ with $0 \neq r_2 r_1 x_2 \in K_2$, and we have

$$0 \neq r_2 r_1 x_1 + r_2 r_1 x_2 \in K_1 \oplus K_2.$$

Thus $K_1 \oplus K_2 \trianglelefteq M_1 \oplus M_2$. $\qquad\qquad\qquad\qquad\qquad\qquad\qquad\qquad\square$

Let N be a submodule of M. If $N' \leq M$ is maximal with respect to $N \cap N' = 0$, then we say that N' is an M-*complement* of N. Using the Maximal Principle we readily see that if $N \leq M$, then the set of those submodules of M whose intersection with N is zero contains a maximal element N'. This proves the first part of

5.21. Proposition. *Every submodule $N \leq M$ has an M-complement. Moreover, if N' is an M-complement of N, then*

(1) $N \oplus N' \trianglelefteq M$;

(2) $(N \oplus N')/N' \trianglelefteq M/N'$.

Proof. (1) If $0 \neq L \leq M$ and $(N \oplus N') \cap L = 0$, then it readily follows that $N \cap (N' + L) = 0$, contrary to the maximality of N'.

(2) Suppose that $L \geq N'$ with $L \cap (N + N') \leq N'$; then by modularity

$$(L \cap N) \oplus N' = L \cap (N + N') \leq N'.$$

Therefore, $L \cap N = 0$ and by maximality of N', $L = N'$. $\qquad\qquad\qquad\square$

5. Exercises

1. Let $_RM$ be a left R-module. Prove that every epimorphism $f:M \to {}_RR$ splits. However, show that there can be monomorphisms $g:{}_RR \to M$ that do not split. [Hint: Let $R = \mathbb{Z}$.]

2. Let M be a non-zero module. In $M^2 = M \times M$, let $M_1 = \{(m,0) \mid m \in M\}$ and $M_2 = \{(0,m) \mid m \in M\}$. For each $\sigma \in End({}_RM)$, set

$$M^\sigma = \{(m, m\sigma) \mid m \in M\}.$$

Then $M^\sigma \le M^2$. Let $K \le M^2$. Prove that
(1) $M^2 = K \oplus M_2$ iff $K = M^\sigma$ for some $\sigma \in End({}_RM)$.
(2) If $K = M^\sigma$ for some automorphism $\sigma \in End({}_RM)$, then

$$M^2 = M_1 \oplus K.$$

3. Prove that if $_RM$ has a distributive lattice (see (0.5)) $\mathscr{S}(M)$ of sub-modules, then each direct summand has a unique complement. For example, show that if R is a Boolean algebra, then $_RR$ has a distributive lattice of submodules.

4. Let $M = K \oplus K' = L \oplus L'$. Prove
(1) $K = L$ implies $K' \cong L'$, but does not imply $K' = L'$.
(2) $K \subseteq H \le M$ implies $H = K \oplus (H \cap K')$.
(3) $K \cap L = 0$ does not imply that $K + L$ is a direct summand of M. [Hint: Consider $\mathbb{Z} \times \mathbb{Z}$ in Exercise (5.2.1).]

5. Let $M = K + L$ and let $f:M \to N$ be an epimorphism. Prove that $N = f(K) \oplus f(L)$ if $K \cap L = Ker f$.

6. (1) In order to give meaning to $M = H \oplus K \oplus L$, prove that if $M = H \oplus H'$ and $H' = K \oplus L$, then $M = (H + K) \oplus L$ and $H + K = H \oplus K$. (I.e., $H \oplus (K \oplus L) = (H \oplus K) \oplus L$.)
(2) Let $H, K, L \le M$. Prove that $M = H \oplus K \oplus L$ iff $H \cap K = 0 = L \cap K$ and $M/K = (H + K)/K \oplus (L + K)/K$.

7. (1) Give an example of an indecomposable module that has a de-composable submodule. [Hint: Try a factor module of $_RR$ where $R = \mathbb{Q}[X, Y]$.]
(2) Give an example of an indecomposable module that has a de-composable factor module.

8. Let $M = M_1 \oplus M_2$ and let $f:M \to N$ be an epimorphism with $K = Ker f$. Then (see Exercise (5.5))

$$N = f(M_1) + f(M_2).$$

(1) Prove that if $K = (K \cap M_1) + (K \cap M_2)$, e.g., if the submodule lattice $\mathscr{S}(M)$ is distributive, then this sum is direct.
(2) Show that in general, however, this sum is not direct. [Hint: Let L be a module, $M = L \times L$ and $f:M \to L$ via $f(l_1, l_2) = l_1 - l_2$.]

9. Let $M = M_1 \oplus M_2$ and let $N \le M$. Then

$$N \ge (N \cap M_1) \oplus (N \cap M_2).$$

If either $M_1 \leq N$ or $M_2 \leq N$, or if the submodule lattice $\mathscr{S}(M)$ is distributive, then equality holds. Show that in general, however, the inequality may be strict.

10. Let $M = M_1 \oplus M_2$ and let p_i $(i = 1, 2)$ be the corresponding projections.
(1) Prove that if $N \leq M$, then $p_1(N)/N \cap M_1 \cong p_2(N)/N \cap M_2$.
(2) Conversely, prove that if $K_i \leq N_i \leq M_i$ $(i = 1, 2)$, and if $N_1/K_1 \cong N_2/K_2$, then there is an $N \leq M$ with $K_i = N \cap M_i$, and $N_i = p_i(N)$ $(i = 1, 2)$. [Hint: Let $\sigma: N_1/K_1 \to N_2/K_2$ be an isomorphism and $N = \{n_1 + n_2 \mid n_2 + K_2 = \sigma(n_1 + K_1)\}$.]
(3) The two extreme cases are of interest; they occur when $N = (N \cap M_1) + (N \cap M_2)$ and when $N \cap M_1 = N \cap M_2 = 0$. Give a non-trivial example of each of these.

11. Let $g: N \to M$ and $f: K \to N$ be homomorphisms. Prove
(1) If f and g are both split monomorphisms (epimorphisms), then gf is a split monomorphism (epimorphism).
(2) Show the converse of (1) is false.
(3) Infer from (1) that a direct summand of a direct summand is a direct summand.

12. Suppose that the following diagram of modules and homomorphisms is commutative

$$\begin{array}{ccccccccc} 0 \to & A & \xrightarrow{f} & B & \xrightarrow{g} & C & \to 0 \\ & \alpha \downarrow & & \beta \downarrow & & \gamma \downarrow & \\ 0 \to & A' & \xrightarrow{f'} & B' & \xrightarrow{g'} & C' & \to 0 \end{array}$$

and that α, β, γ are isomorphisms. Prove that the top row is (split) exact iff the bottom row is (split) exact.

13. (1) Let $e \in R$ be an idempotent. Show that for each $x \in R$, $t = e + (1 - e)xe$ is also an idempotent. Moreover, show that for each such t there is a $y \in R$ with $e = t + (1 - t)yt$. [Hint: Since $et = e$ and $te = t$, it follows that $y = -(1 - e)xe$ works.]
(2) Let $_R M$ be a non-zero module and let $e \in \mathrm{End}(_R M)$ be an idempotent. For each idempotent $t \in \mathrm{End}(_R M)$ prove that $\mathrm{Im}\, t = \mathrm{Im}\, e$ iff $t = e + (1 - e)xe$ for some $x \in \mathrm{End}(_R M)$.

14. Consider the following commutative diagrams in $_R M$:

(1) Prove that if in the first g is monic, then h is an essential monomorphism iff both f and g are. [Hint: (5.13).] Deduce Proposition (5.16.1).
(2) Prove that if in the second g is epic, then h is a superfluous epimorphism iff both f and g are. [Hint: (5.15).] Deduce Proposition (5.17.1).

15. Consider the following commutative diagram in the category $_R M$

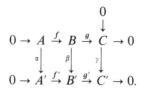

(1) Assume that both rows are exact and α is epic. Prove that if the g is superfluous, so is g'.

(2) Show that (1) is equivalent to (5.18).

16. (1) In Exercise (5.15) assume both rows are exact and γ is monic. Prove that if f' is essential, then so is f.

(2) Deduce that if $K \trianglelefteq M$ and if $x \in M$, then

$$\rho_x^{\leftarrow}(K) = \{r \in R \mid rx \in K\} \trianglelefteq {}_R R.$$

17. Let M be a non-zero module and let K be an $(R, End({}_R M))$-submodule $K \leq {}_R M_{End({}_R M)}$.

(1) Prove that if ${}_R M = M_1 \oplus M_2$, then $K = (K \cap M_1) \oplus (K \cap M_2)$.

(2) Prove that if ${}_R K \ll {}_R M$, and if ${}_R(M/K)$ is indecomposable, then ${}_R M$ is indecomposable.

(3) Prove that if $K \trianglelefteq M$ and if K is indecomposable, then M is indecomposable.

18. Let I be a nilpotent left ideal of R. Prove that for each left R-module M, $IM \ll M$. [Hint: If $IM + N = M$, then $I^2 M + IN + N = M$.]

19. Let M be an abelian group and $K \leq M$. Prove that

(1) Every homomorphism $f: K \to \mathbb{Q}$ has an extension $\bar{f}: M \to \mathbb{Q}$. [Hint: The set $G = \{g_L \mid L \leq M$ and $g_L \in Hom_{\mathbb{Z}}(L, \mathbb{Q})\}$ is partially ordered by set inclusion (each $g \in G$ is a set of ordered pairs). There is a $g_L \in G$ maximal with respect to $K \leq L$ and $(g_L \mid K) = f$. If $x \in M \backslash L$, then for some n, $\mathbb{Z} x \cap L = \mathbb{Z} n x \neq 0$ and there is an $h: \mathbb{Z} x + L \to \mathbb{Q}$ with $h(mx + l) = mn^{-1} g_L(nx) + g_L(l)$.]

(2) Every monomorphism $g: \mathbb{Q} \to M$ splits.

§6. Direct Sums and Products of Modules

In this section we consider two (dual) generalizations of finite products of modules (2.1.6) and of internal direct sums $M = M_1 \oplus M_2$ of a module.

Throughout this section we shall suppose that $(M_\alpha)_{\alpha \in A}$ is an indexed class of left R-modules. Analogues for right modules and bimodules should prove no difficulty.

Direct Products

The cartesian product $\bigtimes_A M_\alpha$ of the sets $(M_\alpha)_{\alpha \in A}$ becomes an R-module with operations defined coordinatewise. That is, if π_α denotes the α-th coordinate map, then for each pair x, y in the product and each $r \in R$

$$\pi_\alpha(x + y) = \pi_\alpha(x) + \pi_\alpha(y) \qquad \pi_\alpha(rx) = r\pi_\alpha(x).$$

That these are (well-defined) operations on the product is immediate from (0.4), and it is elementary to check that they do induce the claimed module structure. In A-tuple notation the operations on the product are given, somewhat imprecisely, by

$$(x_\alpha) + (y_\alpha) = (x_\alpha + y_\alpha), \qquad r(x_\alpha) = (rx_\alpha).$$

The resulting module, called the *direct* (or *cartesian*) *product* of $(M_\alpha)_{\alpha \in A}$, will be denoted by

$$\Pi_A M_\alpha,$$

or some reasonably natural variation such as $\Pi_{i=1}^n M_i$, or $M_1 \times \ldots \times M_n$ in the finite case. If $M_\alpha = M$ for all $\alpha \in A$, we write

$$M^A = \Pi_A M.$$

This is simply the set of all functions from A to M with coordinatewise operations. If $A = \varnothing$, the product has exactly one element (the empty function) and so

$$\Pi_\varnothing M_\alpha = 0 = M^\varnothing.$$

The fundamental property of $\Pi_A M_\alpha$ is given in

6.1. Proposition. *Let $(M_\alpha)_{\alpha \in A}$ be an indexed set of modules. Let N be a module and $(f_\alpha)_{\alpha \in A}$ be homomorphisms $f_\alpha : N \to M_\alpha$. Then there exists a unique homomorphism $f : N \to \Pi_A M_\alpha$ such that for each $\alpha \in A$ the following diagram commutes*

Proof. For each $x \in N$ define $f(x) \in \Pi_A M_\alpha$ coordinatewise (see (0.4)) by

$$\pi_\alpha f(x) = f_\alpha(x) \qquad (\alpha \in A).$$

Since the π_α and the f_α are homomorphisms, it follows also that $f : x \mapsto f(x)$ defines a homomorphism $N \to \Pi_A M$. Moreover $\pi_\alpha f = f_\alpha$ for all $\alpha \in A$. To complete the proof suppose that $g : N \to \Pi_A M_\alpha$ is a homomorphism such that $\pi_\alpha g = f_\alpha$ for all $\alpha \in A$. Then for each $x \in M$ and each $\alpha \in A$ we have $\pi_\alpha g(x) = \pi_\alpha f(x)$ so $g(x) = f(x)$. (See (0.4).) Thus $g = f$. $\qquad \square$

The unique homomorphism $f : N \to \Pi_A M_\alpha$ in (6.1) is called the *direct product* of $(f_\alpha)_{\alpha \in A}$ and is often denoted by $f = \Pi_A f_\alpha$. It is characterized by

$$\pi_\alpha(\Pi_A f_\alpha) = f_\alpha \qquad (\alpha \in A).$$

6.2. Corollary. *Let $f_\alpha : N \to M_\alpha \, (\alpha \in A)$ be an indexed set of homomorphisms. Then*

$$Ker(\Pi_A f_\alpha) = \bigcap_A Ker f_\alpha.$$

Proof. Set $f = \Pi_A f_\alpha$, and let $x \in N$. Then $f(x) = 0$ iff $\pi_\alpha f(x) = 0$ for all $\alpha \in A$ iff $f_\alpha(x) = 0$ for all $\alpha \in A$.   ☐

If $B \subseteq A$, then we have the two products $\Pi_B M_\beta$ and $\Pi_A M_\alpha$. If $x \in \Pi_B M_\beta$, then x is a function with domain B and has a unique extension to an element $\bar{x} \in \Pi_A M_\alpha$ that is zero on all $\alpha \notin B$. So there is a map

$$\iota_B : \Pi_B M_\beta \to \Pi_A M_\alpha$$

defined by $\iota_B : x \mapsto \bar{x}$. Clearly ι_B is an R-monomorphism whose image is the submodule of $\Pi_A M_\alpha$ consisting of those A-tuples that vanish outside of B. On the other hand, for each $x \in \Pi_A M_\alpha$, a function with domain A, its restriction $(x \mid B)$ is an element of $\Pi_B M_\beta$. Clearly the restriction map

$$\pi_B : \Pi_A M_\alpha \to \Pi_B M_\beta$$

defined by $\pi_B : x \mapsto (x \mid B)$ is an R-homomorphism from $\Pi_A M_\alpha$ onto $\Pi_B M_\beta$. With the help of (5.1) and (5.3) it is easy to check the following properties of these maps.

6.3. Proposition. *Let $(M_\alpha)_{\alpha \in A}$ be an indexed set of modules and let A be the disjoint union $A = B \cup C$. Then*

(1) $\pi_B \iota_B = 1_{\Pi_B M_\beta}$;

(2) $\Pi_A M_\alpha = \iota_B(\Pi_B M_\beta) \oplus \iota_C(\Pi_C M_\gamma)$;

(3) $0 \to \Pi_B M_\beta \xrightarrow{\iota_B} \Pi_A M_\alpha \xrightarrow{\pi_C} \Pi_C M_\gamma \to 0$ *is split exact.*   ☐

In practice if $\beta \in A$, we usually identify $\Pi_{\{\beta\}} M_\beta$ with M_β itself, and $\pi_{\{\beta\}}$ with π_β. Also we usually write ι_β for $\iota_{\{\beta\}}$. This monomorphism

$$\iota_\beta : M_\beta \to \Pi_A M_\alpha,$$

called the *β-coordinate injection*, is characterized by

$$\pi_\alpha \iota_\beta = \delta_{\alpha\beta} 1_{M_\beta} \qquad (\alpha \in A).$$

Of course, it is a special case of (6.3) that the sequences

$$0 \to M_\beta \xrightarrow{\iota_\beta} \Pi_A M_\alpha \xrightarrow{\pi_{A \setminus \{\beta\}}} \Pi_{A \setminus \{\beta\}} M_\alpha \to 0$$

$$0 \to \Pi_{A \setminus \{\beta\}} M_\alpha \xrightarrow{\iota_{A \setminus \{\beta\}}} \Pi_A M_\alpha \xrightarrow{\pi_\beta} M_\beta \to 0$$

are split exact.

The "universal mapping property" of $\Pi_A M_\alpha$ described in Proposition 6.1 actually serves to characterize the direct product. Thus a pair $(M, (p_\alpha)_{\alpha \in A})$ consisting of a module M and homomorphisms

$$p_\alpha : M \to M_\alpha \qquad (\alpha \in A)$$

is called a *(direct) product* of $(M_\alpha)_{\alpha \in A}$ in case for each module N and each set of homomorphisms

$$f_\alpha : N \to M_\alpha$$

there exists a unique homomorphism $f: N \to M$ such that

$$f_\alpha = p_\alpha f \qquad (\alpha \in A).$$

Very informally, a product of $(M_\alpha)_{\alpha \in A}$ is a gadget that stores in a single homomorphism any collection of homomorphisms $(f_\alpha: N \to M_\alpha)$ from a module to the M_α and is programmed to sort them out again (via the p_α).

Observe that Proposition (6.1) says, in particular, that the indexed set $(M_\alpha)_{\alpha \in A}$ does have at least one product, namely its cartesian product $(\prod_A M_\alpha, (\pi_\alpha)_{\alpha \in A})$. We now see that all products of the $(M_\alpha)_{\alpha \in A}$ are actually isomorphic in a strong sense.

6.4. Theorem. *Let* $(M, (p_\alpha)_{\alpha \in A})$ *be a product of* $(M_\alpha)_{\alpha \in A}$. *Then a pair* $(M', (p'_\alpha)_{\alpha \in A})$, *where each* $p'_\alpha: M' \to M_\alpha$ *is an R-homomorphism* $(\alpha \in A)$, *is also a product of* $(M_\alpha)_{\alpha \in A}$ *if and only if there exists a (necessarily unique) isomorphism* $p: M' \to M$ *such that* $p_\alpha p = p'_\alpha$ *for each* $\alpha \in A$.

Proof. Since $(M, (p_\alpha)_{\alpha \in A})$ is a product, there is a unique homomorphism $p: M' \to M$ with $p_\alpha p = p'_\alpha$ for each $\alpha \in A$.

($\Rightarrow$). If $(M', (p'_\alpha)_{\alpha \in A})$ is also a product, then there is a unique homomorphism $p': M \to M'$ with $p'_\alpha p' = p_\alpha$ for each α. Then $p'_\alpha = p_\alpha p = p'_\alpha p' p$.

But $p'_\alpha = 1_{M'} p'_\alpha$, so by uniqueness, $p'p = 1_{M'}$. Similarly, $pp' = 1_M$.

($\Leftarrow$). Suppose that $f_\alpha: N \to M_\alpha$ $(\alpha \in A)$ are R-homomorphisms. Since $(M, (p_\alpha)_{\alpha \in A})$ is a product, there is a unique homomorphism h making the outside triangle in

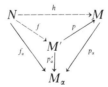

commute for each $\alpha \in A$. So assuming p is an isomorphism and taking $f = p^{-1}h$ we have that $(M', (p'_\alpha)_{\alpha \in A})$ is a product. $\square$

6.5. Examples. (1) Let V be a two-dimensional vector space over a field K, and let (x_1, x_2) be a basis for V. If (p_1, p_2) are the usual linear functionals with kernels (Kx_2, Kx_1), then $(V, (p_1, p_2))$ is a product of (K, K). In particular, a module M may be a product of an indexed set $(M_\alpha)_{\alpha \in A}$ via many different homomorphisms $(p_\alpha)_{\alpha \in A}$.

(2) Consider the abelian group $\mathbb{Z}_{30}$. The residues modulo 2, 3, and 5, respectively, give epimorphisms

$$p_2:\mathbb{Z}_{30} \to \mathbb{Z}_2, \quad p_3:\mathbb{Z}_{30} \to \mathbb{Z}_3, \quad p_5:\mathbb{Z}_{30} \to \mathbb{Z}_5.$$

Therefore, by (6.1) there is a homomorphism p from $\mathbb{Z}_{30}$ to the product $\mathbb{Z}_2 \times \mathbb{Z}_3 \times \mathbb{Z}_5$ such that $p_\alpha = \pi_\alpha p$ ($\alpha = 2, 3, 5$) where the π_α are the coordinate projections of the product. By (6.2)

$$Ker\ p = Ker\ p_2 \cap Ker\ p_3 \cap Ker\ p_5 = 2\mathbb{Z}_{30} \cap 3\mathbb{Z}_{30} \cap 5\mathbb{Z}_{30} = 0.$$

Thus p is a monomorphism. So since $\mathbb{Z}_{30}$ and $\mathbb{Z}_2 \times \mathbb{Z}_3 \times \mathbb{Z}_5$ have the same finite cardinality, p is an isomorphism. Observe that this implies by (6.4) that $(\mathbb{Z}_{30}, (p_2, p_3, p_5))$ is an (abstract) product of $(\mathbb{Z}_2, \mathbb{Z}_3, \mathbb{Z}_5)$; clearly though, $\mathbb{Z}_{30}$ and the cartesian product $\mathbb{Z}_2 \times \mathbb{Z}_3 \times \mathbb{Z}_5$ are quite different sets.

Direct Sums—Coproducts

Recall that a product of $(M_\alpha)_{\alpha \in A}$ is something of a computer to organize sets of homomorphisms into the M_α. We now turn to the dual question of studying gadgets that organize homomorphisms from the M_α. The definition is almost self-evident; we simply reverse the arrows in the definition of a product.

Formally, then, a pair $(M, (j_\alpha)_{\alpha \in A})$ consisting of a module M and homomorphisms

$$j_\alpha:M_\alpha \to M$$

is a *direct sum* (or *a coproduct*) of $(M_\alpha)_{\alpha \in A}$ in case for each module N and each set of homomorphisms

$$f_\alpha:M_\alpha \to N \qquad (\alpha \in A)$$

there is a unique homomorphism $f:M \to N$ such that

$$f_\alpha = f j_\alpha \qquad (\alpha \in A).$$

The next result, whose proof is obtained by reversing the arrows in that of (6.4), establishes that if direct sums do exist, they are essentially unique.

6.6. Theorem. *Let* $(M, (j_\alpha)_{\alpha \in A})$ *be a direct sum of* $(M_\alpha)_{\alpha \in A}$. *Then a pair* $(M', (j'_\alpha)_{\alpha \in A})$, *where each* $j'_\alpha:M_\alpha \to M'$ *is an R-homomorphism* ($\alpha \in A$), *is also a direct sum of* $(M_\alpha)_{\alpha \in A}$ *if and only if there exists a (necessarily unique) isomorphism* $j:M \to M'$ *such that* $jj_\alpha = j'_\alpha$ *for each* $\alpha \in A$.

External Direct Sums

Now we establish that direct sums do exist. An element $x \in \Pi_A M_\alpha$ is *zero for almost all* $\alpha \in A$ (or *almost always zero*) in case its *support*

$$S(x) = \{\alpha \in A \mid x(\alpha) = \pi_\alpha(x) \neq 0\}$$

is finite. Since 0 is almost always zero and since both $S(x + y) \subseteq S(x) \cup S(y)$ and $S(rx) \subseteq S(x)$, it follows that

$$\oplus_A M_\alpha = \{x \in \Pi_A M_\alpha \mid x \text{ is almost always zero}\}$$

is a submodule of $\Pi_A M_\alpha$. This submodule is the (*external*) *direct sum* of $(M_\alpha)_{\alpha \in A}$; as we shall see, the use of "direct sum" is justified. We employ natural variations of this notation, such as $\oplus_{i=1}^n M_i$ in the finite case. Of course, if A is finite, then the external direct sum is the cartesian product. Moreover, if $M_\alpha = M$ for all $\alpha \in A$, then

$$M^{(A)} = \oplus_A M$$

designates the external direct sum of card A copies of M.

In general, for an indexed set $(M_\alpha)_{\alpha \in A}$ and for each $\alpha \in A$, the image $\iota_\alpha(M_\alpha)$ is the set of $x \in \Pi_A M_\alpha$ with $S(x) \subseteq \{\alpha\}$. Moreover, $x \in \Pi_A M_\alpha$ has finite support iff it is a finite sum of elements each of whose support is a singleton. Thus $\oplus_A M_\alpha$ is the submodule of $\Pi_A M_\alpha$ spanned by its submodules $(\iota_\alpha(M_\alpha))_{\alpha \in A}$. Since the images $\iota_\alpha(M_\alpha)$ are in $\oplus_A M_\alpha$, we usually feel free to treat each ι_α also as a monomorphism from M_α to $\oplus_A M_\alpha$. Similarly, we often view each π_α as an epimorphism from $\oplus_A M_\alpha$ to M_α.

Now suppose that N is a module and that $(f_\alpha)_{\alpha \in A}$ is an indexed set of homomorphisms

$$f_\alpha : M_\alpha \to N \qquad (\alpha \in A).$$

For each $x \in \oplus_A M_\alpha$, its support, $S(x) = \{\alpha \in A \mid \pi_\alpha(x) = 0\}$, is finite so there is a function $f : \oplus_A M_\alpha \to N$ defined by

$$f(x) = \Sigma_{\alpha \in S(x)} f_\alpha \pi_\alpha(x)$$

(where we let $f(x) = 0$ if $S(x) = \varnothing$). It is easy to check, since the f_α and π_α are homomorphisms, that f is a homomorphism. We call f the *direct sum* of $(f_\alpha)_{\alpha \in A}$ and write both

$$f = \oplus_A f_\alpha$$

and for each $x = (x_\alpha)_{\alpha \in A} \in \oplus_A M_\alpha$,

$$f(x) = \Sigma_A f_\alpha(x_\alpha).$$

Also it is clear that for each $\alpha \in A$, $f\iota_\alpha = f_\alpha$. This direct sum behaves very much like a regular sum (see Exercise (6.7)). For example, if $g : N \to K$, then $gf_\alpha : M_\alpha \to K$ for each $\alpha \in A$, and

$$g(\oplus_A f_\alpha) = \oplus_A (gf_\alpha).$$

6.7. Proposition. *Let* $(M_\alpha)_{\alpha \in A}$ *be an indexed set of modules. Let* N *be a module and* $(f_\alpha)_{\alpha \in A}$ *be an indexed class of homomorphisms* $f_\alpha : M_\alpha \to N$ *for* $\alpha \in A$. *Then there exists a unique homomorphism* $f : \oplus_A M_\alpha \to N$ *(necessarily* $f = \oplus_A f_\alpha$*) such that*

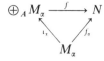

commutes for each $\alpha \in A$. *Thus* $(\oplus_A M_\alpha, (\iota_\alpha)_{\alpha \in A})$ *is a direct sum of* $(M_\alpha)_{\alpha \in A}$.

Proof. In view of our above remarks only the question of uniqueness remains. But that the sum $f = \oplus_A f_\alpha$ is unique with the desired properties follows from (3.1). □

There is one simple fact about direct sums of homomorphisms that we might record now. Suppose $f = \oplus_A f_\alpha$. Then since $f\iota_\alpha = f_\alpha$, it is clear that $Im f_\alpha \leq Im f$. Thus it is immediate from the definition that

6.8. Proposition. *If* $f = \oplus_A f_\alpha$ *is the direct sum of homomorphisms* $f_\alpha : M_\alpha \to N$, *then*

$$Im f = \Sigma_A \, Im f_\alpha.$$ □

Suppose $B \subseteq A$. Then it is easy to check that the restriction of ι_B to $\oplus_B M_\beta$ is a homomorphism into the direct sum $\oplus_A M_\alpha$. Similarly, the restriction of π_B to $\oplus_A M_\alpha$ is an epimorphism onto the direct sum $\oplus_B M_\beta$. Again we conserve notation and denote these restrictions by ι_B and π_B. From (6.3) we infer

6.9. Proposition. *Let* $(M_\alpha)_{\alpha \in A}$ *be an indexed set of modules and let* A *be the disjoint union* $A = B \cup C$. *Then*
(1) $\pi_B \iota_B = 1_{\Pi_B M_\beta}$;
(2) $\oplus_A M_\alpha = \iota_B(\oplus_B M_\beta) \oplus \iota_C(\oplus_C M_\gamma)$;
(3) $0 \to \oplus_B M_\beta \xrightarrow{\iota_B} \oplus_A M_\alpha \xrightarrow{\pi_C} \oplus_C M_\gamma \to 0$ *is split exact.*   □

Internal Direct Sums

Let M_1, M_2 be submodules of a module M. and let $i_1 : M_1 \to M$, and $i_2 : M_2 \to M$ be their inclusion maps. Then (see §5) M is the internal direct sum of M_1 and M_2 iff in our present terminology $i_1 \oplus i_2$ is an isomorphism. By virtue of (6.6) and (6.7) this is equivalent to having $(M, (i_1, i_2))$ a direct sum of (M_1, M_2).

More generally, suppose that $(M_\alpha)_{\alpha \in A}$ is an indexed set of submodules of a module M. Let

$$i_\alpha : M_\alpha \to M \qquad (\alpha \in A)$$

be the corresponding inclusion maps. Generalizing our definition in §5 we say that M is the (*internal*) *direct sum of its submodules* $(M_\alpha)_{\alpha \in A}$ in case the direct sum map

$$i = \oplus_A i_\alpha : \oplus_A M_\alpha \to M$$

is an isomorphism. This condition holds, namely i is an isomorphism, iff each $x \in M$ has a unique representation as a sum

$$x = \Sigma_A x_\alpha$$

with $x_\alpha \in M_\alpha$ zero for almost all $\alpha \in A$. With this summation notation, we have, since i is an R-homomorphism,

$$\Sigma_A x_\alpha + \Sigma_A y_\alpha = \Sigma_A (x_\alpha + y_\alpha) \qquad \text{and} \qquad r(\Sigma_A x_\alpha) = \Sigma_A r x_\alpha.$$

Also if $f : M \to N$ is a homomorphism, then since each of these $\Sigma_A x_\alpha$ is really just a finite sum in M,

$$f(\Sigma_A x_\alpha) = \Sigma_A f(x_\alpha).$$

In other words, if M is the internal direct sum of $(M_\alpha)_{\alpha \in A}$, then we can study M "coordinatewise".

We now have three concepts of "direct sum", the abstract direct sum ($=$ coproduct), the external direct sum of an indexed set of modules, and the internal direct sum of submodules. We assume that the distinctions among them are reasonably clear. (But see the exercises, particularly Exercise (6.4).)

Let $(M_\alpha)_{\alpha \in A}$ be an indexed set of submodules of M with inclusion maps $(i_\alpha)_{\alpha \in A}$. Then by (6.8) we have

$$Im(\oplus_A i_\alpha) = \Sigma_A \, Im \, i_\alpha = \Sigma_A M_\alpha.$$

So whether or not it is epic, if the direct sum map $i = \oplus_A i_\alpha$ is monic, then the submodule $\Sigma_A M_\alpha$ is an internal direct sum of its submodules $(M_\alpha)_{\alpha \in A}$. As for the case of two submodules (§5) we can characterize this in the lattice $\mathscr{S}(M)$ of submodules. We say $(M_\alpha)_{\alpha \in A}$ is *independent* in case for each $\alpha \in A$

$$M_\alpha \cap (\Sigma_{\beta \neq \alpha} M_\beta) = 0.$$

Clearly this is consistent with our earlier definition for two submodules. Of course, it is possible for $(M_\alpha)_{\alpha \in A}$ to be independent in pairs without being independent. However,

6.10. Proposition. *Let $(M_\alpha)_{\alpha \in A}$ be an indexed set of submodules of a module M with inclusion maps $(i_\alpha)_{\alpha \in A}$. Then the following are equivalent:*
(a) *$\Sigma_A M_\alpha$ is the internal direct sum of $(M_\alpha)_{\alpha \in A}$;*
(b) *$i = \oplus_A i_\alpha : \oplus_A M_\alpha \to M$ is monic;*
(c) *$(M_\alpha)_{\alpha \in A}$ is independent;*
(d) *$(M_\alpha)_{\alpha \in F}$ is independent for every finite subset $F \subseteq A$;*
(e) *For every pair $B, C \subseteq A$, if $B \cap C = \varnothing$, then*

$$(\Sigma_B M_\beta) \cap (\Sigma_C M_\gamma) = 0.$$

Proof. Each one of these conditions is clearly equivalent to stating that 0 has a unique representation $0 = \Sigma_A x_\alpha$ with $x_\alpha \in M_\alpha$ zero for almost all $\alpha \in A$.   □

6.11. Corollary. *The module M is the internal direct sum of its submodules $(M_\alpha)_{\alpha \in A}$ if and only if $(M_\alpha)_{\alpha \in A}$ is independent and spans M.*   □

If the submodules $(M_\alpha)_{\alpha \in A}$ of M are independent, we say that the sum $\Sigma_A M_\alpha$ is *direct* and write

$$\Sigma_A M_\alpha = \oplus_A M_\alpha;$$

we may also refer to this as a *direct decomposition* of $\Sigma_A M_\alpha$. As usual we shall allow more or less self-explanatory variations of this notation and terminology.

Now the external direct sum of $(M_\alpha)_{\alpha \in A}$ is the internal direct sum of the images $(\iota_\alpha(M_\alpha))_{\alpha \in A}$ but *not* the internal direct sum of $(M_\alpha)_{\alpha \in A}$. Thus the notation $\oplus$ is being asked to serve two distinct, but related, duties. Only on rare occasions does this cause difficulty. On these we may denote the external direct sum by

$$\dot{\oplus}_A M_\alpha$$

or in the finite case

$$M_1 \dot{\oplus} \ldots \dot{\oplus} M_n.$$

6.12. Proposition. *Let* $(M_1, \ldots, M_n)$ *be a finite sequence of modules. Then*

$$M_1 \times \ldots \times M_n = M_1 \dot{\oplus} \ldots \dot{\oplus} M_n = \iota_1(M_1) \oplus \ldots \oplus \iota_n(M_n). \qquad \square$$

6.13. Examples. (1) Let V be a vector space over a field K, and let $(x_\alpha)_{\alpha \in A}$ be an indexed set in V. Then $(x_\alpha)_{\alpha \in A}$ span V iff $V = \Sigma_A K x_\alpha$. Also $(x_\alpha)_{\alpha \in A}$ is an independent set of vectors iff the indexed set $(K x_\alpha)_{\alpha \in A}$ of cyclic submodules of V is independent. Thus V is the internal direct sum

$$V = \Sigma_A K x_\alpha = \oplus_A K x_\alpha$$

iff $(x_\alpha)_{\alpha \in A}$ is a basis for V.

(2) Consider again the abelian group $\mathbb{Z}_{30}$. (See (6.5).) It has subgroups $15\mathbb{Z}_{30}, 10\mathbb{Z}_{30}, 6\mathbb{Z}_{30}$; let i_2, i_3, i_5 be the corresponding inclusion maps. Then by (6.7) the direct sum $i = i_2 \oplus i_3 \oplus i_5$ is a homomorphism

$$i : (15\mathbb{Z}_{30}) \oplus (10\mathbb{Z}_{30}) \oplus (6\mathbb{Z}_{30}) \to \mathbb{Z}_{30}.$$

Now

$$Im\, i = 15\mathbb{Z}_{30} + 10\mathbb{Z}_{30} + 6\mathbb{Z}_{30} = \mathbb{Z}_{30},$$

so i is epic. By a cardinality argument it follows that i is an isomorphism. Thus $\mathbb{Z}_{30}$ is the internal direct sum of its submodules $(15\mathbb{Z}_{30}, 10\mathbb{Z}_{30}, 6\mathbb{Z}_{30})$. Note also that these modules are isomorphic, respectively, to $\mathbb{Z}_2, \mathbb{Z}_3, \mathbb{Z}_5$. So

$$\mathbb{Z}_{30} = (15\mathbb{Z}_{30}) \oplus (10\mathbb{Z}_{30}) \oplus (6\mathbb{Z}_{30})$$
$$\cong \mathbb{Z}_2 \dot{\oplus} \mathbb{Z}_3 \dot{\oplus} \mathbb{Z}_5 = \mathbb{Z}_2 \times \mathbb{Z}_3 \times \mathbb{Z}_5.$$

Properties of Independence

In addition to the characterizations given in (6.10) there are three other properties of independence having special importance. The first is a generalization of the familiar fact that in a vector space an ordered set of vectors is independent iff no one of the vectors depends on its predecessors.

6.14. Proposition. *A sequence* $M_1, M_2, \ldots$ *of submodules of* M *is independent if and only if for each* $n \geq 1$

$$(M_1 + \ldots + M_n) \cap M_{n+1} = 0.$$

Proof. ($\Rightarrow$). See (6.10.e). ($\Leftarrow$). If $x_i \in M_i$, and $x_1 + \ldots + x_{n+1} = 0$ then $x_{n+1} \in (M_1 + \ldots + M_n) \cap M_{n+1}$. □

The next result is a formal statement of the simple fact that independent sets of submodules of independent submodules form an independent set of submodules.

6.15. Proposition. *Let* $(M_\beta)_{\beta \in B}$ *be independent submodules of a module* M. *For each* $\beta \in B$, *let* $(L_{\alpha_\beta})_{\alpha_\beta \in A_\beta}$ *be an indexed class of submodules of* M_β. *Let* A *be the disjoint union* $A = \cup_B A_\beta$. *If* $(L_{\alpha_\beta})_{\alpha_\beta \in A_\beta}$ *is independent for each* $\beta \in B$, *then* $(L_\alpha)_{\alpha \in A}$ *is independent.*

Proof. Suppose there is a finite set $\alpha_1, \ldots, \alpha_n \in A$ and $x_i \in L_{\alpha_i}$ with

$$x_1 + \ldots + x_n = 0.$$

We may assume there is a k and a β with $\alpha_1, \ldots, \alpha_k \in A_\beta$ and $\alpha_{k+1}, \ldots, \alpha_n \notin A_\beta$. The independence of $(M_\beta)_{\beta \in B}$ then forces

$$x_1 + \ldots + x_k = 0 = x_{k+1} + \ldots + x_n,$$

and the independence of $(L_{\alpha_\beta})_{\alpha_\beta \in A_\beta}$ forces $x_1 = \ldots = x_k = 0$, etc. □

This has the following very useful corollary about internal direct sums.

6.16. Corollary. *Let* $M = \Sigma_B M_\beta$, *let* $M_\beta = \Sigma_{A_\beta} L_{\alpha_\beta}$ *for each* $\beta \in B$, *and let* A *be the disjoint union* $A = \bigcup_B A_\beta$. *Then*

$$M = \oplus_A L_\alpha \quad iff \quad M = \oplus_B M_\beta \quad and \quad M_\beta = \oplus_{A_\beta} L_{\alpha_\beta} \quad (\beta \in B).$$

Proof. ($\Leftarrow$). By (6.15). ($\Rightarrow$). By (6.10.e). □

Our final result on independence is the important one that independence is preserved under "essential changes" of the submodules.

6.17. Proposition. *Suppose that* $(L_\alpha)_{\alpha \in A}$ *is a set of independent submodules of* M. *If* $(M_\alpha)_{\alpha \in A}$ *is a set of submodules of* M *such that* $L_\alpha \trianglelefteq M_\alpha$ *for each* $\alpha \in A$, *then*

(1) $(M_\alpha)_{\alpha \in A}$ *is independent, and*
(2) $\oplus_A L_\alpha \trianglelefteq \oplus_A M_\alpha$.

Proof. Suppose that L_1 and L_2 are independent submodules of M with $L_1 \trianglelefteq M_1$ and $L_2 \trianglelefteq M_2$. Then

$$(L_1 \cap M_2) \cap L_2 = L_1 \cap L_2 = 0$$

so, since $L_2 \trianglelefteq M_2$, we have $L_1 \cap M_2 = 0$. But

$$(M_1 \cap M_2) \cap L_1 \leq L_1 \cap M_2 = 0.$$

So, since $L_1 \trianglelefteq M_1$, $M_1 \cap M_2 = 0$. That is, (M_1, M_2) is an independent set of submodules of M. Moreover, by (5.20.2), we have $L_1 \oplus L_2 \trianglelefteq M_1 \oplus M_2$. Now if (1) and (2) hold for $F = \{\alpha_1, ..., \alpha_n\} \subseteq A$, then for any $\alpha_{n+1} \in A \backslash F$ the sum

$$\Sigma_{i=1}^{n+1} M_i = (\oplus_{i=1}^{n} M_{\alpha_i}) + M_{\alpha_{n+1}}$$

is direct, and

$$\oplus_{i=1}^{n+1} L_{\alpha_i} \trianglelefteq \oplus_{i=1}^{n+1} M_{\alpha_i}.$$

Thus, arguing inductively, (1) and (2) hold for every finite subset of $F \subseteq A$, and by (6.10.d) $(M_\alpha)_{\alpha \in A}$ is independent. But if $0 \neq x \in \oplus_A M_\alpha$, then there exists a finite subset $F \subseteq A$ such that $0 \neq x \in \oplus_F M_\alpha$. So since $\oplus_F L_\alpha \trianglelefteq \oplus_F M_\alpha$, by (5.19) there is an $r \in R$ with

$$0 \neq rx \in \oplus_F L_\alpha \leq \oplus_A L_\alpha$$

and hence $\oplus_A L_\alpha \trianglelefteq \oplus_A M_\alpha$. □

The Idempotents for a Decomposition

Suppose M has an (internal) direct decomposition $M = \oplus_A M_\alpha$. Then (see (6.16)) for each $\alpha \in A$

$$M = M_\alpha \oplus (\Sigma_{\beta \neq \alpha} M_\beta);$$

so by (5.6) and (5.7), there is a unique idempotent $e_\alpha \in End(_R M)$ with

$$M_\alpha = Im\, e_\alpha \qquad \text{and} \qquad \Sigma_{\beta \neq \alpha} M_\beta = Ker\, e_\alpha.$$

We call the idempotents $(e_\alpha)_{\alpha \in A}$ the idempotents for the decomposition $M = \oplus_A M_\alpha$, and for each $\alpha \in A$, we call e_α the idempotent for M_α in this decomposition.

6.18. Proposition. *Let $(M_\alpha)_{\alpha \in A}$ be submodules of a module M. Then $M = \oplus_A M_\alpha$ if and only if there exists an (necessarily unique) indexed set $(e_\alpha)_{\alpha \in A}$ of idempotent endomorphisms of M such that for all $\alpha \in A$*

$$M_\alpha = Im\, e_\alpha \qquad \text{and} \qquad \Sigma_{\beta \neq \alpha} M_\beta = Ker\, e_\alpha.$$

Moreover, if such idempotent endomorphisms of M exist, then e_α is the idempotent for M_α in the decomposition $M = \oplus_A M_\alpha$.

Proof. We need only prove the sufficiency. So assume $(e_\alpha)_{\alpha \in A}$ satisfies the stated condition. Then by (5.6) since $Im\, e_\alpha \cap Ker\, e_\alpha = 0$ for each $\alpha \in A$, it follows that $(M_\alpha)_{\alpha \in A}$ is independent. But also by (5.6), $Im\, e_\alpha + Ker\, e_\alpha = M$. So clearly $(M_\alpha)_{\alpha \in A}$ spans M. The final claim follows from the uniqueness assertion in (5.7). □

A set of idempotents $(e_\alpha)_{\alpha \in A}$ in a ring is said to be *orthogonal* in case the set is "pair-wise" orthogonal, i.e.,

$$e_\alpha e_\beta = \delta_{\alpha\beta} e_\alpha \qquad (\alpha, \beta \in A).$$

6.19. Corollary. *The idempotents* $(e_\alpha)_{\alpha \in A}$ *for a decomposition* $M = \overset{\bullet}{\oplus}_A M_\alpha$ *are orthogonal. Moreover, if* $x \in M$, *then* $xe_\alpha = 0$ *for almost all* $\alpha \in A$ *and*

$$x = \Sigma_A x e_\alpha.$$

Proof. If $\alpha \neq \beta$, then $M_\beta \subseteq \Sigma_{\gamma \neq \alpha} M_\gamma = Ker\, e_\alpha$ so $Me_\beta e_\alpha = M_\beta e_\alpha = 0$. Moreover, writing $x = \Sigma_A x_\alpha$ with $x_\alpha \in M_\alpha = Me_\alpha$ zero for almost all $\alpha \in A$, we have

$$xe_\beta = \Sigma_A x_\alpha e_\beta = x_\beta e_\beta = x_\beta$$

for all $\beta \in A$. □

A finite orthogonal set of idempotents $e_1, \ldots, e_n$ in a ring R is said to be *complete* in case

$$e_1 + \ldots + e_n = 1 \in R.$$

From the following corollary we see that there is a 1–1 correspondence between the finite direct decompositions of a module and the complete sets of orthogonal idempotents in its endomorphism ring.

6.20. Corollary. *Let* $M_1, \ldots, M_n$ *be submodules of* M. *Then*

$$M = M_1 \oplus \ldots \oplus M_n$$

if and only if there exists a (necessarily unique) complete set $e_1, \ldots, e_n$ *of orthogonal idempotents in* $End(_R M)$ *with*

$$M_i = Me_i \qquad (i = 1, \ldots, n).$$

Proof. ($\Rightarrow$). Let $e_1, \ldots, e_n$ be the idempotents for $M = M_1 \oplus \ldots \oplus M_n$. By (6.19) they are orthogonal and $1_M = e_1 + \ldots + e_n$ for

$$x = \Sigma_{i=1}^n xe_i = x(e_1 + \ldots + e_n) \qquad (x \in M).$$

($\Leftarrow$). If $e_1, \ldots, e_n$ are as claimed, then for each $x \in M$

$$x = x(e_1 + \ldots + e_n) = xe_1 + \ldots + xe_n,$$

so clearly $M_i = Im\, e_i$ and $\Sigma_{j \neq i} M_j = Ker\, e_i$. Now apply (6.18). □

A Characterization of Direct Sums

There is a variation of these last results that provides a valuable characterization of abstract direct sums. Its real value will be evident when we begin studying additive functors.

6.21. Proposition. *Let* $(M_\alpha)_{\alpha \in A}$ *be an indexed set of modules, let* M *be a module, and for each* $\alpha \in A$, *let* $j_\alpha : M_\alpha \to M$ *be a homomorphism. Then* $(M, (j_\alpha)_{\alpha \in A})$ *is a direct sum of* $(M_\alpha)_{\alpha \in A}$ *if and only if there exist (necessarily unique) homomorphisms* $q_\alpha : M \to M_\alpha$ $(\alpha \in A)$ *satisfying, for all* $\alpha, \beta \in A$ *and all* $x \in M$,

    (i) $q_\beta j_\alpha = \delta_{\alpha\beta} 1_{M_\alpha}$,

    (ii) $q_\alpha(x) = 0$ *for almost all* $\alpha \in A$,

    (iii) $\Sigma_A j_\alpha q_\alpha(x) = x$.

Moreover, if $(M, (j_\alpha)_{\alpha \in A})$ *is a direct sum of* $(M_\alpha)_{\alpha \in A}$ *and if* $f_\alpha : M_\alpha \to N$ *$(\alpha \in A)$ are homomorphisms, then*

$$f : x \mapsto \Sigma_A f_\alpha q_\alpha(x) \qquad (x \in M)$$

is the unique homomorphism $f : M \to N$ *such that* $f_\alpha = fj_\alpha$ $(\alpha \in A)$.

    Proof. ($\Rightarrow$). Suppose that $(M, (j_\alpha)_{\alpha \in A})$ is a direct sum of $(M_\alpha)_{\alpha \in A}$. Then ((6.6) and (6.7)) the direct sum map $j = \oplus_A j_\alpha : \oplus_A M_\alpha \to M$ is an isomorphism. Let ι_α and π_α be the usual coordinate maps for the external direct sum $\oplus_A M_\alpha$. For each $\alpha \in A$ let

$$q_\alpha = \pi_\alpha j^{-1} : M \to M_\alpha.$$

Then since $j_\alpha = j\iota_\alpha$ for each $\alpha \in A$, it is easy to see that the q_α satisfy (i) and (ii). Now for each $x \in M$,

$$x = jj^{-1}(x) = j(\Sigma_A \iota_\alpha \pi_\alpha j^{-1}(x))$$
$$= \Sigma_A j \iota_\alpha \pi_\alpha j^{-1}(x) = \Sigma_A j_\alpha q_\alpha(x).$$

By (6.8), $M = \Sigma_A \, Im \, j_\alpha$, so in view of (i) and (3.1), the q_α are unique.

    ($\Leftarrow$). If $(q_\alpha)_{\alpha \in A}$ satisfy the three conditions and $f_\alpha : M_\alpha \to N$ $(\alpha \in A)$, then define $f : M \to N$ by

$$f(x) = \Sigma_A f_\alpha q_\alpha(x) \qquad (x \in M).$$

Then for all $\alpha \in A$ and all $x_\alpha \in M_\alpha$,

$$fj_\alpha(x_\alpha) = \Sigma_{\beta \in A} f_\beta q_\beta(j_\alpha(x_\alpha)) = f_\alpha(x_\alpha).$$

Moreover, if $g : M \to N$ with $gj_\alpha = f_\alpha$ for each $\alpha \in A$, then for all $x \in M$

$$g(x) = g(\Sigma_A j_\alpha q_\alpha(x)) = \Sigma_A f_\alpha q_\alpha(x) = f(x). \qquad \square$$

    This important characterization assumes a particularly nice form for the case of finite direct sums.

    6.22. Corollary. *Let* $(M_1, \ldots, M_n)$ *be a finite sequence of modules and let* $j_i : M_i \to M$ $(i = 1, \ldots, n)$ *be homomorphisms. Then* $(M, (j_1, \ldots, j_n))$ *is a direct sum of* $(M_1, \ldots, M_n)$ *if and only if there exist homomorphisms* $q_i : M \to M_i$ $(i = 1, \ldots, n)$ *such that for all* $1 \leq i, k \leq n$

$$q_k j_i = \delta_{ik} 1_{M_i} \qquad and \qquad \Sigma_{i=1}^n j_i q_i = 1_M. \qquad \square$$

Products and Sums of Functions

    6.23. Let $(M_\alpha)_{\alpha \in A}$ be an indexed set of modules. Let $(M, (p_\alpha)_{\alpha \in A})$ be a product of $(M_\alpha)_{\alpha \in A}$. Then by (6.4) and (6.1), there is a unique isomorphism

$p:\Pi_A M_\alpha \to M$ such that $p_\alpha p = \pi_\alpha$, the α-coordinate map of $\Pi_A M_\alpha$. For each $\alpha \in A$ define $i_\alpha = p\iota_\alpha : M_\alpha \to M$. Then for each $\alpha, \beta \in A$

$$p_\alpha i_\beta = p_\alpha(p\iota_\beta) = (p_\alpha p)\iota_\beta = \pi_\alpha \iota_\beta = \delta_{\alpha\beta} 1_{M_\beta}.$$

We infer from (5.1) that the maps

$$i_\alpha : M_\alpha \to M \qquad p_\alpha : M \to M_\alpha$$

are a split monomorphism and a split epimorphism, with

$$M = (Im\, i_\alpha) \oplus (Ker\, p_\alpha).$$

The homomorphisms $(i_\alpha)_{\alpha \in A}$ and $(p_\alpha)_{\alpha \in A}$ are the *injections* and the *projections*, respectively, of the product $(M, (p_\alpha)_{\alpha \in A})$. If N is another module and if $f_\alpha : N \to M_\alpha$ ($\alpha \in A$) are homomorphisms, then by definition there is a unique $f : N \to M$ with $p_\alpha f = f_\alpha$ ($\alpha \in A$). Generalizing our earlier definition, we call f the *direct product of* $(f_\alpha)_{\alpha \in A}$ *relative to the projections* $(p_\alpha)_{\alpha \in A}$. Using (6.2) and the fact that $p_\alpha = \pi_\alpha p^{-1}$, we have

$$Ker\, f = \cap_A Ker\, f_\alpha.$$

6.24. Dually, let $(M, (j_\alpha)_{\alpha \in A})$ be a direct sum of $(M_\alpha)_{\alpha \in A}$. Let $(q_\alpha)_{\alpha \in A}$ be the homomorphisms guaranteed by (6.21). Then since $q_\beta j_\alpha = \delta_{\alpha\beta} 1_{M_\alpha}$, we infer from (5.1) that

$$j_\alpha : M_\alpha \to M \qquad q_\alpha : M \to M_\alpha$$

are a split monomorphism and a split epimorphism with

$$M = (Im\, j_\alpha) \oplus (Ker\, q_\alpha).$$

We call $(j_\alpha)_{\alpha \in A}$ and $(q_\alpha)_{\alpha \in A}$ the *injections* and *projections* of the direct sum $(M, (j_\alpha)_{\alpha \in A})$. Suppose $f_\alpha : M_\alpha \to N$ ($\alpha \in A$) are homomorphisms. Then the unique $f : M \to N$ with $fj_\alpha = f_\alpha$ ($\alpha \in A$) is called the *direct sum of* $(f_\alpha)_{\alpha \in A}$ *relative to the injections* $(j_\alpha)_{\alpha \in A}$. With an assist from (6.8) the characterization (6.21) of f gives

$$Im\, f = \Sigma_A Im\, f_\alpha.$$

6.25. There is yet another useful variation. Let $(M'_\alpha)_{\alpha \in A}$ be a second indexed set of modules indexed by A and for each $\alpha \in A$ let

$$f_\alpha : M_\alpha \to M'_\alpha.$$

If $(P, (p_\alpha)_A)$ and $(P', (p'_\alpha)_A)$ are products respectively of these modules, then it is easy to see that there is a unique homomorphism $P \to P'$, which we also denote by $\Pi_A f_\alpha$ such that

$$p'_\alpha(\Pi_A f_\alpha) = f_\alpha p_\alpha \qquad (\alpha \in A).$$

Dually, if $(S, (j_\alpha)_A)$ and $(S', (j'_\alpha)_A)$ are direct sums, then there is a unique homomorphism $S \to S'$, which we also denote by $\oplus_A f_\alpha$ such that

$$(\oplus_A f_\alpha) j_\alpha = j'_\alpha f_\alpha \qquad (\alpha \in A)$$

$$
\begin{array}{ccc}
P & \xrightarrow{\;\Pi f_\alpha\;} & P' \\
{\scriptstyle p_\alpha}\downarrow & & \downarrow{\scriptstyle p'_\alpha} \\
M_\alpha & \xrightarrow{\;\;J_\alpha\;\;} & M'_\alpha
\end{array}
\qquad\qquad
\begin{array}{ccc}
S & \xrightarrow{\;\oplus f_\alpha\;} & S' \\
{\scriptstyle j_\alpha}\uparrow & & \uparrow{\scriptstyle j'_\alpha} \\
M_\alpha & \xrightarrow{\;\;J_\alpha\;\;} & M_\alpha.
\end{array}
$$

Finally, for cartesian products and for external direct sums, it is easy to prove that under either of these maps

$$
(x_\alpha)_{\alpha \in A} \mapsto (f_\alpha(x_\alpha))_{\alpha \in A}
$$

and that

$$
Im(\Pi_A f_\alpha) = \Pi_A\, Im f_\alpha, \qquad Ker(\Pi_A f_\alpha) = \Pi_A\, Ker f_\alpha
$$

and

$$
Im(\oplus_A f_\alpha) = \oplus_A\, Im f_\alpha, \qquad Ker(\oplus_A f_\alpha) = \oplus_A\, Ker f_\alpha.
$$

(Note that this last statement does not make sense in the cases of the "abstract" direct product and sum. But of course, it does have an obvious interpretation.)

6. Exercises

1. Let $(M_\alpha)_{\alpha \in A}$ be an indexed set of modules, let M be a module and let $j_\alpha : M_\alpha \to M$ and $p_\alpha : M \to M_\alpha$ ($\alpha \in A$) be homomorphisms.
   (1) Assume $(M, (p_\alpha)_{\alpha \in A})$ is a direct product of $(M_\alpha)_{\alpha \in A}$ with injections $(j_\alpha)_{\alpha \in A}$. Prove that $(M, (j_\alpha)_{\alpha \in A})$ is a direct sum of $(M_\alpha)_{\alpha \in A}$ iff A is finite.
   (2) Assume $(M. (j_\alpha)_{\alpha \in A})$ is a direct sum of $(M_\alpha)_{\alpha \in A}$ with projections $(p_\alpha)_{\alpha \in A}$. Prove that $(M, (p_\alpha)_{\alpha \in A})$ is a direct product of $(M_\alpha)_{\alpha \in A}$ iff A is finite.
2. Let $(M_\alpha)_{\alpha \in A}$ be an indexed set of modules, let $K_\alpha \leq M_\alpha$ ($\alpha \in A$), and let $i_\alpha : M_\alpha/K_\alpha \to (\oplus_A M_\alpha)/(\oplus_A K_\alpha)$ and $p_\alpha : (\Pi_A M_\alpha)/(\Pi_A K_\alpha) \to M_\alpha/K_\alpha$ be the canonical maps. Prove that
   (1) $(\oplus_A M_\alpha/\oplus_A K_\alpha, (i_\alpha)_{\alpha \in A})$ is a direct sum of $(M_\alpha/K_\alpha)_{\alpha \in A}$.
   (2) $(\Pi_A M_\alpha/\Pi_A K_\alpha, (p_\alpha)_{\alpha \in A})$ is a direct product of $(M_\alpha/K_\alpha)_{\alpha \in A}$.
3. Let $(M_\alpha)_{\alpha \in A}$ be an indexed set of left R-modules and let I be a left ideal of R. Prove that

   $$
   I(\oplus_A M_\alpha) = \oplus_A I M_\alpha \qquad \text{and} \qquad (\oplus_A M_\alpha)/I(\oplus_A M_\alpha) \cong \oplus_A M_\alpha/I M_\alpha.
   $$

4. Let $(M_\alpha)_{\alpha \in A}$ be an indexed set of modules.
   (1) For each $\alpha \in A$ let $j_\alpha : M_\alpha \to M$ be a monomorphism. Prove that $(M, (j_\alpha)_{\alpha \in A})$ is a direct sum of $(M_\alpha)_{\alpha \in A}$ iff M is the internal direct sum of the images $(Im\, j_\alpha)_{\alpha \in A}$.
   (2) On the other hand suppose that each M_α is a submodule of M with inclusion maps $i_\alpha : M_\alpha \to M$. Prove that M is the internal direct sum of $(M_\alpha)_{\alpha \in A}$ iff $(M, (i_\alpha)_{\alpha \in A})$ is a direct sum. Show that it is possible for M to be isomorphic to the external direct sum $\oplus_A M_\alpha$ yet not be the internal direct sum of these submodules.

5. Assume that M has a decomposition $M = \oplus_A M_\alpha$. Prove that if $f : M \to N$ is an isomorphism, then $N = \oplus_A f(M_\alpha)$.

6. Let $\phi : R \to S$ be a ring homomorphism. Let M be a left S-module with a direct decomposition $M = \oplus_A M_\alpha$. Prove that with the R-module structure induced by ϕ, M is also the internal direct sum of its submodules $(M_\alpha)_{\alpha \in A}$.

7. Let $f_\alpha : M_\alpha \to M$ and $g_\alpha : M_\alpha \to M$ be homomorphisms $(\alpha \in A)$, let $f = \oplus_A f_\alpha$ and $g = \oplus_A g_\alpha$ be their direct sums, let $h : M \to N$. Prove that $f + g = \oplus_A (f_\alpha + g_\alpha)$ and $hf = \oplus_A (hf_\alpha)$.

8. Prove that $\mathbb{Q}/\mathbb{Z} \cong \oplus_p \mathbb{Z}_{p^\infty}$.

9. Let R be a commutative integral domain and let $(M_\alpha)_{\alpha \in A}$ be an indexed set of R-modules. Prove that each M_α is divisible (torsion free) iff $\Pi_A M_\alpha$ $(\oplus_A M_\alpha)$ is divisible (torsion free). [Hint: See Exercises (3.14) and (3.15).]

10. Let M be a torsion free abelian group (Exercise (3.14)). Prove that there is a monomorphism $f : M \to \mathbb{Q}^M$. [Hint: If $0 \neq x \in M$, there is a homomorphism $f_x : M \to \mathbb{Q}$ with $f_x(x) \neq 0$. (Exercise (5.19).)]

11. (1) Prove that $\mathbb{Z}^\mathbb{N}/\mathbb{Z}^{(\mathbb{N})}$ cannot be embedded in a product $\mathbb{Z}^A$. [Hint: Let $x \in \mathbb{Z}^\mathbb{N}$ with $\pi_n(x) = 2^n$ for each $n \in \mathbb{N}$. Its image in $\mathbb{Z}^\mathbb{N}/\mathbb{Z}^{(\mathbb{N})}$ is divisible by every power of 2.]
   (2) Prove that the natural monomorphism $0 \to \mathbb{Z}^{(\mathbb{N})} \to \mathbb{Z}^\mathbb{N}$ does not split.

12. Let $M_1, M_2, \ldots$ be an independent sequence of submodules of M. Show that it is possible for each sum $\Sigma_{i=1}^n M_i$ to be a direct summand of M without $\Sigma_{n=1}^\infty M_n = \oplus_{n=1}^\infty M_n$ being a direct summand of M. [Hint: Exercise (6.11).]

13. Let $M_1 \leq M_2 \leq \ldots$ be a chain of submodules of M with each M_n a direct summand of M. Prove that there exists a sequence $M_1', M_2', \ldots$ of direct summands of M such that $M_n = M_1' \oplus M_2' \oplus \ldots \oplus M_n'$ $(n = 1, 2, \ldots)$. [Hint: See Exercise (5.4.2).]

14. Let $\mathcal{K}$ be a set of non-zero submodules of M and let $N \leq M$.
   (1) Prove that there is an indexed set $(K_\alpha)_{\alpha \in A}$ in $\mathcal{K}$ maximal with respect to $N + \Sigma_A K_\alpha = N \oplus (\oplus_A K_\alpha)$. [Hint: Apply The Maximal Principle to those independent indexed sets $(K_\alpha)_{\alpha \in A}$ with
$$N \cap (\Sigma_A K_\alpha) = 0.]$$
   (2) In particular, if $M = \oplus_A K_\alpha$, there is a subset $B \subseteq A$ maximal with respect to $N \cap (\oplus_B K_\beta) = 0$. Show though that even if N is a direct summand of M, the sum $N + \Sigma_B K_\beta$ need not be. [Hint: Exercise (5.4.3).]

15. Let $_R M$ have a finite spanning set. Prove that every direct decomposition $M = \oplus_A M_\alpha$ of M has at most finitely many non-zero terms. On the other hand, prove that the number of terms need not be bounded.

16. Let M be a module and A be a set. Then by definition,
$$card(M^{(A)}) \leq card(M^A) = (card\ M)^{(card\ A)}$$
   Prove that if $M \neq 0$
   (1) If either A or M is infinite, then $card(M^{(A)}) = (card\ M) \cdot (card\ A)$.

[Hint: Consider the set of finite subsets of $M \times A$ (see (0.10)).]

(2) $card(\mathbb{R}^{\mathbb{N}}) = card(\mathbb{R}^{(\mathbb{N})})$. (But these real vector spaces are not iso-morphic. Can you prove it?)

17. Let M be a left R-module and let A and B be sets. Define maps $\phi:(M^A)^B \to (M^B)^A$ and $\theta:(M^A)^B \to M^{A \times B}$ by

$$[(\phi f)(a)](b) = [f(b)](a) \quad \text{and} \quad [\theta(f)(b)](a) = f(a, b).$$

Prove that

(1) Both ϕ and θ are left R-isomorphisms.

(2) ϕ restricted to $(M^A)^{(B)}$ is a monomorphism into $(M^{(B)})^A$.

(3) The restriction of (2) need not be epic.

18. Let $(M_\alpha)_{\alpha \in A}$ be an indexed set of submodules of M. It is *co-independent* in case $M_\alpha + (\cap_{\beta \neq \alpha} M_\beta) = M$ for each $\alpha \in A$. This gives rise to a notion dual to that of an internal direct sum. Indeed, consider the homo-morphism $f: M \to \Pi_A(M/M_\alpha)$ defined coordinatewise by

$$\pi_\alpha f: x \mapsto x + M_\alpha,$$

and prove

(1) f is monic iff $\cap_A M_\alpha = 0$.

(2) If f is epic, then $(M_\alpha)_{\alpha \in A}$ is co-independent. [Hint: For each $x \in M$ and $\alpha \in A$ there is an x_α such that $\pi_\beta f(x_\alpha) = \delta_{\alpha\beta} x + M_\beta$ $(\beta \in A)$.]

(3) If A is finite and if $(M_\alpha)_{\alpha \in A}$ is co-independent, then f is epic.

19. Let $(M_\alpha)_{\alpha \in A}$ be an indexed set of modules and let $M \leq \Pi_A M_\alpha$. Then M is a *subdirect product* of $(M_\alpha)_{\alpha \in A}$ in case $(\pi_\alpha | M): M \to M_\alpha$ is an epi-morphism for each $\alpha \in A$. Clearly both $\Pi_A M_\alpha$ and $\oplus_A M_\alpha$ are subdirect products of $(M_\alpha)_{\alpha \in A}$.

(1) Let $C(\mathbb{R}) \leq \mathbb{R}^{\mathbb{R}}$ be the $\mathbb{R}$-submodule of all functions $f: \mathbb{R} \to \mathbb{R}$ that are continuous in the usual topology. Prove that $C(\mathbb{R})$ is a subdirect product of $\mathbb{R}$-copies of $\mathbb{R}$. Deduce that a subdirect product need not contain the direct sum.

(2) Prove that a module M is isomorphic to a subdirect product of $(M_\alpha)_{\alpha \in A}$ iff there exists an indexed class $(f_\alpha)_{\alpha \in A}$ of epimorphisms $f_\alpha: M \to M_\alpha$ $(\alpha \in A)$ with $\cap_A Ker f_\alpha = 0$.

(3) Prove that $\mathbb{Z}$ is isomorphic to a subdirect product of $(\mathbb{Z}_n)_{n > 1}$. More-over, prove that if A is an infinite subset of $\mathbb{N}$ (e.g., if $A = \mathbb{P}$), then $\mathbb{Z}$ is isomorphic to a subdirect product of $(\mathbb{Z}_n)_{n \in A}$.

(4) Prove that if $X \subseteq \mathbb{R}$ is a dense subset (e.g., if $X = \mathbb{Q}$), then $C(\mathbb{R})$ is isomorphic to a subdirect product of X copies of $\mathbb{R}$. [Hint: See (1) and (2).]

20. For many purposes the concept of a subdirect product is not a very sharp one. Even simple examples like those of Exercise (6.19) should make it clear that a single module may be representable as a subdirect product in a number of vastly different ways. At one extreme, however, we say that a non-zero module M is *subdirectly irreducible* in case whenever there is a monomorphism $f: M \to \Pi_A M_\alpha$ with each $\pi_\alpha f$ epic, then at least one $\pi_\alpha f$ is an isomorphism. Thus if M is subdirectly irreducible, then whenever it is isomorphic to a subdirect product of modules, it is actually isomorphic to one of them.

(1) Prove that $_RM$ is subdirectly irreducible iff the intersection of its non-zero submodules is non-zero. (See Exercise (6.19.2).)

(2) Using The Maximal Principle prove that if $0 \neq x \in M$, then there is a submodule $N_x < M$ maximal with respect to the property $x \notin N_x$. Then prove that M/N_x is subdirectly irreducible.

(3) Prove that every module is isomorphic to a subdirect product of subdirectly irreducible modules. [Hint: $\cap_{x \neq 0} N_x = 0.$]

21. Let $_RM$ be a non-zero module. Set $T = End(_RM)$. Let $A \neq \varnothing$. Informally each $(x_\alpha)_{\alpha \in A} \in M^{(A)}$ can be viewed as a $1 \times A$ row finite "matrix over M". If $[\![t_{\alpha\beta}]\!]$ is an A-square row finite matrix over T, define

$$(x_\alpha)_{\alpha \in A} [\![t_{\alpha\beta}]\!] = (\Sigma_{\alpha \in A} x_\alpha t_{\alpha\beta})_{\beta \in A}.$$

(1) Prove that relative to this right scalar multiplication, $M^{(A)}$ is a left R-, right $\mathbb{RFM}_A(T)$ bimodule.

(2) Prove that the bimodule of (1) is balanced whenever $_RM$ is balanced.

(3) In particular, prove that

$$End(_RR^{(A)}) \cong \mathbb{RFM}_A(R) \quad \text{and} \quad BiEnd(_RR^{(A)}) \cong R.$$

22. Let $m, n \in \mathbb{N}$. Then $Hom_R(_RR^{(m)}, _RR^{(n)})$ is a left $\mathbb{M}_m(R)$ right $\mathbb{M}_n(R)$ bimodule. (See (4.4) and Exercise (6.21).) Prove that there is a bimodule isomorphism

$$\theta : Hom_R(_RR^{(m)}, _RR^{(n)}) \to \mathbb{M}_{m \times n}(R)$$

such that $[r_1, \ldots, r_m] \, \theta(f) = f(r_1, \ldots, r_m).$

§7. Decomposition of Rings

For each ring R there are the three "regular" modules $_RR$, R_R, and $_RR_R$, and each has its own decomposition theory. The results of the previous sections readily specialize to give us basic information about the decompositions of $_RR$ and of R_R. Indeed, as we saw in (4.11), right multiplication ρ and left multiplication λ are ring isomorphisms.

$$\rho : R \to End(_RR), \qquad \lambda : R \to End(R_R).$$

Thus, (5.8) and (5.6) apply to give a characterization of the direct summands of $_RR$ (and of R_R):

7.1. Proposition. *A left ideal I of a ring R is a direct summand of $_RR$ if and only if there is an idempotent $e \in R$ such that*

$$I = Re.$$

Moreover, if $e \in R$ is an idempotent, then so is $1 - e$, and Re and $R(1 - e)$ are direct complements of each other. That is,

$$_RR = Re \oplus R(1 - e). \qquad \square$$

No direct decomposition of $_RR$ (or of R_R) can have infinitely many non-zero summands. Indeed, suppose that

$$R = \oplus_A I_\alpha$$

is a decomposition of $_RR$ as a direct sum of left ideals $(I_\alpha)_{\alpha \in A}$, and let $(p_\alpha)_{\alpha \in A}$ be the projection maps $p_\alpha : R \to I_\alpha$. By (6.21), $p_\alpha(1) \neq 0$ for at most finitely $\alpha \in A$. But

$$I_\alpha = \operatorname{Im} p_\alpha = Rp_\alpha(1),$$

so $I_\alpha \neq 0$ for at most finitely many $\alpha \in A$. It should be noted, however, that there need be no bound on the number of non-zero terms that can appear in a direct decomposition of $_RR$ (or R_R). (Consider $\mathbb{Z}^{\mathbb{N}}$.) Now recalling that $e_1, e_2, \ldots, e_n \in R$ is a complete set of pairwise orthogonal idempotents if and only if $\qquad\qquad\qquad\qquad\qquad\qquad\qquad\qquad\qquad\qquad\qquad\quad$ □

$$e_i e_j = \delta_{ij} e_i \qquad (i, j = 1, \ldots, n)$$

and

$$1 = e_1 + \ldots + e_n,$$

we have by (6.20) the following extension of (7.1):

7.2. Proposition. *Let $I_1, \ldots, I_n$ be left ideals of the ring R. Then the following statements are equivalent about the left R-module $_RR$:*

(a) $R = I_1 \oplus \ldots \oplus I_n$;
(b) *Each element $r \in R$ has a unique expression*

$$r = r_1 + \ldots + r_n$$

with $r_i \in I_i$ $(i = 1, \ldots, n)$;

(c) *There exists a (necessarily unique) complete set $e_1, \ldots, e_n$ of pairwise orthogonal idempotents in R with*

$$I_i = Re_i \qquad (i = 1, \ldots, n).$$

Note in particular, that if $e_1, \ldots, e_n$ are idempotents in R that satisfy (c), then for each $r \in R$

$$r = re_1 + \ldots + re_n.$$

Thus in the unique expression $r = r_1 + \ldots + r_n$ for r promised in (b) we have $r_i = re_i$ $(i = 1, \ldots, n)$. Since a result similar to (7.2) holds for the decompositions of the right regular module R_R we have

7.3. Corollary.   *If $e_1, \ldots, e_n$ is a complete set of pairwise orthogonal idempotents for the ring R, then*

$$_RR = Re_1 \oplus \ldots \oplus Re_n$$

and

$$R_R = e_1 R \oplus \ldots \oplus e_n R. \qquad\qquad\qquad\qquad\qquad\qquad\qquad\quad □$$

Recall that an idempotent $e \in R$ is primitive in case it is non-zero and cannot be written as a sum $e = e' + e''$ of non-zero orthogonal idempotents. A left (right) ideal of R is *primitive* in case it is of the form Re (eR) for some primitive idempotent $e \in R$. Since the endomorphism ring of Re is isomorphic to eRe, we have by (5.10) and (5.11)

7.4. Corollary. *Let $e \in R$ be a non-zero idempotent. Then the following statements are equivalent:*

(a) *e is a primitive idempotent;*
(b) *Re is a primitive left ideal of R;*
(c) *eR is a primitive right ideal of R;*
(d) *Re is an indecomposable direct summand of $_R R$;*
(e) *eR is an indecomposable direct summand of R_R;*
(f) *The ring eRe has exactly one non-zero idempotent, namely e.*   □

In later sections we shall be very much concerned with the existence and properties of "indecomposable decompositions" of modules (see, for example, §12). For $_R R$ this means simply that $_R R$ is a direct sum of (necessarily) finitely many primitive left ideals, or by (7.2) and (7.4):

7.5. Corollary. *For a ring R the left regular module $_R R$ is a direct sum $I_1 \oplus \ldots \oplus I_n$ of primitive left ideals if and only if there exists a complete set $e_1, \ldots, e_n$ of pairwise orthogonal primitive idempotents in R with*

$$I_i = Re_i \qquad (i = 1, \ldots, n). \qquad \qquad \square$$

Of course, (7.4) and (7.3) show that if $_R R$ has an indecomposable decomposition, so does R_R. The existence of such decompositions for R is far from common, but many rings that are met in practice do have indecomposable left and right decompositions. (See Exercise (7.4).) These ideas will be developed further in Chapter 7.

Suppose now that R has a decomposition as a direct sum of ideals. That is, suppose that

$$_R R = R_1 \oplus \ldots \oplus R_n$$

where each R_i is a non-zero two-sided ideal of R. Then by (7.2) there exists a unique set $u_1, \ldots, u_n$ of non-zero pairwise orthogonal idempotents in R with

$$1 = u_1 + \ldots + u_n$$

and

$$R_i = Ru_i \qquad (i = 1, \ldots, n).$$

For each i since $u_i \in R_i$, and since R_i is an ideal, $u_i R \subseteq R_i$. Thus, if $i \neq j$, then

$$u_i R u_j \subseteq u_i R \cap Ru_j \subseteq R_i \cap R_j = 0.$$

So, for each $r \in R$,

$$u_i r = (u_i r)(1) = u_i r(u_1 + \ldots + u_n)$$

$$= u_i r u_i = (u_1 + \ldots + u_n) r u_i = r u_i.$$

In other words, each u_i is a central idempotent and each

$$R_i = Ru_i = u_iR = u_iRu_i$$

is a ring with identity u_i. Conversely, if $u_1, \ldots, u_n$ is an orthogonal set of non-zero central idempotents of R with $1 = u_1 + \ldots + u_n$, then clearly each

$$R_i = Ru_i \qquad (i = 1, \ldots, n)$$

is an ideal of R and

$$_RR = R_1 \oplus \ldots \oplus R_n.$$

Observe also that this is a decomposition of R as a right module R_R and as a bimodule $_RR_R$. When R has such a decomposition, we say that R is *the ring direct sum of the ideals* $R_1, \ldots, R_n$, we call $R_1, \ldots, R_n$ *ring direct summands of R*, and write

$$R = R_1 \dotplus \ldots \dotplus R_n.$$

We also say that this is a *ring decomposition* of R.

Suppose now that R is the ring direct sum of the ideals $R_1, \ldots, R_n$ and that $u_1, \ldots, u_n$ are the associated central idempotents. Then it is easy to check that the map defined by

$$r \mapsto (ru_1, \ldots, ru_n) \qquad (r \in R)$$

is a ring isomorphism from R onto the cartesian product $R_1 \times \ldots \times R_n$ of the rings $R_1, \ldots, R_n$. Conversely, if $R_1, \ldots, R_n$ is a finite sequence of rings and if $\iota_1, \ldots, \iota_n$ are the canonical injections of these rings into the product $R_1 \times \ldots \times R_n$, then again it is easy to check that

$$R_1 \times \ldots \times R_n = \iota_1(R_1) \dotplus \ldots \dotplus \iota_n(R_n).$$

Here, of course, the central idempotents in $R_1 \times \ldots \times R_n$ of this decomposition are just $\iota_1(1_1), \ldots, \iota_n(1_n)$, the natural images of the identities of the rings $R_1, \ldots, R_n$. Now summarizing we have:

7.6. Proposition. *Let* $R_1, \ldots, R_n$ *be non-zero two-sided ideals of R. Then the following statements are equivalent:*

(a) $R = R_1 \dotplus \ldots \dotplus R_n$;

(b) $_RR = R_1 \oplus \ldots \oplus R_n$;

(c) *As an abelian group R is the direct sum of* $R_1, \ldots, R_n$;

(d) *There exist pairwise orthogonal central idempotents* $u_1, \ldots, u_n \in R$ *with* $1 = u_1 + \ldots + u_n$, *and*

$$R_i = Ru_i \qquad (i = 1, \ldots, n). \qquad \square$$

Our use of the terminology "ring" direct sum may appear to conflict with the terminology of category theory in the sense that it is practically never a direct sum (= coproduct) in the category of rings. (See Exercise (7.18).) Fortunately, in formal category theory the accepted term is "coproduct". Thus we shall allow historical precedent and our frequent desire to view this notion in the context of module decomposition to override this slight conflict.

A ring R is said to be *indecomposable* in case it has no ring decompositions with more than one term.

7.7. Corollary. *A ring R is an indecomposable ring if and only if 1 is the only non-zero central idempotent of R.*

Proof. The sufficiency is clear from (7.6). Conversely, if u is a non-zero central idempotent, then $1 - u$ and u are orthogonal central idempotents, so by (7.6) if R is indecomposable, $1 - u = 0$ and $u = 1$. ☐

In general, a ring need not admit a ring decomposition into indecomposable rings. (See Exercise (7.8).) However, if such a decomposition does exist, then it is unique in a strong sense.

7.8. Proposition. *Let $R = R_1 \dotplus \ldots \dotplus R_n$ be a ring decomposition of R with each $R_1, \ldots, R_n$ indecomposable as a ring. Let $u_1, \ldots, u_n$ be the central idempotents of this decomposition. If $R = S_1 \dotplus \ldots \dotplus S_m$ is a ring decomposition of R with associated central idempotents $v_1, \ldots, v_m$, then there is a partition $A_1, \ldots, A_m$ of $\{1, \ldots, n\}$ such that*

$$v_i = \Sigma_{A_i} u_j \qquad (i = 1, \ldots, m).$$

So, in particular,

$$S_i = \dotplus_{A_i} R_j \qquad (i = 1, \ldots, m).$$

Proof. For each i and j, it is clear that $v_i u_j$ is a central idempotent of R_j; so by (7.7) either $v_i u_j = 0$ or $v_i u_j = u_j$. Set

$$A_i = \{j \mid v_i u_j = u_j\}.$$

Now since $v_1, \ldots, v_m$ are orthogonal, $A_1, \ldots, A_m$ are pairwise disjoint, and since $1 = v_1 \dotplus \ldots + v_m$, we have $A_1 \cup \ldots \cup A_m = \{1, \ldots, n\}$. Finally, since $1 = u_1 + \ldots + u_n$, we have $v_i = v_i u_1 + \ldots + v_i u_n = \Sigma_{A_i} u_j$. ☐

The central idempotents of any ring form a Boolean algebra, and indeed this last result is simply a special case of a proposition about Boolean algebras. (See Exercise (7.7).)

There is one important class of rings that do have decompositions as a direct sum of indecomposable rings—those rings R for which the module $_R R$ has a decomposition as a direct sum of primitive left ideals. Indeed, for such a ring R there is a valuable method for determining the (necessarily unique) indecomposable ring decomposition of R from any one of its (possibly many) left decompositions. Thus suppose that $_R R$ has a decomposition as a direct sum of primitive left ideals. By (7.5) this means there exists a complete set $e_1, \ldots, e_n$ of pairwise orthogonal primitive idempotents in R. Set

$$E = \{e_1, \ldots, e_n\}.$$

On E define a relation $\sim$ by

$$e_i \sim e_j$$

in case there is a $1 \leq k \leq n$ with

$$e_k Re_i \neq 0 \qquad \text{and} \qquad e_k Re_j \neq 0.$$

Then $\sim$ is a reflexive and symmetric relation on E. It can be extended to an equivalence relation $\approx$ defined by

$$e_i \approx e_j$$

in case there is a sequence $i_1, \ldots, i_t$ in $\{1, \ldots, n\}$ such that

$$e_i \sim e_{i_1} \sim \ldots \sim e_{i_t} \sim e_j.$$

Observe that if $u \in R$ is a non-zero central idempotent and $ue_i \neq 0$, then ue_i and $(1 - u)e_1$ are orthogonal idempotents, $e_i = ue_i + (1 - u)e_1$, and e_i is primitive, so $ue_i = e_i$. So if $e_i, e_k \in E$ with $e_k Re_i \neq 0$ and $ue_i \neq 0$, then $ue_k Re_i = e_k Rue_i = e_k Re_i \neq 0$; whence $0 \neq ue_k = e_k$. Arguing thusly, we see that if $e_i \sim e_j$, then $ue_i \neq 0$ iff $ue_j \neq 0$. This extends to give that if $e_i \approx e_j$, then

$$ue_i \neq 0 \qquad \text{iff} \qquad ue_j \neq 0.$$

Let $E_1, \ldots, E_m$ be the $\approx$ equivalence classes of E, and for each $i = 1, \ldots, m$, let

$$u_i = \Sigma E_i$$

be the sum of the idempotents e_j in the class E_i. Then (Exercise (7.5)) each u_i is a non-zero idempotent of R and the set $u_1, \ldots, u_m$ is pairwise orthogonal with $1 = u_1 + \ldots + u_m$. These idempotents $u_1, \ldots, u_m$ are called the *block idempotents* of R and the rings $u_1 Ru_1, \ldots, u_m Ru_m$ are the *blocks* of R determined by E. As one consequence of the next result, these block idempotents and their blocks are independent of the primitive idempotents E.

7.9. The Block Decomposition Theorem. *Let R be a ring whose identity can be written as a sum*

$$1 = e_1 + \ldots + e_n$$

of pairwise orthogonal primitive idempotents; let $u_1, \ldots, u_m$ be the block idempotents of R determined by $E = \{e_1, \ldots, e_n\}$. Then $u_1, \ldots, u_m$ are pairwise orthogonal central idempotents, with

$$1 = u_1 + \ldots + u_m.$$

Moreover, each block $u_i Ru_i$ $(i = 1, \ldots, m)$ is an indecomposable ring, and

$$R = (u_1 Ru_1) \dotplus \ldots \dotplus (u_m Ru_m)$$

is a (necessarily unique) decomposition of R into indecomposable rings.

Proof. As we noted above, $u_1, \ldots, u_m$ are pairwise orthogonal and $1 = u_1 + \ldots + u_m$. If $i \neq j$, then by the way $\sim$ is defined, $u_i Ru_j = 0$. So for each $r \in R$ we have $u_i r = u_i r(u_1 + \ldots + u_m) = u_i ru_i = (u_1 + \ldots + u_m)ru_i = ru_i$. That is, each u_i is central. To complete the proof it will now suffice to prove that u_i is the unique non-zero central idempotent of $u_i R u_i$. If, on the contrary there is a non-zero central idempotent $v \in u_i R u_i$ with $v \neq u_i$, then $w = u_i - v$ and v are orthogonal non-zero central idempotents in R with

$$v = vu_i \quad \text{and} \quad w = wu_i.$$

Thus, if E_i is the class of u_i in $\{e_1, \ldots, e_n\}$, there must be $e_j, e_k \in E_i$ with

$$ve_j \neq 0 \quad \text{and} \quad we_k \neq 0.$$

Since e_j and e_k are primitive, this means

$$ve_j = e_j \quad \text{and} \quad we_k = e_k.$$

But since $vw = 0$, this implies that

$$ve_j \neq 0 \quad \text{and} \quad ve_k = 0$$

which, as we saw in the discussion preceding the theorem, is contrary to $e_j \approx e_k$. □

As we have now seen, the decomposition theory of a ring R and that of its left regular and right regular modules $_R R$ and R_R are simply equivalent to that of its idempotents. Since idempotents are preserved under ring homomorphisms, the direct summands of R yield direct summands of the factor rings of R. Specifically, we have

7.10. Proposition. *Let I be a proper ideal of the ring R. If $e \in R$ is a (central) idempotent of R, then $e + I$ is a (central) idempotent of the factor ring R/I, and both as left R-modules and left R/I-modules.*

$$(R/I)(e + I) = (Re + I)/I \cong Re/Ie.$$

In particular, if $e_1, \ldots, e_n \in R$ is a pairwise orthogonal set of idempotents of R with $1 = e_1 + \ldots + e_n$, then

$$R/I \cong Re_1/Ie_1 \oplus \ldots \oplus Re_n/Ie_n$$

both as left R-modules and left R/I-modules.

Proof. Most of this is a trivial consequence of the fact that the natural map $R \to R/I$ is a surjective ring homomorphism. Thus, finally, since

$$Re \cap I = \{re \in R \mid re \in I\} = Ie$$

we have by (3.7.3) that

$$(Re + I)/I \cong Re/(Re \cap I) = Re/Ie. \quad \square$$

Of course, in a factor ring of R there may be considerably more decomposability than just that inherited from R. Indeed, if I is an ideal of R and $e \in R \backslash I$ is a primitive idempotent, then $e + I$ is certainly a non-zero idempotent of R/I, but it need not be primitive. A problem of great interest that we shall consider later (see §27) is that of determining conditions under which decomposition of R/I "lift" to ones of R.

7. Exercises

1. Let R be the ring of all 2×2 upper triangular matrices over $\mathbb{Z}_2$. (See Exercise (2.7.2).)
   (1) List all direct summands of ${}_R R$ and of R_R.
   (2) For each direct summand in (1) list all idempotents spanning it.
   (3) Find two idempotents $e, f \in R$ with $Re = Rf$ and $eR \neq fR$.

2. Let e and f be idempotents in a ring R. Prove that
   (1) $Re = Rf$ iff $f = e + (1 - e)xe$ for some $x \in R$. [See Exercise (5.13).]
   (2) $Re \cong Rf$ iff there exist $x \in eRf$ and $y \in fRe$ with $xy = e$ and $yx = f$.
   (3) $Re \cong Rf$ iff $eR \cong fR$.
   (4) If $e, f \in Cen\, R$ (in particular, if R is commutative), then $Re \cong Rf$ iff $e = f$.

3. Let V_Q be a vector space over a field Q, and $R = End(V_Q)$.
   (1) Prove that an idempotent $e \in R$ is primitive iff $dim(Im\, e) = 1$.
   (2) Prove that if $e, f \in R$ are primitive idempotents, then $Re \cong Rf$. [See Exercise (7.2.2).]
   (3) Prove that if $dim\, V = n$, then R has a complete set $e_1, \ldots, e_n$ of orthogonal primitive idempotents.

4. Let Q be a field and $n > 1$. Consider $\mathbb{M}_n(Q)$.
   (1) Find a complete set of pairwise orthogonal primitive idempotents in $\mathbb{M}_n(Q)$. [Hint: See Exercise (7.3.2).]
   (2) Find two different complete sets of pairwise orthogonal primitive idempotents in the subring R of all $n \times n$ upper triangular matrices.

5. Let $e_1, \ldots, e_n$ be pairwise orthogonal idempotents in a ring R.
   (1) Prove that $e = e_1 + \ldots + e_n$ is an idempotent of R.
   (2) Prove that $e_1, \ldots, e_n, 1 - e$ is a complete set of pairwise orthogonal idempotents of R if $e \neq 1$.

6. Let e, f be idempotents in a ring R.
   (1) Prove that if either $ef = fe$ or $fe = 0$, then $e + f - ef$ is idempotent and $Re + Rf = R(e + f - ef)$.
   (2) Show that it is possible for $ef = 0$ without e and f being orthogonal.
   (3) Prove that $Re + Rf = Re \oplus R(f - fe)$.
   (4) Show that in general $e + f - ef$ need not be idempotent. [Hint: Consider $\mathbb{M}_2(\mathbb{Z})$.]

7. (1) Let B be a Boolean ring (see Exercise (1.13)). Suppose that there is a set $e_1, \ldots, e_n \in B$ of pairwise orthogonal primitive idempotents with $1 = e_1 + \ldots + e_n$. Prove that if $a \in B$ is non-zero, then there exists a unique subset $\{i_1, \ldots, i_m\} \subseteq \{1, \ldots, n\}$ such that $a = e_{i_1} + \ldots + e_{i_m}$.
   (2) Let $B(R)$ be the set of central idempotents of a ring R. Define an operation $\#$ on $B(R)$ by $e \# f = e + f - ef$. Prove that $B(R)$ is a Boolean ring with the multiplication from R and with addition $\#$. [Hint: See Exercise (7.6).]
   (3) Deduce Proposition (7.8) from (1) and (2).

8. Let R be the ring of all continuous functions $f : \mathbb{Q} \to \mathbb{R}$. Prove that there is a bijection from $B(R)$ (see Exercise (7.7)) to the set of clopen sets of $\mathbb{Q}$.

Deduce that R has infinitely many complete sets of pairwise orthogonal idempotents but not even one primitive idempotent.

9. Let ρ be a binary relation on a set A. Define a relation $\hat{\rho}$ on A by $a\hat{\rho}b$ in case there exists a finite sequence $x_1, \ldots, x_n \in A$ with $a\rho x_1$, $x_n \rho b$, and $x_i \rho x_{i+1}$ $(i = 1, \ldots, n - 1)$. Clearly $\rho = \hat{\rho}$ iff ρ is transitive.
   (1) Prove that $\hat{\rho}$ is transitive; it is the *transitive extension* of ρ.
   (2) Prove that if ρ is reflexive and symmetric, then $\hat{\rho}$ is an equivalence relation.

10. Let $E = \{e_1, \ldots, e_n\}$ be a complete set of pairwise orthogonal primitive idempotents for a ring R. Recall that $e_i \sim e_j$ in case $e_k R e_i \neq 0$ and $e_k R e_j \neq 0$ for some e_k. Define a relation ρ on E by $e_i \rho e_j$ in case $e_i R e_j \neq 0$ or $e_j R e_i \neq 0$.
   (1) Show that ρ and $\sim$ need not be the same. [Hint: Upper triangular matrices.]
   (2) Prove that ρ and $\sim$ have the same transitive extension.

11. Let I be a non-zero left ideal of a ring R. If I is a ring direct summand of R, then it is both a direct summand of $_R R$ and an ideal of R.
   (1) Show that the converse of this last statement is false. [Hint: Consider upper triangular matrices.]
   (2) Show that a non-zero ideal I of R is a ring direct summand iff there is an idempotent $e \in R$ such that $I = eR = Re$.

12. Let R be a ring and I a left ideal of R. Prove that:
   (1) If I is a direct summand of $_R R$, then $I^2 = I$.
   (2) Even if R is a commutative ring, $I^2 = I$ does not force I to be a direct summand. [Hint: An infinite product of fields.]
   (3) If R is commutative, I is finitely generated, and $I^2 = I$, then I is a direct summand. [Hint: Say $I = Rx_1 + \ldots + Rx_n = Ix_1 + \ldots + Ix_n$. Suppose $t_i \in I$ with $(1 - t_i)I \subseteq Ix_i + \ldots + Ix_n$. Then there is a $t_{i+1} \in I$ with $(1 - t_{i+1})I \subseteq Ix_{i+1} + \ldots + Ix_n$. Set $e = t_{n+1}$, and note that $I = Re$.]

13. Let $I_1, \ldots, I_n$ be ideals of a ring R. They are *(pairwise) comaximal* in case $I_i + I_j = R$ whenever $i \neq j$. For example, if each I_i is a maximal ideal, then the ideals are comaximal.
   (1) Prove *The Chinese Remainder Theorem*: If $I_1, \ldots, I_n$ are comaximal ideals, then the natural map $\phi : R \to R/I_1 \times \ldots \times R/I_n$ is a surjective ring homomorphism with kernel $I_1 \cap \ldots \cap I_n$. [Hint: The (R,R) bimodules $I_1, \ldots, I_n$ are co-independent (Exercise (6.18); for example,

$$I_1 + \cap_{i=2}^n I_i \supseteq (I_1 + I_2) \ldots (I_1 + I_n) = R^{n-1} = R.]$$

   (2) From (1) deduce the classical Chinese Remainder Theorem of elementary number theory.

14. Let R be the ring of $n \times n$ upper triangular matrices over a field Q $(n > 1)$. For each k let $I_k = \{[a_{ij}] \in R \mid a_{kk} = 0\}$, and let $J = \cap_{k=1}^n I_k$.
   (1) Prove that each I_k is a maximal ideal of R with $R/I_k \cong Q$.
   (2) Using The Chinese Remainder Theorem (Exercise (7.13)) prove that R/J is isomorphic to a product of n copies of Q.

15. Let R be a ring, let I be an ideal of R, and let $u \in R$ be idempotent modulo I (that is, $u^2 - u \in I$). Then u can be *lifted* to an idempotent in R in case there is an idempotent e in R with $e - u \in I$.

   (1) Let $n > 1$ in $\mathbb{N}$. Prove that if n is not a power of a prime, then there exist idempotents modulo $\mathbb{Z}n$ in $\mathbb{Z}$ that cannot be lifted to idempotents in $\mathbb{Z}$.

   (2) Prove that if R is the ring of $n \times n$ upper triangular matrices over a field Q and if J is the ideal of matrices having zero on the diagonal, then every idempotent modulo J can be lifted to an idempotent in R. [Hint: See Exercises (7.14) and (7.7).]

   (3) Prove that with R as in (2) there are idempotents that are central modulo J that cannot be lifted to central idempotents of R.

16. Let $(R_\alpha)_{\alpha \in A}$ be an indexed set of rings and let R be a subring of $\Pi_A R_\alpha$. Then R is a *subdirect product* of $(R_\alpha)_{\alpha \in A}$ in case the homomorphism $(\pi_\alpha | R) : R \to R_\alpha$ is surjective for each $\alpha \in A$. (See Exercise (6.19).)

   (1) Let $(I_\alpha)_{\alpha \in A}$ be an indexed set of ideals of a ring R. Prove that the image of the natural map $\phi : R \to \Pi_A(R/I_\alpha)$ defined coordinatewise by $\pi_\alpha \phi : r \mapsto r + I_\alpha$ is a subdirect product of $(R/I_\alpha)_{\alpha \in A}$ with $Ker\ \phi = \cap_A I_\alpha$.

   (2) Prove that there exist two indexed sets $(R_\alpha)_{\alpha \in A}$ and $(S_\beta)_{\beta \in B}$ of rings with no pair R_α and S_β isomorphic such that the ring $\mathbb{Z}$ is isomorphic to a subdirect product of $(R_\alpha)_{\alpha \in A}$ and to a subdirect product of $(S_\beta)_{\beta \in B}$.

   (3) Let R be a Boolean ring and let $0 \neq x \in R$. Prove that there is a maximal ideal I_x of R with $x \notin I_x$. Deduce that R is isomorphic to a subdirect product of copies of $\mathbb{Z}_2$. [Hint: Note that $R(1 - x)$ is a proper ideal excluding x. See Exercise (1.13).]

17. Let G be a group of order n, and let K be a commutative ring in which $n = n \cdot 1$ is invertible. Let $R = KG$ be the group ring of G over K (see Exercise (1.15)), and let I be a left ideal of R. Then of course R is a left K-module and I is a K-submodule of $_K R$.

   (1) Suppose that $_K I$ is a direct summand of $_K R$ with a projection map $p :_K R \to _K I$ (along some complement of $_K I$ in $_K R$). Set

   $$e = n^{-1} \Sigma_G g^{-1} p(g).$$

   Prove that for each $h \in G$, $he = n^{-1} \Sigma_G g^{-1} p(gh)$; then deduce that e is an idempotent in R, $e \in I$, and $xe = x$ for each $x \in I$.

   (2) Prove the remarkable fact that $_K I$ is a direct summand of $_K R$ iff $_R I$ is a direct summand of $_R R$.

18. Let $(R_\alpha)_{\alpha \in A}$ be an indexed class of rings.

   (1) A pair $(R, (p_\alpha)_{\alpha \in A})$ consisting of a ring R and ring homomorphisms $p_\alpha : R \to R_\alpha$ ($\alpha \in A$) is a *product* of $(R_\alpha)_{\alpha \in A}$ in case for each ring S and each indexed class $(q_\alpha)_{\alpha \in A}$ of ring homomorphisms $q_\alpha : S \to R_\alpha$ ($\alpha \in A$), there is a unique ring homomorphism $\phi : S \to R$ such that $q_\alpha = p_\alpha \phi$ for each $\alpha \in A$. Prove that

   $$(\Pi_A R_\alpha, (\pi_\alpha)_{\alpha \in A})$$

   is a product of $(R_\alpha)_{\alpha \in A}$ and that if $(R, (p_\alpha)_{\alpha \in A})$ is a product of $(R_\alpha)_{\alpha \in A}$, then there is a ring homomorphism $\phi: \Pi_A R_\alpha \to R$ with $\pi_\alpha = p_\alpha \phi$ for each $\alpha \in A$.

(2) Dualizing the concept of a product define a *coproduct* $(R, (i_\alpha)_{\alpha \in A})$ of $(R_\alpha)_{\alpha \in A}$. Let m and n be natural numbers. Prove that $(\mathbb{Z}_m, \mathbb{Z}_n)$ has a coproduct iff m and n are not relatively prime. (Note, however, that if m and n are relatively prime, then there is, in some sense, no "need" for a coproduct.)

§8. Generating and Cogenerating

The important concept of a spanning set for a module is not categorical and does not have a natural dual. There is, however, an effectively equivalent one that is categorical and that does have a very important dual. These dual concepts of generating and cogenerating are the subjects of this section.

We resume our practice of assuming that all modules and homomorphisms are left R-modules and left R-homomorphisms over a ring R.

Generated and Cogenerated Classes

Let $\mathcal{U}$ be a class of modules. A module M is *(finitely) generated by* $\mathcal{U}$ (or $\mathcal{U}$ *(finitely) generates M*) in case there is a (finite) indexed set $(U_\alpha)_{\alpha \in A}$ in $\mathcal{U}$ and an epimorphism

$$\oplus_A U_\alpha \to M \to 0.$$

If $\mathcal{U} = \{U\}$ is a singleton, then we simply say that U (finitely) generates M; of course this means that there is an epimorphism

$$U^{(A)} \to M \to 0$$

for some (finite) set A. One of the most important examples is given in

8.1. Theorem. *If a module $_RM$ has a spanning set $X \subseteq M$, then there is an epimorphism*

$$R^{(X)} \to M \to 0.$$

Moreover, R finitely generates M if and only if M has a finite spanning set.

Proof. Let $X \subseteq M$ span M. For each $x \in X$, right multiplication $\rho_x: r \mapsto rx$ is a left R-homomorphism $R \to M$. Let $\rho = \oplus_X \rho_x$ be the direct sum of these homomorphisms. Then

$$\rho: R^{(X)} \to M$$

and $\operatorname{Im} \rho = \Sigma_X \operatorname{Im} \rho_x = \Sigma_X Rx = M$. Thus ρ is epic. The final statement is now clear. $\qquad \square$

There is another simple and familiar example that serves very well to illustrate this entire section.

8.2. Example. Recall that a group $_Z M$ is torsion in case each of its elements has finite order. So if $_Z M$ is torsion, then for each $x \in M$ there is an $n(x) > 0$ in $\mathbb{N}$ and a homomorphism $f_x : \mathbb{Z}_{n(x)} \to M$ with $Im f_x = \mathbb{Z}x$, and the direct sum $f = \oplus_M f_x$ is an epimorphism

$$f : \oplus_M \mathbb{Z}_{n(x)} \to M.$$

Conversely, if M is an epimorphic image of a direct sum of finite cyclic groups, it is clearly spanned by elements of finite order, so it is torsion. In other words an abelian group is torsion iff it is generated by

$$\mathcal{U} = \{ \mathbb{Z}_n \mid n = 2, 3, \dots \}.$$

Observe also that this is equivalent to being generated by the single group $\oplus_{\mathbb{N}} \mathbb{Z}_n$. (See (8.9).)

The notion of generating is categorical. It depends only on the objects and the morphisms of the category $_R \mathsf{M}$ and not on the elements of any module M. And of course this concept has a natural dual—simply turn the arrows around.

Let $\mathcal{U}$ be a class of modules. A module M is (*finitely*) *cogenerated by* $\mathcal{U}$ (or $\mathcal{U}$ (*finitely*) *cogenerates* M) in case there is a (finite) indexed set $(U_\alpha)_{\alpha \in A}$ in $\mathcal{U}$ and a monomorphism

$$0 \to M \to \Pi_A U_\alpha.$$

We make the obvious adjustments in terminology if $\mathcal{U} = \{U\}$ is a singleton. Again there is a particularly easy and illuminating example in $_Z \mathsf{M}$.

8.3. Example. If M is a torsion-free abelian group, then there is a monomorphism $M \to \mathbb{Q}^M$ (see Exercise (6.10)); thus, $_Z M$ is cogenerated by $_Z \mathbb{Q}$. On the other hand, any subgroup of $\mathbb{Q}^A$ is certainly torsion free. In other words, the torsion-free abelian groups are precisely the abelian groups cogenerated by $\mathbb{Q}$.

Let $\mathcal{U}$ be a class of modules. The class of all modules generated by $\mathcal{U}$ is denoted $Gen(\mathcal{U})$ and the class cogenerated by $\mathcal{U}$ is denoted $Cog(\mathcal{U})$. Also $FGen(\mathcal{U})$ and $FCog(\mathcal{U})$ denote the classes finitely generated and finitely cogenerated by $\mathcal{U}$, respectively.

For example, if $\mathcal{U} = \{ \mathbb{Z}_n \mid n > 1 \}$, then $Gen(\mathcal{U})$ is the class of torsion groups, whereas $Cog(\mathbb{Q})$ is the class of torsion-free groups. Viewed in the light of these examples, the next two propositions are rather obvious.

8.4. Proposition. *Let $\mathcal{U}$ be a class of modules.*

(1) *If M is in $Gen(\mathcal{U})(FGen(\mathcal{U}))$, then so is every epimorphic image of M.*

(2) *If $(M_\alpha)_{\alpha \in A}$ is a (finite) indexed set in $Gen(\mathcal{U})(FGen(\mathcal{U}))$, then $\oplus_A M_\alpha$ is in $Gen(\mathcal{U})(FGen(\mathcal{U}))$.*

Proof. (1) If $f : \oplus_A U_\alpha \to M$ and $g : M \to M'$ are epimorphisms, then so is $gf : \oplus_A U_\alpha \to M'$.

(2) Let $f_\alpha : \oplus_B U_\beta \to M_\alpha$ be an epimorphism for each $\alpha \in A$. Then (6.25) the direct sum $f = \oplus_A f_\alpha$ is an epimorphism

$$f : \oplus_A (\oplus_{B_\alpha} U_{\beta_\alpha}) \to \oplus_A M_\alpha.$$

But if $C = \cup_A B_\alpha$, then $\oplus_A(\oplus_{B\alpha} U_{\beta\alpha}) \cong \oplus_C U_\gamma$.   □

Stated somewhat informally this last result says that the class of modules (finitely) generated by $\mathscr{U}$ is closed in $_R M$ under isomorphism, forming factor modules, and taking (finite) direct sums. We omit the proof of the following, a simple dual to that of (8.4).

8.5. Proposition. *Let $\mathscr{U}$ be a class of modules.*

(1) *If M is in $Cog(\mathscr{U})(FCog(\mathscr{U}))$, and if $g : M' \to M$ is a monomorphism, then M' is in $Cog(\mathscr{U})(FCog(\mathscr{U}))$;*

(2) *If $(M_\alpha)_{\alpha \in A}$ is a (finite) indexed set in $Cog(\mathscr{U})(FCog(\mathscr{U}))$, then $\Pi_A M_\alpha$ is in $Cog(\mathscr{U})(FCog(\mathscr{U}))$.*   □

There is an easy consequence of these propositions. In effect it says that generating and cogenerating are "transitive".

8.6. Corollary. *Let $\mathscr{U}$ and $\mathscr{V}$ be classes of modules.*

(1) *If $\mathscr{V}$ is contained in $Gen(\mathscr{U})(FGen(\mathscr{U}))$, then so is $Gen(\mathscr{V})(FGen(\mathscr{V}))$.*

(2) *If $\mathscr{V}$ is contained in $Cog(\mathscr{U})(FCog(\mathscr{U}))$, then so is $Cog(\mathscr{V})(FCog(\mathscr{V}))$.*

8.7. Remark. There is another way of describing each of the concepts generating and cogenerating that is related to those of spanning and of subdirect product (Exercise (6.19)). These formulations are also immediate from (6.8) and (6.2).

(1) *The class $\mathscr{U}$ generates M iff M is a sum of submodules each an epimorphic image of some module of $\mathscr{U}$.*

(2) *The class $\mathscr{U}$ cogenerates M iff there is a set $\mathscr{K}$ of submodules of M such that M/K is embedded in some module of $\mathscr{U}$ for each $K \in \mathscr{K}$ and $\cap \mathscr{K} = 0$.*

Generators and Cogenerators

It follows from Corollary (8.6) that if $\mathscr{U}$ and $\mathscr{V}$ are classes of modules that generate each other, then $Gen(\mathscr{U}) = Gen(\mathscr{V})$; in particular, $\mathscr{U}$ and $\mathscr{V}$ could be quite different yet generate the same classes. Thus, given $\mathscr{U}$ it is appropriate to seek out some canonical class that also generates $Gen(\mathscr{U})$.

Of course there is one essentially trivial reduction. A set $\mathscr{U}' \subseteq \mathscr{U}$ is a *class of representatives* (of the isomorphism types) of $\mathscr{U}$ in case each $U \in \mathscr{U}$ is isomorphic to some element of $\mathscr{U}'$; if in addition, no two elements of $\mathscr{U}'$ are isomorphic, then the class of representatives is *irredundant* (see (0.2)). Clearly, if $\mathscr{U}'$ is a class of representatives of $\mathscr{U}$, then $Gen(\mathscr{U}) = Gen(\mathscr{U}')$ and $Cog(\mathscr{U}) = Cog(\mathscr{U}')$.

Given a class $\mathscr{U}$, a module G is a *generator for $Gen(\mathscr{U})$* in case $Gen(\mathscr{U}) = Gen(G)$. A module C is a *cogenerator for $Cog(\mathscr{U})$* in case $Cog(\mathscr{U}) = Cog(C)$. A generator (cogenerator) for the class $_R\mathcal{M}$ of all left R-modules is usually called simply a *(left R-) generator (cogenerator)* without reference to the class. In this terminology the first half of (8.1) can be rephrased.

8.8. Corollary. *The regular module $_RR$ is a generator.*

As we mentioned above, in §18 we shall prove that dually there is a natural left R-cogenerator. Also in this terminology, Example (8.3) says that $\mathbb{Q}$ is a cogenerator for the class of torsion-free groups, and Example (8.2) says that $G = \oplus_{n>1} \mathbb{Z}_n$ is a generator for the class of torsion groups.

It follows immediately from (8.6) that G generates $Gen(\mathcal{U})$ iff $G \in Gen(\mathcal{U})$ and G generates each $U \in \mathcal{U}$. And of course the dual assertion holds for cogenerators. With these observations, it is nearly trivial to prove

8.9. Proposition. *If $\mathcal{U}$ has a set $\{U_\alpha \mid \alpha \in A\}$ of representatives, then*
(1) $\oplus_A U_\alpha$ is a generator for $Gen(\mathcal{U})$;
(2) $\oplus_A U_\alpha$ and $\Pi_A U_\alpha$ are cogenerators for $Cog(\mathcal{U})$.

Proof. (2) By (8.5) both $\Pi_A U_\alpha$ and its submodule $\oplus_A U_\alpha$ are in $Cog(\mathcal{U})$. The injection maps $\iota_\alpha : U_\alpha \to \oplus_A U_\alpha$ are monic, so $\oplus_A U_\alpha$ cogenerates each U_α; hence it cogenerates $Cog(\mathcal{U})$. Trivially $\Pi_A U_\alpha$ cogenerates $\oplus_A U_\alpha$. □

For a module U, the next result gives a very useful characterization of $Gen(U)$ and $Cog(U)$.

8.10. Proposition. *Let U and M be modules. Then*
(1) U (finitely) generates M if and only if there is a (finite) subset $H \subseteq Hom_R(U, M)$ with $M = \Sigma_{h \in H} Im\, h$;
(2) U (finitely) cogenerates M if and only if there is a (finite) subset $H \subseteq Hom_R(M, U)$ with $0 = \cap_{k \in H} Ker\, h$.

Proof. The proof of (1) is an easy variation of that of (8.1). For (2) suppose U cogenerates M. Say $f : M \to U^A$ is a monomorphism. Then for each $\alpha \in A$, $f_\alpha = \pi_\alpha f : M \to U$ is a homomorphism. Since f is the direct product of $(f_\alpha)_{\alpha \in A}$, $\cap_A Ker\, f_\alpha = 0$ (6.2). Conversely, if $H \subseteq Hom_R(M, U)$, then the direct product

$$\Pi_H h : M \to U^H$$

has kernel $\cap_H Ker\, h$. □

A corollary says in effect that U generates (cogenerates) M iff $Hom_R(U, M)$ (resp., $Hom_R(M, U)$) "separates points" in $Hom_R(M, N)$ (resp., $Hom_R(N, M)$) for every module N.

8.11. Corollary. *Let U and M be modules. Then*
(1) U generates M iff for every non-zero homomorphism $f : M \to N$ there exists an $h \in Hom_R(U, M)$ such that $f h \neq 0$;
(2) U cogenerates M iff for every non-zero homomorphism $f : N \to M$ there exists an $h \in Hom_R(M, U)$ such that $hf \neq 0$.

Proof. (1) Set $H = Hom(U, M)$, and $T = \Sigma_H Im\, h \leq M$. If $f : M \to N$, then $f h = 0$ for all $h \in H$ iff $T \leq Ker\, f$. On the other hand T is contained in the kernel of the natural map $M \to M/T$. Now apply (8.10.1). (2) is even easier. □

The Trace and Reject

Let $\mathscr{U}$ be a class of modules. From (8.7) it is clear that whether or not $\mathscr{U}$ generates M, there is a unique largest submodule of M generated by $\mathscr{U}$. And dually, there is a unique largest factor module of M cogenerated by $\mathscr{U}$. The *trace of $\mathscr{U}$ in M* and the *reject of $\mathscr{U}$ in M* are defined by

$$Tr_M(\mathscr{U}) = \Sigma\{Im\, h \mid h:U \to M \quad \text{for some } U \in \mathscr{U}\}$$

and

$$Rej_M(\mathscr{U}) = \cap\{Ker\, h \mid h:M \to U \quad \text{for some } U \in \mathscr{U}\}.$$

Observe that even though the class $\mathscr{U}$ need not be a set, these sums and intersections are taken over sets of submodules of M, and thus are well-defined submodules of M. In the particular case where $\mathscr{U} = \{U\}$ is a singleton, these assume the simpler form.

$$Tr_M(U) = \Sigma\{Im\, h \mid h \in Hom_R(U, M)\},$$

$$Rej_M(U) = \cap\{Ker\, h \mid h \in Hom_R(M, U)\}.$$

As an easy extension of (8.10) we have

8.12. Proposition. *Let $\mathscr{U}$ be a class of modules, and let M be a module. Then*
(1) $Tr_M(\mathscr{U})$ is the unique largest submodule L of M generated by $\mathscr{U}$;
(2) $Rej_M(\mathscr{U})$ is the unique smallest submodule K of M such that M/K is cogenerated by $\mathscr{U}$.

Proof. (1) Let $(U_\alpha)_{\alpha \in A}$ be an indexed set in $\mathscr{U}$ and let $h: \oplus_A U_\alpha \to M$. Then $Im\, h = \Sigma_A Im(h\iota_\alpha) \le Tr_M(\mathscr{U})$, so every submodule of M in $Gen(\mathscr{U})$ is contained in $Tr_M(\mathscr{U})$. On the other hand there is an indexed set $(U_\alpha)_{\alpha \in A}$ and homomorphisms $h_\alpha: U_\alpha \to M$ with $Tr_M(\mathscr{U}) = \Sigma_A Im\, h_\alpha$. Thus $\oplus_A h_\alpha: \oplus_A U_\alpha \to M$ has image $Tr_M(\mathscr{U})$ (see (6.8)), so $Tr_M(\mathscr{U})$ is in $Gen(\mathscr{U})$.

(2) Let $(U_\alpha)_{\alpha \in A}$ be an indexed set in $\mathscr{U}$, let $h: M \to \Pi_A M_\alpha$, and let $K = Ker\, h$. Then $K = \cap_A Ker(\pi_\alpha h) \supseteq Rej_M(\mathscr{U})$, so if M/K is cogenerated by $\mathscr{U}$, $K \supseteq Rej_M(\mathscr{U})$. On the other hand there is an indexed set $(U_\alpha)_{\alpha \in A}$ in $\mathscr{U}$ and $h_\alpha: M \to U_\alpha$ with $Rej_M(\mathscr{U}) = \cap_A Ker\, h_\alpha$. Thus $\Pi_A h_\alpha: M \to \Pi_A U_\alpha$ has kernel $Rej_M(\mathscr{U})$ (see (6.2)), so $M/Rej_M(\mathscr{U})$ is in $Cog(\mathscr{U})$. $\square$

8.13. Corollary. *Let M be a module and $\mathscr{U}$ a class of modules. Then*
(1) $\mathscr{U}$ generates M iff $Tr_M(\mathscr{U}) = M$;
(2) $\mathscr{U}$ cogenerates M iff $Rej_M(\mathscr{U}) = 0$. $\square$

One part of the next corollary says in effect that the trace of the trace is the trace. Another says that if the reject is factored out, there is zero reject.

8.14. Corollary. *Let M be a module and $\mathscr{U}$ a class of modules. Let $K \le M$. Then*
(1) $K = Tr_M(\mathscr{U})$ iff $K \ge Tr_M(\mathscr{U})$ and $Tr_K(\mathscr{U}) = K$;
(2) $K = Rej_M(\mathscr{U})$ iff $K \le Rej_M(\mathscr{U})$ and $Rej_{M/K}(\mathscr{U}) = 0$.
In particular,
$$Tr_{Tr_M(\mathscr{U})}(\mathscr{U}) = Tr_M(\mathscr{U}) \qquad and \qquad Rej_{M/Rej_M(\mathscr{U})}(\mathscr{U}) = 0. \qquad \square$$

8.15. Examples. (1) If $\mathscr{U} = \{\mathbb{Z}_n \mid n = 2, 3, \ldots\}$, then for each abelian group M, the trace $Tr_M(\mathscr{U})$ is simply the torsion subgroup $T(M)$ of M. (See Exercise (8.8).) Of course $T(M)$ is the unique largest torsion subgroup of M, and $T(T(M)) = T(M)$.

(2) If M is an abelian group, then $Rej_M(\mathbb{Q})$ is the intersection of all $K \leq M$ with M/K torsion free. So $Rej_M(\mathbb{Q})$ is again just the torsion subgroup $T(M)$ of M, the unique smallest subgroup with $M/T(M)$ torsion free. And of course $T(M/T(M)) = 0$.

Clearly both $Tr_M(\mathscr{U})$ and $Rej_M(\mathscr{U})$ are left R-submodules of M. Moreover, they are both stable under endomorphisms of M.

8.16. Proposition. *Let $\mathscr{U}$ be a class of modules, let M and N be modules, and let $f : M \to N$ be a homomorphism. Then*

$$f(Tr_M(\mathscr{U})) \leq Tr_N(\mathscr{U}) \qquad and \qquad f(Rej_M(\mathscr{U})) \leq Rej_N(\mathscr{U}).$$

In particular, $Tr_M(\mathscr{U})$ and $Rej_M(\mathscr{U})$ are left R- right $End(_RM)$-bisubmodules of M.

Proof. For the first, simply observe that for each $h \in Hom_R(U, M)$ we have $fh \in Hom_R(U, N)$ and $f(Im\,h) = Im\,fh$. For the second, if $x \in Rej_M(U)$ and $h \in Hom_R(N, U)$, then $hf \in Hom_R(M, U)$ so $h(f(x)) = 0$. □

In general, the image of the trace or reject need not be the trace or reject of the image. For instance, consider the natural map $\mathbb{Z} \to \mathbb{Z}_n$ in the setting of Example (8.15). But with the hypothesis of (8.16) we do have

8.17. Corollary. (1) *If $f : M \to N$ is monic and $Tr_N(\mathscr{U}) \subseteq Im\,f$, then*

$$f(Tr_M(\mathscr{U})) = Tr_N(\mathscr{U});$$

(2) *If $f : M \to N$ is epic and $Ker\,f \subseteq Rej_M(\mathscr{U})$, then*

$$f(Rej_M(\mathscr{U})) = Rej_N(\mathscr{U}).$$

Proof. We shall prove (2). By (8.16), $f(Rej_M(\mathscr{U})) \leq Rej_M(\mathscr{U})$. But if f is epic with $Ker\,f \subseteq Rej_M(\mathscr{U})$, then by The Factor Theorem (3.6), there is an isomorphism

$$M/Rej_M(\mathscr{U}) \to N/f(Rej_M(\mathscr{U})).$$

By (8.14) the rejects of $\mathscr{U}$ in these are both zero, so by (8.14.2), we have $f(Rej_M(\mathscr{U})) = Rej_N(\mathscr{U})$. □

For an indexed set $(M_\alpha)_{\alpha \in A}$ of modules and class of modules $\mathscr{U}$, the direct sum of the traces $Tr_{M_\alpha}(\mathscr{U})$ and the direct sum of the rejects $Rej_{M_\alpha}(\mathscr{U})$ are both contained in $\oplus_A M_\alpha$. In fact, these are the trace and reject of $\mathscr{U}$, respectively in $\oplus_A M_\alpha$.

8.18. Proposition. *If $(M_\alpha)_{\alpha \in A}$ is an indexed set of modules, then for each module M*

$$Tr_{\oplus_A M_\alpha}(\mathscr{U}) = \oplus_A Tr_{M_\alpha}(\mathscr{U})$$

and

$$Rej_{\oplus_A M_\alpha}(\mathscr{U}) = \oplus_A Rej_{M_\alpha}(\mathscr{U}).$$

Proof. We shall do the trace. A similar argument works for the reject. Applying (8.16) to the natural injections ι_α and projections π_α we have

$$Tr_{\oplus_A M_\alpha}(\mathcal{U}) = \Sigma \iota_\alpha \pi_\alpha (Tr_{\oplus_A M_\alpha}(\mathcal{U}))$$

$$\leq \Sigma \iota_\alpha (Tr_{M_\alpha}(\mathcal{U})) \leq Tr_{\oplus_A M_\alpha}(\mathcal{U}).$$

Thus the inequalities are equalities. But

$$\Sigma \iota_\alpha (Tr_{M_\alpha}(\mathcal{U})) = \oplus_A Tr_{M_\alpha}(\mathcal{U}). \qquad \square$$

8.19. Lemma. *Let $\mathcal{U}$ and $\mathcal{V}$ be classes of modules.*
(1) *If $\mathcal{V} \subseteq Gen(\mathcal{U})$, then $Tr_M(\mathcal{V}) \leq Tr_M(\mathcal{U})$;*
(2) *If $\mathcal{V} \subseteq Cog(\mathcal{U})$, then $Rej_M(\mathcal{U}) \leq Rej_M(\mathcal{V})$.*

Proof. We shall do (2). Suppose $x \notin Rej_M(\mathcal{V})$. Then there is a homomorphism $f: M \to V$ with $V \in \mathcal{V}$ and $x \notin Ker f$. Since $V \in Cog(\mathcal{U})$, there is a homomorphism $h: V \to U$ with $U \in \mathcal{U}$ and $f(x) \notin Ker h$. Now $hf: M \to U$ with $x \notin Ker hf$. So $x \notin Rej_M(\mathcal{U})$. $\qquad \square$

8.20. Proposition. *Let G be a generator for $Gen(\mathcal{U})$ and let C be a cogenerator for $Cog(\mathcal{U})$. Then for each M*

$$Tr_M(\mathcal{U}) = Tr_M(G) \qquad and \qquad Rej_M(\mathcal{U}) = Rej_M(C).$$

In particular, if $(U_\alpha)_{\alpha \in A}$ is an indexed set of modules

$$Tr_M(\oplus_A U_\alpha) = \Sigma_A Tr_M(U_\alpha)$$

$$Rej_M(\Pi_A U_\alpha) = \cap_A Rej_M(U_\alpha) = Rej_M(\oplus_A U_\alpha).$$

Proof. By (8.9) and (8.19). $\qquad \square$

Two Special Cases

Since ${}_R R$ is a generator (for ${}_R \mathcal{M}$) it follows from (8.6) that ${}_R M$ is a generator iff it generates ${}_R R$. Whether or not ${}_R M$ is a generator, its trace in R is a measure of how close it comes to being one.

8.21. Proposition. *For each class $\mathcal{U}$ of left R-modules, the trace $Tr_R(\mathcal{U})$ is a two sided ideal. Moreover, a module ${}_R M$ is a generator if and only if $Tr_R(M) = R$.*

Proof. By (4.11), the endomorphisms of ${}_R R$ are the right multiplications $\rho(r)$ by elements of R. Thus by (8.16) $Tr_R(\mathcal{U})$ is a two-sided ideal. The last statement is by (8.8), (8.13.1) and (8.6.1). $\qquad \square$

Recall that if ${}_R M$ is a module, then its (left) annihilator is

$$l_R(M) = \{r \in R \mid rx = 0 \quad (x \in M)\},$$

and that M is faithful in case $l_R(M) = 0$.

8.22. Proposition. *For each left R-module M*

$$Rej_R(M) = l_R(M).$$

In particular, M is faithful if and only if M cogenerates R.

Proof. Using (4.5) we have

$$Rej_R M = \cap \{Ker f \mid f \in Hom(R, M)\}$$
$$= \cap \{Ker \rho(x) \mid x \in M\}$$
$$= \cap_{x \in M} l_R(x) = l_R(M). \qquad \square$$

Motivated by this last fact, we define, for a class $\mathcal{U}$ of left R-modules, its annihilator:

$$l_R(\mathcal{U}) = Rej_R(\mathcal{U}).$$

Thus $l_R(\mathcal{U})$ is simply the intersection of all left ideals I of R such that R/I embeds in some element of $\mathcal{U}$.

8.23. Corollary. *For each class $\mathcal{U}$ of left R-modules, the reject*

$$Rej_R(\mathcal{U}) = l_R(\mathcal{U}) \qquad \square$$

is a two-sided ideal.

8. Exercises

1. Prove that there exist modules U and M such that:
   (1) M is generated by U but not every submodule of M is generated by U. [Hint: Let R be the ring of 2×2 upper triangular matrices over a field and let $U = M$ be a left ideal of R.]
   (2) M is cogenerated by U but not every factor module of M is cogenerated by U.
2. If U generates or cogenerates M, then $l_R(U) \subseteq l_R(M)$. Show, however, that the converse is false.
3. Prove that for a module $_R M$ the following are equivalent: (a) $_R M$ is faithful; (b) M cogenerates R; (c) M cogenerates a generator.
4. Prove that $_R G$ is a generator iff for some natural number n and some module $_R L$ there is an isomorphism $G^{(n)} \cong R \oplus L$. [Hint: See Exercise (5.1).]
5. Prove that if $N \leq M$ and if M either generates N or cogenerates M/N, then N is a $BiEnd(_R M)$ submodule of M.
6. Let M be a left R-module with $S = End(_R M)$. Let $e \in S$ be idempotent. Prove that

$$Tr_M(Me) = (Me)S \quad \text{and} \quad Rej_M(Me) = l_M(Se).$$

7. Let M and U be left R-modules. Prove that

$$Hom_R(M, Tr_U(M)) \cong Hom_R(M, U)$$
$$Hom_R(M/Rej_M(U), U) \cong Hom_R(M, U).$$

8. Let I be a left ideal of R and let M be a left R-module. Prove that $Tr_M(R/I) = Rr_M(I)$.

9. Let $\mathcal{U}$ be the set of all simple abelian groups. Prove that for an abelian group M,
   (1) $Tr_M(\mathcal{U}) = \{x \in M \mid x \text{ has square free order}\}$.
   (2) $Rej_M(\mathcal{U}) = \cap \{N \le M \mid N \text{ is a maximal subgroup of } M\}$.

10. Let R be the ring of $n \times n$ upper triangular matrices over a field Q, and let $\mathcal{U}$ be the class of simple left R-modules. Prove that:
   (1) $Tr_R(\mathcal{U}) = \{[\![a_{ij}]\!] \in R \mid a_{ij} = 0 \ (i \ge 2)\}$.
   (2) $Rej_R(\mathcal{U}) = \{[\![a_{ij}]\!] \in R \mid a_{kk} = 0 \ (k = 1, \ldots, n)\}$.

11. An indexed set $(x_\alpha)_{\alpha \in A}$ of elements of a module is *linearly independent* in case for every finite sequence $\alpha_1, \ldots, \alpha_n$ of distinct elements of A and every $r_1, \ldots, r_n \in R$
$$r_1 x_{\alpha_1} + \ldots + r_n x_{\alpha_n} = 0 \text{ implies } r_1 = \ldots = r_n = 0.$$

An R-module F with a linearly independent spanning set $(x_\alpha)_{\alpha \in A}$ is called a *free R-module (of rank card A) with free basis* $(x_\alpha)_{\alpha \in A}$.
   (1) Let $(x_\alpha)_{\alpha \in A}$ be an indexed set in a left R-module F. For each α let $\rho_\alpha : R \to F$ be the right multiplication $r \mapsto rx_\alpha$. Prove that the following are equivalent:
   (a) F is free with free basis $(x_\alpha)_{\alpha \in A}$;
   (b) $\oplus_A \rho_\alpha : R^{(A)} \to F$ is an isomorphism;
   (c) For every $_R M$ and every indexed set $(y_\alpha)_{\alpha \in A}$ in M there is a unique homomorphism $f : F \to M$ with $f(x_\alpha) = y_\alpha$ $(\alpha \in A)$.
   (2) Prove that a module $_R F$ is free of rank *card A* iff $F \cong R^{(A)}$. Thus there exist free modules of arbitrary rank.
   (3) Prove that every free module is a generator.
   (4) Prove that if $_R F$ is free, then every epimorphism $f : M \to F$ splits. [Hint: Use (c) of part (1).]

12. Let F_R be a free right R-module with free basis $(x_\alpha)_{\alpha \in A}$ and let $S = End(F_R)$. For each $a \in S$ and each x_β there is an indexed set $(a_{\alpha\beta})_{\alpha \in A}$ in R with almost all $a_{\alpha\beta}$ zero such that
$$a(x_\beta) = \Sigma_{\alpha \in A} a_{\alpha\beta} x_\alpha.$$
Prove that $a \mapsto [\![a_{\alpha\beta}]\!]_{(\alpha,\beta) \in A \times A}$ defines a ring isomorphism from $S = End(F_R)$ onto the ring $\mathbb{CFM}_A(R)$ of all $A \times A$ column finite matrices over R.

13. Let F be a free module of rank *card A*. Prove that:
   (1) If A is infinite, every free basis for F has cardinality *card A*.
   (2) If F has a finite spanning set, then A is finite.

14. A ring R is left SBN (for "single basis number") in case all non-zero free modules of finite rank are isomorphic to $_R R$.
   (1) Prove that the following are equivalent:
   (a) R is left SBN;
   (b) $_R R \cong {}_R R^{(2)}$;
   (c) There exist $a, a', b, b' \in R$ with $ab + a'b' = 1$, $ba = b'a' = 1$, and $b'a = ba' = 0$ [Hint: For (b) $\Leftrightarrow$ (c), see (5.3).];

(d) R is right SBN.

(2) Prove that if R is SBN, then so is every factor ring of R.

(3) Prove that if Γ is an infinite set, then the ring R of all Γ-square column finite matrices over a ring S is left SBN. [Hint: Since Γ is infinite, it is the disjoint union of two subsets Γ', Γ'' each with cardinality card Γ. Now consider condition (b) of part (1).]

(4) Give an example of a simple SBN ring. [Hint: Parts (2) and (3).]

15. A ring R is left IBN (for "invariant basis number") in case no two free left R-modules of different ranks are isomorphic. Thus (see Exercise (8.13)), R is IBN iff $R^{(m)} \cong R^{(n)}$ implies $m = n$. Prove that:

  (1) Left IBN implies right IBN. [Hint: Exercise (6.22).]

  (2) If I is an ideal of R and if R/I is IBN, then so is R. [Hint: Exercise (6.3).]

  (3) Every field is IBN.

  (4) Every commutative ring is IBN.

16. Recall that a commutative integral domain R is a *principal ideal domain* (= P.I.D.) in case every ideal of R is principal. Of course $\mathbb{Z}$ is a P.I.D. Essentially all results for abelian groups extend to ones for modules over P.I.D.'s. Show that standard proofs for abelian groups extend to prove the following results for modules over a principal ideal domain R:

  (1) Every finitely spanned torsion-free R-module M is free. (Note: "torsion-free" means $ax = 0$ for $x \in M$ and $a \in R$ implies $a = 0$ or $x = 0$.)

  (2) Every submodule of a free R-module is free.

Chapter 3

Finiteness Conditions for Modules

The first round of generalities is over, and it is now time for us to apply this formal machinery to the study of specific classes of rings and modules. We begin in this chapter with an investigation of the structure of classes of modules having certain natural finiteness properties. In the next chapter we return to the rings themselves.

The lattice of submodules of a module reveals a substantial amount of information about the module and provides a natural means for classifying the module. Our point of departure is the observation that the modules with the simplest possible non-trivial submodule lattices are both simple and indecomposable. We then classify modules with respect to how they are pieced together from simple or from indecomposable modules.

In Section 9 we study those modules that are generated by simple modules. These are precisely the modules that have decompositions as direct sums of simple modules. Such "semisimple" modules form perhaps the most important single class of modules and provide the basic building blocks of much of the theory. In the next two sections we turn to the modules whose submodule lattices satisfy one of the so-called "chain conditions".

Modules satisfying both of these conditions have finite maximal chains—composition series—that generalize the familiar prime factorization of finite abelian groups. Finally, in Section 12 we study modules with yet another "finiteness" property, that is, modules that have indecomposable decompositions. Here, one of our major concerns is the uniqueness of such decompositions and the way such decompositions generalize the notion of a basis for a vector space.

§9. Semisimple Modules—The Socle and the Radical

From the point of view of module theory among the most remarkable features of a vector space are that it has a basis, that the cardinality of this basis is an invariant of the module, and that any independent set can be extended to a basis by adjoining elements from a given basis. These properties can be rephrased in module theoretic terms, and as we shall see in this section, they hold in any module that is generated by simple modules. These are the "semisimple" modules.

Throughout this section we shall continue our agreement that R is a ring, that "module" means "left R-module", etc.

Simple Modules

Recall that a non-zero module $_RT$ is *simple* in case it has no non-trivial submodules. A simple module can be characterized in $_R\mathsf{M}$ as a non-zero module T such that every non-zero homomorphism $T \to N$ ($N \to T$) in $_R\mathsf{M}$ is a monomorphism (epimorphism). (See Exercise (3.2).) From (2.10) and (3.9) we have

9.1. Proposition. *A left R-module T is simple if and only if $T \cong R/M$ for some maximal left ideal M of R.* □

Since the maximal left ideals of R form a set, there is a set $\mathscr{T}$ of representatives of the isomorphism types of simple modules. (See p. 107.) Note, also, that since $_RR$ is cyclic, it does have at least one maximal left ideal (2.8), so there do exist simple modules.

Semisimple Modules

Let $(T_\alpha)_{\alpha \in A}$ be an indexed set of simple submodules of M. If M is the direct sum of this set, then

$$M = \oplus_A T_\alpha$$

is a *semisimple decomposition* of M. A module M is said to be *semisimple* in case it has a semisimple decomposition. Clearly every simple module is semisimple, so for every ring there do exist semisimple modules. As we shall see, semisimple modules need not be plentiful, but since any direct sum of simple modules is semisimple, they are numerous.

If a module M is spanned by simple submodules $(T_\alpha)_{\alpha \in A}$, then the T_α behave very much like the one-dimensional subspaces spanned by a spanning set for a vector space. Evidence of this is given by the fundamental

9.2. Lemma. *Let $(T_\alpha)_{\alpha \in A}$ be an indexed set of simple submodules of the left R-module M. If*

$$M = \Sigma_A T_\alpha,$$

then for each submodule K of M there is a subset $B \subseteq A$ such that $(T_\beta)_{\beta \in B}$ is independent and

$$M = K \oplus (\oplus_B T_\beta).$$

Proof. Let $K \leq M$. By The Maximum Principle there is a subset $B \subseteq A$ maximal with respect to the conditions that $(T_\beta)_{\beta \in B}$ is independent and $K \cap (\Sigma_B T_\beta) = 0$. (See Exercise (6.14).) Then the sum

$$N = K + (\Sigma_B T_\beta) = K \oplus (\oplus_B T_\beta)$$

is direct. We claim $N = M$. For let $\alpha \in A$. Since T_α is simple, either $T_\alpha \cap N = T_\alpha$ or $T_\alpha \cap N = 0$. But $T_\alpha \cap N = 0$ would contradict the maximality of B. Thus $T_\alpha \leq N$ for each $\alpha \in A$, so $M = N$. □

As one consequence of this fundamental lemma we have the following generalization of the fact that in a vector space every spanning set contains a basis.

9.3. Proposition. *If a module M is spanned by an indexed set $(T_\alpha)_{\alpha \in A}$ of simple submodules, then for some $B \subseteq A$*

$$M = \oplus_B T_\beta;$$

that is, M is semisimple.

Proof. In (9.2) let $K = 0$. □

9.4. Proposition. *Let M be a semisimple left R-module with semisimple decomposition $M = \oplus_A T_\alpha$. If*

$$0 \to K \overset{f}{\to} M \overset{g}{\to} N \to 0$$

is an exact sequence of R-modules, then the sequence splits and both K and N are semisimple. Indeed, there is a subset $B \subseteq A$ and isomorphisms

$$N \cong \oplus_B T_\beta \quad \text{and} \quad K \cong \oplus_{A \backslash B} T_\alpha.$$

Proof. Since $\operatorname{Im} f$ is a submodule of M, by (9.2) there is a subset $B \subseteq A$ such that $M = (\operatorname{Im} f) \oplus (\oplus_B T_\beta)$. Thus the sequence splits and $N \cong M/\operatorname{Im} f \cong \oplus_B T_\beta$. But also $M = (\oplus_{A \backslash B} T_\alpha) \oplus (\oplus_B T_\beta)$, so that (see (5.5))

$$K \cong \operatorname{Im} f \cong \oplus_{A \backslash B} T_\alpha. \quad □$$

This is a very significant result. Every submodule and every factor module of a semisimple module are semisimple. Moreover, every submodule is a direct summand. As we shall see below (9.6) this property actually characterizes semisimple modules.

9.5. Corollary. *Let $(T_\alpha)_{\alpha \in A}$ be an indexed set of simple submodules of M. If T is a simple submodule of M such that*

$$T \cap (\Sigma_A T_\alpha) \neq 0,$$

then there is an $\alpha \in A$ such that $T \cong T_\alpha$.

Proof. If T is simple and $T \cap (\Sigma_A T_\alpha) \neq 0$, then $T \leq \Sigma_A T_\alpha$. So clearly we may assume that $M = \Sigma_A T_\alpha$. Then by (9.3) we have that M is semisimple and $M = \oplus_B T_\beta$ for some $B \subseteq A$. Finally, apply (9.4). □

Now we have the following fundamental characterizations of semisimple modules.

9.6. Theorem. *For a left R-module the following statements are equivalent:*
(a) *M is semisimple;*
(b) *M is generated by simple modules;*
(c) *M is the sum of some set of simple submodules;*
(d) *M is the sum of its simple submodules;*
(e) *Every submodule of M is a direct summand;*
(f) *Every short exact sequence*

$$0 \to K \to M \to N \to 0$$

of left R-modules splits.

Proof. The implication (a) $\Rightarrow$ (f) is by (9.4), (f) $\Rightarrow$ (e) is by (5.2), and (b) $\Rightarrow$ (a) is by (9.3). Also (b) $\Leftrightarrow$ (c) $\Leftrightarrow$ (d) are all trivial. Finally,

(e) $\Rightarrow$ (d). Assume that M satisfies (e). We claim that every non-zero submodule of M has a simple submodule. Indeed, let $x \neq 0$ in M. Then (2.8) Rx has a maximal submodule, say H. By (e), we have that $M = H \oplus H'$ for some $H' \leq M$. Thus, by modularity (2.5), $Rx = Rx \cap M = H \oplus (Rx \cap H')$ and $Rx \cap H' \cong Rx/H$ is simple (2.10). So Rx has a simple submodule. Let N be the sum of all simple submodules of M. Then $M = N \oplus N'$ by (e) for some $N' \leq M$. Since $N \cap N' = 0$, N' has no simple submodule. But as we have just seen, this means $N' = 0$. So $N = M$. $\square$

It is clear then that if R is a division ring, then every R-vector space ${}_R M$ is semisimple, for M is generated by its cyclic modules and every non-zero cyclic R-module is simple. Also by (9.6.d) an abelian group M is semisimple iff it is spanned by its elements of prime order. (See Exercise (9.1).)

The Socle

The equivalence (a) $\Leftrightarrow$ (b) in (9.6) says that the class of semisimple left R-modules is precisely the class $Gen(\mathscr{S})$ of modules generated by the simple modules $\mathscr{S}$. Therefore, each module M has a (unique) largest semisimple submodule, the trace of $\mathscr{S}$ in M. This submodule, usually called the *socle* of M (from the French word for "pedestal"), and abbreviated

$$Soc\, M = Tr_M(\mathscr{S}),$$

is of fundamental importance. Clearly, M is semisimple iff $M = Soc\, M$. An important multiple characterization of the socle is

9.7. Proposition. *If M is a left R-module, then*
$$Soc\, M = \Sigma\{K \leq M \mid K \text{ is minimal in } M\}$$
$$= \bigcap\{L \leq M \mid L \text{ is essential in } M\}.$$

Proof. The first equality is trivial. To prove the final equality, let $T \leq M$ be simple. If $L \trianglelefteq M$, then $T \cap L \neq 0$, so $T \leq L$. Thus $Soc\, M$ is contained in every essential submodule of M. On the other hand, set $H = \bigcap\{L \leq M \mid L \trianglelefteq M\}$. We claim that H is semisimple. For let $N \leq H$ and let $N' \leq M$ be a complement of N. (See (5.21).) Then $N + N' = N \oplus N' \trianglelefteq M$. But then $N \leq H \leq N \oplus N'$, and by modularity

$$H = H \cap (N \oplus N') = N \oplus (H \cap N').$$

Thus N is a direct summand of H. Therefore (9.6.e), H is semisimple; so $H \leq Soc\, M$. $\square$

Many properties of the socle are immediate from the fact that $Soc\, M$ is just the trace in M of some class of modules. For example, $Soc(_RR)$ is an ideal of R. (See (8.21).) More generally,

9.8. Proposition. *Let M and N be left R-modules and let $f: M \to N$ be an R-homomorphism. Then*

$$f(Soc\ M) \leq Soc\ N.$$

In particular, $Soc\ M$ is a left R-right $End(_RM)$ submodule of M.

Proof. By (8.16).   □

9.9. Corollary. *Let M be a module and let $K \leq M$. Then*

$$Soc\ K = K \cap Soc\ M.$$

In particular,

$$Soc(Soc\ M) = Soc\ M.$$

Proof. By (9.8) $Soc\ K \leq Soc\ M$. By (9.4) $K \cap Soc\ M$ is semisimple so contained in $Soc\ K$.   □

Now the socle $Soc\ M$ of M is the largest submodule of M that is contained in every essential submodule of M. In general, though, $Soc\ M$ need not be essential in M; in fact, non-zero modules can have zero socles. (See Exercise (9.2).) However, we do have

9.10. Corollary. *Let M be a left R-module. Then $Soc\ M \trianglelefteq M$ if and only if every non-zero submodule of M contains a minimal submodule.*

Proof. This follows from (9.7) and (9.9).   □

As we have noted, the class of simple left R-modules has a set $\mathcal{T}$ of representatives. So from (8.20) we have

9.11. Proposition. *Let $\mathcal{T}$ be a set of representatives of the simple left R-modules. Then for each $_RM$*

$$Soc\ M = Tr_M(\mathcal{T}) = Tr_M(\oplus_{\mathcal{T}}\ T) = \Sigma_{\mathcal{T}}\ Tr_M(T).$$ □

Note that one consequence of (9.11) is that the class of semisimple R-modules has a semisimple generator, namely $\oplus_{\mathcal{T}}\ T$. If T is simple, then the trace $Tr_M(T)$ of T in M is called the *T-homogeneous* component of $Soc\ M$. Of course, $Tr_M(T)$ is generated by a simple module, so it is semisimple and in $Soc\ M$. By (9.5) every simple submodule of $Tr_M(T)$ is isomorphic to T. For example, the $\mathbb{Z}_p$-homogeneous component of the socle of an abelian group M is simply the set of elements of order p. (See Exercise (9.1).)

A semisimple module H is *T-homogeneous* in case

$$H = Tr_H(T).$$

Thus it is clear that for any module M, the T-homogeneous component of $Soc\ M$ is the unique largest T-homogeneous semisimple submodule of M. Of course, if M has no simple submodules isomorphic to T, then the T-homogeneous component of its socle is zero. By (9.11) the homogeneous components of $Soc\ M$ span $Soc\ M$; by (9.5) they are independent (see Exercise (9.8)); and by (8.16) they are stable under endomorphisms of M. Thus we have

9.12. Proposition. *The socle of a left R-module M is, as a left R- right* $End(_RM)$*-bimodule, a direct sum of its homogeneous components.* $\square$

The Radical

The socle of a module M is the largest submodule of M generated by the class $\mathscr{S}$ of simple modules. There is a dual: for every module M there is a unique "largest" factor module of M cogenerated by $\mathscr{S}$. However we focus less on this factor module of M, called the *capital* of M, than on the corresponding reject of $\mathscr{S}$ in M.

Let $\mathscr{S}$ be the class of simple left R-modules. For each left R-module M the (*Jacobson*) *radical* of M is the reject in M of $\mathscr{S}$:

$$Rad\ M = Rej_M(\mathscr{S}).$$

A dual version of (9.7) is now given by the following characterization of the radical.

9.13. Proposition. *Let M be a left R-module. Then*
$$Rad\ M = \bigcap \{K \leq M \mid K \text{ is maximal in } M\}$$
$$= \Sigma\{L \leq M \mid L \text{ is superfluous in } M\}.$$

Proof. Since $K \leq M$ is maximal in M iff M/K is simple, the first equality is immediate from the definition of the reject in M of a class. For the second equality, let $L \ll M$. If K is a maximal submodule of M, and if $L \nleq K$, then $K + L = M$; but then since $L \ll M$, we have $K = M$, a contradiction. We infer that every superfluous submodule of M is contained in $Rad\ M$. On the other hand, let $x \in M$. If $N \leq M$ with $Rx + N = M$, then either $N = M$ or there is a maximal submodule K of M with $N \leq K$ and $x \notin K$. (See Exercise (2.9).) If $x \in Rad\ M$, then the latter cannot occur; thus $x \in Rad\ M$ forces $Rx \ll M$ and the second equality is proved. $\square$

Since the radical of M is simply the reject in M of a class of modules, we infer many properties of $Rad\ M$ from those of rejects. For example (8.23) $Rad(_RR)$ is an ideal of R. More generally, by (8.16) we have

9.14. Proposition. *Let M and N be left R-modules and let* $f: M \to N$ *be an R-homomorphism. Then*

$$f(Rad\ M) \leq Rad\ N.$$

In particular, Rad M is a left R- right $End(_RM)$*-submodule of M.* $\square$

Given a homomorphism $f: M \to N$ we have just seen that $f(Rad\ M) \leq Rad\ N$. Even if f is an epimorphism, we cannot expect $f(Rad\ M)$ to be the radical of N. (See Exercise (9.2).) However, an immediate consequence of (8.17.2) is

9.15. Proposition. *If $f: M \to N$ is an epimorphism and if $Ker f \leq Rad\, M$, then $Rad\, N = f(Rad\, M)$. In particular,*

$$Rad(M/Rad\, M) = 0. \qquad \square$$

Recall that $Soc\, M = M$ iff M is semisimple. The dual statement is

9.16. Proposition. *Let M be a left R-module. Then $Rad\, M = 0$ if and only if M is cogenerated by the class of simple modules. In particular, if M is semisimple, then $Rad\, M = 0$.*

Proof. The first assertion is by (8.13.2) and the second by the fact that a direct sum of simples is contained in a product of simples. $\square$

The dual of the T-homogeneous component of the socle of M is the reject $Rej_M(T)$. And dual to (9.11) we have by (8.20)

9.17. Proposition. *Let $\mathcal{T}$ be a set of representatives of the simple left R-modules. Then for each $_R M$,*

$$Rad\, M = Rej_M(\Pi_{\mathcal{T}}\, T) = Rej_M(\oplus_{\mathcal{T}}\, T) = \cap_{\mathcal{T}}\, Rej_M(T).$$

Now the radical of M is the smallest submodule of M that contains all superfluous submodules. However, the radical need not be superfluous. (See Exercise (9.2).) We do have an important sufficient condition for $Rad\, M \ll M$, but surprisingly the condition is not necessary. (See (9.10) and Exercise (9.4).)

9.18. Proposition. *If every proper submodule of M is contained in a maximal submodule of M, then $Rad\, M$ is the unique largest superfluous submodule of M.*

Proof. Let L be a proper submodule of M and let K be a maximal submodule with $L \leq K$. Then by (9.13) $L + Rad\, M \leq K \neq M$. $\square$

We conclude this section by noting that both the socle and the radical behave well toward direct sums. (But see Exercise (9.12) for the product.) For by (8.18) we have

9.19. Proposition. *If $(M_\alpha)_{\alpha \in A}$ is an indexed set of submodules of M with $M = \oplus_A M_\alpha$, then*

$$Soc\, M = \oplus_A Soc\, M_\alpha \qquad and \qquad Rad\, M = \oplus_A Rad\, M_\alpha. \qquad \square$$

9. Exercises

1. Let $_Z M$ be an abelian group.
   (1) Prove that $Soc\, M$ is the subgroup spanned by the elements of prime order. (See Exercise (8.8).)
   (2) Prove that the $\mathbb{Z}_p$-homogeneous component of $Soc\, M$ is $r_M(p)$.
   (3) Let $n \in \mathbb{N}$. Prove that $\mathbb{Z}_n$ is semisimple iff n is square free (i.e., n is divisible by no square other than 1).
2. Let $_Z M$ be an abelian group. Prove that: (1) If $_Z M$ is torsion-free, then

$Soc\ M = 0$; (2) If $_{\mathbb{Z}}M$ is torsion, then $Soc\ M \trianglelefteq M$; (3) If $_{\mathbb{Z}}M$ is divisible, then $Rad\ M = M$.

3. Compute both the socle and the radical of each of the following left $\mathbb{Z}$-modules:
   (1) $\mathbb{Z}$; (2) $\mathbb{Z}_n$; (3) $\mathbb{R}$; (4) $\mathbb{Z}_{(p)}$; (5) $\mathbb{Z}_{p^\infty}$.

4. Since $_{\mathbb{Z}}\mathbb{Z}$ generates $_{\mathbb{Z}}\mathscr{M}$ (see (8.1)), there exists an epimorphism $f: \mathbb{Z}^{(A)} \to \mathbb{Q}$. Prove that $Ker\ f = K$ is contained in no maximal subgroup of $\mathbb{Z}^{(A)}$. Deduce that for a module M, $Rad\ M \ll M$ does not imply that every proper submodule is contained in a maximal one.

5. Let R be the ring of 2×2 upper triangular matrices over a field.
   (1) Compute the socle and the radical of both $_R R$ and R_R. [Note that each of these four is an ideal of R; also note any similarities or dissimilarities.]
   (2) Show that R has two non-isomorphic simple modules, but that $Soc(_R R)$ has only one non-zero homogeneous component.

6. Let D be a division ring and let M_D be a finite dimensional vector space over D. Let $R = End(M_D)$. Prove that
   (1) $Rad(_R R) = Rad(R_R) = 0$.
   (2) $Soc(_R R) = Soc(R_R) = R$.

7. Let M be a left R-module. Prove that the following are equivalent:
   (a) M is semisimple; (b) For every $K \leq M$ and every R-homomorphism $f: K \to H$ there is an extension $\bar{f}: M \to H$ of f; (c) For every $K \leq M$ and every R-homomorphism $g: H \to M/K$ there is a homomorphism $\bar{g}: H \to M$ with $g = n_K \bar{g}$.

8. Let $(T_\alpha)_{\alpha \in A}$ and $(S_\beta)_{\beta \in B}$ be indexed sets of simple submodules of a module $_R M$. Prove that if $(\Sigma_A T_\alpha) \cap (\Sigma_B S_\beta) \neq 0$, then $T_\alpha \cong S_\beta$ for some $\alpha \in A$ and $\beta \in B$.

9. Let M and N be left R-modules and let $f: M \to N$ be an epimorphism. Show that it is possible for $Rad\ N \nsubseteq f(Rad\ M)$. But prove that if $M/Rad\ M$ is semisimple, then $Rad(N) = f(Rad\ M)$.

10. Let M be a left R-module and let $K \leq M$. Prove that:
    (1) $K = Rad\ M$ iff $K \leq Rad\ M$ and $Rad(M/K) = 0$.
    (2) $K = Soc\ M$ iff $K \geq Soc\ M$ and $Soc\ K = K$.
    (3) If $K \ll M$ and $Rad\ M/K = 0$, then $K = Rad\ M$.
    (4) If $K \trianglelefteq M$ and $Soc\ K = K$, then $K = Soc\ M$.

11. Prove that $Rad\ M = 0$ iff M is a subdirect product of simples. But show that $Rad\ M = 0$ is possible even though M is neither a sum nor product of simple modules.

12. Show that a product $\Pi_A M_\alpha$ of simple modules M_α $(\alpha \in A)$ need not be semisimple. [Hint: Let K be a field and let $R = K^A$. Then $_R R$ is a product of simple modules. Compute the socle.]

13. Let T be a simple left R-module. Assume that $_R R = Soc\ _R R$. Prove that the T-homogeneous component of $_R R$ is a ring direct summand of R, and deduce that as a ring, R is the direct sum of its homogeneous components.

14. A module M is *co-semisimple* in case every submodule of M is the

intersection of maximal submodules. Prove that:

(1) M is co-semisimple iff $Rad(M/K) = 0$ for all $K \leq M$.

(2) Every submodule and every factor module of a co-semisimple module is co-semisimple.

(3) Every semisimple module is co-semisimple.

(4) If R is a Boolean ring, then $_R R$ is co-semisimple. Infer that co-semisimple modules need not be semisimple. [Hint: Let K be a submodule of R. Then R/K is a Boolean ring as well as a factor module of R. Apply Exercise (7.16.3) to get that $_R R$ is co-semisimple. For the final assertion, let $K = \mathbb{Z}_2$ in the Hint for Exercise (9.12).]

§10. Finitely Generated and Finitely Cogenerated Modules—Chain Conditions

As we have noted, the concepts of spanning sets, and finite spanning sets are not categorical and do not have duals. Here, however, we reformulate the concept of finitely spanned both lattice theoretically and categorically, and we do obtain an important dual.

Finitely Generated Modules

A module M is *finitely generated* in case for every set $\mathscr{A}$ of submodules of M that spans M, there is a finite set $\mathscr{F} \subseteq \mathscr{A}$ that spans M; that is,

$$\Sigma \mathscr{A} = M \qquad \text{implies} \qquad \Sigma \mathscr{F} = M$$

for some finite $\mathscr{F} \subseteq \mathscr{A}$. This is nothing really new; it is simply a reformulation of a familiar concept.

10.1. Proposition. *The following statements about a left R-module are equivalent:*

(a) *M is finitely generated;*

(b) *For every set $f_\alpha : U_\alpha \to M$ $(\alpha \in A)$ with $M = \Sigma_A Im f_\alpha$, there is a finite set $F \subseteq A$ with $M = \Sigma_F Im f_\alpha$;*

(c) *For every indexed set $(U_\alpha)_{\alpha \in A}$ and epimorphism $\bigoplus_A U_\alpha \to M \to 0$, there is a finite set $F \subseteq A$ and an epimorphism $\bigoplus_F U_\alpha \to M \to 0$;*

(d) *Every module that generates M finitely generates M;*

(e) *M contains a finite spanning set.*

Proof. The implications (a) $\Rightarrow$ (b) and (c) $\Rightarrow$ (d) are both clear.

(b) $\Rightarrow$ (c). By (6.8) we have that $f : \bigoplus_A U_\alpha \to M$ is an epimorphism iff $\Sigma_A Im f\iota_\alpha = M$. And of course $f\iota_\alpha : U_\alpha \to M$ $(\alpha \in A)$.

(d) $\Rightarrow$ (e). This follows from (8.1).

(e) $\Rightarrow$ (a). Suppose that $\{x_1, \ldots, x_n\}$ is a finite spanning set in M and suppose that $\mathscr{A}$ is a set of submodules of M with $M = \Sigma \mathscr{A}$. Then for each x_i there is a finite subset $\mathscr{F}_i \subseteq \mathscr{A}$ with $x_i \in \Sigma \mathscr{F}_i$. Set $\mathscr{F} = \mathscr{F}_1 \cup \ldots \cup \mathscr{F}_n$.

Then $\mathscr{F}$ is finite, and since $\Sigma\mathscr{F}$ is a submodule of M that contains a spanning set for M, $\Sigma\mathscr{F} = M$. That is, M is finitely generated.   □

Finitely Cogenerated Modules

The definition of a finitely generated module has an obvious if not so familiar dual. A module M is *finitely cogenerated* in case for every set $\mathscr{A}$ of submodules of M

$$\cap\mathscr{A} = 0 \qquad \text{implies} \qquad \cap\mathscr{F} = 0$$

for some finite $\mathscr{F} \subseteq \mathscr{A}$.

For example, the abelian group $\mathbb{Z}$ is finitely generated but not finitely cogenerated. The group $\mathbb{Z}_{p^\infty}$ is finitely cogenerated but not finitely generated.

Only four of the conditions of Proposition 10.1 have duals and surprisingly only three of these are equivalent. Although we state these equivalences now, we have no immediate need of one of the implications and the proof of this one will be postponed until §18 (see (18.17)).

10.2. Proposition. *The following statements about a left R-module M are equivalent:*

(a) *M is finitely cogenerated;*

(b) *For every set $f_\alpha : M \to U_\alpha$ $(\alpha \in A)$ with $\cap_A \operatorname{Ker} f_\alpha = 0$, there is a finite set $F \subseteq A$ with $\cap_F \operatorname{Ker} f_\alpha = 0$;*

(c) *For every indexed set $(U_\alpha)_{\alpha \in A}$ and monomorphism $0 \to M \to \Pi_A U_\alpha$, there is a finite set $F \subseteq A$ and a monomorphism $0 \to M \to \Pi_F U_\alpha$.*

Proof. (a) ⇒ (b). This is clear.

(b) ⇒ (a). Let $\{M_\alpha \mid \alpha \in A\}$ be submodules of M with $\cap_A M_\alpha = 0$. Apply (b) to the natural maps $f_\alpha : M \to M/M_\alpha$ $(\alpha \in A)$ to get (a).

(b) ⇒ (c). Suppose $f : M \to \Pi_A U_\alpha$ is a monomorphism. Then by (6.2), $\cap_A \operatorname{Ker} \pi_\alpha f = 0$. So by (b) there is a finite set $F \subseteq A$ with $\cap_F \operatorname{Ker} \pi_\alpha f = 0$. So again by (6.2) $\pi_F f : M \to \Pi_F U_\alpha$ is a monomorphism.

(c) ⇒ (b). This implication is proved in (18.17). However, see Exercise (10.4).   □

10.3. Corollary. *If M is finitely cogenerated, then every module that cogenerates M finitely cogenerates M.*

Proof. By the implication (b) ⇒ (c) of (10.2).   □

The property of finitely cogenerated modules stated in (10.3) is the dual of (10.1.d); however, it does not characterize finitely cogenerated modules. For example, the abelian group $\oplus_p \mathbb{Z}_p$ is not finitely cogenerated yet every group that cogenerates it finitely cogenerates it. (See Exercise (10.2).) This fact does not flaw the Principle of Duality that the dual of a theorem is a theorem. The implication (d) ⇒ (a) in (10.1) is simply not a theorem in the category $_R\mathsf{M}$, for to obtain it requires some version of the non-categorical statement (10.1.e). One version of (10.1.e) is that $_R R$ is a finitely generated generator in $_R M$. And sure enough whenever $_R M$ has a finitely cogenerated cogenerator, then the converse of (10.3) is true in $_R M$. (See Exercise (10.3).)

The Roles of the Radical and the Socle

Next we state fundamental characterizations of finitely generated and finitely cogenerated modules. They show that "finitely generated" and "finitely cogenerated" are determined by the radical and the socle, respectively.

10.4. Theorem. *Let M be a left R-module. Then*
(1) M is finitely generated if and only if $M/\mathrm{Rad}\, M$ is finitely generated and the natural epimorphism

$$M \to M/\mathrm{Rad}\, M \to 0$$

is superfluous (i.e., $\mathrm{Rad}\, M \ll M$);
(2) M is finitely cogenerated if and only if $\mathrm{Soc}\, M$ is finitely cogenerated and the inclusion map

$$0 \to \mathrm{Soc}\, M \to M$$

is essential (i.e., $\mathrm{Soc}\, M \trianglelefteq M$).

Proof. We shall prove (2). The proof of (1) is dual.

($\Rightarrow$). Clearly a submodule of a finitely cogenerated module is finitely cogenerated. So it will suffice to show that if M is finitely cogenerated, then $\mathrm{Soc}\, M \trianglelefteq M$. But suppose $K \leq M$ with $(\mathrm{Soc}\, M) \cap K = 0$. Now $\mathrm{Soc}\, M$ is the intersection of all essential submodules of M (see (9.7)), so since M is finitely cogenerated, there exist essential submodules $L_1, \ldots, L_n$ of M with $L_1 \cap \ldots \cap L_n \cap K = 0$. But $(L_1 \cap \ldots \cap L_n) \trianglelefteq M$ (see (5.16.2)), whence $K = 0$.

($\Leftarrow$). Let $\mathrm{Soc}\, M$ be finitely cogenerated and essential in M. Let $\mathscr{A}$ be any set of submodules of M with $\bigcap \mathscr{A} = 0$. Then $\bigcap \{(A \cap \mathrm{Soc}\, M) \mid A \in \mathscr{A}\} = 0$. This forces

$$(A_1 \cap \ldots \cap A_n) \cap (\mathrm{Soc}\, M) = (A_1 \cap \mathrm{Soc}\, M) \cap \ldots \cap (A_n \cap \mathrm{Soc}\, M) = 0$$

for some $A_1, \ldots, A_n \in \mathscr{A}$. But $\mathrm{Soc}\, M \trianglelefteq M$ whence $A_1 \cap \ldots \cap A_n = 0$.   □

10.5. Corollary. *Let M be a non-zero module.*
(1) If M is finitely generated, then M has a maximal submodule;
(2) If M is finitely cogenerated, then M has a minimal submodule.   □

For semisimple modules the two concepts are equivalent and we have

10.6. Proposition. *The following statements about a semisimple module M are equivalent:*
(a) M is finitely cogenerated;
(b) $M = T_1 \oplus \ldots \oplus T_n$ with T_i simple $(i = 1, \ldots, n)$;
(c) M is finitely generated.

Proof. (a) $\Rightarrow$ (b). Assume (a). Then since M clearly can be embedded in a product of simple modules and since (10.2.c) holds, M can be embedded in a product of finitely many simples. Now apply (9.4).

(b) $\Rightarrow$ (c). Assume (b). Then clearly M has a finite spanning set. Apply (10.1).

(c) $\Rightarrow$ (a). Assume (c). Since M is semisimple, it is spanned by simple submodules. So by (c) it is spanned by a finite set $T_1, \ldots, T_n$ of simple submodules. We shall prove (a) by induction on n. Certainly if $n = 1$, then M is simple, and finitely cogenerated. Assume inductively that $n > 1$ and that any module spanned by fewer than n simple modules is finitely cogenerated. Now suppose $\mathscr{A}$ is a set of submodules of M with $\cap \mathscr{A} = 0$. Then $T_n \cap L = 0$ for some $L \in \mathscr{A}$. By (9.4), $L = S_1 \oplus \ldots \oplus S_m$ with each S_i simple and $m < n$. Set $\mathscr{A}' = \{N \cap L \mid N \in \mathscr{A}\}$, so $\mathscr{A}'$ is a set of submodules of L with $\cap \mathscr{A}' = 0$. So for some finite set $\{N_1, \ldots, N_k\} \subseteq \mathscr{A}$,

$$L \cap N_1 \cap \ldots \cap N_k = 0,$$

and M is finitely cogenerated. $\qquad\square$

This last Proposition and (10.4.2) combine to establish the following characterization of finitely cogenerated modules.

10.7. Proposition. *A module is finitely cogenerated if and only if its socle is essential and finitely generated.* $\qquad\square$

It is clear from the definitions that if M is finitely generated (finitely cogenerated), then so is every factor module (submodule) of M. Thus we have at once the necessity of the conditions in the next result.

10.8. Proposition. *Let $M = M_1 \oplus \ldots \oplus M_n$. Then M is finitely generated (finitely cogenerated) if and only if each M_i ($i = 1, \ldots, n$) is finitely generated (finitely cogenerated).*

Proof. Since the union of spanning sets for the M_i ($i = 1, \ldots, n$) is a spanning set for M, the finitely generated case is settled by (10.1). So it suffices to show that if M_i ($i = 1, \ldots, n$) is finitely cogenerated, then so is M. But we know (9.19)

$$Soc\ M = (Soc\ M_1) \oplus \ldots \oplus (Soc\ M_n).$$

Since each M_i is finitely cogenerated, each $Soc\ M_i$ is finitely generated by (10.7). Thus $Soc\ M$ is finitely generated by the other part of this Proposition. Also by (10.7) each $Soc\ M_i \trianglelefteq M_i$, whence by (6.17) $Soc\ M \trianglelefteq M$. Finally another application of (10.7) gives that M is finitely cogenerated. $\qquad\square$

The Chain Conditions

Modules for which every submodule (every factor module) is finitely generated (finitely cogenerated) can be characterized in terms of certain "chain conditions". In general, neither of these finiteness conditions implies the other although in some very special settings they may be equivalent. Note for example, that the submodules of $\mathbb{Z}$ are finitely generated and the factor modules of $\mathbb{Z}_{p^\infty}$ are finitely cogenerated. A set $\mathscr{L}$ of submodules of M satisfies the *ascending* chain condition in case for every chain

$$L_1 \leq L_2 \leq \ldots \leq L_n \leq \ldots$$

in $\mathscr{L}$, there is an n with $L_{n+i} = L_n$ $(i = 1, 2, \ldots)$. Turn the inequalities around for the *descending chain condition*. (See Exercise (10.9).)

A module M is *noetherian* in case the lattice $\mathscr{S}(M)$ of all submodules of M satisfies the ascending chain condition. It is *artinian* in case $\mathscr{S}(M)$ satisfies the descending chain condition.

10.9. Proposition. *For a module M the following statements are equivalent:*
(a) *M is noetherian;*
(b) *Every submodule of M is finitely generated;*
(c) *Every non-empty set of submodules of M has a maximal element.* $\square$

The proof of this proposition is dual to that of the next proposition and therefore it will be omitted.

10.10. Proposition. *For a module M the following statements are equivalent.*
(a) *M is artinian;*
(b) *Every factor module of M is finitely cogenerated;*
(c) *Every non-empty set of submodules of M has a minimal element.*

Proof. (a) $\Rightarrow$ (c). Let $\mathscr{A}$ be a non-empty set of submodules of M and suppose that $\mathscr{A}$ does not have a minimal element. Then for each $L \in \mathscr{A}$ the set $\{L' \in \mathscr{A} \mid L' < L\}$ is not empty. Thus, by the Axiom of Choice (0.2), there is a function $L \mapsto L'$ with $L > L'$ for each $L \in \mathscr{A}$. Let $L \in \mathscr{A}$. Then

$$L > L' > L'' > \ldots$$

is an infinite descending chain of submodules of M.

(c) $\Rightarrow$ (b). Assume (c). Then by (2.9) it will suffice to show that if $K \leq M$ and if $\mathscr{A}$ is a collection of submodules of M with $K' = \cap \mathscr{A}$, then $K = \cap \mathscr{F}$ for some finite subset $\mathscr{F} \subseteq \mathscr{A}$. But set $\mathscr{P} = \{\cap \mathscr{F} \mid \mathscr{F} \subseteq \mathscr{A} \text{ is finite}\}$. Then by (c), $\mathscr{P}$ has a minimal element, $\cap \mathscr{F}$. Clearly, $K = \cap \mathscr{F}$.

(b) $\Rightarrow$ (a). Assume (b) and suppose that M has a descending chain

$$L_1 \geq L_2 \geq \ldots \geq L_n \geq \ldots$$

of submodules. Set $K = \cap_{\mathbb{N}} L_n$. Then since M/K is finitely cogenerated, there must be some n with $K = L_n$ whence $L_{n+1} = L_n$ $(i = 1, 2, \ldots)$. $\square$

10.11. Corollary. *Let M be a non-zero module.*
(1) *If M is artinian, then M has a simple submodule; in fact, Soc M is an essential submodule;*
(2) *If M is noetherian, then M has a maximal submodule; in fact, Rad M is a superfluous submodule.* $\square$

10.12. Proposition. *Let*

$$0 \to K \to M \to N \to 0$$

be an exact sequence of left R-modules. Then M is artinian (noetherian) if and only if both K and N are artinian (noetherian).

Proof. Let M be artinian. Then since K is isomorphic to a submodule of M, K is artinian by the definition. Also every factor module of N is isomorphic to a factor module of M (3.7), so by (10.10) N is artinian.

Conversely, suppose K and N are both artinian; we claim that M is artinian. Clearly we may assume that $K \le M$ and that $M/K = N$. Now suppose that

$$L_1 \ge L_2 \ge \dots \ge L_n \ge \dots$$

is a descending chain of submodules of M. Since $M/K \cong N$ is artinian, there is an integer m such that

$$L_m + K = L_{m+i} + K \qquad (i = 1, 2, \dots).$$

Since K is artinian, there is an integer $n \ge m$ such that

$$L_n \cap K = L_{n+i} \cap K \qquad (i = 1, 2, \dots).$$

Thus using modularity and the fact that $L_n \ge L_{n+i}$, we have for each $i = 1, 2, \dots,$

$$L_n = L_n \cap (L_n + K) = L_n \cap (L_{n+i} + K)$$
$$= L_{n+i} + (L_n \cap K) = L_{n+i} + (L_{n+i} \cap K) = L_{n+i}.$$

Therefore M is artinian. The proof of the noetherian case is dual. $\square$

10.13. Corollary. *Let $M = M_1 \oplus \dots \oplus M_n$. Then M is artinian (noetherian) if and only if each $M_i (i = 1, \dots, n)$ is artinian (noetherian).* $\square$

One of the most significant properties of artinian and noetherian modules is that each such module admits a finite indecomposable direct decomposition. Note, however, that modules that are just finitely generated need not have such a decomposition; for example, if R is a product of infinitely many copies of a field, then $_R R$ is cyclic but has no indecomposable decomposition (also see Exercise (7.8)).

10.14. Proposition. *Let M be a non-zero module that has either the ascending or the descending chain condition on direct summands (e.g., if M is artinian or noetherian). Then M is the direct sum*

$$M = M_1 \oplus \dots \oplus M_n$$

of a finite set of indecomposable submodules.

Proof. For each non-zero module M that does not have a finite indecomposable decomposition choose a proper decomposition

$$M = N' \oplus M'$$

such that M' has no finite indecomposable decomposition. Suppose M is non-zero and is not a finite direct sum of indecomposable modules. Then

$$M = N' \oplus M', \qquad M' = N'' \oplus M'', \dots$$

is a sequence of proper decompositions. So there exist infinite chains

$$N' < N' \oplus N'' < \dots \qquad \text{and} \qquad M > M' > M'' > \dots$$

of direct summands of M.   □

The four finiteness conditions are equivalent for semisimple modules.

10.15. Proposition. *For each module M the following statements are equivalent:*
  (a) *$Rad\, M = 0$ and M is artinian;*
  (b) *$Rad\, M = 0$ and M is finitely cogenerated;*
  (c) *M is semisimple and finitely generated;*
  (d) *M is semisimple and noetherian;*
  (e) *M is the direct sum of a finite set of simple submodules.*

Proof. The implications (a) $\Rightarrow$ (b) and (d) $\Rightarrow$ (c) are immediate from Propositions (10.10) and (10.9), respectively.
  (b) $\Rightarrow$ (e). Assume (b). Then by (9.16) and (10.2.c) M is isomorphic to a submodule of a finite product P of simple modules. Since such a product is necessarily a direct sum (6.12), P is semisimple. Now apply (9.4).
  (c) $\Leftrightarrow$ (e). By Proposition 10.6.
  (e) $\Rightarrow$ (a) and (e) $\Rightarrow$ (d). Assume (e). Then M is semisimple, and by (9.16) we have $Rad\, M = 0$. Clearly a simple module is both artinian and noetherian. Now apply (10.13).   □

10.16. Corollary. *For a semisimple module M the following statements are equivalent:*
  (a) *M is artinian;*
  (b) *M is noetherian;*
  (c) *M is finitely generated;*
  (d) *M is finitely cogenerated.*   □

Chain Conditions for Rings

A ring R is *left artinian* (*right artinian*) in case the left (right) regular module $_R R$ (R_R) is an artinian module. The ring is *artinian* in case it is both left artinian and right artinian; i.e., in case $_R R$ and R_R are both artinian modules. The concepts *left noetherian, right noetherian* or simply *noetherian* for a ring are similarly defined in terms of the regular modules $_R R$ and R_R.

It is easy to see that the ring R of all 2×2 upper triangular matrices

$$\begin{bmatrix} a & b \\ 0 & \gamma \end{bmatrix}$$

with $a, b \in \mathbb{R}$ and $\gamma \in \mathbb{Q}$ is both left artinian and left noetherian, but it is neither right artinian nor right noetherian. Of course, $\mathbb{Z}$ is a noetherian ring that is not artinian. However, in §15 we shall prove the remarkable fact that every (left) artinian ring is (left) noetherian.

10.17. Proposition. *If R is either left or right artinian or noetherian, then R has a block decomposition*

$$R = R_1 \dotplus \ldots \dotplus R_n$$

as a ring direct sum of indecomposable rings.

Proof. By (10.14) and (7.5), R has a complete set of pairwise orthogonal primitive idempotents. Now apply (7.9).   □

Observe that if R is left or right artinian, then certainly $_R R_R$ is artinian; i.e., the ring R has the descending chain condition on (two-sided) ideals. On the other hand, a ring R can have the descending chain condition on ideals yet be neither left nor right artinian; indeed (see Exercise (10.14)) there are simple rings that are not artinian.

10.18. Proposition. *For each ring R the following statements are equivalent:*
(a) *R is left artinian;*
(b) *R has a generator $_R G$ that is artinian;*
(c) *Every finitely generated left R-module is artinian;*
(d) *Every finitely generated left R-module is finitely cogenerated.*

Proof. (a) $\Rightarrow$ (b). This follows since $_R R$ is a generator (8.8).
(b) $\Rightarrow$ (c). Assuming (b), we have that for each finite set F, $G^{(F)}$ is artinian by (10.13). But if M is finitely generated, then by (10.1.d), M is isomorphic to a factor of $G^{(F)}$ for some finite set F. Now apply (10.12) to deduce that M is artinian.
(c) $\Rightarrow$ (d). Immediate from (10.10).
(d) $\Rightarrow$ (a). Assume (d). Since $_R R$ is finitely generated, so is every factor module of $_R R$. So by (d) every factor module of $_R R$ is finitely cogenerated. Now apply (10.10).   □

The proof of the following result, similar to that of (10.18), will be omitted.

10.19. Proposition. *For each ring R the following statements are equivalent:*
(a) *R is left noetherian;*
(b) *R has a generator $_R G$ that is noetherian;*
(c) *Every finitely generated left R-module is noetherian;*
(d) *Every submodule of every finitely generated left R-module is finitely generated.*   □

10. Exercises

1. (1) Prove that if $_R M$ is finitely generated (finitely cogenerated), then so is every factor module (submodule) of M.

   (2) Give an example of a finitely generated module (in fact, a cyclic one) with submodules that are not finitely generated.

2. Prove that every $\mathbb{Z}$-module that cogenerates $M = \oplus_p \mathbb{Z}_p$ finitely cogenerates M, but that M is not finitely cogenerated.

3. (1) Let R be a ring that has a finitely cogenerated cogenerator $_R C$. Prove that for $_R M$ the following are equivalent:

   (a) M is finitely cogenerated; (b) Every module that cogenerates M finitely cogenerates M; (c) There is an $n \in \mathbb{N}$ and a monomorphism $M \to C^{(n)}$.

   (2) Generalize Exercise (10.2) by proving that if R is a ring with an infinite set $(T_n)_{n \in \mathbb{N}}$ of pairwise non-isomorphic simple modules, then $M = \oplus_{\mathbb{N}} T_n$ satisfies (1.b) but is not finitely cogenerated.

4. A slight variation of the condition of Corollary (10.3) does characterize finitely cogenerated modules. Prove that $_R M$ is finitely cogenerated iff for every module U and every set A, if there is a monomorphism $f : M \to U^A$, then there is a finite subset $F \subseteq A$ such that $\pi_F \circ f : M \to U^F$ is a monomorphism. [Hint: ($\Leftarrow$) Suppose $M_\alpha \leq M$ and $\cap_A M_\alpha = 0$. Set $U = \Pi_A M/M_\alpha$ and consider some monomorphism $M \to U^A$.]

5. Prove that $_R M$ is finitely generated iff for every chain $\mathscr{C}$ of proper submodules of M, its union $\bigcup \mathscr{C}$ is also a proper submodule. [Hint: Assume that the submodules of M satisfy the condition. Consider the set $\mathscr{D} = \{K \leq M \mid M/K \text{ is not finitely generated}\}$. If $\mathscr{D} \neq \varnothing$, then the condition on chains of submodules implies that $\mathscr{D}$ has a maximal element, say N (why?). But if $x \in M \backslash N$, and $N \in \mathscr{D}$, then $N + Rx$ is also in $\mathscr{D}$, a contradiction. So since $M \notin \mathscr{D}$, we must conclude that $\mathscr{D} = \varnothing$.]

6. Prove that $_R M$ is finitely cogenerated iff for every chain $\mathscr{C}$ of non-zero submodules of M its intersection $\cap \mathscr{C}$ is not zero. [Hint: Suppose M is not finitely cogenerated. Then there is a set $\mathscr{A}$ of submodules maximal with respect to $\cap \mathscr{A} = 0$ and $\cap \mathscr{F} \neq 0$ for all finite $\mathscr{F} \subseteq \mathscr{A}$. Let $\mathscr{C}$ be a maximal chain in $\mathscr{A}$. If $\cap \mathscr{C} \neq 0$, then since $\mathscr{A}$ is closed under finite intersections, $\cap \mathscr{C} \in \mathscr{A}$.]

7. Let $\phi : Q \to R$ be a ring homomorphism and let M be a left R-module. Then via ϕ, M is a left Q-module (see Exercise (4.15)). Prove that if $_Q M$ is artinian or noetherian, then so is $_R M$. Deduce that if R is a finite dimensional algebra (via ϕ) over a field Q, then the following are equivalent: (a) $_R M$ is artinian and noetherian; (b) $_R M$ is finitely generated; (c) $_Q M$ is finite dimensional.

8. Let M_R be a non-zero homogeneous semisimple module (e.g., a vector space) and let $S = End(M_R)$. Prove that

   (1) The set $U = \{\gamma \in S \mid Im \gamma \text{ is finitely generated}\}$ is the unique smallest non-zero ideal of S. [Hint: If T_1, T_2 are simple submodules of M, then by

the homogeneity of M and by Exercise (9.7) there exist $e_i = e_i^2 \in S$ and $f \in S$ with $e_i M = T_i$ and $(e_2 f e_1 \mid T_1): T_1 \to T_2$ an isomorphism.]

(2) $Soc(_S S) = Soc(S_S) = U$ and $Rad(_S S) = Rad(S_S) = 0$.

9. A poset $(P, \leq)$ satisfies the *ascending (descending) chain condition* in case there is no infinite properly ascending (descending) chain $a_1 < a_2 < a_3 < \ldots$ ($a_1 > a_2 > a_3 > \ldots$) in P.

(1) Prove that a poset P satisfies the ascending (descending) chain condition iff it satisfies the maximum (minimum) condition (i.e., every non-empty subset of P contains a maximal (minimal) element).

(2) Apply (1) to obtain another proof of (10.14). [Hint: If M has the a.c.c. on direct summands, let $\mathscr{P}$ be the set of direct summands of M having finite indecomposable decompositions. Let $N \in \mathscr{P}$ be maximal and suppose $M = N \oplus N'$ with $N' \neq 0$. The set $\mathscr{P}'$ of proper direct summands of N' has a maximal element N'' and $N' = N'' \oplus N'''$ for some $N''' \neq 0$. Consider $N \oplus N'''$.]

10. (1) A lattice L with greatest element u has the *finite join property* (abbreviated FJP) in case each subset $\mathscr{A}$ with join u has a finite subset $\mathscr{F}$ with join u. Prove that L has the ascending chain condition iff for each $a \in L$ the sublattice

$$a^- = \{x \in L \mid x \leq a\}$$

has the FJP.

(2) A lattice L with least element 0 has the *finite meet property* (FMP) in case its dual has the FJP. Prove that L has the descending chain condition iff for each $a \in L$ the sublattice

$$a^+ = \{x \in L \mid x \geq a\}$$

has the FMP. [Hint: This should follow from (1)!]

11. Prove that the following statements about a non-zero module $_R M$ are equivalent:

(a) The set of direct summands of M has the ascending chain condition;
(b) The set of direct summands of M has the descending chain condition;
(c) $End(_R M)$ has no infinite orthogonal set of non-zero idempotents.

[Hint: For (a) $\Leftrightarrow$ (b) consider Exercise (10.9). For (c) $\Rightarrow$ (b) suppose $M = L_0 \geq L_1 \geq L_2 \geq \ldots$ is a chain of direct summands. Then for each n, $M = K_1 \oplus \ldots \oplus K_n \oplus L_n$ with $K_n \oplus L_n = L_{n-1}$. Let e_n be the idempotent of K_n in this decomposition.]

12. Over a field Q let R be the set of all $\mathbb{N}$-square row finite matrices $A = \llbracket \alpha_{mn} \rrbracket$ such that $\alpha_{mm} = \alpha_{nn}$ for all m and n (i.e., with constant diagonal) and $\alpha_{mn} = 0$ if $m \neq 1$ and $m \neq n$ (i.e., only the first row can be non-zero off the diagonal).

(1) Prove that R is a subalgebra of the Q-algebra $\mathbb{RFM}_{\mathbb{N}}(Q)$ of $\mathbb{N}$-square row finite matrices over Q. [Note that R is isomorphic to the ring $Q[X_1, X_2, \ldots]$ of polynomials in "$\mathbb{N}$ indeterminants" modulo the ideal generated by all $X_i X_j$ $(i, j = 1, 2, \ldots)$.]

(2) Prove that R is a commutative local ring (i.e., R has a unique maximal ideal).

(3) Prove that $_R R$ is finitely generated but not noetherian. [Hint: If J is the unique maximal ideal, then $_R J$ is not noetherian.]

(4) Let M be the left R-module $Hom_Q(R_R, Q)$. (See (4.4).) Prove that M is finitely cogenerated but not artinian. [Hint: The set K of all $f \in M$ with $J \leq Ker f$ is the unique minimal submodule of $_R M$ and $K \trianglelefteq M$.]

13. Let R be the ring of all 2×2 upper triangular matrices

$$\begin{bmatrix} a & b \\ 0 & \gamma \end{bmatrix}$$

with $a, b \in \mathbb{R}$ and $\gamma \in \mathbb{Q}$. Prove that R is left artinian and left noetherian but neither right noetherian nor right artinian.

14. Let Q be a field and let $n \in \mathbb{N}$. For each $A \in \mathbb{M}_n(Q)$ let $D(A)$ be the $\mathbb{N} \times \mathbb{N}$ matrix over Q given in block form:

$$D(A) = \begin{bmatrix} A & & & & \\ & A & & 0 & \\ & & A & & \\ 0 & & & \cdot & \\ & & & & \cdot \\ & & & & & \cdot \end{bmatrix}$$

Let R be the set of all $D(A)$ for $A \in \mathbb{M}_n(Q)$ and $n \in \mathbb{N}$.

(1) Prove that R is a simple subring of the ring $\mathbb{CFM}_\mathbb{N}(Q)$ of column finite matrices over Q.

(2) Prove that $_R R$ satisfies neither the ascending nor the descending chain conditions for direct summands. [Hint: In the descending case, consider the idempotents E_n with (i, j) entry $\delta_{i\,2n}\delta_{j\,2n}$.]

(3) Deduce that R is a simple Q-algebra that is not finite dimensional and not a division ring.

15. Prove that every finitely cogenerated module has a finite indecomposable decomposition.

16. (1) Prove that for a Boolean ring R, the following are equivalent: (a) R is artinian; (b) R is noetherian; (c) R is finite; (d) R is semisimple. [Hint: If R is not finite, then for each $0 \neq a \in R$, one of the rings Ra or $R(1 - a)$ is not finite.]

(2) Prove that if $_R M$ is an artinian or noetherian module over a Boolean ring R, then M is semisimple.

§11. Modules with Composition Series

Suppose that M is a non-zero module with the property that every non-zero submodule of M has a maximal submodule. For example, by (10.11) and

(10.12) we have that every non-zero noetherian module has this property. In any event, given such a module M it has a maximal submodule M_1, and either $M_1 = 0$ or in turn it has a maximal submodule M_2. Then clearly every such process leads to an infinite descending chain

$$M > M_1 > M_2 > \dots$$

of submodules, each maximal in its predecessor, or there is finite chain

$$M > M_1 > M_2 > \dots > M_n = 0$$

with each term maximal in its predecessor. Observe that if in addition M is artinian, then only the latter option can occur.

Similarly, if M is a non-zero module with the property that every non-zero factor module has a simple submodule (e.g., if M is artinian), then there is an ascending chain

$$0 < L_1 < L_2 < \dots$$

of submodules of M each maximal in its successor. Again, if M is noetherian, the chain terminates at M after finitely many terms; i.e., $L_n = M$ for some n.

From the existence of such chains of submodules it is possible to prove a substantial number of the familiar arithmetic properties of dimension for vector spaces.

Composition Series

Let M be a non-zero module. A finite chain of $n + 1$ submodules of M

$$M = M_0 > M_1 > \dots > M_n = 0$$

is called a *composition series* of *length n* for M provided that M_{i-1}/M_i is simple $(i = 1, 2, \dots, n)$; i.e., provided each term in the chain is maximal in its predecessor. We have just noted that if a module is both artinian and noetherian, then it has such a series. Indeed those are the only modules with composition series.

11.1. Proposition. *A non-zero module M has a composition series if and only if M is both artinian and noetherian.*

Proof. In view of the above remarks it suffices to prove the necessity of the condition. So suppose that M has a composition series; we shall induct on the minimum length, say n, of all such series. Certainly if $n = 1$, then M is simple and we are done. Otherwise, if

$$M = M_0 > M_1 > \dots > M_n = 0$$

is a composition series of minimal length for M, then M_1 has a composition series of length $n - 1$ and M/M_1 is simple. Now apply (10.12).   □

11.2. Corollary. *Let K, M, and N be non-zero modules and suppose there is* *an exact sequence*

$$0 \to K \to M \to N \to 0$$

of homomorphisms. Then M has a composition series if and only if K and N both have composition series.

Proof. This is immediate from (10.12) and (11.1).   □

We shall return to this corollary later in the section and obtain a sharpened form of one direction of it, a form that is the basis for some of the arithmetic properties of such modules.

Now let M be an arbitrary module and let $L \leq M$. Then whether or not L is a term in a composition series for M, if L has a maximal submodule K, the simple module L/K is called a *composition factor* of M. Moreover, if M has a composition series

$$M = M_0 > M_1 > \ldots > M_n = 0,$$

then the simple modules

$$M_0/M_1, M_1/M_2, \ldots, M_{n-1}/M_n$$

are called the *composition factors* of the series. If M has a second composition series

$$M = N_0 > N_1 > \ldots > N_p = 0$$

then the two series are *equivalent* in case $n = p$ and there is a permutation σ of $\{1, 2, \ldots, n\}$ such that

$$M_i/M_{i+1} \cong N_{\sigma(i)}/N_{\sigma(i)+1} \qquad (i = 1, 2, \ldots, n).$$

Observe that equivalence simply means that for each simple R-module T the number of isomorphic copies of T in the sequence of composition factors for the one composition series equals the number of isomorphic copies of T in the other.

11.3. The Jordan–Hölder Theorem. *If a module M has a composition series, then every pair of composition series for M are equivalent.*

Proof. If M has a composition series, then denote by $c(M)$ the minimum length of such a series for M. We shall induct on $c(M)$. Clearly, if $c(M) = 1$, there is no challenge. So assume that $c(M) = n > 1$ and that any module with a composition series of smaller length has all of its composition series equivalent. Let

(1) $$M = M_0 > M_1 > \ldots > M_n = 0$$

be a composition series of minimal length for M and let

(2) $$M = N_0 > N_1 > \ldots > N_p = 0$$

be a second composition series for M. If $M_1 = N_1$, then by the induction hypothesis, since $c(M_1) \leq n - 1$, the two series are equivalent. So we may assume that $M_1 \neq N_1$. Then since M_1 is a maximal submodule of M, we have $M_1 + N_1 = M$, so by (3.7.3).

(3) $$M/M_1 = (M_1 + N_1)/M_1 \cong N_1/(M_1 \cap N_1),$$

and

(4) $$M/N_1 = (M_1 + N_1)/N_1 \cong M_1/(M_1 \cap N_1).$$

Thus $M_1 \cap N_1$ is maximal in both M_1 and N_1. Now by (11.2), $M_1 \cap N_1$ has a composition series

$$M_1 \cap N_1 = L_0 > L_1 > \ldots > L_k = 0.$$

So

$$M_1 > L_0 > \ldots > L_k = 0$$

and

$$N_1 > L_0 > \ldots > L_k = 0$$

are composition series for M_1 and N_1. Since $c(M_1) < n$, every two composition series for M_1 are equivalent, so the two series

$$M = M_0 > M_1 > M_2 > \ldots > M_n = 0$$

and

$$M = M_0 > M_1 > L_0 > \ldots > L_k = 0$$

are equivalent. In particular, $k < n - 1$, so clearly $c(N_1) < n$. Thus by our induction hypothesis, every two composition series for N_1 are equivalent. Thus the two series

$$M = N_0 > N_1 > N_2 > \ldots > N_p = 0$$

and

$$M = N_0 > N_1 > L_0 > \ldots > L_k = 0$$

are equivalent. But as we noted in (3) and (4)

$$M/M_1 \cong N_1/L_0 \quad \text{and} \quad M/N_1 \cong M_1/L_0;$$

thus the series (1) and (2) are equivalent, and we are done.   □

Composition Length

It is an immediate consequence of the Jordan-Hölder Theorem that for any module having a composition series, all composition series for that module have the same length. A module M that is both artinian and noetherian is said to be of *finite length*; as we have just noted, for such a module M we can define its *(composition) length* $c(M)$ unambiguously by

$$c(M) = \begin{cases} 0 & \text{if } M = 0 \\ n & \text{if } M \text{ has a composition series of length } n. \end{cases}$$

If a module M is not of finite length, we say it is of *infinite length* and write

$$c(M) = \infty.$$

A finite dimensional vector space clearly has a composition length and this length is simply the dimension of the space. Indeed the function c of composition length behaves on modules of finite length very much like the dimension function behaves on finite dimensional vector spaces.

Now we prove the promised revision of one half of Corollary (11.2). Let K, M, and N be modules and let

$$0 \to K \xrightarrow{f} M \xrightarrow{g} N \to 0$$

be an exact sequence. Suppose further that

$$K = K_0 > K_1 > ... > K_n = 0$$

and

$$N = N_0 > N_1 > ... > N_p = 0$$

are composition series for K and N, respectively. For each $i = 0, 1, ..., n$, let $K_i' = f(K_i)$ and for each $j = 0, 1, ..., p$, let $N_j' = g^{\leftarrow}(N_j)$. Then by (3.8) the series

$$M = N_0' > N_1' > ... > N_p' = K_0' > K_1' > ... > K_n' = 0$$

is a composition series for M. Thus, in view of the uniqueness of length of such composition series, we have

11.4. Corollary. *Let K, M, and N be modules and suppose there is an exact sequence*

$$0 \to K \to M \to N \to 0$$

of homomorphisms. Then

$$c(M) = c(N) + c(K). \qquad \square$$

From this Corollary we deduce easily the following fundamental result:

11.5. Corollary. [The Dimension Theorem.] *Let M be a module of finite length and let K and N be submodules of M. Then*

$$c(K + N) + c(K \cap N) = c(K) + c(N).$$

Proof. By (3.7), $(K + N)/N \cong K/(K \cap N)$. Then apply (11.4) to the two exact sequences

$$0 \to N \to K + N \to (K + N)/N \to 0$$

and

$$0 \to K \cap N \to K \to K/(K \cap N) \to 0$$

to get

$$c(K + N) - c(N) = c(K) - c(K \cap N). \qquad \square$$

Fitting's Lemma

An endomorphism f of a finite dimensional vector space induces a direct decomposition of the space into two subspaces, on one of which f is nilpotent and on the other of which f is invertible. This fact has a generalization of fundamental importance to the study of modules of finite length. Its proof depends on

11.6. Lemma. *Let M be a module and let f be an endomorphism of M.*

(1) *If M is artinian, then $Im f^n + Ker f^n = M$ for some n, whence f is an automorphism if and only if it is monic;*

(2) *If M is noetherian, then $Im f^n \cap Ker f^n = 0$ for some n, whence f is an automorphism if and only if it is epic.*

Proof. For (1) observe that

$$Im f \geq Im f^2 \geq \ldots .$$

Assume that M is artinian. Then this descending chain is finite, and there is an n such that $Im f^{2n} = Im f^n$.

Let $x \in M$. Then $f^n(x) \in Im f^{2n}$, so $f^n(x) = f^{2n}(y)$ for some $y \in M$. Clearly

$$x = f^n(y) + (x - f^n(y)) \in Im f^n + Ker f^n.$$

Finally, if f is monic, then $Ker f^n = 0$, so that $Im f^n = M$ whence $Im f = M$.
We omit the proof of (2).   □

11.7. Proposition. [Fitting's Lemma.] *If M is a module of finite length n and if f is an endomorphism of M, then*

$$M = Im f^n \oplus Ker f^n.$$

Proof. By (11.1), M is both artinian and noetherian, so by the Lemma, there is an m with $M = Im f^m \oplus Ker f^m$. But since M has length n, both $Im f^n = Im f^m$ and $Ker f^n = Ker f^m$.   □

11.8. Corollary. *Let M be an indecomposable module of finite length. Then the following statements about an endomorphism f of M are equivalent:*

(a) *f is a monomorphism;*

(b) *f is an epimorphism;*

(c) *f is an automorphism;*

(d) *f is not nilpotent.*   □

11. Exercises

1. Let n be a positive integer.
   (1) Determine the composition length $c(\mathbb{Z}_n)$ of the $\mathbb{Z}$-module $\mathbb{Z}_n$.
   (2) Characterize those n for which $\mathbb{Z}_n$ has a unique composition series.
2. Give examples of modules M such that $c(M) = 2$ and such that:
   (1) M has exactly one composition series.

(2) M has exactly two composition series.

(3) M has infinitely many composition series.

3. Give an example of a module M that does not have a composition series but for which every non-zero submodule has a maximal submodule and every non-zero factor module has a minimal submodule.

4. Let $M_1, \ldots, M_n$ be submodules of M such that each M/M_i has finite length. Prove that $M/(M_1 \cap \ldots \cap M_n)$ has finite length. Moreover determine a formula for computing this length.

5. (1) Let M be a module of finite length and let $(M_\alpha)_{\alpha \in A}$ be an indexed set of submodules with $M = \Sigma_A M_\alpha$. Prove that $c(M) = \Sigma_A c(M_\alpha)$ iff $M = \oplus_A M_\alpha$.

   (2) Let M be semisimple. Prove that $c(M)$ is finite iff M is finitely generated.

6. Prove the Schreier Refinement Theorem: If M is a module of finite length and if

$$M = N_0 > N_1 > \ldots > N_p = 0$$

is a chain of submodules of M, then there is a composition series for M whose terms include $N_0, N_1, \ldots, N_p$.

7. Prove that if $L \cong M/K$ and T is isomorphic to a composition factor of M then T is isomorphic to a composition factor of either K or L (even if M isn't of finite length).

8. Let M be noetherian and let f be an endomorphism of M. Suppose that $\text{Coker} f$ has finite length. Prove that both $\text{Coker} f^n$ and $\text{Ker} f^n$ have finite length $(n = 1, 2, \ldots)$.

   [Hint: By (11.6.2) there is an m with $\text{Ker} f^m \cap \text{Im} f^m = 0$.]

9. (1) Prove that if $_RM$ is either artinian or noetherian and if $m, n \in \mathbb{N}$ with $M^{(m)} \cong M^{(n)}$, then $m = n$. [Hint: (11.6).]

   (2) Deduce that if R has an ideal I such that R/I is left noetherian or left artinian, then R is IBN. (See Exercise (8.15).)

   (3) Find a simple ring that is neither left artinian, right artinian, left noetherian nor right noetherian.

10. Let $(L, \leq)$ be a complete modular lattice. Prove that if L has a maximal chain of finite length, then every two maximal chains have the same length. [Hint: Use an induction argument similar to that in the proof of (11.3). Also see Exercise (2.6.2).]

11. Prove that if M has two semisimple decompositions $M = \oplus_A T_\alpha = \oplus_B S_\beta$ then these two decompositions are equivalent, i.e., there is a bijection $\sigma : A \to B$ such that $T_\alpha \cong S_{\sigma(\alpha)}$ $(\alpha \in A)$. [Hint: One may assume that M is homogeneous. (Why?) If A is finite use the Jordan-Hölder Theorem. If A is infinite, argue as in Exercise (2.18).]

12. Prove the following version of Fitting's Lemma: If M is a module of finite length and $f : M \to M$ is an endomorphism, then there exist submodules I and K such that $M = I \oplus K$, $(f \mid I) : I \to I$ is an automorphism and $(f \mid K) : K \to K$ is nilpotent. [Hint: If $c(M) = n$, let $I = \text{Im} f^n$ and $K = \text{Ker} f^n$.]

§12. Indecomposable Decompositions of Modules

Recall that a module is indecomposable in case it is non-zero and has no non-trivial direct summands. A direct decomposition

$$M = \oplus_A M_\alpha$$

of a module M as a direct sum of indecomposable submodules $(M_\alpha)_{\alpha \in A}$ is an *indecomposable decomposition*. For example, semisimple modules (§9), artinian modules and noetherian modules (10.14) all have such decompositions. Indeed, as we have observed in §9 it is the existence of such indecomposable decompositions with *simple* terms that allowed us to prove, for semisimple modules, analogues of the standard properties of vector spaces.

Not every module admits an indecomposable decomposition. Indeed, if R is the ring of all continuous functions from $\mathbb{Q}$ to $\mathbb{R}$, then the left regular module $_R R$ has no indecomposable direct summands, so certainly no indecomposable decompositions (see Exercise (7.8)). Nevertheless a significant number of the modules met in practice do have indecomposable decompositions, and the study of these modules and their decomposition theories is one of the most important in ring theory. There are two main directions this study takes, the study of the structure of indecomposable modules and the study of the behavior of the decompositions themselves. Anything even resembling a definitive study of these awaits the work of future generations. The structure of indecomposable modules, even over comparatively simple rings, can be staggeringly complex. In this section we concern ourselves with the decompositions and shall see that even if there is an indecomposable decomposition, there is no guarantee that it is particularly well behaved.

Equivalent Decompositions

We begin our study with an important concept for decompositions that generalizes one of the fundamental properties of bases in vector spaces. Let M be a module. Two direct decompositions

$$M = \oplus_A M_\alpha = \oplus_B N_\beta$$

of M are said to be *equivalent* in case there is a bijection, called an *equivalence map*, $\sigma : A \to B$ such that

$$M_\alpha \cong N_{\sigma(\alpha)} \qquad (\alpha \in A).$$

For example, every two indecomposable decompositions of a semisimple module are equivalent (Exercise (11.11)).

It is easy to check that in the set of all direct decompositions of a module the property of being equivalent defines an equivalence relation.

12.1. Proposition. *Let $(M_\alpha)_{\alpha \in A}$ and $(N_\beta)_{\beta \in B}$ be indexed sets of non-zero submodules of M. Suppose*

$$M = \oplus_A M_\alpha = \oplus_B N_\beta.$$

Let $\sigma : A \to B$ be a map. These two decompositions are equivalent via σ if and only if there is an automorphism f of M with $f(M_\alpha) = N_{\sigma(\alpha)}$ for each $\alpha \in A$.

Proof. ($\Rightarrow$). For each $\alpha \in A$, let $f_\alpha : M_\alpha \to N_{\sigma(\alpha)}$ be an isomorphism. Then since σ is bijective, the direct sum (see (6.25)) $f = \oplus_A f_\alpha : M \to M$ is an automorphism with $f(M_\alpha) = f_\alpha(M_\alpha) = N_{\sigma(\alpha)}$.

($\Leftarrow$). It will suffice to show that $\sigma : A \to B$ is a bijection. But $\alpha \neq \alpha'$ in A implies $M_\alpha \neq M_{\alpha'}$ whence $N_{\sigma(\alpha)} \neq N_{\sigma(\alpha')}$ and $\sigma(\alpha) \neq \sigma(\alpha')$. Now

$$f(M) = \oplus_A f(M_\alpha) = \oplus_A N_{\sigma(\alpha)} = M.$$

So if $\beta \in B$ and $\beta \notin \sigma(A)$, then $N_\beta = N_\beta \cap M = N_\beta \cap (\Sigma_A N_{\sigma(\alpha)}) = 0$ which is not the case. $\qquad \square$

It is immediate that any decomposition of a module that is equivalent to an indecomposable one is indecomposable. On the other hand two indecomposable decompositions of a module need not be equivalent. For one example of this phenomenon see Exercise (12.4). Thus it is important, although non-trivial, to devise meaningful sufficient conditions for indecomposable decompositions to be equivalent.

Decompositions that Complement Direct Summands

We consider next a generalization of a fundamental property of semisimple modules (9.2). First recall that if M is a module, then a direct summand K of M is a maximal direct summand of M if and only if K has an indecomposable direct complement N in M. Now a decomposition

$$M = \oplus_A M_\alpha$$

of a module M as a direct sum of non-zero submodules $(M_\alpha)_{\alpha \in A}$ is said to *complement direct summands* (*complement maximal direct summands*) in case for every (every maximal) direct summand K of M there is a subset $B \subseteq A$ with

$$M = (\oplus_B M_\beta) \oplus K.$$

Of course, a decomposition that complements direct summands complements maximal direct summands. The converse fails, for as we have seen (Exercise (7.8)) there are modules having no indecomposable direct summands and for such a module every decomposition complements maximal direct summands. A decomposition that complements (all) direct summands is necessarily indecomposable. (See Exercise (12.2).)

Now suppose that a module M has a direct decomposition

$$M = \oplus_A M_\alpha$$

that complements (maximal) direct summands. If M' is a second module and if $f : M \to M'$ is an isomorphism, then

$$M' = \oplus_A f(M_\alpha)$$

is a direct decomposition of M' that complements (maximal) direct summands. In particular, by Proposition 12.1 if one of two equivalent decompositions of M complements (maximal) direct summands, then so does the other. (See Exercise (12.1).)

12.2. Lemma. *Let* $M = \oplus_A M_\alpha$ *be a decomposition that complements maximal direct summands. If*

$$M = N_1 \oplus \ldots \oplus N_n \oplus K$$

with each $N_1, \ldots, N_n$ *indecomposable, then there exist* $\alpha_1, \ldots, \alpha_n \in A$ *such that*

$$M_{\alpha_i} \cong N_i \qquad (i = 1, \ldots, n),$$

and for each $1 \leq l \leq n$,

$$M = M_{\alpha_1} \oplus \ldots \oplus M_{\alpha_l} \oplus N_{l+1} \oplus \ldots \oplus N_n \oplus K.$$

Proof. We induct on n. If $n = 1$, then K is a maximal direct summand, so the result is immediate. Suppose

$$M = N_1 \oplus \ldots \oplus N_n \oplus N_{n+1} \oplus K$$

with the N_i indecomposable, and let $M_{\alpha_1}, \ldots, M_{\alpha_n}$ satisfy the conclusion of the lemma through $l = n$. Then

$$M_{\alpha_1} \oplus \ldots \oplus M_{\alpha_n} \oplus K$$

is a maximal direct summand of M with direct complement N_{n+1}. So there is an $M_{\alpha_{n+1}}$, necessarily isomorphic to N_{n+1}, such that

$$M = M_{\alpha_1} \oplus \ldots \oplus M_{\alpha_n} \oplus M_{\alpha_{n+1}} \oplus K. \qquad \square$$

Our first main goal is to show that if M has an indecomposable decomposition complementing maximal direct summands, then every two indecomposable decompositions are equivalent, whence every indecomposable decomposition complements maximal direct summands. To show this we require the following lemma.

12.3. Lemma. *Let* $M = \oplus_A M_\alpha$ *be a decomposition that complements* (*maximal*) *direct summands. Let* $A' \subseteq A$ *and set* $M' = \Sigma_{A'} M_{\alpha'}$. *Then*

$$M' = \oplus_{A'} M_{\alpha'}$$

is a decomposition of M' *that complements* (*maximal*) *direct summands. Moreover, if* M *has a decomposition that complements direct summands, then so does every direct summand of* M.

Proof. It is clear that $M' = \oplus_{A'} M_{\alpha'}$ is a decomposition of M'. Suppose that K is a (maximal) direct summand of M'. Then

$$(\oplus_{A \setminus A'} M_\alpha) \oplus K$$

is a (maximal) direct summand of M. So by hypothesis there is a subset $B' \subseteq A$ such that

$$M = (\oplus_{A \setminus A'} M_\alpha) \oplus (\oplus_{B'} M_{\beta'}) \oplus K.$$

But then clearly we must have $B' \subseteq A'$ and

$$M' = (\oplus_{B'} M_{\beta'}) \oplus K.$$

This proves the first statement. The final statement follows from the first in view of the fact that if $M = \oplus_A M_\alpha$, if N is a direct summand of M, and if $B \subseteq A$ with $M = (\oplus_B M_\beta) \oplus N$, then $N \cong \oplus_{A \setminus B} M_\alpha$. $\qquad\square$

Incidentally, it is apparently not known whether the last assertion of (12.3) holds for decompositions just complementing maximal direct summands.

12.4. Theorem. *If a module M has an indecomposable decomposition that complements maximal direct summands, then all indecomposable decompositions of M are equivalent.*

Proof. Suppose that $M = \oplus_A M_\alpha$ and $M = \oplus_C N_\gamma$ are indecomposable decompositions and that the first complements maximal direct summands. For each indecomposable direct summand L of M set

$$A(L) = \{\alpha \in A \mid M_\alpha \cong L\} \quad\text{and}\quad C(L) = \{\gamma \in C \mid N_\gamma \cong L\}.$$

Then to complete the proof it will suffice to show that for each L, there is a bijection from $A(L)$ onto $C(L)$ or equivalently that

$$card\ A(L) = card\ C(L).$$

This will involve several steps.

First, suppose that $A(L)$ is finite. Then by (12.2) for each finite subset $F = \{\gamma_1, \ldots, \gamma_n\} \subseteq C(L)$, there is an injection $\tau_F : F \to A$ such that

$$L \cong N_{\gamma_i} \cong M_{\tau_F(\gamma_i)} \qquad (i = 1, \ldots, n).$$

Thus $Im\ \tau_F \subseteq A(L)$ and in this case $card\ C(L) \leq card\ A(L)$.

Next suppose that $A(L)$ is infinite. Let $(p_\gamma)_{\gamma \in C}$ be the projections for the decomposition $M = \oplus_C N_\gamma$. For each $\alpha \in A$, set

$$F_\alpha = \{\gamma \in C \mid M = M_\alpha \oplus (\oplus_{\beta \neq \gamma} N_\beta)\}.$$

Clearly by (5.5), $\gamma \in F_\alpha$ iff $(p_\gamma \mid M_\alpha) : M_\alpha \to N_\gamma$ is an isomorphism. Also $M = \oplus_A M_\alpha$ complements each $\oplus_{\beta \neq \gamma} N_\beta$, so clearly

$$C(L) = \cup_{A(L)} F_\alpha.$$

Let $\alpha \in A$. Since $(N_\gamma)_{\gamma \in C}$ spans M, there exist $\gamma_1, \ldots, \gamma_n \in C$ with

$$M_\alpha \cap (N_{\gamma_1} + \ldots + N_{\gamma_n}) \neq 0.$$

Thus, $Ker(p_\gamma \mid M_\alpha) = 0$ only if $\gamma \in \{\gamma_1, \ldots, \gamma_n\}$; hence each F_α is finite. But this means $\alpha \mapsto F_\alpha$ is a mapping from $A(L)$ to a set of finite subsets of $C(L)$ that cover $C(L)$. Therefore $card\ C(L) \leq card\ (\mathbb{N} \times A(L))$. But since $A(L)$ is infinite, $card\ (\mathbb{N} \times A(L)) = card\ A(L)$. (See (0.10).)

We now have that for each indecomposable direct summand L of M, *card* $C(L) \leq$ *card* $A(L)$. That is, there is an injection $\sigma : C \to A$ such that $N_\gamma \cong M_{\sigma(\gamma)}$ for each $\gamma \in C$. Therefore there is an isomorphism

$$f : M = \oplus_C N_\gamma \to \oplus_C M_{\sigma(\gamma)}$$

such that $f(N_\gamma) = M_{\sigma(\gamma)}$ for each $\gamma \in C$. Thus, by (12.3) the decomposition $M = \oplus_C N_\gamma$ also complements maximal direct summands. So we can reverse the roles of A and C and infer that for each indecomposable module L,

$$\textit{card } A(L) \leq \textit{card } C(L)$$

So (0.10) there is a bijection $C(L) \to A(L)$. $\qquad\square$

12.5. Corollary. *If a module M has an indecomposable decomposition that complements (maximal) direct summands, then every indecomposable decomposition of M complements (maximal) direct summands.* $\qquad\square$

Azumaya's Decomposition Theorem

A ring R is said to be *local* in case for each pair $a, b \in R$ if $a + b$ is invertible, then either a or b is invertible. (Also see Exercise (2.12).) We shall have more to say about such rings in §15. For now we simply observe (see Exercise (12.9)) that if R is local, then 0 and 1 are its only idempotents. So, in particular, a module with a local endomorphism ring must be indecomposable (5.10). This establishes the first assertion of the following important theorem of Azumaya.

12.6. Theorem [Azumaya]. *If a module has a direct decomposition*

$$M = \oplus_A M_\alpha$$

where each endomorphism ring $End(M_\alpha)$ is local, then this is an indecomposable decomposition and

(1) *Every non-zero direct summand of M has an indecomposable direct summand;*

(2) *The decomposition $M = \oplus_A M_\alpha$ complements maximal direct summands and thus is equivalent to every indecomposable decomposition of M.*

Proof. Throughout this proof we shall treat the elements of the various endomorphism rings as right operators. Now suppose that

$$(1) \qquad\qquad M = \oplus_A M_\alpha$$

is a decomposition whose terms have local endomorphism rings, and that $M = N \oplus N'$ is a decomposition of M with N non-zero. Let e and $e' = 1 - e$ be the orthogonal idempotents in $End(M)$ such that

$$N = Me \qquad \text{and} \qquad N' = Me'.$$

We claim that N has a decomposition $N = K \oplus N''$ such that, for some $\alpha \in A$, e restricts to an isomorphism $(e \,|\, M_\alpha) : M_\alpha \to K$. First observe that since the submodules $(M_\alpha)_{\alpha \in A}$ span M, there is a finite set $\alpha_1, \ldots, \alpha_n \in A$ such that

$$N \cap (M_{\alpha_1} \oplus \ldots \oplus M_{\alpha_n}) \neq 0.$$

Next let e_1 be the idempotent for M_{α_1} in the decomposition (1). Then $e_1 End(M)e_1$, which is isomorphic to $End(M_{\alpha_1})$ by (5.9), is a local ring with identity e_1. Thus since

$$e_1 = e_1 e e_1 + e_1 e' e_1,$$

one of these terms must be invertible in $e_1 End(M)e_1$. Thus for some $f_1 \in \{e, e'\}$, $e_1 f_1 e_1$ is invertible in $e_1 End(M)e_1$. Set

(2) $$K_1 = Im(e_1 f_1).$$

Since both $e_1 f_1 e_1$ and e_1 are isomorphisms from M_{α_1} to M_{α_1},

(3) $$(f_1 \mid M_{\alpha_1}): M_{\alpha_1} \to K_1 \quad \text{and} \quad (e_1 \mid K_1): K_1 \to M_{\alpha_1}$$

are isomorphisms. Now the second of these together with (5.5) gives a decomposition

(1)$_1$ $$M = K_1 \oplus (\oplus_{\alpha \neq \alpha_1} M_\alpha)$$

in which each term has a local endomorphism ring. If $n > 1$, then let e_2 be the idempotent for M_{α_2} in the decomposition (1)$_1$. Repeating this last argument we obtain $f_2 \in \{e, e'\}$ with

(2)$_1$ $$K_2 = Im(e_2 f_2)$$

and isomorphisms

(3)$_1$ $$(f_2 \mid M_{\alpha_2}): M_{\alpha_2} \to K_2 \quad \text{and} \quad (e_2 \mid K_2): K_2 \to M_{\alpha_2}$$

so that

(1)$_2$ $$M = K_1 \oplus K_2 \oplus (\oplus_{\alpha \neq \alpha_1, \alpha_2} M_\alpha).$$

We can continue this until we have

(1)$_n$ $$M = K_1 \oplus \ldots \oplus K_n \oplus (\oplus_{\alpha \neq \alpha_1, \ldots, \alpha_n} M_\alpha)$$

and a sequence $f_1, f_2, \ldots, f_n$ from $\{e, e'\}$ with each

$$(f_i \mid M_{\alpha_i}): M_{\alpha_i} \to K_i$$

an isomorphism. At least one of the f_i must be e for if all the f_i are e', then $e' = 1 - e$ would restrict to an isomorphism

$$M_{\alpha_1} \oplus \ldots \oplus M_{\alpha_n} \to K_1 \oplus \ldots \oplus K_n;$$

this is impossible, however, because

$$(Ker\, e') \cap (M_{\alpha_1} \oplus \ldots \oplus M_{\alpha_n}) = N \cap (M_{\alpha_1} \oplus \ldots \oplus M_{\alpha_n}) \neq 0.$$

Therefore for some $1 \leq i \leq n, f_i = e$, and we have that K_i is a direct summand of M that is contained in $N = Me$ such that $(e \mid M_{\alpha_i}): M_{\alpha_i} \to K_i$ is an isomorphism. Thus, taking $\alpha = \alpha_i$ and $K = K_i$ we have

(4) $$N = K \oplus N''$$

such that e restricts to an isomorphism

(5) $$(e \mid M_\alpha) : M_\alpha \to K.$$

This essentially completes the proof. For first $K \cong M_\alpha$ is an indecomposable direct summand of N, and second if N is indecomposable, then we must have $K = N$ so that, by (5) and (5.5), M_α is a complement of the maximal direct summand N' of M. Of course the last assertion of part (2) now follows from (12.4). $\qquad\Box$

12.7. Corollary. *If M has a finite direct decomposition*

$$M = M_1 \oplus \dots \oplus M_n$$

where each endomorphism ring $End(M_i)$ is local, then this decomposition complements direct summands.

Proof. Let $M = N \oplus K$ where $N \neq 0$. Then by (12.6) there is a $1 \leq i_1 \leq n$ and an $N_1 \leq N$ with $M = M_{i_1} \oplus N_1 \oplus K$. If $N_1 \neq 0$, then again by (12.6) there is a $1 \leq i_2 \leq n$ and an $N_2 \leq N_1$ with $M = M_{i_1} \oplus M_{i_2} \oplus N_2 \oplus K$, and clearly $i_1 \neq i_2$. Continuing by induction and noting that this can continue for at most n steps, we conclude that there exist $i_1, \dots, i_k$ for some $k \leq n$ with

$$M = M_{i_1} \oplus \dots \oplus M_{i_k} \oplus K. \qquad\Box$$

The condition of local endomorphism rings is not necessary for a decomposition to complement maximal direct summands. Indeed any indecomposable module, with or without a local endomorphism ring, has a decomposition that complements direct summands. Of course a module that has no indecomposable direct summands has a decomposition that complements maximal direct summands.

The Krull–Schmidt Theorem

The classical Krull-Schmidt Theorem is now an easy consequence of Azumaya's Theorem and the following

12.8. Lemma. *If M is an indecomposable module of finite length, then $End(M)$ is a local ring.*

Proof. Let M be an indecomposable module of finite length $c(M) = n$. Let $f, g \in End(M)$, and suppose that $f + g$ is invertible in $End(M)$. It will suffice to show that if g is not invertible, then f is. But if $f + g$ is invertible, then for some automorphism h,

$$(f + g)h = 1_M$$

in $End(M)$. If g is not invertible, then by (11.8) neither is gh, so also by (11.8) gh is nilpotent; in fact, $(gh)^n = 0$. Thus

$$(1 - gh)(1 + gh + \ldots (gh)^{n-1}) = 1.$$

In other words fh is invertible whence f is invertible.   □

12.9. The Krull–Schmidt Theorem. *Let M be a non-zero module of finite length. Then M has a finite indecomposable decomposition*

$$M = M_1 \oplus \ldots \oplus M_n$$

such that for every indecomposable decomposition

$$M = N_1 \oplus \ldots \oplus N_k,$$

$n = k$ *and there is a permutation σ of $\{1, \ldots, n\}$ such that*

$$M_{\sigma(i)} \cong N_i \qquad (i = 1, \ldots, n),$$

and for each $1 \le l \le n$,

$$M = M_{\sigma(1)} \oplus \ldots \oplus M_{\sigma(l)} \oplus N_{l+1} \oplus \ldots \oplus N_n.$$

In fact the decomposition $M = M_1 \oplus \ldots \oplus M_n$ complements direct summands.

Proof. Since M has finite length, we know that it does have a finite indecomposable decomposition

$$M = M_1 \oplus \ldots \oplus M_n.$$

(See (11.1) and (10.14).) We have from (12.8) then that each $End(M_i)$ is local; thus, the corollary (12.7) to Azumaya's Theorem applies, and the decomposition complements direct summands. The other assertions follow at once from (12.2).   □

As we noted earlier, the hypothesis of local endomorphism rings is not necessary for an indecomposable decomposition to complement maximal direct summands. However, one consequence of the following result is that if both M and $M^{(2)} = M \times M$ have indecomposable decompositions that complement maximal direct summands, then the endomorphism rings of the terms in these decompositions must be local.

12.10. Proposition. *Let $M = \bigoplus_A M_\alpha$ be an indecomposable decomposition that complements maximal direct summands. If M_α appears at least twice in this decomposition (i.e., there is a $\beta \ne \alpha$ in A such that $M_\beta \cong M_\alpha$), then $End(M_\alpha)$ is a local ring.*

Proof. In view of Lemma 12.3 it will suffice to show that if M is an indecomposable module and if the decomposition

$$M^{(2)} = M \times M = M_1 \oplus M_2$$

where

$$M_1 = \{(m, 0) \mid m \in M\} \qquad \text{and} \qquad M_2 = \{(0, m) \mid m \in M\},$$

complements maximal direct summands, then $End(M)$ is a local ring. So suppose that we have these hypotheses and let

$$\pi_i : M^{(2)} \to M \qquad (i = 1, 2)$$

be the natural coordinate projections with $Ker\ \pi_i = M_j\ (i \neq j)$. Now let $f, g \in End(_R M)$ with

$$f - g = 1_M.$$

It will suffice to prove that either f or g is an automorphism. So set

$$M' = \{(mf, mg) \mid m \in M\} \qquad \text{and} \qquad M_d = \{(m, m) \mid m \in M\}.$$

Then, from the fact that $(mf, mg) = (n, n)$ implies $m = m(f - g) = n - n = 0$, and the identity

$$(m, n) = ((m - n)f, (m - n)g) + (m - (m - n)f, m - (m - n)f),$$

we have $M^{(2)} = M_d \oplus M'$.

But also we see at once that $M \cong M_d$. Whence M' is a maximal direct summand of $M^{(2)}$. Thus, either $M^{(2)} = M_1 \oplus M'$ or $M^{(2)} = M_2 \oplus M'$, and therefore, either

$$(\pi_2 \mid M') : M' \to M \qquad \text{or} \qquad (\pi_1 \mid M') : M' \to M$$

is an isomorphism. Finally it is easy to check that this means either f or g is an automorphism, as desired.   ☐

12. Exercises

1. Let $f : M \to N$ be an isomorphism and let $M = \oplus_A M_\alpha$. Prove that this decomposition complements (maximal) direct summands iff $N = \oplus_A f(M_\alpha)$ complements (maximal) direct summands.
2. Prove that if $M = \oplus_A M_\alpha$ is a decomposition that complements direct summands, then each M_α is indecomposable.
3. Let A be an infinite set and let $S = \mathbb{R}^A$.
   (1) Prove that $_S S$ does not have an indecomposable decomposition.
   (2) The constant functions in S form a subring of S isomorphic to $\mathbb{R}$. Prove that the resulting module $_\mathbb{R} S$ has an indecomposable decomposition.
   (3) Give an example of an indecomposable $\mathbb{R}$-direct summand of S that is not an S-direct summand of S.
4. (1) Let I and J be left ideals of a ring R such that $I + J = R$. Prove that as left R-modules, $I \oplus J \cong R \oplus (I \cap J)$. [Hint: The natural epimorphism $I \oplus J \to R$ splits. (Exercise (5.1).)]
   (2) Let $R = \mathbb{Z}[\sqrt{-5}]$. Prove that R has a module M having inequivalent indecomposable decompositions. [Hint: For each $r = a + b\sqrt{-5}$ in R define $\bar{r} = a - b\sqrt{-5}$, and

$$\|r\| = r\bar{r} = a^2 + 5b^2.$$

Show that $\|rs\| = \|r\| \|s\|$. Deduce then that the ideal I generated by $\{3, 2 + \sqrt{-5}\}$ is not principal and similarly that the ideal J generated by $\{3, 2 - \sqrt{-5}\}$ is not principal.]

5. An indexed set $(M_\alpha)_{\alpha \in A}$ is *homologically independent* in case $\alpha \neq \beta$ implies $Hom_R(M_\alpha, M_\beta) = 0$. Let $M = \oplus_A M_\alpha$ with $(M_\alpha)_{\alpha \in A}$ homologically independent.

   (1) Prove that if each M_α is indecomposable and if K is a non-zero direct summand of M, then $K = \oplus_B M_\beta$ for some (necessarily unique) $B \subseteq A$. [Hint: Let $(e_\alpha)_{\alpha \in A}$ be the idempotents in $End(_R M)$ for the given decomposition. Let $e = e^2 \in End(_R M)$. Then $\alpha \neq \beta$ implies $e_\beta e e_\alpha = 0$.]

   (2) If each M_α has an indecomposable decomposition that complements (maximal) direct summands, then so does M.

6. Give an example of an indecomposable decomposition $M = \oplus_A M_\alpha$ with A infinite that complements direct summands and no $End(_R M_\alpha)$ local. [Hint: See Exercise (12.5).]

7. Let M be a left R-module and set $B = BiEnd(_R M)$. Let K, L, M_α $(\alpha \in A)$ and N_γ $(\gamma \in C)$ be submodules of $_R M$. Prove that

   (1) K is an (indecomposable) direct summand of $_R M$ iff K is an (indecomposable) direct summand of $_B M$. [Hint: See Proposition (4.12).]

   (2) If K is a direct summand of M, then $End(_R K) = End(_B K)$.

   (3) K and L are R-isomorphic direct summands of $_R M$ iff they are B-isomorphic direct summands of $_B M$.

   (4) $M = \oplus_A M_\alpha = \oplus_C N_\gamma$ are equivalent decompositions of $_R M$ iff they are equivalent decompositions of $_B M$.

   (5) $M = \oplus_A M_\alpha$ complements (maximal) direct summands in $_R M$ iff it complements (maximal) direct summands in $_B M$.

   (6) If $_R M$ is simple (semisimple), then $_B M$ is simple (semisimple).

   (7) If $_B M$ is semisimple, then $_R M$ has a decomposition that complements direct summands.

8. Let M have the property that every direct summand has an indecomposable decomposition. Prove that if M has a decomposition that complements maximal direct summands, then so does every direct summand.

9. Prove that if R is a local ring, then 0 and 1 are its only idempotents. Show that the converse is false.

10. (1) Deduce from Proposition 12.10 that the $\mathbb{Z}$-module $\mathbb{Z} \oplus \mathbb{Z}$ does not have an indecomposable decomposition that complements maximal direct summands. Observe, however, that every two indecomposable decompositions of $\mathbb{Z} \oplus \mathbb{Z}$ are equivalent. [Hint: Exercise (8.16).]

    (2) Determine a maximal direct summand of $\mathbb{Z} \oplus \mathbb{Z}$ that is not complemented by the decomposition $\mathbb{Z}(1, 1) \oplus \mathbb{Z}(1, 2)$.

Chapter 4

Classical Ring-Structure Theorems

As we saw in the last chapter semisimple modules play a distinguished role in the theory of modules. Classically, the most important class of rings consists of those rings R whose category $_R\mathsf{M}$ has a semisimple generator. A characteristic property of such a ring R, called a "semisimple" ring, is that each left R-module is semisimple. These rings are the objects of study in Section 13 where we prove the fundamental Wedderburn-Artin characterization of these rings as direct sums of matrix rings over division rings. In particular, a semisimple ring is a direct sum of rings each having a simple faithful left module. In Section 14 we study rings characterized by this latter property—the "(left) primitive" rings. Here we prove Jacobson's important generalization of the semisimple case characterizing left primitive rings as "dense rings" of linear transformations.

If R is a ring, then the radical of the regular module $_RR$ is an ideal, the "radical" of the ring R. This ideal, an object of considerable importance, is the focus of attention in Section 15. It is characterized as the unique smallest ideal of R modulo which R can be suitably represented as a subring of a product of left primitive rings.

§13. Semisimple Rings

As we have noted several times, the good behavior of vector spaces is often a consequence of their special decomposition theory. It is more than that vector spaces are direct sums of simple modules; it is that they are direct sums of copies of the *same* simple module. Module theoretically this property of division rings D is just that the category of left D-modules has a simple generator. It is not restricted to division rings—indeed any endomorphism ring of a finite dimensional vector space also has this property. We begin by considering this from the point of view of matrices.

A Simple Example

13.1. Let D be a division ring and $n \in \mathbb{N}$. Let $\mathbb{C}_n(D)$ be the set of all $n \times 1$ column matrices over D and let $\mathbb{R}_n(D)$ be the set of all $1 \times n$ row matrices over D. Then $\mathbb{C}_n(D)$ is an n-dimensional right D-vector space and $\mathbb{R}_n(D)$ is an n-dimensional left D-vector space:

$$\mathbb{C}_n(D) = (D_D)^{(n)} \qquad \text{and} \qquad \mathbb{R}_n(D) = (_DD)^{(n)}.$$

Moreover, the usual matrix multiplications λ and ρ are ring isomorphisms

$$\lambda : \mathbb{M}_n(D) \to End(\mathbb{C}_n(D)_D)$$

$$\rho : \mathbb{M}_n(D) \to End({}_D\mathbb{R}_n(D)).$$

So $\mathbb{C}_n(D)$ and $\mathbb{R}_n(D)$ are left and right $\mathbb{M}_n(D)$ modules respectively. But notice that $\mathbb{C}_n(D)$ is a simple left $\mathbb{M}_n(D)$-module and $\mathbb{R}_n(D)$ is a simple right $\mathbb{M}_n(D)$-module. (See Exercise (13.3).) Let $E_1, E_2, \ldots, E_n$ be the primitive diagonal idempotents of $\mathbb{M}_n(D)$. Then as a left $\mathbb{M}_n(D)$-module

$$\mathbb{M}_n(D) = \mathbb{M}_n(D)E_1 \oplus \ldots \oplus \mathbb{M}_n(D)E_n$$

$$\cong \mathbb{C}_n(D) \oplus \ldots \oplus \mathbb{C}_n(D)$$

and as a right $\mathbb{M}_n(D)$-module

$$\mathbb{M}_n(D) = E_1 \mathbb{M}_n(D) \oplus \ldots \oplus E_n \mathbb{M}_n(D)$$

$$\cong \mathbb{R}_n(D) \oplus \ldots \oplus \mathbb{R}_n(D).$$

In particular, $\mathbb{M}_n(D)$ is generated both as a left module and as a right module over itself by a simple module. So by (8.8) and (8.6) *every* left $\mathbb{M}_n(D)$-module is generated by the simple $\mathbb{M}_n(D)$ module $\mathbb{C}_n(D)$ and *every* right $\mathbb{M}_n(D)$ is generated by $\mathbb{R}_n(D)$.

This rather inelegant looking example (but see Exercise (13.2)) is really the whole story. For as we shall see, the property of having a simple generator characterizes (to within isomorphism) such matrix rings. Thus, in particular, from this assumption on one side, we can deduce it on the other side. The first step for the converse of this example will deal with endomorphism rings of finite direct sums of a module.

Simple Artinian Rings

Let R be an arbitrary ring, ${}_RM$ a non-zero left R-module, and $n > 0$ a natural number. In what follows we shall write the endomorphisms of M and of $M^{(n)}$ as right operators, and we shall also write the natural injections and projections

$$\iota_i : M \to M^{(n)} \qquad \text{and} \qquad \pi_i : M^{(n)} \to M$$

on the right. Now for each $\alpha = [\![\alpha_{ij}]\!] \in \mathbb{M}_n(End(M))$ define $\rho(\alpha) \in End(M^{(n)})$ coordinatewise by

$$(x\rho(\alpha))\pi_j = \Sigma_i \, x\pi_i \alpha_{ij}.$$

Then $x\rho(\alpha)$ is simply the usual matrix product

$$x\rho(\alpha) = [x_1, \ldots, x_n][\![\alpha_{ij}]\!]$$

where the elements x of $M^{(n)}$ are considered as $1 \times n$ row matrices $x = [x_1, \ldots, x_n]$ over M. Thus it follows from computations just as in ordinary matrix multiplication that $M^{(n)}$ is a bimodule

$$_R M^{(n)}{}_{\mathbb{M}_n(End(M))}$$

via ρ. That is,

$$\rho : \mathbb{M}_n(End(M)) \to End(M^{(n)})$$

is a ring homomorphism. (See Proposition 4.10.)

13.2. Proposition. *Let M be a non-zero left R-module and let $n > 0$ be a natural number. Then*

$$\rho : \mathbb{M}_n(End(M)) \to End(M^{(n)})$$

is a ring isomorphism.

Proof. If $\alpha \in Ker\,\rho$, then for each i, j, $\alpha_{ij} = \iota_i \rho(\alpha) \pi_j = 0$, so ρ is injective. Finally, if $\gamma \in End(M^{(n)})$, then

$$(x\rho([\![\iota_k \gamma \pi_l]\!]))\pi_j = \Sigma_i\, x\pi_i(\iota_i \gamma \pi_j) = (x\gamma)\pi_j$$

and ρ is an isomorphism. $\qquad\qquad\qquad\qquad\qquad\qquad\qquad\qquad\square$

13.3. Schur's Lemma. *If $_R T$ is a simple module, then $End(_R T)$ is a division ring.*

Proof. Every non-zero endomorphism $T \to T$ is an isomorphism. $\qquad\square$

Now we have the very fundamental Wedderburn characterization of simple artinian rings (see (13.5)) phrased in terms of simple generators.

13.4. Theorem [Wedderburn]. *The ring R has a simple left generator if and only if R is isomorphic to the full matrix ring $\mathbb{M}_n(D)$ for some division ring D and some natural number n. Moreover, if $_R T$ is a simple left generator for R, then as a ring*

$$R \cong \mathbb{M}_n(D)$$

where $D = End(_R T)$ and $n = c(_R R)$.

Proof. With the notation of (13.1) $C_n(D)$, a simple left $\mathbb{M}_n(D)$-module, generates every left $\mathbb{M}_n(D)$-module (see (8.8) and (8.6)), so $\mathbb{M}_n(D)$ has a simple left generator.

For the rest of the Theorem it will suffice to prove the final assertion. So suppose $_R T$ is a simple generator for R. Since $_R R$ is finitely generated and since T generates R, there is an integer m and an epimorphism $T^{(m)} \to\, _R R \to 0$. So by (9.4) $_R R \cong T^{(n)}$ for some natural number n. Therefore $_R R$ has a composition series of length n (see Exercise (11.5)), so $c(_R R) = n$ (see §11). Now by (4.11) and (13.2), as rings

$$R \cong End(_R R) \cong End(_R T^{(n)}) \cong \mathbb{M}_n(End(T)).$$

Finally, by Schur's Lemma (13.3), $End(T) = D$ is a division ring. $\qquad\square$

Observe that this Theorem implies that if R has a simple generator $_R T$, then T_D is a finite dimensional vector space over the division ring $D = End(_R T)$ and $R \cong End(T_D)$. (See Exercise (13.1).) In §14 we shall return to consider the Wedderburn Theorem from this point of view as a special case of a general result on biendomorphism rings.

There are other important characterizations of rings having simple left generators. Of particular interest are that they are precisely the simple left artinian rings and that they are also, symmetrically, the rings having simple right generators.

13.5. Proposition. *For a ring R the following statements are equivalent:*
(a) *R has a simple left generator;*
(a') *R has a simple right generator;*
(b) *R is simple and left artinian;*
(b') *R is simple and right artinian;*
(c) *For some simple $_R T$, $_R R \cong T^{(n)}$ for some n;*
(c') *For some simple T_R, $R_R \cong T^{(n)}$ for some n;*
(d) *R is simple and $_R R$ is semisimple;*
(d') *R is simple and R_R is semisimple.*

Proof. (a) $\Leftrightarrow$ (c). This is clear.

(a) $\Rightarrow$ (d). Assume R has a simple left generator T. Let I be a proper ideal of R. Then I is contained in a maximal left ideal L of R, and we have $R/L \cong T$. But clearly $_R T$ is faithful. So, since $IR \subseteq L$,

$$I \le l_R(R/L) = l_R(T) = 0$$

and R is simple. Since (a) $\Rightarrow$ (c), R is semisimple.

(d) $\Rightarrow$ (b). If $_R R$ is a direct sum of simples, it must be a finite direct sum of simples (see §7), so (see (10.15)), $_R R$ is artinian.

(b) $\Rightarrow$ (a). If R is left artinian, then (10.11) R has a minimal non-zero left ideal T. Now the trace $Tr_R(T) \ne 0$ of T in R is an ideal (8.21) of R, so if R is simple, then $Tr_R(T) = R$. That is, $_R R$ is generated by T.

(a) $\Leftrightarrow$ (a'). Since $\mathbb{M}_n(D)$ has a simple left generator and a simple right generator, (13.4) establishes this equivalence.

(a') $\Leftrightarrow$ (b') $\Leftrightarrow$ (c') $\Leftrightarrow$ (d') are now clear.   □

In particular, from this proposition we see that for simple rings, the conditions, left artinian, right artinian, and artinian are equivalent. A ring satisfying the equivalent conditions of (13.5) (i.e., a ring that is isomorphic to an $n \times n$ matrix ring over a division ring) is usually referred to as a *simple artinian ring*.

The Wedderburn–Artin Theorem

A ring R is said to be *semisimple* in case the left regular module $_R R$ is semisimple. By (13.5) we have that every simple artinian ring is semisimple. Also it follows that any ring direct sum of semisimple rings is also semisimple. (See Exercise (13.6).) Thus, we have one implication in the following result—one of the most important theorems in all of algebra.

13.6. Theorem [Wedderburn–Artin]. *A ring R is semisimple if and only if it is a (ring) direct sum of a finite number of simple artinian rings.*

This version actually understates the case. Indeed, to prove the remaining implication of the Wedderburn-Artin Theorem we shall make the following analysis of the

13.7. Structure of a Semisimple Ring [Wedderburn–Artin]. *Let R be a semisimple ring. Then R contains a finite set $T_1, T_2, \ldots, T_m$ of minimal left ideals which comprise an irredundant set of representatives of the simple left R-modules. Moreover for each such set the homogeneous components*

$$Tr_R(T_i) = RT_iR \qquad (i = 1, 2, \ldots, m)$$

are simple artinian rings and R is the ring direct sum

$$R = RT_1R \dotplus \ldots \dotplus RT_mR.$$

Finally, T_i is a simple generator for the ring RT_iR and

$$RT_iR \cong \mathbb{M}_{n_i}(D_i)$$

where

$$n_i = c(RT_iR) \qquad and \qquad D_i = End(_RT_i)$$

$(i = 1, 2, \ldots, m)$.

Proof. By (9.1) and (9.4) every simple left R-module is isomorphic to a minimal left ideal of R. In particular, for each simple $_RT$ the trace $Tr_R(T) \neq 0$. Now $_RR$ is the direct sum of these traces (9.12); so (see §7) there is a finite set $T_1, T_2, \ldots, T_m$ of minimal left ideals of R that is an irredundant set of representatives of simple left R-modules. By (8.21) each of the traces $Tr_R(T_i)$ is an ideal of R and hence

$$_RR_R = Tr_R(T_1) \oplus \ldots \oplus Tr_R(T_m).$$

So, by (7.6) each $Tr_R(T_i)$ is a ring and this latter is a ring direct sum

$$R = Tr_R(T_1) \dotplus \ldots \dotplus Tr_R(T_m).$$

Certainly $T_i \subseteq Tr_R(T_i)$ and so by (7.6) it follows that T_i is a simple left ideal of the ring $Tr_R(T_i)$. Since T_i generates $Tr_R(T_i)$ as an R-module (8.12) it generates it as a $Tr_R(T_i)$-module. Thus by (13.5), $Tr_R(T_i)$ is a simple ring, hence a minimal two-sided ideal of R, so

$$Tr_R(T_i) = RT_iR.$$

The rest of the proof is now an easy application of (13.4). □

13.8. Corollary. *A ring R is semisimple if and only if R_R is semisimple.*

Proof. This is clear from (13.5) and (13.6). □

Now we easily deduce the following important characterizations of semisimple rings.

13.9. Proposition. *For a ring R the following statements are equivalent:*
(a) *R is semisimple;*
(b) *R has a semisimple left generator;*
(c) *Every short exact sequence*

$$0 \to K \to M \to N \to 0$$

of left R-modules splits;
(d) *Every left R-module is semisimple.*
Moreover, these statements are equivalent if throughout "left" is replaced by "right".

Proof. In view of (13.8) it will clearly suffice to prove the equivalence of the "left"-hand version of the conditions.

(a) $\Rightarrow$ (b). By (8.8), $_R R$ is a left generator.

(b) $\Rightarrow$ (d). Every module is an epimorphic image of a direct sum of copies of any generator. Now apply (9.4).

(d) $\Rightarrow$ (c) $\Rightarrow$ (a). This is by Theorem (9.6).   □

This result implies immediately the following characterization of categories of $_R M$ for which R is a semisimple ring.

13.10. Corollary. *For a ring R the following statements are equivalent:*
(a) *R is semisimple;*
(b) *Every monomorphism in $_R M$ splits;*
(c) *Every epimorphism in $_R M$ splits.*   □

13. Exercises

1. Let R have a simple generator $_R T$ and let $D = End(_R T)$.
   (1) Prove that if $_R M$ is simple, then $M \cong T$.
   (2) For some n, $_R R \cong T^{(n)}$ and $R \cong M_n(D)$. (See (13.4).) Prove that $dim(T_D) = n$ and that $\lambda: R \to BiEnd(_R T)$ is an isomorphism. [Hint: With the notation of (13.1), $C_n(D)$ is a simple left R-module.]
   (3) Prove that $Cen\, R \cong Cen(End(_R T))$. [Hint: Exercise (4.6).]
2. Let V be an n-dimensional right D-vector space over the division ring D. Let $v_1, \ldots, v_n$ be a basis for V. Set $R = End(V_D)$.
   (1) Prove that $_R V$ is a simple generator (for $_R M$).
   (2) For each $1 \leq i, j \leq n$, let $e_{ij} \in R$ with

$$e_{ij}(v_k) = \delta_{ik} v_j \qquad (k = 1, \ldots, n).$$

   Prove that $e_{ii} = e_i$ is an idempotent of R and that $Re_i \cong {}_R V$.
   (3) Prove that $_R R = Re_1 \oplus \ldots \oplus Re_n$.
   (4) Deduce that $R_R = e_1 R \oplus \ldots \oplus e_n R$, whence $e_i R$ is a simple generator for R_R.
3. Let D be a division ring, $n \in \mathbb{N}$, $1 \leq k \leq n$, and set

$$C(k) = \{ [\![\alpha_{ij}]\!] \in \mathbb{M}_n(D) \,|\, \alpha_{ij} = \delta_{jk}\alpha_{ij} \}$$

$$R(k) = \{ [\![\alpha_{ij}]\!] \in \mathbb{M}_n(D) \,|\, \alpha_{ij} = \delta_{ik}\alpha_{ij} \}.$$

(1) Prove that $C(k)$ is a simple left ideal of $\mathbb{M}_n(D)$ and $R(k)$ is a simple right ideal of $\mathbb{M}_n(D)$. [Hint: Exercise (13.2).]

(2) Prove that as left $\mathbb{M}_n(D)$ modules $C(k) \cong \mathbb{C}_n(D)$ (see 13.1)) and as right $\mathbb{M}_n(D)$ modules, $R(k) \cong \mathbb{R}_n(D)$.

(3) Prove that the left (right) regular module over $\mathbb{M}_n(D)$ is the direct sum of $C(1), \ldots, C(n)$ (resp. $R(1), \ldots, R(n)$).

4. (1) Let R be a semisimple ring and let I be a proper ideal of R. Prove that R/I is also a semisimple ring.

(2) Show that subrings of semisimple rings need not be semisimple.

5. (1) Let $\phi : R \to S$ be a surjective ring homomorphism. Prove that S is a semisimple ring iff $_R S$ is semisimple.

(2) State and prove necessary and sufficient conditions in order that $\mathbb{Z}_n$ be a semisimple ring. [Hint: Exercise (9.3).]

6. Let $(R_\alpha)_{\alpha \in A}$ be an indexed class of rings. Prove that the product $\Pi_A R_\alpha$ is semisimple iff A is finite and each R_α is semisimple.

7. (1) Prove that if R is isomorphic to a subdirect product of a finite set $(R_k)_{k=1}^n$ of simple rings, then R is a ring direct sum of simple rings. [Hint: Exercises (7.13) and (7.16).]

(2) Prove that if R is isomorphic to a subdirect product of a finite set of semisimple rings, then R is semisimple.

8. (1) Let I be a minimal left ideal of a ring R. Prove that I is a direct summand of $_R R$ iff $I^2 \neq 0$. [Hint: If $I^2 \neq 0$, then $I = Ix$ for some $x \in I$. So there is an $e \in I$ such that $ex = x$. Now suppose $0 \neq R(e - e^2)$.]

(2) Prove that for a left artinian ring R the following are equivalent: (a) R is semisimple; (b) R contains no non-zero nilpotent (nil) left ideals; (c) R contains no non-zero left ideals with square zero; (d) for all $x \in R$, $xRx = 0$ implies $x = 0$.

9. (1) Prove that the converse of Schur's Lemma is false. That is, show that there exists a non-simple module whose endomorphism ring is a division ring. [Hint: Consider a ring of upper triangular matrices.]

(2) Let R be a ring having no non-zero nilpotent left ideals. Prove that if I is a left ideal of R such that $End(_R I)$ is a division ring, then $_R I$ is simple. [Hint: Let $0 \neq x \in I$. Then Rx is not nilpotent, so for some $r \in R$, $\rho_{rx} : I \to I$ is a non-zero endomorphism of I.]

10. A ring R is *left cosemisimple* (or a *left V ring*) in case $_R M$ has a semisimple cogenerator.

(1) Prove that for a ring R the following are equivalent: (a) R is left cosemisimple; (b) $Rad(M) = 0$ for all left R-modules M; (c) Every left R-module is cosemisimple; (d) $_R M$ has a cosemisimple cogenerator. [Hint: See Exercise (9.14).]

(2) Prove that every semisimple ring is cosemisimple.

(3) Prove that if $_R M$ has a simple cogenerator, then R is a simple ring.

(4) Prove that for a ring R the following are equivalent:

(a) R is simple artinian; (b) Every non-zero left R-module is a generator; (c) Every non-zero left R-module is a cogenerator.

11. Let G be a finite group of order n and let K be a field whose characteristic does not divide n. Thus $n = n \cdot 1$ is invertible in K. Using the results of Exercise (7.17) prove

    Maschke's Theorem *If G is a group of order n and if K is a field whose characteristic does not divide n, then the group ring KG is semisimple.*

12. Let R be a finite dimensional algebra over an algebraically closed field. Prove that if R is a simple ring, then $R \cong \mathbb{M}_n(K)$. [Hint: First R is artinian (Exercise (10.7)). If $R \cong \mathbb{M}_l(D)$ where D is a division ring, then $Cen\, D = Cen\, R = K$ (see Exercises (4.4) and (4.5)). So D is finite dimensional over K. Thus $K = D$.]

13. (1) Using the fact that every finite division ring is a field, prove

    Theorem. *Let R be a simple ring of m elements and let K be its center. Then: $K = GF(p^n)$ for some prime p and some n; $m = (p^n)^{k^2}$ for some k; and*

    $$R \cong \mathbb{M}_k(GF(p^n)).$$

    [Here $GF(p^n)$ is the unique (to within isomorphism) finite field of p^n elements.]

    (2) Deduce that there is a natural bijection between the finite semisimple rings (to within isomorphism) and the set of all finite sequences

    $$(p_1, n_1, k_1), \ldots, (p_l, n_l, k_l)$$

    of triples of natural numbers with $p_1, \ldots, p_l$ prime.

§14. The Density Theorem

Recall that if T is a faithful left R-module, then the natural map $\lambda: R \to BiEnd(_R T)$ is an injective ring homomorphism. One consequence of the last section (see Exercise (13.1)) is that if R has a simple generator T, then $_R T$ is faithful, the mapping λ is an isomorphism, and $BiEnd(_R T)$ is the endomorphism ring of a finite dimensional vector space T_D over a division ring $D = End(_R T)$. More generally, in this section we consider those rings R having a faithful simple module T. For such a ring $BiEnd(_R T)$ is the endomorphism ring of a (possibly infinite dimensional) vector space. Then the classical Jacobson-Chevalley Density Theorem asserts that the canonical image of R in $BiEnd(_R T)$ is a "dense" subring. The first step toward proving this is a lemma concerning biendomorphism rings.

Biendomorphism Rings of Direct Sums

Suppose that the module M has a direct summand M'. Then M' is stable under $BiEnd(_RM)$. Indeed, if $M = M' \oplus M''$ and if $e \in End(_RM)$ is the idempotent for M' in this decomposition, then for each $b \in BiEnd(_RM)$

$$b(M') = b(Me) = (bM)e \subseteq M'.$$

Also every endomorphism of M' extends to one of M, so the restriction to M' of a biendomorphism of M is a biendomorphism of M' and the restriction map

$$BiEnd(_RM) \xrightarrow{Res} BiEnd(_RM')$$

is a ring homomorphism.

14.1. Lemma. *Let the left R-module M be the direct sum* $M = M' \oplus M''$ *of submodules M' and M''. Then the restriction map Res is a ring homomorphism making the diagram*

$$
\begin{array}{ccc}
 & R & \\
\lambda \swarrow & & \searrow \lambda \\
BiEnd(_RM) & \xrightarrow{Res} & BiEnd(_RM')
\end{array}
$$

commute. Moreover,

(1) *If M' generates or cogenerates M'', then Res is injective;*

(2) *If M' generates and cogenerates M'', then Res is an isomorphism.*

Proof. The first assertion is an immediate consequence of the preceding remarks. For the others, let

$$S = End(_RM),$$

and let $e \in S$ be the idempotent for M' in the direct decomposition $M = M' \oplus M''$. Then by (5.9) there is a ring isomorphism $\rho : eSe \to End(_RM')$ where for each $s \in S$ and each $x \in M'$, $\rho(ese) : x \mapsto xese$. Thus, it follows that M' is a right eSe module, and

$$BiEnd(_RM') = End(M'_{eSe}).$$

So now let $b \in BiEnd(_RM)$ and suppose that $(b \mid M') = 0$. If M' generates M'', then clearly it generates M (8.4) whence $M'S = Tr_M(M') = M$ (see Exercise (8.6)), so $b(M) = b(M'S) = (bM')S = 0$. On the other hand suppose M' cogenerates M''. Then M' cogenerates M whence (see Exercise (8.6)), $l_M(Se) = Rej_M(M') = 0$. But $(bM)Se = b(MSe) \leq bM' = 0$ so that $bM = 0$. In either case $b = 0$ and (1) is established.

Now for part (2) suppose that M' generates and cogenerates M''; then we need only show that $Res : BiEnd(_RM) \to BiEnd(_RM')$ is surjective. But let $a \in BiEnd(_RM') = End(M'_{eSe})$. We claim that there is an S-homomorphism $\bar{a} : M'S \to M$ such that

$$\bar{a} : \Sigma\, x_i s_i \mapsto \Sigma\, (ax_i)s_i$$

for all $x_i \in M'$ and $s_i \in S$. Indeed, suppose $\Sigma\, x_i s_i = 0$. Then for each $s \in S$, $(\Sigma(ax_i)s_i)se = \Sigma(ax_i)es_i se = a(\Sigma\, x_i es_i se) = 0$. But since M' cogenerates M, $l_M(Se) = 0$, so $\Sigma(ax_i)s_i = 0$, and our claim follows. Also since M' generates

M, $M'S = M$, so $\bar{a} \in BiEnd(_R M)$. Finally, it is clear that $(\bar{a} \mid M') = a$, and thus (2) is proved. □

Now let M be a non-zero left R-module and let A be a non-empty set. Since M is a left $BiEnd(_R M)$-module, the direct sum $M^{(A)}$ is not only a left R-module but also a left $BiEnd(_R M)$ module with respect to "coordinatewise" scalar multiplication. That is, for each $b \in BiEnd(_R M)$ and each $x = (x_\alpha)_{\alpha \in A} \in M^{(A)}$,

$$bx = (bx_\alpha)_{\alpha \in A}.$$

Equivalently (see §2), there is a ring homomorphism μ from $BiEnd(_R M)$ to to $\mathbb{Z}$-endomorphism ring of $M^{(A)}$ such that

$$\mu(b)(x) = (bx_\alpha)_{\alpha \in A}.$$

We claim that in fact these $\mathbb{Z}$-endomorphisms $\mu(b)$ of $M^{(A)}$ are the biendomorphisms of $_R M^{(A)}$. To see this, first let us denote, for each $\alpha \in A$, the α-coordinate injection and projection of $M^{(A)}$ by ι_α and π_α, respectively, and let us view ι_α and π_α as right operators. It is clear that ι_α and π_α are also the α-coordinate injection and projection of $M^{(A)}$ viewed as a $BiEnd(_R M)$-module. (See Exercise (12.7).) Let $\gamma \in A$. Then $\iota_\gamma : M \to M\iota_\gamma$ is an R-isomorphism so (see Exercise (4.14)) there is a ring isomorphism $\phi : BiEnd(_R M) \to BiEnd(_R M\iota_\gamma)$ such that

$$\phi(b)(m\iota_\gamma) = (bm)\iota_\gamma = b(m\iota_\gamma).$$

By (14.1) $Res : BiEnd(_R M^{(A)}) \to BiEnd(_R M\iota_\gamma)$ is an isomorphism. For each $b \in BiEnd(_R M)$ let $\bar{b} = Res^{-1}(\phi(b)) \in BiEnd(_R M^{(A)})$. Then

$$\bar{b}(m\iota_\gamma) = \phi(b)(m\iota_\gamma) = (bm)\iota_\gamma.$$

Then for each $\alpha \in A$, since $\pi_\gamma \iota_\alpha \in End(_R M^{(A)})$, we have

$$\bar{b}(m\iota_\alpha) = \bar{b}(m\iota_\gamma \pi_\gamma \iota_\alpha) = (\bar{b}(m\iota_\gamma))\pi_\gamma \iota_\alpha$$
$$= ((bm)\iota_\gamma)\pi_\gamma \iota_\alpha = b(m\iota_\alpha) = \mu(b)(m\iota_\alpha).$$

Thus, since $(Im\ \iota_\alpha)_{\alpha \in A}$ spans $M^{(A)}$ over $\mathbb{Z}$, we see that $\mu(b) = \bar{b}$. Therefore $\mu = Res^{-1} \circ \phi$ is a ring isomorphism $\mu : BiEnd(_R M) \to BiEnd(_R M^{(A)})$ and we have proved

14.2. Proposition. *Let M be a non-zero left R-module and let A be a non-empty set. Then there is a ring isomorphism*

$$\mu : BiEnd(_R M) \to BiEnd(_R M^{(A)})$$

defined coordinatewise by $\mu(b)(x_\alpha)_{\alpha \in A} = (bx_\alpha)_{\alpha \in A}$. □

Now we are ready to prove

The Density Theorem

14.3. The Density Theorem. *Let M be a semisimple left R-module. If*

$x_1, \ldots, x_n \in M$ and $b \in BiEnd(_RM)$, then there is an $r \in R$ such that

$$bx_i = rx_i \qquad (i = 1, \ldots, n).$$

Proof. Since M is semisimple, $M^{(n)}$ is also semisimple. Thus the cyclic submodule $R(x_1, \ldots, x_n)$ of $M^{(n)}$ is a direct summand of $M^{(n)}$; so $R(x_1, \ldots, x_n)$ is also a $BiEnd(_RM^{(n)})$ submodule of $M^{(n)}$. Then by (14.2), $R(x_1, \ldots, x_n)$ is a $BiEnd(_RM)$-submodule of $M^{(n)}$; in particular,

$$(BiEnd(_RM))(x_1, \ldots, x_n) = R(x, \ldots, x_n).$$

Thus, for $b \in BiEnd(_RM)$ there is an $r \in R$ such that

$$(bx_1, \ldots, bx_n) = b(x_1, \ldots, x_n)$$
$$= r(x_1, \ldots, x_n) = (rx_1, \ldots, rx_n). \qquad \square$$

There is sound topological justification for the name of this theorem. Indeed, consider the cartesian product M^M. Then the product topology on M^M induced by the discrete topology on M is called the "finite topology" on M^M. For $f \in M^M$ a neighborhood base for f in this topology consists of the sets

$$\{g \in M^M \mid f(x_i) = g(x_i) \text{ for all } x_1, \ldots, x_n\}$$

as $\{x_1, \ldots, x_n\}$ ranges over the finite subsets of M. Now suppose that M is an abelian group, and let R and S be subrings of $End(M)$. In particular, R and S inherit the finite topology from M^M. Thus, if R is a subring of S we say that R is *dense* in S (over M) in case in the finite topology R is a dense subset of S. Of course this means that for every finite set $x_1, \ldots, x_n \in M$ and every $s \in S$ there is an $r \in R$ such that

$$rx_i = sx_i \qquad (i = 1, \ldots, n).$$

Suppose next that M is a left R-module. Then the image $\lambda(R)$ of R under the natural map $\lambda: R \to End(M_{\mathbb{Z}})$ is a subring of $BiEnd(_RM)$, and The Density Theorem states that if $_RM$ is semisimple, then $\lambda(R)$ is dense in $BiEnd(_RM)$.

Now we turn to Jacobson's generalization of simple artinian rings and the Wedderburn Structure Theorem for these rings. A ring R is *left primitive* in case it has a simple faithful left module. Since a simple artinian ring has a simple left generator and since a generator is faithful, every simple artinian ring is left primitive. The Wedderburn Theorem asserts that a simple artinian ring is isomorphic to the ring of endomorphisms of a finite dimensional vector space. The generalization for primitive rings is the following.

14.4. The Density Theorem for Primitive Rings. *Let R be a left primitive ring with simple faithful module $_RT$, and let*

$$D = End(_RT).$$

Then D is a division ring, T_D is a D-vector space and, via left multiplication λ, R is isomorphic to a dense subring of $End(T_D)$. In particular, for every finite D-linearly independent set $x_1, \ldots, x_n \in T$ and every $y_1, \ldots, y_n \in T$ there is an $r \in R$ such that

$$rx_i = y_i \qquad (i = 1, \ldots, n).$$

Proof. By Schur's Lemma (13.2), D is a division ring, whence T_D is a D-vector space. Since $_RT$ is faithful and simple, the ring homomorphism $\lambda: R \to End(T_D) = BiEnd(_RT)$ is injective and The Density Theorem (14.3) establishes that the image is dense in $End(T_D)$. For the final statement suppose $x_1, \ldots, x_n \in T$ are D-linearly independent and $y_1, \ldots, y_n \in T$. Then there is a linear transformation $b \in End(T_D)$ such that $b(x_i) = y_i$ $(i = 1, \ldots, n)$. Now apply (14.3). □

The converse is also true and thus there is the following important characterization of left primitive rings.

14.5. Corollary. *A ring is left primitive if and only if it is isomorphic to a dense ring of linear transformations of a vector space. In other words, a ring R is left primitive if and only if there is a division ring D and a bimodule $_RT_D$ with $_RT$ faithful such that for every finite D-linearly independent set $x_1, \ldots, x_n \in T$ and every $y_1, \ldots, y_n \in T$ there is an $r \in R$ such that*

$$rx_i = y_i \qquad (i = 1, \ldots, n).$$

Proof. In view of (14.4) it will suffice to prove that if D is a division ring and $_RT_D$ satisfies the final condition, then R is left primitive. But by hypothesis $_RT$ is faithful. Moreover, it is simple. For if $x \in T$ is non-zero, then $\{x\}$ is D-linearly independent, so again by hypothesis $Rx = T$. Hence R is left primitive. □

14.6. Remarks.

(1) We claimed above that Theorem (14.4) is a generalization of the Wedderburn Theorem for simple artinian rings. For as we noted then a simple artinian ring R is left primitive, so by (14.4) it is isomorphic to a dense subring of $End(M_D)$ for some D-vector space M_D. Using the fact that R is left artinian, it is easy to show (see Exercise (14.4)) that M_D is finite dimensional. Then using density we have another proof of the fact that R is isomorphic to $End(M_D)$.

(2) Every simple ring is primitive (see Exercise (14.1)). The converse fails, however. For example, by (14.5), $End(M_D)$ is primitive for every vector space M_D, but unless M_D is finite dimensional, then $End(M_D)$ is not simple. On the other hand there are simple rings that are not artinian. (See Exercise (11.9).)

(3) One notable feature of the structure theorems for simple artinian rings in §13 was the left-right symmetry of these theorems. Such symmetry does not extend to primitive rings. By a *right primitive* ring we mean a ring having a simple faithful right module. Then it can be seen that left primitive rings need not be right primitive. (See Bergman [64].) In this connection we shall let *primitive ring* mean *left primitive ring*.

Matrix Representation

If D is a division ring and M_D is a right D-vector space, then M_D is free (see

Exercises (8.11) and (2.17)) so the ring $End(M_D)$ of endomorphisms of M_D is isomorphic to a ring of column finite matrices over D. (Exercise (8.12).) That is, if $(x_\alpha)_{\alpha \in \Omega}$ is a basis for M_D, then the mapping

$$a \mapsto [\![a_{\alpha\beta}]\!]$$

from $End(M_D)$ to the ring $\mathbb{CFM}_\Omega(D)$ of all column-finite $\Omega \times \Omega$-matrices over D defined by

$$a(x_\beta) = \Sigma_\alpha x_\alpha a_{\alpha\beta}$$

is an isomorphism. Because of the help this matrix representation can provide, particularly in the study of examples, it deserves a bit of attention here.

Just as in elementary linear algebra, the vector space M_D can be viewed as the set of all column finite $\Omega \times 1$-column vectors over D. Moreover, M_D then has a basis $(e_\alpha)_{\alpha \in \Omega}$ where the column vector e_α is 1 in the "α-row" and 0 elsewhere. Then with the isomorphism

$$a \mapsto [\![a_{\alpha\beta}]\!]$$

from $End(M_D)$ onto $\mathbb{CFM}_\Omega(D)$ determined by this basis, we have

$$a(e_\beta) = [\![a_{\alpha\beta}]\!]e_\beta$$

for each $a \in End(M_D)$ and $\beta \in \Omega$. In other words, we can view M_D as Ω-column vectors, $End(M_D)$ as the ring $\mathbb{CFM}_\Omega(D)$ of $\Omega \times \Omega$-column finite matrices over D, and the action of $End(M_D)$ as given by matrix multiplication.

Now it is easy to check that a subring R of $End(M_D)$ is dense (over M_D) if and only if for each finite set $\beta_1, \ldots, \beta_n \in \Omega$ and finite set $y_1, \ldots, y_n \in M_D$, there is an $a \in R$ with

$$a(e_{\beta_i}) = y_i \qquad (i = 1, \ldots, n).$$

(See Exercise (14.3).) That is, the density condition only needs to be tested on finite subsets of a given basis. For matrix rings this implies that if D is a division ring, if Ω is a non-empty set, and if R is a subring of $\mathbb{CFM}_\Omega(D)$ such that for every finite set $\Gamma \subseteq \Omega$ the restriction of R to $\Omega \times \Gamma$ is $\mathbb{CFM}_{\Omega \times \Gamma}(D)$, then R is primitive. Of course the converse is true in the sense that every primitive ring is isomorphic to such a subring of some $\mathbb{CFM}_\Omega(D)$. In particular, if D is a division ring and if R is a subring of $\mathbb{CFM}_\mathbb{N}(D)$, then R is dense in $\mathbb{CFM}_\mathbb{N}(D)$ (and hence is primitive) iff for each $n \in \mathbb{N}$ and each $U \in \mathbb{M}_n(D)$, there is a matrix in R of the form

$$\begin{bmatrix} U & X \\ 0 & Y \end{bmatrix}.$$

14. Exercises

1. (1) Prove that every simple ring is both left and right primitive.

(2) Prove that every commutative primitive ring is a field.

2. Let Q be a field. For each $n \in \mathbb{N}$, each $A \in \mathbb{M}_n(Q)$, and each $s \in Q$, let $[A, s] \in \mathbb{M}_\mathbb{N}(Q)$ be the matrix

$$
[A, s] = \begin{bmatrix} A & & & \\ & s & & \\ & & s & 0 \\ 0 & & & s \\ & & & & \ddots \\ & & & & & \ddots \end{bmatrix}.
$$

Let S be a subring of Q and $R = \{[A, s] \mid s \in S,\ A \in \mathbb{M}_n(Q),\ n \in \mathbb{N}\}$.

(1) Prove that R is a primitive subring of $\mathbb{CFM}_\mathbb{N}(Q)$ with $Cen\ R \cong S$. Thus each subring of a field is the center of some primitive ring. (See Exercise (1.9).)

(2) Prove that if S is not a field, then R has a non-primitive factor ring.

3. Let D be a division ring and let M_D be a right D-vector space with basis $(x_\alpha)_{\alpha \in A}$. Let R be a subring of $End(M_D)$. Prove that R is dense in $End(M_D)$ iff for each finite subset $F \subseteq A$ and every set $(m_\gamma)_{\gamma \in F}$ of elements of M, there is an $r \in R$ with $r(x_\gamma) = m_\gamma$ for all $\gamma \in F$.

4. Let D be a division ring, let M_D be a right D-vector space, and let R be a dense subring of $End(M_D)$. Prove that if $x_1, x_2, x_3, \ldots$ are D-linearly independent in M, then

$$
l_R(x_1) > l_R(x_1, x_2) > l_R(x_1, x_2, x_3) > \ldots.
$$

5. Let M_D be an infinite dimensional vector space over a division ring D and let R be a subring of $End(M_D)$. Prove that if R is dense in $End(M_D)$, then for each $n \in \mathbb{N}$ there is a subring S_n of R and a surjective ring homomorphism $\phi_n : S_n \to \mathbb{M}_n(D)$.

6. Let $n > 1$ and let R be a primitive ring such that $x^n = x$ for each $x \in R$. Prove that R is a division ring. (Actually, R is a field. Can you prove it?)

7. (1) Prove that if R is primitive and $e \in R$ is a non-zero idempotent, then eRe is primitive. [Hint: If ${}_R M$ is simple, then eM is either 0 or eRe-simple.]

(2) Let R be a ring and let $n > 1$. Prove that R is left primitive iff $\mathbb{M}_n(R)$ is left primitive. [Hint: Exercise (1.8).]

8. Let M and N be left R-modules. Prove that

(1) If M is balanced and either generates or cogenerates N, then $M \oplus N$ is balanced.

(2) If M generates and cogenerates N and N generates and cogenerates M, then

$$
BiEnd({}_R M) \cong BiEnd({}_R N) \quad \text{and} \quad Cen(End({}_R M)) \cong Cen(End({}_R N)).
$$

[Hint: Exercise (4.6).]

9. Let M be a left R-module and let λ be the canonical ring homomorphism $R \to BiEnd(_R M)$.

   (1) Prove that the following assertions are equivalent: (a) $\lambda(R)$ is dense in $BiEnd(_R M)$; (b) For each $n > 0$, each R-submodule of $M^{(n)}$ is a $BiEnd(_R M)$ submodule of $M^{(n)}$; (c) For each $n > 0$, each R-submodule of $M^{(n)}$ is a $BiEnd(_R M^{(n)})$ submodule of $M^{(n)}$.

   (2) Prove that if $_R M$ is a cogenerator, then $\lambda(R)$ is dense in $BiEnd(_R M)$. [Hint: Part (1) and Exercise (8.5).]

10. A ring R is *prime* in case each non-zero left ideal is faithful. Prove that

    (1) A commutative ring is prime iff it is an integral domain.

    (2) For a ring R, these are equivalent:

    (a) R is prime;

    (b) Every non-zero right ideal is (right) faithful;

    (c) For each pair I_1, I_2 of non-zero ideals, $I_1 I_2 \neq 0$;

    (d) For all $x, y \in R$, $xRy = 0$ implies $x = 0$ or $y = 0$.

    (3) Every primitive ring is prime.

    (4) Every left artinian prime ring is simple.

11. (1) Let R be a prime ring. Prove that if $Soc \,_R R \neq 0$, then R is primitive, $Soc \,_R R$ is homogeneous, and $Soc \,_R R \trianglelefteq \,_R R$.

    (2) Prove that if R is prime and $Soc \,_R R$ is non-zero and of finite length, then R is simple artinian.

    (3) Prove that if R is prime and $Soc \,_R R$ is a simple left ideal, then R is a division ring.

    (4) Show that there exist primitive rings R with $Soc \,_R R = 0$. [Hint: $End(M_D)$ for M_D a vector space has a simple factor ring.]

12. (1) Prove that if R is a prime ring and $e \in R$ is a non-zero idempotent, then eRe is a prime ring. [Hint: Exercise (14.10.2).]

    (2) Let R be a ring and $n > 1$. Prove that R is prime iff $\mathbb{M}_n(R)$ is prime. [Hint: Exercise (1.8).]

13. Let V be an infinite dimensional vector space over a division ring D. For each $f \in End(V_D)$ define the *rank* of f by $rank f = dim(Im f)$.

    (1) Let c be an infinite cardinal. Prove that

    $$I_c = \{f \in End(V_D) \mid rank f < c\}$$

    is an ideal of $End(V_D)$.

    (2) Let I be an ideal of $End(V_D)$ and let $f \in I$. Prove that

    $$\{g \in End(V_D) \mid rank g \leq rank f\}$$

    is contained in I.

    (3) Prove that $I_{\aleph_0}$ is the unique minimal ideal of $End(V_D)$ and $I_{dim V}$ is the unique maximal ideal of $End(V_D)$.

    (4) It can be shown that the collection of cardinal numbers that are less than or equal to a given cardinal is well ordered by $\leq$. (See Stoll [63].) Using this fact prove that $I \neq 0$ is a proper ideal of $End(V_D)$ iff $I = I_c$ for some cardinal c such that $\aleph_0 \leq c \leq dim V$. Conclude that the ideal lattice of $End(V_D)$ is well ordered.

§15. The Radical of a Ring—Local Rings and Artinian Rings

Let R be a ring. Then $End(_RR)$ is simply the ring of right multiplications by elements of R (see (4.11)). Thus, by (9.14) the radical $Rad(_RR)$ of $_RR$ is a (two-sided) ideal of R. This ideal of R is called the (*Jacobson*) *radical of* R, and we usually abbreviate

$$J(R) = Rad(_RR).$$

One consequence of the first theorem of this section (15.3) is that this radical is also $Rad(R_R)$; therefore we have to contend with no left-right ambiguity.

Primitive Ideals

The first goal of this section is to obtain several characterizations of the radical $J(R)$ of a ring. One of the more important of these is that $J(R)$ is the smallest ideal modulo which R is "residually primitive". We say that an ideal P of R is a (*left*) *primitive ideal* in case R/P is a (left) primitive ring. Similarly, a *right primitive ideal* is an ideal P of R such that R/P is a right primitive ring. Since every simple ring is primitive (see Exercise (14.1)), every maximal ideal is both left and right primitive. However, although a left primitive ideal of R is a two-sided ideal, it need not be right primitive (see Bergman [64]). Of course, the primitive ideals of R are simply the kernels of the ring homomorphisms of R onto dense rings of linear transformations of vector spaces. Another easy characterization:

15.1. Proposition. *An ideal P of a ring R is a primitive ideal if and only if there exists a maximal left ideal M of R such that*

$$P = l_R(R/M) = Rej_R(R/M).$$

Proof. The factor ring R/P is primitive iff R/P has a faithful simple module iff (2.12) P is the annihilator of a simple left R-module. The second equality is by (8.22). □

Characterizations of the Radical

Since R is finitely generated as a left R-module, its radical $J(R)$ is the unique largest superfluous left R-submodule of R. (See (9.18) and (10.5.1).) Also since every nilpotent left ideal of R is superfluous in $_RR$, the radical $J(R)$ contains all nilpotent left ideals. (See Exercise (5.18).) Indeed, if R is left artinian, then $J(R)$ is the unique largest nilpotent left ideal of R. (See (15.19).) In general, however, $J(R)$ is not nilpotent, or even nil. But there is a useful generalization of nilpotence that leads to a generalization of the above characterization of $J(R)$ for left artinian rings.

An element $x \in R$ is *left quasi-regular* in case $1 - x$ has a left inverse in R. Similarly $x \in R$ is *right quasi-regular* (*quasi-regular*) in case $1 - x$ has a right (two-sided) inverse in R. Of course, an element of R can be left quasi-regular but not right quasi-regular. A subset of R is *left quasi-regular* (etc.) in case each element of R is left quasi-regular (etc.). This does generalize nilpotence. For if $x \in R$ with $x^n = 0$, then

$$(1 + x + \ldots + x^{n-1})(1 - x) = 1 = (1 - x)(1 + x + \ldots + x^{n-1})$$

so x is quasi-regular.

15.2. Proposition. *For a left ideal I of R the following statements are equivalent:*

(a) *I is left quasi-regular;*
(b) *I is quasi-regular;*
(c) *I is superfluous in R.*

Proof. (a) $\Rightarrow$ (b). Assume (a) and let $x \in I$. Then x is left quasi-regular, so $x'(1 - x) = 1$ for some $x' \in R$. Thus since $x'x \in I$ is left quasi-regular and since $x' = 1 + x'x = 1 - (-x'x)$, there is a $y \in R$ such that $yx' = 1$. But then x' is invertible and $y = 1 - x$. So $(1 - x)x' = 1$ and x is quasi-regular.

(b) $\Rightarrow$ (c). Assume (b) and let K be a left ideal of R with $R = I + K$. Then there exist $x \in I$ and $k \in K$ with $1 = x + k$. So $k = 1 - x$ is invertible whence $1 \in K$ and $K = R$.

(c) $\Rightarrow$ (a). Assume (c) and let $x \in I$. Then $Rx \ll R$. But $R = Rx + R(1 - x)$, whence $R(1 - x) = R$, so $1 - x$ has a left inverse. $\qquad\square$

Now we come to an important multiple characterization of $J(R)$. One consequence of it is that there is a fourth condition equivalent to the three of (15.2), namely, that I is right quasi-regular.

15.3. Theorem. *Given a ring R each of the following subsets of R is equal to the radical $J(R)$ of R.*

(J_1) *The intersection of all maximal left (right) ideals of R;*
(J_2) *The intersection of all left (right) primitive ideals of R;*
(J_3) *$\{x \in R \mid rxs \text{ is quasi-regular for all } r, s \in R\}$;*
(J_4) *$\{x \in R \mid rx \text{ is quasi-regular for all } r \in R\}$;*
(J_5) *$\{x \in R \mid xs \text{ is quasi-regular for all } s \in R\}$;*
(J_6) *The union of all the quasi-regular left (right) ideals of R;*
(J_7) *The union of all the quasi-regular ideals of R;*
(J_8) *The unique largest superfluous left (right) ideal of R.*

Moreover, (J_3), (J_4), (J_5), (J_6) and (J_7) also describe the radical $J(R)$ if "quasi-regular" is replaced by "left quasi-regular" or by "right quasi-regular".

Proof. To denote the right-hand version of (J_1), (J_2), (J_6) and (J_8) we shall append an asterisk. Thus J_1^* is the intersection of all maximal right ideals of R. Now by (9.17) and (15.1), we have that

$$J_1 = Rad(_RR) = \cap\{Rej_R(T) \mid {}_RT \text{ is simple}\} = J_2.$$

Also since $_RR$ is finitely generated, we know from (9.13) and (10.4.1) that $J_1 = J_8$ and by (15.2) we see that $J_6 = J_8$. Of course, also

$$J_1^* = J_2^* = J_6^* = J_8^*.$$

But since J_2 and J_2^* are ideals, J_6 and J_6^* are ideals. So clearly $J_6 = J_6^* = J_7$. Now it is immediate that

$$J_6 \subseteq J_3 \subseteq J_4 \subseteq J_6,$$

and

$$J_6^* \subseteq J_3 \subseteq J_5 \subseteq J_6^*,$$

so we do have the equality claimed by the first assertion of the theorem. Also in their left quasi-regular versions

$$J_7 \subseteq J_3 \subseteq J_4 \subseteq J_6.$$

But thanks to (15.2) the left quasi-regular version and the quasi-regular version of J_6 are equal as are the two versions of J_7. Similarly, the right quasi-regular versions of J_3, J_5, J_6^* and J_7 equal $J(R)$. Now we have all the claimed sets equal $J(R)$ except for the left quasi-regular versions of J_5 ($= J_6^*$ with left quasi-regular) and the right quasi-regular versions of J_4 ($= J_6$ with right quasi-regular). We shall show that the right quasi-regular version of J_4 is the radical and let symmetry handle the other. Clearly, in their right quasi-regular forms $J_3 \subseteq J_4$ and as we have seen, $J_3 = J_1^*$; so in order to accomplish this it only remains to show that every right quasi-regular left ideal is contained in $J_1^* = Rad(R_R)$. Suppose then that Rx is right quasi-regular and that $x \notin J_1^*$; then there exists a maximal right ideal K of R with $x \notin K$. Thus for some $r \in R$ and $k \in K$

$$1 = xr + k.$$

Now rx is right quasi-regular, so there is a $u \in R$ with

$$(1 - rx)u = 1.$$

Therefore,

$$x = x(1 - rx)u = (x - xrx)u$$

$$= xu - (1 - k)xu = kxu \in K.$$

This contradiction means that $Rx \subseteq J_1^*$ as claimed, and the proof is complete. □

15.4. Corollary. *If R is a ring, then*

$$Rad(_R R) = J(R) = Rad(R_R). \qquad □$$

In view of (8.22), (15.1), and the fact (15.3) that $J(R) = J_1 = J_2$, we have immediately

15.5. Corollary. *If R is a ring, then $J(R)$ is the annihilator in R of the class of simple left (right) R-modules.* □

A key fact about the Jacobson radical $J(R)$ of R (or indeed about any "radical"; see p. 174) is

15.6. Corollary. *If I is an ideal of a ring R, and if $J(R/I) = 0$, then $J(R) \subseteq I$.*

Proof. If $x \notin I$, there exists a maximal left ideal M of R with $I \subseteq M$ and $x \notin M$ (see (15.3) and (3.8)), so $x \notin J(R)$. ☐

15.7. Corollary. *For an ideal I of a ring R the following are equivalent:*
(a) $I = J(R)$;
(b) *I is left quasi-regular and $J(R/I) = 0$;*
(c) *I is left quasi-regular and $J(R) \subseteq I$;*
(d) $_RI$ *is superfluous in $_RR$ and $J(R/I) = 0$;*
(e) $_RI$ *is superfluous in $_RR$ and $J(R) \subseteq I$.* ☐

The radical of a factor ring of R is at least as big as the corresponding factor of $J(R)$, but they need not be equal. Indeed, the ring $\mathbb{Z}$ has zero radical, but $\mathbb{Z}_4$ does not. (See Exercise (15.1).)

15.8. Corollary. *If R and S are rings and if $\phi : R \to S$ is a surjective ring homomorphism, then $\phi(J(R)) \subseteq J(S)$. Moreover, if $\mathrm{Ker}\,\phi \subseteq J(R)$, then $\phi(J(R)) = J(S)$. In particular,*

$$J(R/J(R)) = 0.$$

Proof. Clearly since ϕ is surjective, $\phi(J(R))$ is a quasi-regular ideal of S; thus $\phi(J(R)) \subseteq J(S)$ by (15.3). On the other hand suppose $\mathrm{Ker}\,\phi \subseteq J(R)$. If M is a maximal left ideal of R, then $\mathrm{Ker}\,\phi \subseteq J(R) \subseteq M$, so by (3.11), $\phi(M)$ is a maximal left ideal of S and by (15.3), $J(S) \subseteq \phi(M)$. But also by (15.3) and (3.11), $\phi(J(R))$ is the intersection of all $\phi(M)$ for M a maximal left ideal of R. So $J(S) \subseteq \phi(J(R))$ ☐

15.9. Corollary. *If R is the ring direct sum of ideals $R_1, R_2, \ldots, R_n$, then*

$$J(R) = J(R_1) + J(R_2) + \ldots + J(R_n).$$

Proof. Let $1 = u_1 + u_2 + \ldots + u_n$ where $u_1, u_2, \ldots, u_n$ are pairwise orthogonal central idempotents. Then it is easy to see that I is a quasi-regular ideal in R if and only if $I = I_1 + \ldots + I_n$ where, for each $k = 1, \ldots, n$, I_k is a quasi-regular ideal in the ring $u_k R u_k = R_k$. ☐

As we have already noted, $J(R)$ contains every nilpotent left ideal of R. Recall that an ideal (left, right, or two-sided) is *nil* in case each of its elements is nilpotent. Thus, more generally,

15.10. Corollary. *If R is a ring, then every nil left, right, or two-sided ideal of R is left quasi-regular, whence every nil left, right, or two-sided ideal of R is contained in $J(R)$.*

Proof. Every nilpotent element $x \in R$ is left quasi-regular, for if $x^n = 0$, $(1 + x + \ldots + x^{n-1})(1 - x) = 1$. ☐

15.11. Corollary. *If R is a ring, then $J(R)$ contains no non-zero idempotent.*

Proof. If $e \in R$ is idempotent and if $e \in J(R)$, then Re is a superfluous direct summand of $_RR$. Thus, $e = 0$. ☐

15.12. Corollary. *Let I be an ideal of the ring R. If I is nil and if $J(R/I) = 0$, then $I = J(R)$. On the other hand, if $I \subseteq J(R)$ and if every non-zero left ideal of R/I contains a non-zero idempotent, then $I = J(R)$.*

Proof. The first assertion is immediate from (15.7) and (15.10). The second follows from (15.11) and (15.8).   □

Recall (9.18) and (10.5) that if $_RM$ is non-zero and finitely generated, then *Rad M* $\neq M$. Using this fact we have the following very useful characterization of $J(R)$. It is often called Nakayama's Lemma.

15.13. Corollary. *For a left ideal I of a ring R, the following are equivalent:*

(a) *$I \leq J(R)$;*
(b) *For every finitely generated left R-module M, if $IM = M$, then $M = 0$;*
(c) *For every finitely generated left R-module M, IM is superfluous in M.*

Proof. (a) $\Rightarrow$ (b). Suppose $M \neq 0$ is finitely generated. Then M has a maximal submodule K. (See (10.5).) So by (15.5), $J(R)M \leq K$.

(b) $\Rightarrow$ (c). Suppose $N \leq M$ and $IM + N = M$. Then

$$I(M/N) = (IM + N)/N = M/N.$$

So if M is finitely generated, (b) implies $M/N = 0$.

(c) $\Rightarrow$ (a). Assume (c). Then since $_RR$ is finitely generated, $IR \ll R$. Thus, $I \leq IR \leq J(R)$.   □

Semiprimitive Rings

A ring R is *semiprimitive* in case $J(R) = 0$. In particular, a primitive ring is semiprimitive. The left-right symmetry that holds for simple artinian and semisimple rings but that fails for primitive rings (see (14.6.3)) reappears for semiprimitive ones.

15.14. Proposition. *For a ring R the following are equivalent:*

(a) *R is semiprimitive;*
(b) *$_RR$ is cogenerated by the class of simple left R-modules;*
(c) *R_R is cogenerated by the class of simple right R-modules;*
(d) *R has a faithful semisimple module.*

Proof. This is immediate by (15.5) and (9.17).   □

Since $J(R) = J_2$ (see (15.3)), a ring R is semiprimitive iff it is isomorphic to a subdirect product of primitive rings. (See Exercise (7.16).)

Remark. The term "semisimple" is often used for rings R with $J(R) = 0$. This is confusing because a semiprimitive ring need not be semisimple. However (see (15.16)), the semisimple rings are precisely the artinian semiprimitive rings.

Local Rings

Recall from §12 that a ring R is a local ring in case the set of non-invertible elements of R is closed under addition. Using the radical we have the following characterization of this important class of rings.

15.15. Proposition. *For a ring R the following statements are equivalent:*
(a) *R is a local ring;*
(b) *R has a unique maximal left ideal;*
(c) *$J(R)$ is a maximal left ideal;*
(d) *The set of elements of R without left inverses is closed under addition;*
(e) *$J(R) = \{x \in R \mid Rx \neq R\}$;*
(f) *$R/J(R)$ is a division ring;*
(g) *$J(R) = \{x \in R \mid x \text{ is not invertible}\}$;*
(h) *If $x \in R$ then either x or $1 - x$ is invertible.*

Proof. (b) $\Leftrightarrow$ (c). This is immediate from the definition of $J(R)$.

(c) $\Rightarrow$ (d). Assume (c). Then by (15.3) $J(R)$ is the unique maximal left ideal of R. Let $x, y \in R$ be non left invertible. Then since every proper left ideal is contained in a maximal one (10.5), $Rx, Ry \leq J(R)$, whence $x + y \in J(R)$. So $x + y$ is not left invertible.

(d) $\Rightarrow$ (e). Assume (d). Since $J(R)$ is a proper left ideal it will clearly suffice to prove that if $x \in R$ with $Rx \neq R$, then $x \in J(R)$. But then for each $r \in R$, rx does not have a left inverse and $1 = rx + (1 - rx)$, so by (d), $1 - rx$ does have a left inverse. Thus by (15.3), $x \in J_4 = J(R)$.

(e) $\Rightarrow$ (f). Assuming (e) it follows that every non-zero element of $R/J(R)$ has a left inverse there. Thus $R/J(R)$ is a division ring. (See Exercise (1.2).)

(f) $\Rightarrow$ (b). Since a division ring has no non-trivial left ideals, if $R/J(R)$ is a division ring, then $J(R)$ is a maximal left ideal. (See (3.8).)

(h) $\Rightarrow$ (g). Assume (h). Let $x \in R$ be non-invertible, say x has no left inverse. Then no rx is invertible, so by (h) each rx is quasi-regular. Thus $x \in J(R)$.

(f) $\Rightarrow$ (g). Assume (f). Suppose $x \in R$ and $x \notin J(R)$. Then by hypothesis x is invertible modulo $J(R)$. That is, $Rx + J(R) = R$ and $xR + J(R) = R$. But since $J(R) = J_8$ in (15.3), $Rx = R$ and $xR = R$. So x is invertible.

(g) $\Rightarrow$ (f) and (g) $\Rightarrow$ (a) $\Rightarrow$ (h). These are clear.   □

Rings Semisimple Modulo the Radical

One of the most significant of these characterizations is that R is local iff it is a division ring modulo its radical. In particular, a local ring is semisimple modulo its radical. Another class of rings with this property is the class of artinian rings.

15.16. Proposition. *Let R be left artinian. Then R is semisimple if and only if $J(R) = 0$. In particular, $R/J(R)$ is semisimple.*

Proof. The first assertion is just (10.15). The second follows from the first and (15.8) since $R/J(R)$ is a left artinian ring.   □

In general rings do not have any semisimple factor rings. For example, no simple non-semisimple ring can have a semisimple factor ring (see Exercise (10.14)). However, rings R for which $R/J(R)$ is semisimple are of considerable interest. Some of the reason for their importance is given in:

15.17. Proposition. *For a ring R with radical $J(R)$ the following statements are equivalent:*

(a) $R/J(R)$ *is semisimple;*

(b) $R/J(R)$ *is left artinian;*

(c) *Every product of simple left R-modules is semisimple;*

(d) *Every product of semisimple left R-modules is semisimple;*

(e) *For every left R-module M, $Soc\ M = r_M(J(R))$.*

Proof. (a) ⇔ (b) is immediate from (15.16), (10.15), and the fact (15.8) that $J(R/J(R)) = 0$.

(a) ⇒ (e). By (15.5), $J(R)$ annihilates every simple left R-module. Thus, $Soc\ M \subseteq r_M(J(R))$ for every $_RM$. But $J(R)r_M(J(R)) = 0$. Therefore $r_M(J(R))$ is an $R/J(R)$ module. So assuming (a), we have that $r_M(J(R))$ is semisimple and contained in $Soc\ M$.

(e) ⇒ (d). Since $J(R)$ annihilates all semisimple modules it annihilates all products of semisimple modules. Thus, assuming (e) we have that every product of semisimple modules is its own socle and hence is semisimple.

(d) ⇒ (c). This is clear.

(c) ⇒ (a). We know that $R/J(R)$ is cogenerated by simple R-modules. Thus by (9.4) it follows that (c) implies (a).   □

For any left R-module M the factor module $M/Rad\ M$ is cogenerated by the simple left R-modules. But by (15.5) we know that $J(R)$ annihilates all simple modules, so certainly it annihilates $M/Rad\ M$. In other words,

$$J(R)M \leq Rad\ M.$$

In general, equality does not hold. But if R is semisimple modulo its radical, then not only does $J(R)$ determine the socle of each $_RM$, by $Soc\ M = r_M(J(R))$, but also, it determines the radical of M. Specifically,

15.18. Corollary. *Let R be a ring with radical $J = J(R)$. Then for every left R-module M,*

$$JM \leq Rad\ M.$$

If R is semisimple modulo its radical, then for every left R-module M,

$$JM = Rad\ M$$

and M/JM is semisimple.

Proof. The first inequality has been established above. Now assume that R/J is semisimple. Let M be a left R-module. Then M/JM is a semisimple

R/J-module, and hence (2.12) a semisimple R-module. Therefore

$$Rad(M/JM) = 0$$

and by (9.15) it follows that $Rad\ M \leq JM$.   □

The Radical of an Artinian Ring

If R is left artinian, then its radical $J(R)$ is the unique smallest ideal modulo which R is semisimple. Now we can also characterize $J(R)$ for artinian rings as the unique largest nilpotent ideal.

15.19. Theorem. *If R is a left artinian ring, then its radical $J(R)$ is the unique largest nilpotent left, right, or two-sided ideal in R.*

Proof. In view of (15.10) it will suffice to prove that for a left artinian ring R its radical $J = J(R)$ is nilpotent. But since we do have

$$J \supseteq J^2 \supseteq J^3 \supseteq \ldots,$$

if R is left artinian, then $J^n = J^{n+1}$ for some $n > 0$. Suppose $J^n \neq 0$. Then the collection of left ideals of R that are not annihilated by J^n is not empty. So (10.10) there is a left ideal I of R minimal with respect to the property $J^n I \neq 0$. Let $x \in I$ with $J^n x \neq 0$. Then $Jx \leq Rx \leq I$ and $J^n(Jx) = J^{n+1}x = J^n x \neq 0$. So by the minimality of I, we have $Jx = Rx$, contrary to (15.13).   □

Now it is easy to prove the following very remarkable result:

15.20. Theorem [Hopkins]. *Let R be a ring with $J = J(R)$. Then R is left artinian if and only if R is left noetherian, J is nilpotent, and R/J is semisimple.*

Proof. If R is left artinian, then by (15.19) J is nilpotent and by (15.16) R/J is semisimple. So we may assume that R/J is semisimple and that J is nilpotent, say $J^n = 0$. We induct on n. If $n = 1$, then $R = R/J$ is semisimple, so (10.16) gives the proof. So let $n > 1$ and assume the result for every ring of nilpotency index less than n. By (15.12), $J(R/J^{n-1}) = J/J^{n-1}$, so our inductive assumption implies that R/J^{n-1} is left artinian iff it is left noetherian. Now there is a short exact sequence of left R-modules:

$$0 \to J^{n-1} \to R \to R/J^{n-1} \to 0.$$

So by (10.12) R is left artinian (noetherian) iff both J^{n-1} and R/J^{n-1} are. But since $J^n = J(J^{n-1}) = 0$ and since R/J is semisimple, J^{n-1} is semisimple. Thus by (10.16), J^{n-1} is artinian iff it is noetherian.   □

15.21. Corollary. *Let R be left artinian. If M is a left R-module, then*

$$Soc\ M = r_M(J) \trianglelefteq M \qquad and \qquad Rad\ M = JM \ll M.$$

Moreover, for M the following statements are equivalent:
  (a) *M is finitely generated;*
  (b) *M is noetherian;*

(c) *M has a composition series;*
(d) *M is artinian;*
(e) *M/JM is finitely generated.*

Proof. Let $0 \neq x \in M$. Then Rx is a factor of R, hence, Rx is artinian and $Soc\ Rx = Rx \cap (Soc\ M) \neq 0$ by (10.11). By (15.16) R/J is semisimple, so by (15.17.e), $Soc\ M = r_M(J)$. By (15.16) and (15.18), $Rad\ M = JM$. By (15.19), J is nilpotent, so $JM \ll M$. (See Exercise (5.18).)

(a) $\Rightarrow$ (c). If M is finitely generated, then there is an R-epimorphism $R^{(n)} \to M \to 0$. But then since $_RR$ is both artinian and noetherian, so is M ((10.12) and (10.13)), i.e., M has a composition series (11.1).

(c) $\Rightarrow$ (b) and (d). By (11.1).

(b) $\Rightarrow$ (a). By (10.9).

(d) $\Rightarrow$ (e). If M is artinian, then so is M/JM (10.12); hence by (15.18) and (10.15) M/JM is finitely generated.

(e) $\Leftrightarrow$ (a). Since $JM \ll M$, this follows from (15.18) and (10.4). $\qquad\square$

Levitzki's Theorem

It follows from (15.10) and (15.19) that if R is left artinian, then every nil one-sided ideal of R is actually nilpotent. This fact admits the following generalization to left noetherian rings.

15.22. Theorem [Levitzki]. *If R is left noetherian, then every nil one-sided ideal of R is nilpotent.*

Proof. Let R be left noetherian. Then by (10.9) R has a maximal nilpotent ideal, say N. Let $S = R/N$. Then 0 is the only nilpotent ideal in S. We claim that 0 is the only nil right ideal in S. To see this, suppose $0 \neq I \leq S_S$ is nil. Since S is left noetherian, the set $\{l_S(x) \mid 0 \neq x \in I\}$ has a maximal element, say $l_S(x)$. Let $s \in S$ with $xs \neq 0$. Now $xs \in I$ is nilpotent, say $(xs)^{k+1} = 0$ and $(xs)^k \neq 0$. Clearly $l_S(x) \subseteq l_S((xs)^k)$, so by maximality $l_S(x) = l_S((xs)^k)$. Thus $xsx = 0$. Therefore $(SxS)^2 = 0$, $x = 0$, and the claim is established. Thus if I is a nil right ideal in R, then $(I + N)/N = 0 \in R/N$ and $I \subseteq N$, i.e., N contains every nil right ideal of R. But if $a \in R$ and Ra is nil, then aR is also nil (if $(ra)^n = 0$, then $(ar)^{n+1} = 0$) so we see that N also contains every nil left ideal of R. Since N is nilpotent, this completes the proof. $\qquad\square$

Combining (15.20) and (15.22) we have

15.23. Corollary. *Let R be left noetherian. If $R/J(R)$ is semisimple and if $J(R)$ is nil, then R is left artinian.*

General Radicals

Given any non-empty class $\mathscr{P}$ of rings, there is for each ring R an associated $\mathscr{P}$-radical, $Rad_{\mathscr{P}}(R)$, the intersection of all kernels of the surjective homomorphisms $R \to P$ for $P \in \mathcal{P}$. Many of the fundamental properties of these general radicals are very easy to prove. For example, the $\mathcal{P}$-radical is an ideal, the ring modulo the $\mathcal{P}$-radical has zero $\mathcal{P}$-radical, the $\mathcal{P}$-radical is 0 iff R is isomorphic to a subdirect product (see Exercise (7.16)) of rings in $\mathcal{P}$, etc. The Jacobson radical $J(R)$ is just the $\mathcal{P}$-radical for $\mathcal{P}$ the class of primitive rings. In the exercises we shall look at another important radical, namely the one induced by the class of "prime rings".

15. Exercises

1. Compute the Jacobson radical of each of the following rings:
   (1) $\mathbb{Z}$.
   (2) $\mathbb{Z}_n$.
   (3) R where R is a Boolean ring.
   (4) The ring of all $n \times n$ upper triangular matrices over a field K.
2. Let R be the ring of all upper triangular $\mathbb{N} \times \mathbb{N}$-matrices over a field that are 0 a.e. off the main diagonal.
   (1) Show that the Jacobson radical $J(R)$ of R is the subset of all strictly upper triangular matrices (i.e., all $[\![a_{ij}]\!] \in R$ with $a_{ii} = 0$ for all i).
   (2) Show that $J(R)$ is nil but not nilpotent.
3. Let $(M_\alpha)_{\alpha \in A}$ be an indexed class of left R-modules and let I be a right ideal of R.
   (1) Prove that $I(\Pi_A M_\alpha) \subseteq \Pi_A(I M_\alpha)$ and that equality holds whenever I is finitely generated.
   (2) Show that if I is not finitely generated, then equality need not hold in part (1). [Hint: Try $R = \mathbb{Z}^{\mathbb{N}}$ and $I = \mathbb{Z}^{(\mathbb{N})}$.]
   (3) Prove that if R is right artinian, then $Rad(\Pi_A M_\alpha) = \Pi_A(Rad\, M_\alpha)$.
4. Let $(R_\alpha)_{\alpha \in A}$ be an indexed class of rings. Prove that
$$J(\Pi_A R_\alpha) = \{r \in \Pi_A R_\alpha \mid \pi_\alpha(r) \in J(R_\alpha) \quad (\alpha \in A)\}.$$
   That is, with a few notational liberties, $J(\Pi_A R_\alpha) = \Pi_A J(R_\alpha)$.
5. Let R be a commutative ring, and let $\mathcal{I}$ be the set of all maximal ideals of R. Prove that $Rad(M) = \cap_{I \in \mathcal{I}} IM$. [Hint: If L is a maximal submodule of M, then $M/L \cong R/I$ for some $I \in \mathcal{I}$ and $IM \subseteq L$.]
6. Let R be left artinian with radical J and with $T_1, \ldots, T_n$ a complete set of representatives of the simple left R-modules. Prove that
   (1) A two-sided ideal of R is primitive iff it is maximal.
   (2) $Rej_R(T_1), \ldots, Rej_R(T_n)$ is the set of maximal ideals of R.
   (3) $J = \cap_{i=1}^n Rej_R(T_i)$.
   (4) If $_R T$ is simple, then $R/Rej_R(T) \cong Tr_{R/J}(T)$.
7. Let $p \in \mathbb{P}$ be a prime and let J be the radical of $\mathbb{Z}_{(p)}$. (See Exercise (2.12).) Prove that J is not nil but that $\cap_{n=1}^\infty J^n = 0$.
8. Let Ω be an uncountable well-ordered set, let Q be a field, and let V be a right Q-vector space with ordered basis $(x_\alpha)_{\alpha \in \Omega}$. For each $\alpha \in \Omega$, set

$$V_\alpha = \Sigma_{\beta < \alpha} Q x_\beta.$$

Let R be the subset of $End(V_Q)$ of those f such that for some scalar a_f

(i) $\dim_K Im(f - a_f 1_V) < \infty$

(ii) $(f - a_f 1_V)(x_\alpha) \in V_\alpha$ $(\alpha \in \Omega)$.

(1) Prove that R is a subring of $End(V_Q)$. [Note: R can be represented as the ring of all $\Omega \times \Omega$ upper triangular matrices over Q, constant $(= a_f)$ on the diagonal and with at most finitely many non-zero rows above the diagonal.]

(2) Prove that $J = J(R) = \{f \in R \mid a_f = 0\}$.

(3) Prove that if $(f_1, f_2, \ldots)$ is any sequence in J, then there exists an n such that $f_n f_{n-1} \ldots f_1 = 0$. [Hint: For each $\alpha \in \Omega$, let $\alpha_n \in \Omega$ be least such that $Im(f_n f_{n-1} \ldots f_1) \leq V_{\alpha_n}$. If $f_n f_{n-1} \ldots f_1 \neq 0$ for all n, then $\alpha_1 > \alpha_2 > \ldots > \alpha_n > \ldots$, a contradiction.]

(4) In particular, deduce from (3) that J is nil.

(5) Prove, however, that $\cap_{n=1}^{\infty} J^n \neq 0$. [Hint: Let ω be the first element of Ω such that $\{\alpha \in \Omega \mid \alpha < \omega\}$ is uncountable. Define e_n and f_n by

$$e_n(x_\alpha) = \begin{cases} 0 & \text{if } \alpha \leq n \\ x_n & \text{if } \alpha > n \end{cases} \qquad f_n(x_\alpha) = \begin{cases} 0 & \text{if } \alpha < \omega \\ x_n & \text{if } \alpha \geq \omega \end{cases}$$

Then $f_1 = e_1 f_2 = e_1 e_2 f_3 = \ldots$.]

9. A ring R with $J = J(R)$ is *semiprimary* in case R/J is semisimple and J is nilpotent. Prove that if M is a left module over a semiprimary ring R, then $Soc M = r_M(J) \trianglelefteq M$, $Rad M = JM \ll M$, and M is artinian iff M is noetherian.

10. (1) Prove that if R is left artinian, then there exists a finite sequence $M_1, \ldots, N_n$ of distinct maximal ideals of R with $M_1 \cap \ldots \cap M_n$ nilpotent.

(2) Use part (1), Exercises (13.7) and (14.4), and the Density Theorem to obtain another version of the Wedderburn-Artin Theorem (13.7): If R is left artinian with no non-zero nilpotent ideals, then R is isomorphic to a direct sum of rings $R_1, \ldots, R_n$ each the endomorphism ring of a finite dimensional vector space V_i over a division ring D_i.

11. Let e and f be idempotents in a ring R and let J be the Jacobson radical of R.

(1) Prove that $Rad(Re) = Je$.

(2) Prove that if $e \neq 0$, then $J(eRe) = eJe$.

(3) Deduce that the following statements are equivalent:

(a) $Re \cong Rf$; (b) $Re/Je \cong Rf/Jf$; (c) $eR \cong fR$; (d) $eR/eJ \cong fR/fJ$.

12. Prove that the following are equivalent for an element a in a ring R:

(a) Ra is a direct summand of $_R R$; (b) $a = axa$ for some $x \in R$; (c) aR is a direct summand of R_R.

13. A ring R is *von Neumann regular* in case $a \in aRa$ for each $a \in R$. It follows from Exercise (15.12) that R is von Neumann regular iff every principal left ideal is a direct summand iff every principal right ideal is a direct summand.

(1) Prove that R is von Neumann regular iff every finitely generated left (right) ideal is a direct summand. [Hint: Let $e = e^2 \in R$ and $a \in R$. Then by hypothesis there exists $f = f^2 \in R$ such that $Rf = Ra(1 - e) \subseteq Ra + Re$. Show that $Re + Ra = Re + Rf$ and apply Exercise (7.6.1).]

(2) Prove that every von Neumann regular ring is semiprimitive.

(3) Prove that every factor ring of a von Neumann regular ring is von Neumann regular.

(4) Prove that if e is a non-zero idempotent in a von Neumann regular ring R, then eRe is von Neumann regular.

(5) Prove that a commutative ring is von Neumann regular iff $I^2 = I$ for each ideal I of R.

(6) Prove that for a von Neumann regular ring the following properties are equivalent: (a) semisimple; (b) left artinian; (c) right artinian; (d) left noetherian; (e) right noetherian.

(7) Prove that if M_S is semisimple, then $End(M_S)$ is von Neumann regular. [Hint: If $a \in R$, then $M = Ker\ a \oplus L = K \oplus Im\ a$. Let $x = 0 \oplus (a|L)^{-1}$. Consider axa.]

(8) Let R be the primitive ring of Exercise (14.2). Prove that R is von Neumann regular iff the subring S is a subfield of Q.

14. An ideal P of a ring R is a *prime ideal* in case R/P is a prime ring (see Exercise (14.10)). Thus an ideal P is prime iff for each $x, y \in R$, $xRy \subseteq P$ implies $x \in P$ or $y \in P$. Note that for commutative rings this agrees with the definition given in Exercise (2.11). The *prime radical* or the *lower nil radical* $N(R)$ of a ring R is the intersection of all prime ideals of R. A ring R is *semiprime* in case $N(R) = 0$, that is, in case R is isomorphic to a subdirect product of prime rings. (See Exercise (7.16).)

(1) Let I be an ideal of R. Prove that $N(R/I) = 0$ iff I is an intersection of prime ideals of R. In particular, $N(R/N(R)) = 0$.

(2) Prove that $N(R) = 0$ iff R has no non-zero nilpotent left ideals. [Hint: ($\Rightarrow$). See Exercise (14.10.2). ($\Leftarrow$). Let Rx be non-nilpotent and let P be an ideal of R maximal with respect to $(Rx)^n \nsubseteq P$ for all $n \in \mathbb{N}$. By (14.10.2.c) P is a prime ideal.]

(3) Let $U(R)$ be the unique largest nil ideal of R. (See Exercise (2.5). This ideal is called the *upper nil radical* of R.) Prove that $N(R) \subseteq U(R)$; thus $N(R)$ is a nil ideal of R. [Hint: By (2), $N(R/U(R)) = 0$; now apply (1).]

(4) Prove that if R is left noetherian, then $N(R) = U(R)$ is the unique largest nilpotent ideal of R.

(5) Prove that $J(R) \supseteq N(R)$, but that they need not be equal.

(6) Prove that if R is artinian, then $J(R) = N(R)$, in fact, that every prime ideal is maximal.

(7) Prove that if R is commutative, then $N(R)$ is just the set of nilpotent elements of R, so R is semiprime iff R has no non-zero nilpotent elements.

(8) Prove that a commutative ring R is semiprime iff it can be embedded in a product of fields.

Chapter 5

Functors Between Module Categories

It should now be clear that the structure of the category $_R\mathsf{M}$ determines to a significant extent the structure of the ring R. Thus in this chapter we turn to the direct study of these categories $_R\mathsf{M}$. Our starting point will be the study of certain natural "functors" or "homomorphisms" between pairs of these categories.

The various module categories that we study have one important feature distinct from many other classes of categories. For each pair M, N of R modules, the set $Hom_R(M, N)$ is an abelian group. Since this property is an integral part of the structure of these categories, we shall study only functors that respect it. Thus suppose C is a full subcategory of R-modules and that D is a full subcategory of S-modules. Then a functor T from C to D is *additive* in case for each M, N, modules in C, and each pair $f, g : M \to N$ in C,

$$T(f + g) = T(f) + T(g).$$

In particular, if T is additive and covariant, then the restriction

$$T : Hom_R(M, N) \to Hom_S(T(M), T(N))$$

is an abelian group homomorphism, whereas if T is additive and contra-variant, then the restriction

$$T : Hom_R(M, N) \to Hom_S(T(N), T(M))$$

is an abelian group homomorphism.

In Sections 16 and 19 we study the two most important classes of additive functors, the "Hom" and "tensor" functors, between module categories. Certain pathologies of these functors vanish for some distinguished classes of modules. These modules, the projective, injective, and flat modules, are the centers of attention in Sections 17, 18, and part of 19. Projective and injective modules, duals of each other, are particularly important for their universal splitting properties—for example, an injective module is a direct summand of each of its extensions.

Finally, in Section 20 we investigate the notion of a natural transformation between functors. Here we show the naturality of many of the homo-morphisms of earlier sections. As we shall see, it is this naturality that allows us to make some very significant comparisons of rings and their categories by means of the Hom and tensor functors.

§16. The Hom Functors and Exactness—Projectivity and Injectivity

Let R and S be rings and let $U = {}_RU_S$ be a bimodule. Then (see (4.4)) for each left R-module ${}_RM$ there are two S-modules,

$$Hom_R(U_S, M) \in {}_S\mathcal{M} \quad \text{and} \quad Hom_R(M, U_S) \in \mathcal{M}_S.$$

Thus $M \mapsto Hom_R(U_S, M)$ and $M \mapsto Hom_R(M, U_S)$ define functions from ${}_R\mathcal{M}$ to ${}_S\mathcal{M}$ and to $\mathcal{M}_S$, respectively. These two functions can be extended to additive functors from ${}_R\mathsf{M}$ to ${}_S\mathsf{M}$ and to M_S. The resulting functors are of fundamental importance in our analysis of module categories.

Definition of the Hom Functors

As above, let $U = {}_RU_S$ be a bimodule. Let

$$f : {}_RM \to {}_RN$$

be an R-homomorphism in ${}_R\mathsf{M}$. Then for each $\gamma \in Hom_R(U, M)$, we have $f\gamma \in Hom_R(U, N)$. We claim that

$$Hom(U, f) : \gamma \mapsto f\gamma$$

is an S-homomorphism

$$Hom_R(U, f) : Hom_R(U, M) \to Hom_R(U, N).$$

For if $\gamma_1, \gamma_2 \in Hom_R(U, M)$ and $s_1, s_2 \in S$, then for all $u \in U$,

$$f \circ (s_1\gamma_1 + s_2\gamma_2)(u) = f(\gamma_1(us_1) + \gamma_2(us_2))$$

$$= f\gamma_1(us_1) + f\gamma_2(us_2)$$

$$= (s_1(f\gamma_1) + s_2(f\gamma_2))(u).$$

Thus, we do have a function $Hom_R(U, \_) : {}_R\mathsf{M} \to {}_S\mathsf{M}$ defined by

$$Hom_R(U, \_) : M \mapsto Hom_R(U, M)$$

$$Hom_R(U, \_) : f \mapsto Hom_R(U, f).$$

The notation $Hom_R(U, f)$ can be awkward, so if there is no ambiguity with the module U, we are likely to abbreviate

$$f_* = Hom_R(U, f).$$

Note that if $f : M \to N$ in ${}_R\mathsf{M}$, then f_* is characterized by

Now it is an easy matter to check that this function $Hom_R(U, \_)$ is actually an additive covariant functor from ${}_R\mathsf{M}$ to ${}_S\mathsf{M}$.

On the other hand, for the R-homomorphism $f: {}_RM \rightarrow {}_RN$ we can define a mapping

$$f^* = Hom_R(f, U): Hom_R(N, U) \rightarrow Hom_R(M, U)$$

via

$$Hom_R(f, U): \gamma \mapsto \gamma f.$$

It is straightforward to show that $f^* = Hom_R(f, U)$ is an S-homomorphism. For f^* we have

$$M \xrightarrow{\;\;f\;\;} N$$
$$f^*(\gamma) \searrow \qquad \swarrow \gamma$$
$$U$$

Here then we have a function $Hom_R(\_, U): {}_RM \rightarrow M_S$ defined by

$$Hom_R(\_, U): M \mapsto Hom_R(M, U)$$
$$Hom_R(\_, U): f \mapsto Hom_R(f, U).$$

Now finally,

16.1. Theorem. *Let R and S be rings and let $U = {}_RU_S$ be a bimodule. Then*

$$Hom_R(U, \_): {}_RM \rightarrow {}_SM$$

is an additive covariant functor and

$$Hom_R(\_, U): {}_RM \rightarrow M_S$$

is an additive contravariant functor.

Proof. We shall show that $Hom_R(\_, U)$ reverses composition and preserves addition. The rest of the proof is at least as routine and will be omitted. Suppose then that $g: M \rightarrow M'$ and $f: M' \rightarrow M''$ are morphisms in ${}_RM$. If $\gamma \in Hom_R(M'', U)$, then

$$(f \circ g)^*(\gamma) = \gamma \circ f \circ g = g^*(f^*(\gamma)) = (g^* \circ f^*)(\gamma).$$

Next suppose that $f: M \rightarrow N$ and $g: M \rightarrow N$ are morphisms in ${}_RM$. If $\gamma \in Hom(N, U)$, then

$$(f + g)^*(\gamma) = \gamma \circ (f + g) = (\gamma \circ f) + (\gamma \circ g) = f^*(\gamma) + g^*(\gamma). \qquad \square$$

Appealing to opposite rings we can deduce from this Theorem the existence of a variety of additive functors. For example, a bimodule ${}_{R-S}V$ yields a covariant functor

$$Hom_R({}_SV, \_): {}_RM \rightarrow M_S,$$

and a contravariant functor

$$Hom_R(\_, {}_SV): {}_RM \rightarrow {}_SM.$$

We shall frequently refer to this class of functors as the "Hom functors".

General Properties of Additive Functors

Before proceeding to the explicit study of these Hom functors, we shall prove a pair of results about additive functors that we shall need subsequently.

16.2. Proposition. *Let* C *and* D *be full subcategories of the categories of left (or right) modules over rings R and S. Let $F : C \to D$ ($G : C \to D$) be an additive covariant (contravariant) functor. If*

$$0 \to K \overset{f}{-\oplus\to} M \overset{g}{-\oplus\to} N \to 0$$

is split exact in C, *then both*

$$0 \to F(K) \xrightarrow{F(f)} F(M) \xrightarrow{F(g)} F(N) \to 0$$

$$0 \to G(N) \xrightarrow{G(g)} G(M) \xrightarrow{G(f)} G(K) \to 0$$

are split exact in D. *In particular, if $g : M \to N$ is an isomorphism, then both $F(g)$ and $G(g)$ are isomorphisms.*

We shall prove this result jointly with

16.3. Proposition. *Let* C, D, F, *and* G *be as in* (16.2). *If $M, M_1, \ldots, M_n$ are modules in* C *and if $M = M_1 \oplus \ldots \oplus M_n$ is a direct sum with injections $\iota_1, \ldots, \iota_n$ and projections $\pi_1, \ldots, \pi_n$, then*
(1) *$F(M)$ is a direct sum of $F(M_1), \ldots, F(M_n)$ with injections $F(\iota_1), \ldots, F(\iota_n)$ and projections $F(\pi_1), \ldots, F(\pi_n)$;*
(2) *$G(M)$ is a direct sum of $G(M_1), \ldots, G(M_n)$ with injections $G(\pi_1), \ldots, G(\pi_n)$ and projections $G(\iota_1), \ldots, G(\iota_n)$.*

Proof (of (16.2) and (16.3)). For each pair M, N in C, $F : Hom_R(M, N) \to Hom_S(F(M), F(N))$ is a group homomorphism. Thus F of the zero map is the zero map. Now for (16.3.1) we have from (6.22), and the additivity and covariance of F:

$$\Sigma_{i=1}^n F(\iota_i)F(\pi_i) = F(\Sigma_{i=1}^n \iota_i \pi_i) = F(1_M) = 1_{F(M)};$$

$$F(\pi_i)F(\iota_j) = F(\pi_i \iota_j) = F(\delta_{ij}) = \delta_{ij} 1_{F(M_i)}.$$

A similar argument takes care of (16.3.2). Finally, (16.2) now follows as a special case of (16.3) by virtue of (5.3.b) and (6.22).   □

Suppose again that C, D, F and G are as in (16.2). Let $f_i : M_i \to N$ ($i = 1, \ldots, n$) be homomorphisms in C. Applying F to the appropriate diagrams we have for each $i = 1, \ldots, n$,

Thus by (16.3) and the uniqueness of the direct sum map,

$$F(\oplus_{i=1}^n f_i) = \oplus_{i=1}^n F(f_i);$$

hence, relative to the injections $F(\iota_1), \ldots, F(\iota_n)$, F preserves finite direct sum. of homomorphisms as well as of modules. Of course we can also write

$$F(\Pi_{i=1}^n g_i) = \Pi_{i=1}^n F(g_i),$$

$$G(\oplus_{i=1}^n f_i) = \Pi_{i=1}^n G(f_i)$$

and

$$G(\Pi_{i=1}^n g_i) = \oplus_{i=1}^n G(g_i).$$

Direct Sums and Products under Hom

Given a bimodule ${}_RU_S$ the functors $Hom_R(\_, U)$ and $Hom_R(U, \_)$ are both additive, so by (16.3) they "preserve" finite direct sums. In fact, they do even better as the next proposition shows.

16.4. Proposition. *Let ${}_RU_S$ be a bimodule and let $(M_\alpha)_{\alpha \in A}$ be an indexed set of left R-modules.*
(1) *If $(M, (q_\alpha)_{\alpha \in A})$ is a direct product of $(M_\alpha)_{\alpha \in A}$, then*

$$(Hom_R(U_S, M), (Hom_R(U_S, q_\alpha))_{\alpha \in A})$$

is a direct product of the left S-modules $(Hom_R(U_S, M_\alpha))_{\alpha \in A}$;
(2) *If $(M, (j_\alpha)_{\alpha \in A})$ is a direct sum of $(M_\alpha)_{\alpha \in A}$, then*

$$(Hom_R(M, U_S), (Hom_R(j_\alpha, U_S))_{\alpha \in A})$$

is a direct product of the right S-modules $(Hom_R(M_\alpha, U_S))_{\alpha \in A}$.

Proof. We leave (1) as an exercise. To prove (2) let $(\pi_\alpha)_{\alpha \in A}$ be the projections for the direct product $\Pi_A Hom_R(M_\alpha, U_S)$. Then by (6.1) there is an S-homomorphism η making the diagrams

$$Hom_R(M, U_S) \xrightarrow{\quad\eta\quad} \Pi_A Hom_R(M_\alpha, U_S)$$

$$\underset{Hom_R(j_\alpha, U_S)}{\searrow} \qquad \underset{\pi_\alpha}{\swarrow}$$

$$Hom_R(M_\alpha, U_S)$$

commute for all $\alpha \in A$. If $\gamma \in Ker\,\eta$ then

$$0 = \pi_\alpha \eta(\gamma) = Hom_R(j_\alpha, U)(\gamma) = \gamma j_\alpha$$

for all $\alpha \in A$. But this, since $M = \Sigma_A\, Im\, j_\alpha$ (see (6.21.iii)), forces $\gamma = 0$. Thus η is monic. If $(\gamma_\alpha)_{\alpha \in A} \in \Pi_A Hom_R(M_\alpha, U_S)$, then the direct sum map $\oplus_A \gamma_\alpha$, making the diagrams

$$M \xrightarrow{\oplus_A \gamma_\alpha} U$$

$$\underset{j_\alpha}{\nwarrow} \qquad \underset{\gamma_\alpha}{\nearrow}$$

$$M_\alpha$$

co _ute $(\alpha \in A)$, satisfies

$$\pi_\alpha \eta(\oplus_A \gamma_\alpha) = Hom(j_\alpha, U_S)(\oplus_A \gamma_\alpha)$$

$$= (\oplus_A \gamma_\alpha) j_\alpha = \gamma_\alpha \qquad (\alpha \in A).$$

Thus η is an isomorphism, and by (6.4) the proof is complete. $\qquad\square$

Reversing the variables, we see that Proposition (16.4) relates the functors $Hom_R(\oplus_A U_\alpha, \_)$ and $Hom_R(\_, \Pi_A U_\alpha)$ to the functors $Hom_R(U_\alpha, \_)$ and $Hom_R(\_, U_\alpha)$. That is,

$$Hom_R(\oplus_A U_\alpha, M) \cong \Pi_A Hom_R(U_\alpha, M),$$

$$Hom_R(M, \Pi_A U_\alpha) \cong \Pi_A Hom_R(M, U_\alpha).$$

(See (6.4).) The next corollary asserts that these relationships are "natural". (See §20.)

16.5. Corollary. *Let $(U_\alpha)_{\alpha \in A}$ be an indexed set of left R-modules. If M and N are left R-modules, then there exist $\mathbb{Z}$-isomorphisms η_M, η_N, v_N and v_M such that for all $f: {}_R M \to {}_R N$ the diagrams*

(1)
$$
\begin{array}{ccc}
Hom_R(\oplus_A U_\alpha, M) & \xrightarrow{Hom(\oplus_A U_\alpha, f)} & Hom_R(\oplus_A U_\alpha, N) \\
\downarrow{\eta_M} & & \downarrow{\eta_N} \\
\Pi_A Hom_R(U_\alpha, M) & \xrightarrow{\Pi_A Hom(U_\alpha, N)} & \Pi_A Hom_R(U_\alpha, N)
\end{array}
$$

and

(2)
$$
\begin{array}{ccc}
Hom_R(N, \Pi_A U_\alpha) & \xrightarrow{Hom(f, \Pi_A U_\alpha)} & Hom_R(M, \Pi_A U_\alpha) \\
\downarrow{v_N} & & \downarrow{v_M} \\
\Pi_A Hom_R(N, U_\alpha) & \xrightarrow{\Pi_A(Hom(f, U_\alpha))} & \Pi_A Hom_R(N, U_\alpha)
\end{array}
$$

are commutative.

Proof. Again we omit the proof of (1). To prove (2) let $(q_\alpha)_{\alpha \in A}$ be the projections for $\Pi_A U_\alpha$ and let $(\pi_\alpha)_{\alpha \in A}$ and $(\pi'_\alpha)_{\alpha \in A}$ be the projections for $\Pi_A Hom_R(N, U_\alpha)$ and $\Pi_A Hom_R(M, U_\alpha)$, respectively. Then (see (16.4.1) and (6.4)) there are isomorphisms

$$v_N: Hom_R(N, \Pi_A U_\alpha) \to \Pi_A Hom_R(N, U_\alpha)$$

and

$$v_M: Hom_R(M, \Pi_A U_\alpha) \to \Pi_A Hom_R(M, U_\alpha)$$

such that

$$\pi_\alpha v_N = Hom_R(N, q_\alpha) \qquad (\alpha \in A)$$

and

$$\pi'_\alpha v_M = Hom_R(M, q_\alpha) \qquad (\alpha \in A).$$

If $\gamma \in Hom_R(N, \Pi_A U_\alpha)$ then for all $\alpha \in A$

$$\pi'_\alpha(\Pi_A Hom(f, U_\alpha)(\nu_N(\gamma))) = Hom(f, U_\alpha)(\pi_\alpha \nu_N(\gamma))$$
$$= Hom(f, U_\alpha)(Hom(N, q_\alpha)(\gamma)) = q_\alpha \gamma f$$
$$= Hom(M, q_\alpha)(Hom(f, \Pi_A U_\alpha)(\gamma))$$
$$= \pi'_\alpha(\nu_M(Hom(f, \Pi_A U_\alpha)(\gamma))).$$

Thus the diagram commutes, as desired. $\qquad\square$

Although the *Hom* functors preserve split exact sequences, in general they need not preserve all exactness. For the remainder of this section we shall be primarily concerned with the behavior of the *Hom* functors on exact sequences.

Exact Functors

Let **C** and **D** be the full subcategories of categories of modules and let $F : \mathbf{C} \to \mathbf{D}$ be a covariant functor. If for every short exact sequence in **C**

$$0 \to K \to M \to N \to 0$$

the sequence

$$0 \to F(K) \to F(M) \to F(N)$$

is exact in **D**, then F is said to be *left exact*. On the other hand if

$$F(K) \to F(M) \to F(N) \to 0$$

is exact in **D**, then F is said to be *right exact*. In the contravariant case the defining diagrams are

$$0 \to G(N) \to G(M) \to G(K)$$

for left exact and

$$G(N) \to G(M) \to G(K) \to 0$$

for right exact. A functor that is both left and right exact is called simply an *exact functor*. This name is well chosen, for (see Exercise (16.4)) the image of any exact sequence under an exact functor on $_R\mathbf{M}$ is exact.

16.6. Proposition. *The Hom functors are left exact. Thus, in particular if* $_R U$ *is a module, then for every exact sequence* $0 \to K \overset{f}{\to} M \overset{g}{\to} N \to 0$ *in* $_R\mathbf{M}$ *the sequences*

$$0 \to Hom_R(U, K) \overset{f_*}{\to} Hom_R(U, M) \overset{g_*}{\to} Hom_R(U, N)$$

and

$$0 \to Hom_R(N, U) \overset{g^*}{\to} Hom_R(M, U) \overset{f^*}{\to} Hom_R(K, U)$$

are exact.

Proof. We shall do the contravariant case. If $\gamma \in Hom_R(N, U)$ and $0 = g^*(\gamma) = \gamma g$ then $\gamma = 0$ because g is epic. Thus g^* is monic. Since the *Hom* functors are additive we have $f^*g^* = (gf)^* = 0^* = 0$; so that $Im\, g^* \subseteq Ker\, f^*$. If $\beta \in Ker\, f^*$, then $\beta f = f^*(\beta) = 0$, so $Ker\, \beta \subseteq Im\, f = Ker\, g$. So the Factor Theorem tells us that β factors through g. Therefore $\beta = \gamma g = g^*(\gamma) \in Im\, g^*$ and we have proved that $Im\, g^* = Ker\, f^*$. $\square$

The left exactness of the *Hom* functors $Hom_R(V, \_)$ and $Hom_R(\_, V)$ on M_R can be established by considering opposite rings.

M-projective and M-injective Modules

Let $_RU$ be a module. Then in general the functors $Hom_R(U, \_)$ and $Hom_R(\_, U)$ are not exact. For example, it is easy to see that neither $Hom_{\mathbb{Z}}(\mathbb{Z}_2, \_)$ nor $Hom_{\mathbb{Z}}(\_, \mathbb{Z}_2)$ preserves the exactness of the natural short exact sequence

$$0 \to \mathbb{Z} \to \mathbb{Z} \to \mathbb{Z}_2 \to 0.$$

Nevertheless, the functors $Hom_R(U, \_)$ and $Hom_R(\_, U)$ do preserve the exactness of some short exact sequences. We now look at some aspects of this "local exactness".

Let $_RU$ be a module. If $_RM$ is a module, then U is *projective relative to M* (or U is *M-projective*) in case for each epimorphism $g : _RM \to _RN$ and each homomorphism $\gamma : _RU \to _RN$ there is an R-homomorphism $\bar{\gamma} : U \to M$ such that the diagram

$$
\begin{array}{c}
U \\
{\scriptstyle \bar{\gamma}} \swarrow \quad \downarrow {\scriptstyle \gamma} \\
M \xrightarrow[g]{} N \to 0
\end{array}
$$

commutes. On the other hand U is *injective relative to M* (or U is *M-injective*) in case for each monomorphism $f : K \to M$ and each homomorphism $\gamma : _RK \to _RU$ there is an R-homomorphism $\bar{\gamma} : M \to U$ such that the diagram

$$
\begin{array}{c}
U \\
{\scriptstyle \gamma} \uparrow \quad \nwarrow {\scriptstyle \bar{\gamma}} \\
0 \to K \xrightarrow[f]{} M
\end{array}
$$

commutes. The next two results assert that U is M-projective (M-injective) if and only if $Hom_R(U, \_)$ (respectively $Hom_R(\_, U)$) preserves the exactness of all short exact sequences with middle term M.

16.7. Proposition. *Let U and M be left R-modules. Then the following are equivalent:*

   (a) *U is M-projective;*
   (b) *For every short exact sequence with middle term M*

$$0 \to K \xrightarrow{f} M \xrightarrow{g} N \to 0$$

in $_R$M, *the sequence*

$$0 \to Hom_R(U, K) \xrightarrow{f_*} Hom_R(U, M) \xrightarrow{g_*} Hom_R(U, N) \to 0$$

is exact;

    (c) *For each submodule* $_RK \le {}_RM$, *every R-homomorphism* $h: U \to M/K$ *factors through the natural epimorphism* $n_K: M \to M/K$.

    Proof. (a) $\Leftrightarrow$ (b). By (16.6) condition (b) holds if and only if for every epimorphism $M \xrightarrow{f} N \to 0$ the sequence $Hom_R(U, M) \xrightarrow{f_*} Hom_R(U, N) \to 0$ is exact. But f_* is epic if and only if for each $\gamma \in Hom(U, N)$ there is a $\bar{\gamma} \in Hom_R(U, M)$ such that $\gamma = f_*(\bar{\gamma}) = f\bar{\gamma}$.

    (a) $\Rightarrow$ (c). This is clear.

    (c) $\Rightarrow$ (a). Suppose we have an epimorphism $g: M \to N$ with $K = Ker\, g$. Then by The Factor Theorem (3.6.1) there is an isomorphism $h: N \to M/K$ such that $hg = n_K$. By hypothesis, if $\gamma: u \to N$ then $h\gamma$ factors through n_K; that is, there is a $\bar{\gamma}$ such that

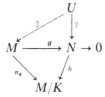

commutes. Thus we have $hg\bar{\gamma} = n_K\bar{\gamma} = h\gamma$. So since h is an isomorphism, $g\bar{\gamma} = \gamma$ and U is M-projective. $\quad\square$

    Similar methods prove

16.8. Proposition. *Let U and M be left R-modules. Then the following are equivalent:*

    (a) *U is M-injective;*

    (b) *For every short exact sequence with middle term M*

$$0 \to K \xrightarrow{f} M \xrightarrow{g} N \to 0$$

in $_R$M, *the sequence*

$$0 \to Hom_R(N, U) \xrightarrow{g^*} Hom_R(M, U) \xrightarrow{f^*} Hom_R(K, U) \to 0$$

is exact;

    (c) *For each submodule* $_RK \le {}_RM$, *every R-homomorphism* $h: K \to U$ *can be extended to an R-homomorphism* $\bar{h}: M \to U$ *(i.e., every $h: K \to U$ factors through the natural monomorphism $i_K: K \to M$).* $\quad\square$

    A module $_RP$ is said to be *projective* in case it is projective relative to every module $_RM$. And a module $_RQ$ is *injective* in case it is injective relative to every module $_RM$. Thus,

16.9. Corollary. *A module $_RP$ is projective if and only if the additive covariant functor $Hom_R(P, {}_-)$ is exact on $_R$M. A module $_RQ$ is injective if and only if the additive contravariant functor $Hom_R({}_-, Q)$ is exact on $_R$M.*

Projectivity and Injectivity Classes

Although there exist modules that are neither projective nor injective, we shall see in later sections that over a ring R there do exist many projective and injective modules. Of course it is trivial that the zero module is both projective and injective. Now let $_RM$ be a module. Denote by $\mathscr{P}\imath(M)$ (respectively, $\mathscr{I}\!n(M)$) *the class of all M-projective (M-injective) modules.* Thus $\mathscr{P}\imath(M)$ and $\mathscr{I}\!n(M)$ contain all the projective and injective modules, respectively, so in particular, they both contain 0. Our next result implies that $\mathscr{P}\imath(M)$ is "closed" under direct sums and that $\mathscr{I}\!n(M)$ is "closed" under direct products.

16.10. Proposition. *Let M be a left R-module and let $(U_\alpha)_{\alpha \in A}$ be an indexed set of left R-modules. Then*
(1) $\oplus_A U_\alpha$ *is M-projective if and only if each U_α is M-projective;*
(2) $\Pi_A U_\alpha$ *is M-injective if and only if each U_α is M-injective.*

Proof. This follows from (16.5). For example, $Hom(\oplus U_\alpha, f)$ is epic if and only if each $Hom(U_\alpha, f)$ is epic. (See also Remark (6.25).) $\qquad\square$

16.11. Corollary. *Let $(U_\alpha)_{\alpha \in A}$ be an indexed set of left R-modules. Then*
(1) $\oplus_A U_\alpha$ *is projective if and only if each U_α is projective;*
(2) $\Pi_A U_\alpha$ *is injective if and only if each U_α is injective.*

Projectivity and Injectivity Domains

The *projectivity domain* of a left R-module U is $\mathscr{P}\imath^{-1}(U)$, the collection of all modules $_RM$ such that U is M-projective. The *injectivity domain* of U is $\mathscr{I}\!n^{-1}(U)$ which consists of those modules M such that U is M-injective. Again it is trivial that 0 belongs to both $\mathscr{P}\imath^{-1}(U)$ and $\mathscr{I}\!n^{-1}(U)$. Also it is immediate that U is projective or injective, respectively, if and only if $\mathscr{P}\imath^{-1}(U)$ or $\mathscr{I}\!n^{-1}(U)$ contains all left R-modules. The most important properties of the classes $\mathscr{P}\imath^{-1}(U)$ and $\mathscr{I}\!n^{-1}(U)$ are that they are both closed under submodules and epimorphic images, that $\mathscr{P}\imath^{-1}(U)$ is closed under finite direct sums and that $\mathscr{I}\!n^{-1}(U)$ is closed under arbitrary direct sums.

16.12. Proposition. *Let U be a left R-module.*
(1) *If $0 \to M' \overset{h}{\to} M \overset{k}{\to} M'' \to 0$ is exact in $_RM$ and U is M-projective, then U is projective relative to both M' and M''.*
(2) *If U is projective relative to each of $M_1, \ldots, M_n$, then U is $\oplus_{i=1}^n M_i$-projective.*
   Moreover, if U is finitely generated and M_α-projective ($\alpha \in A$), then U is projective relative to $\oplus_A M_\alpha$.

Proof. (1) Let U be M-projective and let
$$0 \to M' \overset{h}{\to} M \overset{k}{\to} M'' \to 0$$
be exact. If $g : M'' \to N''$ is epic, then since U is M-projective and gk is epic

$$g_* k_* = (gk)_*$$

is epic. But then g_* is epic so U is M''-projective. To prove that M' is U-projective, we may assume that $M' \leq M$. If $K' \leq M'$ then there is a commutative diagram

$$
\begin{array}{ccccccccc}
& & & & 0 & & & & \\
& & & & \downarrow & & & & \\
0 & \to & M' & \to & M & \to & M/M' & \to & 0 \\
& & {\scriptstyle n_K}\downarrow & & \downarrow & & \downarrow & & \\
0 & \to & M'/K' & \to & M/K' & \to & M/M' & \to & 0 \\
& & \downarrow & & \downarrow & & \downarrow & & \\
& & 0 & & 0 & & 0 & &
\end{array}
$$

with exact rows and columns. Applying $Hom_R(U, \_)$ to this diagram we have the commutative diagram

$$
\begin{array}{ccccccc}
& & & & 0 & & \\
& & & & \downarrow & & \\
0 \to & Hom_R(U, M') & \to & Hom_R(U, M) & \to & Hom_R(U, M/M') & \\
& {\scriptstyle (n_K)_*}\downarrow & & \downarrow & & \downarrow & \\
0 \to & Hom_R(U, M'/K') & \to & Hom_R(U, M/K') & \to & Hom_R(U, M/M') & \\
& \downarrow & & \downarrow & & & \\
& 0 & & 0 & & &
\end{array}
$$

with exact rows and columns. But now it follows from (3.14.4) that $(n_{K'})_*$ is epic. Thus U is M'-projective.

(2) Clearly it is sufficient to show that if U is projective relative to M_1 and M_2, then U is $M_1 \oplus M_2$-projective. So suppose $K \leq M_1 \oplus M_2$. Then the obvious maps yield the commutative diagram

$$
\begin{array}{ccccccccc}
0 & \longrightarrow & M_1 & \overset{\oplus}{\longrightarrow} & M_1 \oplus M_2 & \overset{\oplus}{\longrightarrow} & M_2 & \longrightarrow & 0 \\
& & \downarrow & & {\scriptstyle n_K}\downarrow & & \downarrow & & \\
0 \to & (M_1 + K)/K & \to & (M_1 \oplus M_2)/K & \to & (M_1 \oplus M_2)/(M_1 + K) & \to & 0 \\
& \downarrow & & \downarrow & & \downarrow & & \\
& 0 & & 0 & & 0 & &
\end{array}
$$

with exact rows and columns. To prove that $(n_K)_*$ is epic, apply $Hom_R(U, \_)$ and The Five Lemma (3.15).

Finally, suppose that U is finitely generated and projective relative to each M_α ($\alpha \in A$). If we have the solid part of

$$
\begin{array}{ccc}
& U & \\
{\scriptstyle \bar{\gamma}}\swarrow & \downarrow{\scriptstyle \gamma} & \\
\oplus_A M_\alpha \overset{g}{\to} & N & \to 0
\end{array}
$$

then, since $Im\,\gamma$ is finitely generated and g is epic, there exist $x_1, \ldots, x_n \in \oplus_A M_\alpha$ such that $\{g(x_1), \ldots, g(x_n)\}$ is a spanning set for $Im\,\gamma$. Let $M' = Rx_1 + \ldots +$

$Rx_n \leq \oplus_A M_\alpha$. Then M' is contained in a finite direct sum $\oplus_F M_\alpha \leq \oplus_A M_\alpha$; and hence by (1) and (2) U is M'-projective. Thus there exists a $\bar{\gamma}$ making

$$
\begin{array}{ccc}
 & & U \\
 & \nearrow^{\bar{\gamma}} & \downarrow^{\gamma} \\
M' & \xrightarrow{(g\,|\,M')} & Im\,\gamma \to 0
\end{array}
$$

(and hence the first diagram) commute.   □

16.13. Proposition. *Let U be a left R-module.*

(1) *If $0 \to M' \xrightarrow{h} M \xrightarrow{k} M'' \to 0$ is exact in $_R M$ and U is M-injective, then U is injective relative to both M' and M''.*

(2) *If U is injective relative to each of the R-modules $M_\alpha (\alpha \in A)$, then U is $\oplus_A M_\alpha$-injective.*

Proof. (1) This proof is similar to that of (16.12.1): If $f : K' \to M'$ is monic, then so is $hf : K' \to M$. Thus

$$f^*h^* = (hf)^*$$

is epic, f^* is epic, and hence U is M'-injective. To see that U is M''-injective, assume that $M' \leq K \leq M$ and that $M'' = M/M'$. Then apply $Hom_R(\_, U)$ to the canonical diagram

$$
\begin{array}{ccccccccc}
 & & & & 0 & & & & \\
 & & & & \uparrow & & & & \\
0 & \to & M' & \to & M & \to & M/M' & \to & 0 \\
 & & \uparrow & & \uparrow & & \uparrow & & \\
0 & \to & M' & \to & K & \to & K/M' & \to & 0 \\
 & & \uparrow & & \uparrow & & \uparrow & & \\
 & & 0 & & 0 & & 0 & &
\end{array}
$$

(2) Suppose that $M = \oplus_A M_\alpha$ and U is M_α-injective for all $\alpha \in A$. Let $K \leq M$ and $h : _R K \to _R U$. Let

$$\mathscr{F} = \{f : L \to U \mid K \leq L \leq M \text{ and } (f \mid K) = h\}.$$

Then $\mathscr{F}$ is ordered by set inclusion (i.e., $f \leq g$ if and only if $f \subseteq g \subseteq M \times U$) and $\mathscr{F}$ is clearly inductive. Let

$$\bar{h} : \quad N \to U$$

be a maximal element in $\mathscr{F}$. To complete the proof we need only show that each M_α is contained in N. Let

$$K_\alpha = M_\alpha \cap N.$$

Then

$$(\bar{h} \mid K_\alpha) : K_\alpha \to U,$$

so since $K_\alpha \leq M_\alpha$ and U is M_α-injective, there is a map

$$\bar{h}_\alpha : M_\alpha \to U$$

with

$$(\bar{h}_\alpha \mid K_\alpha) = (\bar{h} \mid K_\alpha).$$

If $m_\alpha \in M_\alpha$ and $n \in N$ such that $m_\alpha + n = 0$, then $m_\alpha = -n \in K_\alpha$ and $\bar{h}_\alpha(m_\alpha) + \bar{h}(n) = \bar{h}(-n) + \bar{h}(n) = 0$. Thus

$$f : m_\alpha + n \mapsto \bar{h}_\alpha(m_\alpha) + \bar{h}(n)$$

is a well-defined R-map

$$f : M_\alpha + N \to U.$$

But $(f \mid N) = \bar{h}$. So by the maximality of $\bar{h}$, $M_\alpha \subseteq N$.   □

16.14. Corollary. *Let U be a left R-module and suppose that G is a generator in ${}_R\mathsf{M}$.*

(1) *If U is G-injective, then U is injective.*

(2) *If U is finitely generated and G-projective, then U is projective.*

Proof. Every left R-module is an epimorph of a direct sum of copies of G. Thus (16.13) and (16.12) apply.   □

16. Exercises

1. (1) Let $\phi : R \to S$ be a ring homomorphism. Prove that the functor $T_\phi : {}_S\mathsf{M} \to {}_R\mathsf{M}$ of Exercise (4.15) is exact.
   (2) Let $e \in R$ be a non-zero idempotent. Prove that the functor $T_e : {}_R\mathsf{M} \to {}_{eRe}\mathsf{M}$ of Exercise (4.17) is exact.
2. Let $F : {}_R\mathsf{M} \to {}_S\mathsf{M}$ be an additive functor, and let ${}_RM \in {}_R\mathscr{M}$. For each $K \leq M$ let $i_{K \leq M}$ denote the inclusion map. Prove that
   (1) If F is covariant, then $K \mapsto Im\, F(i_{K \leq M})$ defines an order preserving map from the lattice $\mathscr{S}(M)$ of submodules of M to the lattice $\mathscr{S}(F(M))$ of submodules of $F(M)$.
   (2) If F is contravariant, then $K \mapsto Ker\, F(i_{K \leq M})$ defines an order reversing map $\mathscr{S}(M) \to \mathscr{S}(F(M))$.
3. Let ${}_RU$ be finitely generated and let $(M_\alpha)_{\alpha \in A}$ be an indexed set of left R-modules. Prove that $Hom_R(U, \oplus_A M_\alpha) \cong \oplus_A Hom_R(U, M_\alpha)$. [Hint: There is a natural monomorphism one way.]
4. Let C be a full subcategory of ${}_R\mathsf{M}$ that contains with each M all submodules and factor modules of M. Let $F : \mathsf{C} \to {}_S\mathsf{M}$ be a covariant additive functor (there are obvious variations of this exercise for contravariant functors). Prove that
   (1) F is exact iff for each exact sequence $M' \xrightarrow{f} M \xrightarrow{g} M''$ in C the induced sequence

$$F(M') \xrightarrow{F(f)} F(M) \xrightarrow{F(g)} F(M'')$$

is exact in $_S\mathsf{M}$. [Hint: See Exercise (3.10).]

(2) F is left exact iff for each exact sequence $0 \to M' \to M \to M''$ in $\mathbf{C}$ the induced sequence $0 \to F(M') \to F(M) \to F(M'')$ is exact in $_S\mathsf{M}$.

(3) State and prove the "right exact" version of part (2).

5. (1) Let E be (U-)injective and let U finitely cogenerate V. Prove that $Tr_E(V) \leq Tr_E(U)$. In particular, if V generates E, then so does U.

(2) Let P be (U-)projective and let U generate V. Prove that $Rej_P(V) \geq Rej_P(U)$. In particular, if V cogenerates P, then so does U.

6. Let R be the ring of all 2×2 upper triangular matrices over $\mathbb{Q}$. Using Exercise (16.5) prove that $Soc\, R$ is not injective and $R/J(R)$ is not projective.

7. Prove that an exact sequence $0 \to K \to M \to N \to 0$ of R-homomorphisms splits if either K is M-injective or N is M-projective.

8. Prove that the following statements about a left R-module M are equivalent: (a) Every left R-module is M-projective; (b) Every left R-module is M-injective; (c) Every simple left R-module is M-projective; (d) M is semisimple. [Hint: (a) $\Rightarrow$ (d) and (b) $\Rightarrow$ (d) by Exercise (16.7). For (c) $\Rightarrow$ (d), first assume that M is cyclic.]

9. Prove that the following statements about a ring R are equivalent: (a) R is semisimple; (b) Every left R-module is projective; (c) Every left R-module is injective; (d) Every simple left R-module is projective. [Hint: Exercise (16.8).]

10. (1) Prove that every divisible abelian group is projective relative to the regular module $_{\mathbb{Z}}\mathbb{Z}$ but no non-zero divisible group is a projective $\mathbb{Z}$-module. So $_{\mathbb{Z}}\mathbb{Q}$ is $_{\mathbb{Z}}\mathbb{Z}$-projective but not projective. Deduce that $\mathscr{P}_r^{-1}(\mathbb{Q})$ is not closed under direct sums. [Hint: If $_{\mathbb{Z}}D$ is divisible, then $Hom_{\mathbb{Z}}(D, \mathbb{Z}_n) = 0$.]

(2) Prove that every torsion-free abelian group F is injective relative to every torsion group T, but that the torsion-free group $_{\mathbb{Z}}\mathbb{Z}$ is not an injective $\mathbb{Z}$-module. Deduce that $\mathscr{I}_n^{-1}(\mathbb{Z})$ is not closed under direct products. [Hint: $\Pi_p \mathbb{Z}_p$ has a torsion-free subgroup.]

11. Let R be a P.I.D. Prove that if a module $_R M$ is divisible (see Exercise (3.15)), then it is injective relative to R and hence is injective over R.

12. Let $_R G$ be a generator, let $_R U$ have a spanning set X, and let U be $G^{(X)}$-projective. Prove that $_R U$ is projective.

13. Let U, V, M be R-modules with $V \leq U$. Prove that:

(1) If U is M-injective and $Tr_U(M) \leq V$, then V is M-injective.

(2) If U is M-projective and $V \leq Rej_U(M)$, then U/V is M-projective.

(3) If $V \trianglelefteq U$ and V is M-injective, then $Tr_U(M) \leq V$ and U is M-injective. [Hint: If $f : M \to U$, let $N = f^{\leftarrow}(V)$. By hypothesis there is a $g : M \to V$ with $(g \mid N) = (f \mid N)$. Observe that $Im(f - g) \cap V = 0$.]

(4) If $V \ll U$ and U/V is M-projective, then $V \leq Rej_U(M)$ and U is M-projective. [Hint: Try for a dual.]

(5) $\mathscr{I}_n(U)(\mathscr{P}_r(U))$ is closed under essential (superfluous) extensions.

14. Let I be an ideal of R and let $_R P$ and $_R E$. Prove that:

(1) If P is projective, then P/IP is R/I-projective.

(2) If E is injective, then $r_E(I)$ is R/I-injective.

15. A module $_RU$ is *projective (injective) modulo its annihilator* in case U is a projective (injective) $R/l_R(U)$ module.

(1) Prove that these are equivalent: (a) $_RU$ is projective modulo its annihilator; (b) U is a projective R/I module for some ideal $I \subseteq l_R(U)$; (c) U is U^A-projective for every set A.

(2) Prove that these are equivalent: (a) $_RU$ is injective modulo its annihilator; (b) U is an injective R/I module for some ideal $I \subseteq l_R(U)$; (c) U is U^A-injective for every set A.

16. Prove that a module that is projective (injective) modulo its annihilator need not be projective (injective). [Hint: See Exercise (16.6).]

17. A module $_RU$ is *quasi-projective (quasi-injective)* in case it is U-projective (U-injective). Thus from Exercise (16.15) we infer that if U is projective (injective) modulo its annihilator, then it is quasi-projective (quasi-injective). Prove that

(1) Every semisimple module is quasi-projective and quasi-injective.

(2) The abelian group $\oplus_\mathbb{P} \mathbb{Z}_p$ is quasi-projective and quasi-injective but neither projective nor injective modulo its annihilator.

(3) Every quasi-projective (quasi-injective) module is quasi-projective (quasi-injective) over its biendomorphism ring.

(4) $U_1 \oplus \ldots \oplus U_n$ is quasi-projective (quasi-injective) iff U_i is U_j-projective (U_j-injective) for all i, j.

(5) The following are equivalent for a ring R: (a) R is semisimple; (b) Every left R-module is quasi-projective; (c) Every left R-module is quasi-injective.

18. Let $S = End(_RU)$ and let $_RT$ be simple. Prove that:

(1) If U is quasi-projective, then the left S-module $Hom_R(U_S, T)$ is simple or zero.

(2) If U is quasi-injective, then the right S-module $Hom_R(T, U_S)$ is simple or zero.

[Hint: For (1) suppose $0 \neq \gamma \in Hom_R(U, T)$. Then γ is epic and U is quasi-projective, so $\gamma_* : {}_SS \to Hom_R(U_S, T)$ is epic. Thus

$$S\gamma = Hom_R(U, T).]$$

19. Let $_RU$ be quasi-injective with $I = l_R(U)$. Prove that the following are equivalent: (a) U finitely cogenerates R/I; (b) U is finitely generated over $End(_RU)$; (c) U finitely cogenerates $BiEnd(_RU)$. Moreover, prove that each of these conditions implies that U is injective both modulo its annihilator and as a $BiEnd(_RU)$ module.

§17. Projective Modules and Generators

Recall that a left R-module P is projective in case P is projective relative to every left R-module. That is, whenever there is given the solid part of a diagram

$$P$$

$$\bar{\gamma} \diagup \quad \Big| \gamma$$

$$M \xrightarrow{g} N \to 0$$

in $_R\mathsf{M}$ with exact row, there is an R-homomorphism $\bar{\gamma}$ such that the whole diagram commutes, i.e., $g\bar{\gamma} = \gamma$. We begin with several

Characterizations of Projective Modules

17.1. Proposition. *The following statements about a left R-module P are equivalent:*

(a) *P is projective;*

(b) *For each epimorphism $f:{}_RM \to {}_RN$ the map*

$$Hom_R(P,f):Hom_R(P, M) \to Hom_R(P, N)$$

is an epimorphism;

(c) *For each bi-module structure $_RP_S$ the functor*

$$Hom_R(P_S, -):{}_R\mathsf{M} \to {}_S\mathsf{M}$$

is exact;

(d) *For every exact sequence*

$$M' \xrightarrow{j} M \xrightarrow{g} M''$$

in $_R\mathsf{M}$ the sequence

$$Hom_R(P, M') \xrightarrow{f_*} Hom_R(P, M) \xrightarrow{g_*} Hom_R(P, M'')$$

is exact.

Proof. (a) $\Leftrightarrow$ (b) and (a) $\Leftrightarrow$ (c) both follow from (16.7). Finally (c) $\Leftrightarrow$ (d) is immediate from Exercise (16.4). $\square$

It is an easy consequence of (4.5) and (17.1) that $_RR$ is projective. A direct proof is also easy. Indeed, suppose we have the solid part of the diagram

$$R$$

$$\rho(m) \diagup \quad \Big| \gamma$$

$$M \xrightarrow{g} N \to 0$$

and that the row is exact. Then there is an $m \in M$ with $g(m) = \gamma(1)$. Clearly $\rho(m):r \mapsto rm$ defines an R-homomorphism $R \to M$ making the entire diagram commute. Recall that a module is free in case it is isomorphic to $R^{(A)}$ for some set A. So the important fact (16.11) that direct sums of projective modules are projective establishes that every free module is projective. This gives rise to the following characterization.

17.2. Proposition. *The following statements about a left R-module P are equivalent:*

(a) P *is projective;*

(b) *Every epimorphism* $_RM \to {}_RP \to 0$ *splits;*

(c) P *is isomorphic to a direct summand of a free left R-module.*

Proof. (a) $\Rightarrow$ (b). Suppose $f: M \to P$ is an epimorphism. If P is projective, then there is a homomorphism g such that $fg = 1_P$, so (§5), the epimorphism f splits.

(b) $\Rightarrow$ (c). This follows from the fact (8.1) that every module is an epimorph of a free module.

(c) $\Rightarrow$ (a). Every free module is projective. Apply (16.11). $\square$

17.3. Corollary. *A left R-module P is finitely generated and projective if and only if for some module $_RP'$ and some integer $n > 0$ there is an R-isomorphism*

$$P \oplus P' \cong R^{(n)}.$$

Proof. A module $_RP$ is finitely generated if and only if for some natural number n, there is an epimorphism

$$R^{(n)} \to P \to 0.$$

But by (17.2) this epimorphism splits if and only if P is projective. $\square$

In §13 we saw that modules over semisimple rings behave very much like modules over division rings. Of course modules over division rings are projective. The parallel behavior of division rings and semisimple rings is seen further in

17.4. Corollary. *A ring R is semisimple if and only if every left R-module is projective.*

Proof. By (13.10) R is semisimple iff every epimorphism in $_RM$ splits. But, by (17.2.b) this condition holds iff every left R-module is projective. $\square$

Characterization of Generators

Recall (see §8) that a module $_RG$ is a *generator* in case G generates every module in $_RM$. That is, G is a generator if and only if for every $_RM$ there is a set A and an R-epimorphism

$$G^{(A)} \to M \to 0.$$

A sort of duality exists between generators and projectives. The projectives are the modules P for which $Hom_R(P, \_)$ maps every epimorphism to an epimorphism; the generators are those modules G for which $Hom_R(G, \_)$ only maps epimorphisms to epimorphisms. Other examples of this phenomenon can be found by comparing (17.1) with the following characterizations of generators.

17.5. Proposition. *For a left R-module G the following statements are equivalent:*

(a) G *is a generator;*

(b) *For every homomorphism f in $_R M$ if $Hom_R(G, f) = 0$, then $f = 0$;*

(c) *For every $f: {}_R M \to {}_R N$ in $_R M$, if $f_*: Hom_R(G, M) \to Hom_R(G, N)$ is epic, then f is epic;*

(d) *A sequence*

$$M' \overset{f}{\to} M \overset{g}{\to} M''$$

is exact in $_R M$ if the sequence

$$Hom_R(G, M') \overset{f_*}{\to} Hom_R(G, M) \overset{g_*}{\to} Hom_R(G, M'')$$

is exact.

Proof. (a) $\Leftrightarrow$ (b). This follows at once from (8.11.1).

(a) $\Rightarrow$ (d). Let $_R G$ be a generator. Suppose

$$_R M' \overset{f}{\to} {}_R M \overset{g}{\to} {}_R M''$$

is a sequence in $_R M$ such that the sequence

$$Hom_R(G, M') \overset{f_*}{\to} Hom_R(G, M) \overset{g_*}{\to} Hom_R(G, M'')$$

is exact. Then $0 = g_* f_* = Hom(G, gf)$. Thus, since (a) $\Leftrightarrow$ (b), we have $gf = 0$. I.e., $Im f \leq Ker g$. Let $x \in Ker g$. Then, since G generates $Ker g$, there exist homomorphisms $\beta_i: G \to Ker g \leq M$ and $y_i \in G$ such that

$$x = \Sigma_{i=1}^n \beta_i(y_i).$$

Then, for each i, $g\beta_i = 0$; so $\beta_i \in Ker g_* = Im f_*$. That is, for each i there is an $\alpha_i \in Hom(G, M')$ with $\beta_i = f_*(\alpha_i) = f\alpha_i$. Therefore,

$$x = \Sigma_{i=1}^n \beta_i(y_i) = \Sigma_{i=1}^n f\alpha_i(y_i) \in Im f.$$

(d) $\Rightarrow$ (c). This implication is clear.

(c) $\Rightarrow$ (a). Assuming (c) it will suffice to show (see (8.13)) that for each $_R M$ the trace $Tr_M(G)$ is M. Consider then the canonical exact sequence

$$0 \to Tr_M(G) \overset{i}{\to} M \overset{n}{\to} M/Tr_M(G) \to 0$$

which yields an exact sequence

$$0 \to Hom_R(G, Tr_M(G)) \overset{i_*}{\to} Hom_R(G, M) \overset{n_*}{\to} Hom_R(G, M/Tr_M(G)).$$

If $\beta \in Hom(G, M)$, then $[n_*(\beta)](G) = n(\beta(G)) = 0$ because

$$\beta(G) \subseteq Tr_M(G) = Ker n.$$

Thus $Im i_* = Ker n_* = Hom_R(G, M)$ and i_* is surjective. So (c) implies that i must be surjective; that is, $Tr_M(G) = Im i = M$. $\square$

Since $_R R$ is a generator (8.8), a module generates $_R R$ if and only if it generates every module. However, $_R R$ is also finitely generated, so any module that generates $_R R$ must finitely generate $_R R$ (see (10.1)). Finally, since $_R R$ is projective, it follows that $_R G$ is a generator if and only if there is a split epimorphism $G^{(n)} \to R \to 0$. In other words

17.6. Proposition. *A left R-module G is a generator if and only if for some module R' and some integer n > 0 there is an R-isomorphism*

$$G^{(n)} \cong R \oplus R'. \qquad \square$$

The dual behavior of generators and finitely generated projective modules is illustrated by (17.3) and (17.6). The next two important results provide further evidence.

17.7. Lemma. *Let $_RQ_S$ be a faithfully balanced bimodule. Then $_RQ$ is a generator if and only if Q_S is finitely generated and projective.*

Proof. ($\Rightarrow$). Since right multiplication $\rho: S \to End(_RQ)$ is a ring isomorphism, as right S-modules

$$Hom_R(Q, Q_S) \cong S_S.$$

Also by (17.6), since $_RQ$ is a generator,

$$_RQ^{(n)} \cong R \oplus R'$$

for some left R-module R'. Now applying (16.3) and (4.5) we get right S-isomorphisms

$$S_S^{(n)} \cong Hom_R(Q, Q_S)^{(n)} \cong Hom_R(Q^{(n)}, Q_S)$$

$$\cong Hom_R(R \oplus R', Q_S) \cong Hom_R(R, Q_S) \oplus Hom_R(R', Q_S)$$

$$\cong Q \oplus Q';$$

so that, by (17.3), Q_S is finitely generated and projective.

($\Leftarrow$). This follows from (17.3) and (17.6) since

$$_RQ^{(n)} \cong Hom_S(S, {_RQ})^{(n)} \cong Hom_S(S^{(n)}, {_RQ})$$

$$\cong Hom_S(Q \oplus Q', {_RQ}) \cong Hom_S(Q, {_RQ}) \oplus Hom_S(Q', {_RQ})$$

$$\cong R \oplus R'. \qquad \square$$

17.8. Theorem *A left R-module G is a generator if and only if*
(i) $_RG$ *is faithful and balanced,*
(ii) $G_{End(_RG)}$ *is finitely generated and projective.*

Proof. ($\Rightarrow$). Since $_RG$ is a generator, by (17.6)

$$_RG^{(n)} \cong R \oplus R'$$

for some R'. Applying (14.2), Exercise (4.14) and (14.1.1) we have a commutative diagram of ring homomorphisms

$$\begin{array}{ccccccc}
R & \xrightarrow{1_R} & R & \xrightarrow{1_R} & R & \xrightarrow{1_R} & R \\
\lambda_1 \downarrow & & \lambda_2 \downarrow & & \lambda_3 \downarrow & & \lambda_4 \downarrow \\
BiEnd(G) & \to & BiEnd(G^{(n)}) & \to & BiEnd(R \oplus R') & \to & BiEnd(_RR)
\end{array}$$

where the composite of maps in the bottom row is injective and, by (4.11), λ_4

is bijective. Thus λ_1 is an isomorphism. That is, making the obvious identifications,

$$R \le BiEnd(G) = BiEnd(G^{(n)}) = BiEnd(R \oplus R')$$

$$\le BiEnd(R) = R.$$

Therefore $_RG$ is faithful and balanced. That is, $_RG_{End\ (_RG)}$ is a faithfully balanced bimodule. Now (17.7) applies.

($\Leftarrow$). This implication follows at once from (17.7). $\square$

Of course, the regular module $_RR$, or indeed any free left R-module, is both a projective module and a generator. In general not all projective modules are generators. The next result, however, shows that the important class of projective generators can be characterized as those projective modules that generate all simple modules.

17.9. Proposition. *Let P be a projective left R-module. Then the following statements are equivalent:*

(a) *P is a generator;*

(b) *$Hom_R(P, T) \ne 0$ for all simple left R-modules T;*

(c) *P generates every simple left R-module.*

Proof. The implications (a) $\Rightarrow$ (c) and (c) $\Rightarrow$ (b) are trivial. Finally for (b) $\Rightarrow$ (a), assume that P satisfies condition (b). It will suffice to prove that P generates R, or (see (8.21)) that $Tr_R(P) = R$. But if $Tr_R(P) \ne R$, then since it is a left ideal, $Tr_R(P)$ is contained in some maximal left ideal I of R. Then R/I is simple, so there is a non-zero R-homomorphism $\gamma : P \to R/I$. Since P is projective, there is a commutative diagram

$$
\begin{array}{ccc}
 & P & \\
 {\scriptstyle \bar\gamma}\swarrow & \downarrow{\scriptstyle \gamma} & \\
R \xrightarrow[n_I]{} & R/I & \longrightarrow 0.
\end{array}
$$

This produces a contradiction since $Im\,\bar\gamma \subseteq Tr_R(P) \subseteq I$. $\square$

Radicals of Projective Modules

It is not surprising, in view of (17.2), that the behavior of projective modules parrots much of that of the regular module $_RR$. Next we observe this in the case of radicals of projective modules.

17.10. Proposition. *Let R be a ring with radical $J(R) = J$. If P is a projective left R-module, then*

$$Rad\ P = JP.$$

Proof. Proposition (17.2) allows us to assume that P is a direct summand of a free module $P \oplus P' = R^{(A)}$. Then by (9.19)

$$Rad\ P \oplus Rad\ P' = Rad(R^{(A)}) = (Rad\ R)^{(A)}$$

$$= J^{(A)} = J \cdot R^{(A)} = JP \oplus JP'.$$

So, since $JP \leq Rad\,P$ and $JP \leq Rad\,P'$ (see (15.18)), we must have $Rad\,P = JP$. $\qquad\square$

Next we calculate the radical of the endomorphism ring of a projective module.

17.11. Proposition. *Let P be a projective left R-module with endomorphism ring* $S = End(_R P)$. *Let* $a \in S$. *Then*

$$a \in J(S) \qquad iff \qquad Im\,a \ll P.$$

Proof. ($\Leftarrow$). Suppose $Im\,a \ll {}_R P$. Then, by (15.3), it will suffice to show that $Sa \ll {}_S S$. So suppose that $I \leq {}_S S$ and $Sa + I = S$. So $1_P = sa + b$ for some $s \in S$ and $b \in I$. Then $P = P1_P \leq Psa + Pb \leq Im\,a + Pb$, so that $Pb = P$. But then b is an epimorphism $b : P \to P$. So, since P is projective, this epimorphism splits and there is some $c \in S$ with $1_P = cb \in I$. Thus $I = S$ and $Sa \ll S$.

($\Rightarrow$). Let $a \in J(S)$ and suppose that $K \leq P$ with $Im\,a + K = P$. Then we readily see that if $n_K : P \to P/K$ is the natural epimorphism, $an_K : P \to P/K$ is epic. So, choosing $s \in S$ such that

commutes, we have $(1 - sa)n_K = 0$. But, since $a \in J(S)$, $1 - sa$ is invertible and $n_K = 0$. Therefore, $K = P$. $\qquad\square$

17.12. Corollary. *Let* $J = J(R)$. *If P is a projective left R-module such that* $JP \ll P$ *(e.g., if* $_R P$ *is finitely generated), then*

$$J(End(_R P)) = Hom_R(P, JP) \quad and \quad End(_R P)/J(End_R P) \cong End(_R P/JP).$$

Proof. Since, by (17.10) we have $Rad\,P = JP$, the hypothesis $JP \ll P$ insures that a submodule of P is superfluous iff it is contained in JP. (See (9.13).) In particular then, by (17.11), an endomorphism a of P belongs to $J(End(_R P))$ iff $Im\,a \leq JP$. Thus $J(End(_R P)) = Hom_R(P, JP)$.

Now observe that, since JP is stable under endomorphisms of P,

$$\phi(s) : (p + JP) \mapsto ps + JP$$

defines a ring homomorphism $\phi : End(_R P) \to End(_R P/JP)$; and that, since P is projective, this homomorphism ϕ is surjective.

But clearly we have $Ker\,\phi = Hom_R(P, JP)$, so that

$$End(_R P)/J(End(_R P)) \cong End(_R P/JP). \qquad\square$$

17.13. Corollary. *Let R be a ring with radical J, let $n \in \mathbb{N}$, and let $e \in R$ be a non-zero idempotent. Then*

$$J(\mathbb{M}_n(R)) = \mathbb{M}_n(J), \qquad and \qquad J(eRe) = eJe.$$

Proof. By (4.11) and (13.2) there is a natural ring isomorphism

$$\rho : \mathbb{M}_n(R) \to End(_R R^{(n)}).$$

Now $J R^{(n)} = J^{(n)}$ and clearly $\rho([r_{ij}]) \in Hom_R(R^{(n)}, J^{(n)})$ iff $[r_{ij}] \in \mathbb{M}_n(J)$. So apply (17.12). For the other assertion, recall (4.15) that there is a natural isomorphism

$$\rho : eRe \to End(_R Re)$$

and clearly $\rho(ere) \in Hom_R(Re, Je)$ iff $ere \in J$. Again apply (17.12). (Or see Exercise (15.11).) $\qquad\qquad\qquad\qquad\qquad\qquad\qquad\qquad\qquad\qquad\qquad \Box$

Now we can prove the important fact that no non-zero projective module is its own radical; that is,

17.14. Proposition. *Every non-zero projective module contains a maximal submodule.*

Proof. Let $_R P$ be projective. Then by (17.3) we may assume that there is a free left R-module F with $F = P \oplus P'$. If P contains no maximal submodule, then by (17.10) we have

$$P = JP \subseteq JF.$$

To prove the proposition we show that this forces $P = 0$. To this end, let $x \in P$. Let e be an idempotent endomorphism of F such that $Fe = P$ and let $(x_\alpha)_{\alpha \in A}$ be a free basis for F. Then, for some finite subset $H \subseteq A$, and some $r_\alpha \in R \; (\alpha \in H)$,

$$x = \Sigma_{\alpha \in H} r_\alpha x_\alpha.$$

Also, for each $\alpha \in H$, there are finite sets $H_\alpha \subseteq A$ and $a_{\alpha\beta} \in J \; (\beta \in H_\alpha)$ such that

$$x_\alpha e = \Sigma_{\beta \in H_\alpha} a_{\alpha\beta} x_\beta.$$

Now, inserting 0's where necessary, we may assume that all of these sums are taken over a common finite subset $K \subseteq A$ to get

$$0 = x - xe = (\Sigma_{\alpha \in K} r_\alpha x_\alpha) - (\Sigma_{\alpha \in K} r_\alpha x_\alpha e)$$

$$= (\Sigma_{\alpha \in K} r_\alpha(\Sigma_{\beta \in K} \delta_{\alpha\beta} x_\beta)) - (\Sigma_{\alpha \in K} r_\alpha(\Sigma_{\beta \in K} a_{\alpha\beta} x_\beta))$$

$$= \Sigma_{\beta \in K}(\Sigma_{\alpha \in K} r_\alpha(\delta_{\alpha\beta} - a_{\alpha\beta})) x_\beta.$$

Since the x_β are independent this equation yields the matrix equation

$$[r_\alpha](I_n - [a_{\alpha\beta}]) = [0] \in \mathbb{M}_{1 \times n}(R)$$

where $n = card(K)$ and I_n is the identity matrix in $\mathbb{M}_n(R)$. But by (17.13) $[a_{\alpha\beta}] \in J(\mathbb{M}_n(R))$ and hence is quasi-regular. Thus $I_n - [a_{\alpha\beta}]$ has an inverse in $\mathbb{M}_n(R)$ and $[r_\alpha] = [0] \in \mathbb{M}_{1 \times n}(R)$. This means

$$x = \Sigma_{\alpha \in K} r_\alpha x_\alpha = 0. \qquad\qquad\qquad\qquad\qquad\qquad\qquad \Box$$

Projective Covers

Every free module is projective (17.2) and every module is generated by $_RR$ (8.1). Thus trivially

17.15. Proposition. *Every module is an epimorphic image of a projective module.*   □

For some modules M an even stronger assertion is possible: there is a projective module P and an epimorphism $f: P \to M$ "minimal" in the sense that $(f \mid L): L \to M$ is epic for no proper submodule of P. It is clear from (5.15) that this minimality condition simply says that $Ker\, f \ll P$. This leads to a formal definition.

A pair (P, p) is a *projective cover* of the module $_RM$ in case P is a projective left R-module and

$$P \xrightarrow{p} M \to 0$$

is a superfluous epimorphism $(Ker\, p \ll P)$. We also employ natural variations and abbreviations of this terminology; for example, we may well call P itself a projective cover of M.

17.16. Examples. (1) If e is an idempotent in R, then by (17.10) and (10.4), $Je = Rad(Re) \ll Re$. So since Re is projective, the pair (Re, n) is a projective cover of Re/Je where $n: Re \to Re/Je$ is the natural map.

(2) The pair $(\mathbb{Z}, r_2)$ where $r_2: \mathbb{Z} \to \mathbb{Z}_2$ is the natural map is not a projective cover since $2\mathbb{Z}$ is not superfluous in $\mathbb{Z}$. In fact, using (17.17) it is easy to prove that $\mathbb{Z}_2$ has no projective covers. (See Exercise (17.14).)

(3) Let R be a local ring and $_RM$ finitely generated. Then R/J is a division ring and M/JM is a finite dimensional vector space over R/J. Say M/JM is k-dimensional; set $P = R^{(k)}$. Then clearly there is an R-epimorphism $\bar{p}: P \to M/JM$ with $Ker\, \bar{p} = JP$. Since $_RP$ is projective and the natural map $n: M \to M/JM$ is epic, there is a homomorphism $p: P \to M$ with $np = \bar{p}$. By Nakayama's Lemma (15.13), $JM \ll M$, so n is a superfluous epimorphism. Thus, by (5.15), p is epic. But $Ker\, p \leq Ker\, \bar{p} = JP \ll P$, whence (P, p) is a projective cover of M.

Now we prove The Fundamental Lemma for Projective Covers. One of its consequences is that if a module does have a projective cover, then it has (essentially) only one.

17.17. Lemma. *Suppose $_RM$ has a projective cover $p: P \to M$. If $_RQ$ is projective and $q: Q \to M$ is an epimorphism, then Q has a decomposition*

$$Q = P' \oplus P''$$

such that

(1) $P' \cong P$;
(2) $P'' \leq Ker\, q$;
(3) $(q \mid P'): P' \to M$ *is a projective cover for M.*

Moreover, if $f:M_1 \to M_2$ *is an isomorphism and if* $p_1:P_1 \to M_1$ *and* $p_2:P_2 \to M_2$ *are projective covers, then there is an isomorphism* $\bar{f}:P_1 \to P_2$ *such that* $p_2 \bar{f} = f p_1$.

Proof. By the projectivity of Q there is a commutative diagram

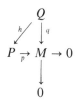

with exact row and column. Since p is a superfluous epimorphism and $ph = q$ is epic, h is also epic by (5.15). But P is projective, so h splits, i.e., there is a monomorphism $g:P \to Q$ such that $hg = 1_P$, and hence $Q = Im\,g \oplus Ker\,h$. Now, setting

$$P' = Im\,g \qquad \text{and} \qquad P'' = Ker\,h,$$

we see that (1) holds because g is monic, and that (2) holds because $ph = q$. But now we have $q(P') = q(Q) = M$, so that

$$P' \xrightarrow{(q\,|P')} M \longrightarrow 0$$

is exact; and this is a projective cover because from $qg = phg = p$, it follows that $Ker(q\,|\,P') = g(Ker\,p)$, a superfluous submodule of $g(P) = P'$. Thus (3) also holds.

To prove the last statement, let $p = p_2$, $q = f p_1$ and $\bar{f} = h$. Then $p_2 \bar{f} = f p_1$. Also $\bar{f} = h$ is epic, $Ker\,\bar{f} = Ker\,p_1$ is a superfluous direct summand of P_1 and $\bar{f}$ is an isomorphism. ☐

17.18. Proposition. *Let e and f be idempotents in a ring R. Then the following are equivalent:*
    (a) $Re \cong Rf$;
    (b) $Re/Je \cong Rf/Jf$;
    (c) $eR/eJ \cong fR/fJ$;
    (d) $eR \cong fR$.

Proof. (a) $\Leftrightarrow$ (b). That (a) implies (b) is clear. If $h:Re/Je \to Rf/Jf$ is an isomorphism, then since the natural maps $Re \to Re/Je$ and $Rf \to Rf/Jf$ are projective covers, it follows from Lemma (17.17) that there is an isomorphism $\bar{h}:Re \to Rf$.

(c) $\Leftrightarrow$ (d). This is symmetric to the proof of (a) $\Leftrightarrow$ (b).

(a) $\Leftrightarrow$ (d). If $Re \cong Rf$, then (see (4.6))

$$eR \cong Hom(Re, R) \cong Hom(Rf, R) \cong fR.$$ ☐

In Chapter 7 we shall deal extensively with projective covers. There projective covers of simple modules will be of particular interest. We now have the resources to give

17.19. Proposition. *Let R be a ring with radical $J = J(R)$. Then the following statements about a projective left R-module P are equivalent:*
  (a) *P is the projective cover of a simple left R-module;*
  (b) *JP is a superfluous maximal submodule of P;*
  (c) *$End(_RP)$ is a local ring.*
Moreover, if these conditions hold, then $P \cong Re$ for some idempotent $e \in R$.

Proof. (a) $\Rightarrow$ (b). Clearly P is the projective cover of a simple module iff P contains a superfluous maximal submodule. But JP is contained in every maximal submodule of P; and JP contains every superfluous submodule of P ((9.13) and (17.10)).

  (b) $\Rightarrow$ (c). If JP is a superfluous maximal submodule of P, then by (17.12) and Schur's Lemma (13.1)

$$End(_RP)/J(End(_RP)) \cong End_R(P/JP)$$

is a division ring. Thus by (15.15), (b) implies that $End(_RP)$ is local.

  (c) $\Rightarrow$ (a). Suppose that $End(_RP)$ is a local ring. Then $P \neq 0$. By (17.14) there is a maximal submodule $K < P$. We claim that the natural epimorphism $P \to P/K \to 0$ is a projective cover, i.e., $K \ll P$. Suppose that $K + L = P$ for some $L \leq P$. Then

$$P/K \cong (L + K)/K \cong L/(L \cap K);$$

so there is a non-zero homomorphism $f : P \to L/(L \cap K)$. Thus, since P is projective there is an endomorphism $s : P \to L \leq P$ such that

$$
\begin{array}{ccc}
 & P & \\
{}^{s}\swarrow & & \downarrow{f} \\
L \underset{\pi}{\to} & L/(L \cap K) & \to 0
\end{array}
$$

commutes. Since $0 \neq f = sn$, $Im\, s \nleq K$; from which it follows that $Im\, s$ is not superfluous in P. Therefore $s \notin J(End(_RP))$ (17.11), s is an invertible endomorphism of P (15.15.g), $L = P$; and we have shown that $K \ll P$.

  Moreover, every simple module is an epimorph of R, so by (17.17), a projective cover P of a simple module must be isomorphic to a direct summand of $_RR$. That is, $P \cong Re$ for some $e = e^2 \in R$. $\square$

17.20. Corollary. *The following statements about an idempotent e in a ring R are equivalent:*
  (a) *Re/Je is simple;*
  (b) *Je is the unique maximal submodule of Re;*
  (c) *eRe is a local ring;*
  (d) *eJ is the unique maximal submodule of eR;*
  (e) *eR/eJ is simple.* $\square$

17. Exercises

1. Prove that every projective module over a P.I.D. is free. [Hint: Exercise (8.16).]

2. Prove that if R is a local ring, then every finitely generated projective module is free. [Hint: See (12.7).]

3. Prove that the abelian group $\mathbb{Z}^{\mathbb{N}}$ is not projective; hence products of projectives need not be projective. [Hint: Let K consist of those $x \in \mathbb{Z}^{\mathbb{N}}$ such that each power of 2 divides all but finitely many $\pi_n(x)$. Show that if $\mathbb{Z}^{\mathbb{N}}$ is free, then K is free of uncountable rank. Now consider $K/2K$.]

4. Let $e \in R$ be idempotent. Prove that:
   (1) Re is simple and faithful iff eR is simple and faithful. [Hint: By (17.20), if $J(R) = 0$, then Re is simple iff eR is. If Re is simple and faithful and $0 \neq r \in R$, then there exist s, $t \in R$ with $srte = e$.]
   (2) If Re is simple and faithful, then $Soc(_R R) = ReR = Soc(R_R)$.

5. Recall that for a ring R the properties x-primitive and y-artinian for $x, y \in \{\text{left}, \text{right}\}$ are equivalent, but that in general a primitive ring need not be right primitive.
   (1) Now prove that for a ring R the following are equivalent: (a) R is primitive and has a minimal left ideal; (b) R is primitive and has a simple projective left module; (c) R is primitive and has a projective simple right module; (d) R is primitive and has a minimal right ideal; (e) R is right primitive and has a minimal right ideal. [Hint: Exercises (13.8) and (17.4).]
   (2) Prove that if R is the endomorphism ring of a vector space, then R is primitive and has a minimal left ideal.
   (3) Prove that if R is a simple ring and has a minimal left ideal, then R is artinian.
   (4) Prove that the primitive ring of Exercise (14.2) has a minimal left ideal.

6. Prove that $_R M$ is faithful iff $_R M$ cogenerates every projective left R-module. (See Exercise (8.3).)

7. Prove that each finitely generated non-zero projective module is a generator over its endomorphism ring and finitely generated projective over its biendomorphism ring.

8. For each left R-module M, let M^* denote the right R-module $M^* = Hom_R(M, R)$. Prove that if M is finitely generated and projective (a generator), then so is M^*.

9. Prove that a module $_R M$ is a projective generator iff $M^{(C)}$ is free for some $C \neq \varnothing$.

10. A generator $_R G$ is a *minimal* generator in case it is an epimorphic image of every generator in $_R M$.
    (1) Show that if R is either semisimple, local, a P.I.D., or the endomorphism ring of an infinite dimensional vector space, then R has a minimal generator. [Hint: In the last three cases $_R R$ is a minimal generator.]
    (2) Let $R = \Pi_{n=1}^{\infty} \mathbb{M}_n(Q)$ where Q is a division ring. Prove that R has no minimal generator. [Hint: For each n let G_n be a generator for $\mathbb{M}_n(Q)$. Then $G_1 \oplus \ldots \oplus G_m \oplus \Pi_{n>m} \mathbb{M}_n(Q)$ is a generator for R.]

11. Let P be a left R-module. A pair of indexed sets $(x_\alpha)_{\alpha \in A}$ in P and

$(f_\alpha)_{\alpha \in A}$ in $Hom_R(P, R)$ is a *dual basis* for P in case for all $x \in P$.

(i) $f_\alpha(x) = 0$ for almost all $\alpha \in A$.

(ii) $x = \Sigma_A f_\alpha(x)x_\alpha$.

Prove

(1) **The Dual Basis Lemma.** P *is projective iff it has a dual basis.* [Hint: (17.2).]

(2) P is finitely generated projective iff there exist $x_1, \ldots, x_n \in P$ and $f_1, \ldots, f_n \in Hom_R(P, R)$ such that for each $x \in P$

$$x = \Sigma_{i=1}^n f_i(x)x_i.$$

12. Recall that if $_RP$ is projective, then $Rad\, P = JP$. Dually prove that $Soc\, P = (Soc\, _RR)P$.

13. Let P be a non-zero projective module. Prove that P is the projective cover of a simple module iff every non-zero factor module of P is indecomposable.

14. (1) Prove that if R is a ring with $J(R) = 0$, then no non-projective R-module has a projective cover. In particular, no finite $\mathbb{Z}$-module has a projective cover.

    (2) Let n be a natural number and let p be a prime that divides n, say p^m is the largest p-power that divides n. Prove that if $1 \le k \le m$, then the epimorphism $\mathbb{Z}_{p^m} \to \mathbb{Z}_{p^k}$ is a projective cover in $_{\mathbb{Z}_n}M$.

15. (1) Prove that if $p_i : P_i \to M_i$ $(i = 1, \ldots, n)$ are projective covers, then $(\oplus_i p_i) : \oplus_i P_i \to \oplus_i M_i$ is a projective cover.

    (2) Prove that if M and $M \oplus N$ have projective covers, then so does N. [Hint: If $p : P \to M$ and $p' : P' \to M \oplus N$ are projective covers, then $f^\leftarrow(N) \to N$ is a projective cover where $f : P' \to P \oplus N$ satisfies $(p \oplus 1_N)f = p'$.]

16. (1) Prove that if $_RU$ has a projective cover, then $\mathcal{P}\imath^{-1}(U)$ is closed under direct products. [Hint: Exercise (16.13).]

    (2) Prove that if $p : P \to U$ is a projective cover, then the following are equivalent: (a) U is quasi-projective; (b) U is projective modulo its annihilator; (c) $Ker\, p = I_R(U)P$.

    (3) Prove that if P is projective and K is a left R- right $End(_RP)$-submodule of P, then P/K is a quasi-projective left R-module.

17. Let U be a quasi-projective module with a projective cover $p : P \to U$. Prove that if $P = P_1 \oplus P_2$, then U is a direct sum $U = U_1 \oplus U_2$ of quasi-projective modules U_1, U_2 such that $(p|P_i) : P_i \to U_i$ are projective covers $(i = 1, 2)$. In particular, if U is indecomposable, then so is P.

18. Let $p_i : P_i \to U_i$ $(i = 1, 2)$ be projective covers. Prove that if $P_1 \cong P_2$ and if $U_1 \oplus U_2$ is quasi-projective, then $U_1 \cong U_2$.

19. Prove that a finite abelian group M is quasi-projective iff for each prime p the p-primary components of M all have the same length. [Hint: Exercises (17.14.2) and (17.18).]

20. Let R be a left artinian ring with $J = J(R)$. Then by (10.14) and (7.2) $R = Re_1 \oplus \ldots \oplus Re_n$ where $e_1, \ldots, e_n$ is a complete set of pairwise

orthogonal primitive idempotents. By (12.8) and (17.20) each Re_i/Je_i is simple and $Re_i \to Re_i/Je_i$ is a projective cover.

(1) Prove that every simple left R-module has a projective cover. [Hint: $R/J \cong Re_1/Je_1 \oplus \ldots \oplus Re_n/Je_n$.]

(2) Prove that every semisimple module has a projective cover. [Hint: If M is semisimple, then there exists an indexed set $(f_\alpha)_{\alpha \in A}$ in $\{e_1, \ldots, e_n\}$ such that $M \cong \oplus_A Rf_\alpha/Jf_\alpha \cong (\oplus_A Rf_\alpha)/(\oplus_A Jf_\alpha)$. By Exercise (5.18), $\oplus_A Jf_\alpha \ll \oplus_A Rf_\alpha$.]

(3) Prove that for each left R-module M, there is an indexed set $(f_\alpha)_{\alpha \in A}$ in $\{e_1, \ldots, e_n\}$ and a projective cover $\oplus_A Rf_\alpha \to M$. [Hint: Since M/JM is semisimple, it has a projective cover $\bar{p}: \oplus_A Rf_\alpha \to M/JM$. Since $M \to M/JM$, $\bar{p}$ lifts to $p: \oplus_A Rf_\alpha \to M$. Since $JM \ll M$ (Exercise (5.18)), p is a projective cover.]

(4) Prove that if $_RP$ is projective, then there exists an indexed set $(f_\alpha)_{\alpha \in A}$ in $\{e_1, \ldots, e_n\}$ such that $P \cong \oplus_A Rf_\alpha$.

§18. Injective Modules and Cogenerators

Recall that a left R-module E is *injective* in case E is injective relative to every left R-module. That is, E is injective in case whenever there is given the solid part of a diagram

$$
\begin{array}{c}
E \\
\gamma \uparrow \ \ \nwarrow \bar{\gamma} \\
0 \to K \underset{f}{\to} M
\end{array}
$$

in $_R\mathsf{M}$ with exact row, there is an R-homomorphism $\bar{\gamma}$ such that the whole diagram commutes; i.e., $\bar{\gamma}f = \gamma$. In other words, the injective modules are the arrow-theoretic or categorical duals of the projective modules.

Characterizations of Injective Modules

The injective and the projective modules have dual effect on the *Hom* functors; in particular, dual to (17.1) we have:

18.1. Proposition. *The following statements about a left R-module E are equivalent:*

(a) *E is injective;*

(b) *For each monomorphism $f: {}_RK \to {}_RM$ the map*

$$Hom(f, E): Hom_R(M, E) \to Hom_R(K, E)$$

is an epimorphism;

(c) *For each bimodule structure $_RE_S$ the functor*

$$Hom_R(-, E_S): {}_R\mathsf{M} \to \mathsf{M}_S$$

is exact;

(d) *For every exact sequence*

$$M' \xrightarrow{f} M \xrightarrow{g} M''$$

in $_R M$ *the sequence*

$$Hom_R(M'', E) \xrightarrow{g^*} Hom_R(M, E) \xrightarrow{f^*} Hom_R(M', E)$$

is exact. ☐

Until we establish the existence of sufficiently many injective modules, we cannot prove the proper dual to (17.2). As a temporary substitute, though, we have at once from (16.11) the important fact: ☐

18.2. Proposition. *Direct products and direct summands of injective modules are injective.*

Every module is an epimorph of a projective (even free) module. One of our tasks is to prove the dual result that every module can be embedded in an injective module. First, however, we establish a very useful test for injectivity. This test (sometimes called "The Baer Criterion") says that injectivity of a module E can be determined by its behavior in the set of diagrams

$$
\begin{array}{c}
E \\
\uparrow \\
0 \to I \to R
\end{array}
$$

when the row is restricted to inclusion maps of left ideals.

18.3. The Injective Test Lemma. *The following statements about a left R-module E are equivalent:*

(a) E *is injective;*

(b) E *is injective relative to R;*

(c) *For every left ideal $I \leq {}_R R$ and every R-homomorphism $h: I \to E$ there exists an $x \in E$ such that h is right multiplication by x*

$$h(a) = ax \qquad (a \in I).$$

Proof. (a) $\Leftrightarrow$ (b). This is by (16.14), since $_R R$ is a generator.

(b) $\Rightarrow$ (c). If E is $_R R$-injective and $I \leq {}_R R$ with $h: I \to E$, then there is an $\bar{h}: R \to E$ such that $(\bar{h} \mid I) = h$. Let $x = \bar{h}(1)$. Then $h(a) = \bar{h}(a) = a\bar{h}(1) = ax$ for all $a \in I$.

(c) $\Rightarrow$ (b). If $I \leq {}_R R$, $x \in E$ and $h(a) = ax$ for all $a \in I$, then right multiplication by x, $\rho(x): R \to E$, extends h. Thus (c) implies that E is $_R R$-injective. ☐

Recall that an abelian group Q is divisible in case $nQ = Q$ for each non-zero integer n. (See Exercise (3.15).)

18.4. Lemma. *An abelian group Q is divisible if and only if Q is injective as a $\mathbb{Z}$-module.*

Proof. ($\Rightarrow$). Every non-zero ideal of $\mathbb{Z}$ is of the form $\mathbb{Z}n$, $n \neq 0$. If Q is a divisible abelian group and $h: \mathbb{Z}n \to Q$, then there is a $b \in Q$ with $h(n) = nb$

and $h(jn) = jh(n) = (jn)b$ for all $jn \in \mathbb{Z}n$. Thus The Injective Test Lemma applies.

($\Leftarrow$). If $_{\mathbb{Z}}Q$ is injective, $a \in Q$ and $0 \neq n \in \mathbb{Z}$, then $h: jn \mapsto ja$ defines a homomorphism $h: \mathbb{Z}n \to Q$ which, by The Injective Test Lemma, must be multiplication by some $b \in Q$. But then $a = h(n) = nb$. $\qquad\square$

18.5. Lemma. *If Q is a divisible abelian group, then the left R-module $Hom_{\mathbb{Z}}(R_R, Q)$ is injective.*

Proof. By (4.4.1), $Hom_{\mathbb{Z}}(R_R, Q)$ is a left R-module. Let $I \leq {_R}R$ and suppose $h: I \to Hom_{\mathbb{Z}}(R_R, Q)$ is an R-homomorphism. Then $\gamma: a \mapsto [h(a)](1)$ defines an abelian group homomorphism $\gamma: {_{\mathbb{Z}}}I \to {_{\mathbb{Z}}}Q$. Thus, since $_{\mathbb{Z}}Q$ is injective, there is a $\bar{\gamma} \in Hom_{\mathbb{Z}}(R, Q)$ such that $(\bar{\gamma} \,|\, I) = \gamma$. Now we have, for all $a \in I$, $r \in R$,

$$(a\bar{\gamma})(r) = \bar{\gamma}(ra) = \gamma(ra) = [h(ra)](1)$$
$$= [r \cdot h(a)](1) = [h(a)](r);$$

so, $h(a) = a\bar{\gamma}$ for all $a \in I$. Therefore, by The Injective Test Lemma, $Hom_{\mathbb{Z}}(R_R, Q)$ is an injective left R-module. $\qquad\square$

18.6. Proposition. *Every left R-module can be embedded in an injective left R-module.*

Proof. Let M be a left R-module. Then by (8.1) there is a set A and a $\mathbb{Z}$-epimorphism $f: \mathbb{Z}^{(A)} \to M$. Thus, since

$$_{\mathbb{Z}}M \cong \mathbb{Z}^{(A)}/Ker f \leq \mathbb{Q}^A/Ker f,$$

and since direct products and factor groups of divisible abelian groups are divisible (see Exercises (3.15) and (6.9)), we may assume that $_{\mathbb{Z}}M \leq {_{\mathbb{Z}}}Q$ with Q divisible. Finally, apply (18.5) to

$$_RM \cong Hom_R(R_R, M) \leq Hom_{\mathbb{Z}}(R_R, M) \leq Hom_{\mathbb{Z}}(R_R, Q). \qquad\square$$

The following partial dual to (17.2) is an immediate consequence of (18.2), (18.6), and Exercise (16.7).

18.7. Proposition. *A left R-module E is injective if and only if every monomorphism*

$$0 \to {_R}E \to {_R}M \qquad\qquad\square$$

splits.

And dual to (17.4) we have from (13.9):

18.8. Corollary. *A ring R is semisimple if and only if every left R-module is injective.* $\qquad\square$

Injective Envelopes

As we have seen (18.6) every R-module M can be embedded in an injective R-module. This leads to a notion dual to that of a projective cover, namely a "minimal" embedding of M in an injective module. A pair (E, i) is an *injective envelope* of M in case E is an injective left R-module and

$$0 \to M \xrightarrow{i} E$$

is an essential monomorphism. Again we shall allow obvious variations in our terminology.

Since $\mathbb{Q}$ is divisible as a $\mathbb{Z}$-module, it is $\mathbb{Z}$-injective. Clearly the inclusion map $i : \mathbb{Z} \to \mathbb{Q}$ is essential. Thus $(\mathbb{Q}, i)$ is an injective envelope of $\mathbb{Z}$. (Also see Exercise (18.2).)

Dual to (17.17) we have the following Fundamental Lemma for Injective Envelopes:

18.9. Lemma. *Let M be a left R-module and suppose that $i : M \to E$ is an injective envelope of M. If $_R Q$ is injective and $q : M \to Q$ is a monomorphism, then Q has a decomposition*

$$Q = E' \oplus E''$$

such that

(1) $E' \cong E$;

(2) $\operatorname{Im} q \leq E'$;

(3) $q : M \to E'$ *is an injective envelope of M.*

Moreover, if $f : M_1 \to M_2$ is an isomorphism and $i_1 : M_1 \to E_1$ and $i_2 : M_2 \to E_2$ are injective envelopes, then there is an isomorphism $\bar{f} : E_1 \to E_2$ such that $\bar{f} i_1 = i_2 f$.

$$
\begin{array}{ccc}
E_1 & \xrightarrow{\bar{f}} & E_2 \\
{\scriptstyle i_1}\uparrow & & \uparrow{\scriptstyle i_2} \\
M_1 & \xrightarrow{f} & M_2
\end{array}
\qquad \square
$$

Not every module has a projective cover (see Exercise (17.14)). Thus the next very important result is especially remarkable.

18.10. Theorem. *Every module has an injective envelope. It is unique to within isomorphism.*

Proof. Let M be a left R-module. Then by (18.6) there is an injective module $_R Q$ with $M \leq Q$. The set of $N \leq Q$ such that $M \trianglelefteq N$ is clearly inductive. So by the Maximal Principle there is a maximal member E of this set. Now choose $E' \leq Q$ maximal with respect to $E \cap E' = 0$, (i.e., let E' be a Q-complement of E) so

$$(E \oplus E')/E' \trianglelefteq Q/E'$$

(see (5.21)). The fact is that $E \oplus E' = Q$. To see this let $g : (E \oplus E')/E' \to E$ be the obvious isomorphism. Then using the injectivity of Q we have a commutative diagram with exact row and column

$$
\begin{array}{ccc}
 & Q & \\
 & {\scriptstyle g}\uparrow \ \ \ \ \searrow{\scriptstyle h} & \\
0 \longrightarrow (E \oplus E')/E' & \xrightarrow[\trianglelefteq]{} & Q/E' \\
 & {\scriptstyle}\uparrow & \\
 & 0 &
\end{array}
$$

By (5.13), h is monic, so

$$M \trianglelefteq E = Im\,g = h((E \oplus E')/E') \trianglelefteq h(Q/E').$$

Therefore $M \trianglelefteq h(Q/E')$ by (5.16), so by the maximality of E

$$h((E \oplus E')/E') = h(Q/E').$$

Then since h is monic, $Q = E \oplus E'$. Now by (18.2) we have that E is injective so the inclusion $M \to E$ is an injective envelope. That it is unique up to isomorphism follows from (18.9).   □

18.11. Corollary. *The following statements about an R-monomorphism*
$i : M \to E$ *are equivalent:*

(a) $i : M \to E$ *is an injective envelope of* M;

(b) *E is an injective module and for every R-monomorphism* $f : M \to Q$ *with*
Q injective there is a monomorphism $g : E \to Q$ *making the following diagram*
commute:

(c) *i is an essential monomorphism and for every essential monomorphism*
$f : M \to N$ there is a monomorphism $g : N \to E$ *making the following diagram*
commute:

Proof. To prove that (a) implies (b) and (c) use injectivity to get g and then use (5.13) to see that g is monic.

On the other hand, assume (b). By Theorem (18.10) there is an injective envelope $f : M \to Q$ for M. Then (b) gives a monomorphism $g : E \to Q$ with $f = gi$. Since E is injective, this monomorphism splits (18.7); say $Q = (Im\,g) \oplus E'$. But f is an essential monomorphism, so $Im\,f \trianglelefteq Q$ and $Im\,f = Im\,gi \leq Im\,g$. Thus, $E' = 0$, and g is an isomorphism; so $i : M \to E$ is also essential.

Finally to prove that (c) implies (a), we use (18.10) to get an injective envelope $f : M \to N$, and then apply (c). We omit the details.   □

It will be a very great convenience for us to take some liberties with our notation for injective envelopes. Every module has an injective envelope but no non-zero module has a unique one. Nevertheless, if $i : {}_R M \to {}_R Q$ is an injective envelope for M, we shall often write $Q = E({}_R M)$, or simply $Q = E(M)$, and say that $E(M)$ is "the injective envelope" of M. Moreover, we shall frequently identify M with its image in $E(M)$ and shall thus think of M as a submodule of $E(M)$. In this guise $E(M)$ is an essential injective extension of M. Then (18.11) can be rephrased loosely to characterize $E(M)$ (to within isomorphism) simultaneously as the unique minimal injective extension and also the unique maximal essential extension of M. Indeed $E(M)$ appears as a direct summand (though not necessarily uniquely; see Exercise (18.6)) of every injective module that contains M, and $E(M)$ contains a copy of every essential extension of M.

Among the more important other properties of the injective envelope we have the following:

18.12. Proposition. *In the category of left R-modules over a ring R:*
(1) *M is injective if and only if $M = E(M)$;*
(2) *If $M \trianglelefteq N$, then $E(M) = E(N)$;*
(3) *If $M \leq Q$, with Q injective, then $Q = E(M) \oplus E'$;*
(4) *If $\oplus_A E(M_\alpha)$ is injective (for instance, if A is finite) then*

$$E(\oplus_A M_\alpha) = \oplus_A E(M_\alpha).$$

Proof. Part (1) is immediate from the definition of the injective envelope. For (2), since $N \trianglelefteq E(N)$, if $M \trianglelefteq N$, then $M \trianglelefteq E(N)$ and $E(N)$ is injective, so the inclusion $M \to E(N)$ is an injective envelope of M. For (3) apply (18.11) to the inclusion map $f : M \to Q$ and then use (18.2). Finally, for (4), suppose $\oplus_A E(M_\alpha)$ is injective. Let

$$f : \oplus_A M_\alpha \to \oplus_A E(M_\alpha)$$

be the direct sum of the injective envelopes $M_\alpha \to E(M_\alpha)$. Since f is monic (6.25) it will suffice to show that it is essential. But this is just (6.17.2). □

Direct Sums of Injectives

It is not true that every direct sum of injective modules is injective. Indeed it is precisely the noetherian rings over which every direct sum of injectives is injective, and over these rings injective envelopes commute with direct sums.

18.13. Proposition. *For a ring R the following are equivalent:*
(a) *Every direct sum of injective left R-modules is injective;*
(b) *If $(M_\alpha)_{\alpha \in A}$ is an indexed set of left R-modules, then*

$$E(\oplus_A M_\alpha) = \oplus_A E(M_\alpha).$$

(c) *R is a left noetherian ring.*

Proof. (a) $\Leftrightarrow$ (b). The one implication is by (18.12.4) and the other by (18.12.1).

(a) $\Rightarrow$ (c). Suppose that (a) holds and that

$$I_1 \leq I_2 \leq \dots$$

is an ascending chain of left ideals in R. Let $I = \cup_{i=1}^{\infty} I_i$. Observe that if $a \in I$, then $a \in I_i$ for all but finitely many $i \in \mathbb{N}$. So there is an

$$f: I \to \oplus_{i=1}^{\infty} E(R/I_i)$$

defined via

$$\pi_i f(a) = a + I_i \qquad (a \in I).$$

By The Injective Test Lemma there is an $x \in \oplus_{i=1}^{\infty} E(R/I_i)$ such that $f(a) = ax$ for all $a \in I$. Now choose n such that $\pi_{n+k}(x) = 0$, $k = 0, 1, \dots$. So

$$I/I_{n+k} = \pi_{n+k}(f(I)) = \pi_{n+k}(Ix) = I\pi_{n+k}(x) = 0$$

or, equivalently, $I_n = I_{n+k}$ for all $k = 0, 1, 2, \dots$.

(c) $\Rightarrow$ (a). If R is left noetherian, $I \leq {}_R R$ and $f: I \to \oplus_A E_\alpha$, then since I is finitely generated, $Im f$ is contained in $\oplus_F E_\alpha$ for some finite subset $F \subseteq A$. Now apply (18.2) and The Injective Test Lemma. $\qquad\square$

Cogenerators

A module C in ${}_R\mathsf{M}$ is a *cogenerator* (see §8) in case C cogenerates every left R-module; that is, in case each left R-module M can be embedded in a product of copies of C

$$0 \to M \to C^A$$

(i.e., $Rej_M(C) = 0$). In terms of the functor $Hom_R(\_, C)$ we have

18.14. Proposition. *For a left R-module C the following statements are equivalent:*

(a) *C is a cogenerator;*

(b) *For every homomorphism f in ${}_R\mathsf{M}$ if $Hom_R(f, C) = 0$, then $f = 0$;*

(c) *For every $f: {}_R M \to {}_R N$ in ${}_R\mathsf{M}$, if $f^*: Hom_R(N, C) \to Hom_R(M, C)$ is epic, then f is monic;*

(d) *A sequence*

$$M' \xrightarrow{f} M \xrightarrow{g} M''$$

is exact in ${}_R\mathsf{M}$ if the sequence

$$Hom_R(M'', C) \xrightarrow{g^*} Hom_R(M, C) \xrightarrow{f^*} Hom_R(M', C)$$

is exact.

Proof. (a) $\Leftrightarrow$ (b). This is by (8.11.2).

(a) $\Rightarrow$ (d). Let C be a cogenerator. Suppose that $f: M' \to M$, and $g: M \to M''$ are such that

$$Hom_R(M'', C) \xrightarrow{g^*} Hom_R(M, C) \xrightarrow{f^*} Hom_R(M', C)$$

is exact. Then since $Hom_R(gf, C) = Hom_R(f, C)Hom_R(g, C) = 0$, we see that $Im f \leq Ker g$. Let $n: M \to M/Im f$ be the natural epimorphism. Then for each $h: M/Im f \to C$

$$[f^*(hn)](M') = h(n(Im f)) = 0,$$

so that $hn \in Ker f^* = Im g^*$. Equivalently $hn = \alpha g$ for some $\alpha \in Hom_R(M'', C)$. But now we have

$$h(Ker g/Im f) = hn(Ker g) = \alpha g(Ker g) = 0.$$

Thus $Ker g/Im f \leq Rej_{M/Im f}(C) = 0$ and hence

$$M' \xrightarrow{f} M \xrightarrow{g} M''$$

is exact.

(d) $\Rightarrow$ (c). This is clear.

(c) $\Rightarrow$ (a). It is not difficult to see that if $n : M \rightarrow M/Rej_M(C)$ is the natural map, then $n^* : Hom_R(M/Rej_M(C), C) \rightarrow Hom_R(M, C)$ is an isomorphism. Thus, under the hypothesis (c), n must always be monic; i.e., $Rej_M(C) = 0$. $\square$

Dual to (17.9) we have

18.15. Proposition. *Let E be an injective left R-module. Then the following are equivalent:*

(a) *E is a cogenerator;*

(b) *$Hom_R(T, E) \neq 0$ for all simple left R-modules T;*

(c) *E cogenerates every simple left R-module.*

Proof. The implications (a) $\Rightarrow$ (c) and (c) $\Rightarrow$ (b) are trivial. For (b) $\Rightarrow$ (a) assume that E satisfies (b). Let M be a left R-module and let $0 \neq x \in M$. Since Rx is cyclic, it contains a maximal submodule, so by (b) there is a nonzero homomorphism $h : Rx \rightarrow E$. But E is injective, so h can be extended to a homomorphism $\bar{h} : M \rightarrow E$ with $\bar{h}(x) = h(x) \neq 0$. Thus, $Rej_M(E) = 0$. $\square$

Now we shall see that there exist cogenerators in the category $_R$M. In fact, $_R$M contains a cogenerator C_0, which we call the *minimal cogenerator*, that embeds in every cogenerator in $_R$M.

18.16. Corollary. *Let $\mathscr{S}_0$ denote an irredundant set of representatives of the simple modules in $_R$M. Then*

$$C_0 = \oplus_{T \in \mathscr{S}_0} E(T)$$

is a cogenerator in $_R$M. Moreover, for a left R-module C, the following are equivalent:

(a) *C is a cogenerator;*

(b) *$E(T)$ is isomorphic to a direct summand of C for every simple left R-module T;*

(c) *C_0 is isomorphic to a submodule of C.*

Proof. It follows from (18.15) that the injective module $\Pi_{T \in \mathscr{S}_0} E(T)$ is a cogenerator. But this module is clearly cogenerated by $\oplus_{T \in \mathscr{S}_0} E(T)$. Thus, by (8.6.2), the first statement holds and (c) $\Rightarrow$ (a). To see that (a) $\Rightarrow$ (b) observe that if T is simple and is not contained in the kernel of $f : E(T) \rightarrow C$, then $Ker f \cap T = 0$; but since $T \trianglelefteq E(T)$, it follows that f is monic. Finally, to prove that (b) $\Rightarrow$ (c) observe that an irredundant set $\mathscr{S}_0$ of simple submodules of C must be independent; hence, by (6.17.1), the set $\{E(T) \mid T \in \mathscr{S}_0\}$ of their essential extensions is also independent in C. □

As promised in §9 we have

18.17. Corollary. *A left R-module M is finitely cogenerated if and only if for every indexed set of left R-modules $(U_\alpha)_{\alpha \in A}$ and every monomorphism*

$$0 \to M \to \Pi_A U_\alpha$$

there is a monomorphism

$$0 \to M \to \Pi_F U_\alpha$$

for some finite subset $F \subseteq A$.

Proof. We need only prove sufficiency (see (10.2)). By (18.16) M is cogenerated by injective envelopes of simple modules. Thus by hypothesis there is a finite set of simple modules $T_1, \ldots, T_n$ such that M is isomorphic to a submodule of $E(T_1) \oplus \ldots \oplus E(T_n) = E(T_1 \oplus \ldots \oplus T_n)$. But by (10.7) this module (whose socle is $T_1 \oplus \ldots \oplus T_n$), is finitely cogenerated; hence M is finitely cogenerated. □

While on the subject we observe

18.18. Proposition. *A module M is finitely cogenerated if and only if $E(M) \cong E(T_1) \oplus \ldots \oplus E(T_n)$ for some finite set $T_1, \ldots, T_n$ of simple modules.*

Proof. Clearly $E(T_1) \oplus \ldots \oplus E(T_n) \cong E(T_1 \oplus \ldots \oplus T_n)$ has a finitely generated essential socle. Thus each of its submodules is finitely cogenerated. Conversely, if $Soc\, M = T_1 \oplus \ldots \oplus T_n \trianglelefteq M$ then $E(M) \cong E(T_1) \oplus \ldots \oplus E(T_n)$.
□

Finally, we see that the injective cogenerators are distinguished in the class of injective modules as are the projective generators distinguished in the class of projective modules. The class of injective cogenerators is closed under the formation of direct products, and a module is injective if and only if it is a direct summand of an injective cogenerator. There is, however, one notable difference. Every ring R possesses a unique (to within isomorphism) minimal injective cogenerator, namely, $E(C_0)$, but in general, a ring need not have a minimal projective generator. (See Exercise (17.10).)

18.19. Corollary. *Let $\mathscr{S}_0$ denote an irredundant set of representatives of the simple modules in $_R\mathsf{M}$. Then*

$$Q = E(\oplus_{T \in \mathscr{S}_0} T)$$

is an injective cogenerator in $_R\mathsf{M}$. Moreover,

(1) $Q = E(C_0)$;

(2) *If Q' is an injective cogenerator in $_R\mathsf{M}$, then there is a (split) monomorphism $Q \to Q'$.*

Proof. Clearly $C_0 \trianglelefteq Q$ so (1) holds and Q is an injective cogenerator. Finally (2) follows from (18.16). □

Endomorphism Rings of Injective Modules

We conclude this section with the duals of (17.11) and (17.12).

18.20. Proposition. *Let E be an injective left R-module with endomorphism ring $S = End(_RE)$. Let $a \in S$. Then*

$$a \in J(S) \qquad iff \qquad Ker\, a \trianglelefteq E.$$

Proof. If $Ker\, a \trianglelefteq E$, then it will suffice to prove that $aS \ll S_S$. (See (15.3).) The rest of the proof is entirely dual to that of (17.11) and will be omitted. □

18.21. Corollary. *Let E be an injective left R-module such that $Soc\, E \trianglelefteq E$ (e.g., if $_RE$ is finitely cogenerated). Let $S = End(_RE)$. Then*

$$J(S) = r_S(Soc\, E) \qquad and \qquad S/J(S) \cong End(_R Soc\, E).$$

Proof. This is dual to the proof of (17.12).   □

18. Exercises

1. A ring R is *left (right) self-injective* in case $_RR$ (R_R) is injective. Prove that if R is a P.I.D., and $I \neq 0$ is an ideal of R, then R/I is self-injective.

2. Let R be a commutative integral domain. Prove that:
   (1) If Q is the field of quotients of R, then $_RQ = E(_RR)$. [Hint: If $_RI \leq {}_RR$ and $f: {}_RI \to {}_RQ$, then $\bar{f}: \Sigma\, q_i a_i \mapsto \Sigma\, q_i f(a_i)$ defines a Q-homomorphism.]
   (2) If $_RE$ is injective, then E is divisible (see Exercise (3.15)).
   (3) If R is a P.I.D., then $_RE$ is injective iff $_RE$ is divisible.

3. Let D be a division ring, $Q = \mathbb{M}_n(D)$, and R the subring of upper triangular matrices. Prove that $_RQ$ is an injective envelope of $_RR$. [Hint: Let S be the set of all $a \in Q$ zero off the first row. Then S is an ideal of R, $Sq = 0$ implies $q = 0$, and $SQ \leq R$. In particular $_RR \trianglelefteq {}_RQ$. Let I be a left ideal of R and $\phi: I \to Q$ be an R-homomorphism. If $\Sigma\, q_i a_i = 0$ with $a_i \in I$, then $s\Sigma\, q_i \phi(a_i) = 0$ for each $s \in S$; thus there is a Q-homomorphism $\bar{\phi}: QI \to Q$ such that $(\bar{\phi}\,|\,I) = \phi$. By (18.8) $_QQ$ is injective so $\bar{\phi}$ extends to an endomorphism of $_QQ$.]

4. Let M_D be a non-zero vector space over a division ring D and let $R = End(M_D)$. Prove that:
   (1) $Soc\,_R R = \{f \in R \,|\, rank\, f < \infty\} = Soc\, R_R$, and $(Soc\,_R R)^2 = Soc\,_R R$. [Hint: Exercises (14.13) and (17.4).]
   (2) If $\phi: Soc(R_R) \to R_R$ is a right R-homomorphism, then there exists a unique extension $\bar{\phi}: R_R \to R_R$ of ϕ. [Hint: Let $(x_\alpha)_{\alpha \in A}$ be a basis for M_D and let $e_\alpha \in R$ be defined by $e_\alpha(x_\beta) = \delta_{\alpha\beta} x_\alpha$. Then $\bar{\phi}$ is defined by

   $$\bar{\phi}(f)(x) = \Sigma_A [\phi(e_\alpha)](e_\alpha(f(x))).$$

   (3) If I is a right ideal of R, then $Soc(R_R) \leq I \oplus I'$ for some right ideal I' of R. [Hint: See (5.21).]

(4) The ring R is right self-injective.

(5) If M_D is infinite dimensional, then R is not left self-injective, in fact R has a primitive idempotent e such that Re is not injective. [Hint: There is an R-homomorphism $\phi: Soc(_RR) \to {}_RR$ such that $\phi(Re_\alpha) = Re$ for each $\alpha \in A$.]

5. Let M be a non-zero module. Prove that $E(M^{(A)}) = E(M)^A$ iff A is finite.

6. Let E denote the injective $\mathbb{Z}_4$-module $\mathbb{Z}_4 \oplus \mathbb{Z}_4$ (see Exercise (18.1)). Let $M = \{(0,0), (2,2)\} \le E$. Prove that:

   (1) M is not injective.

   (2) M is the intersection of injective submodules of E.

   (3) E contains more than one copy of $E(M)$.

7. If M is a left R-module, then it is contained in an injective module E (see (18.6)). Forming injective envelopes is not a "closure" operation in the usual sense, for $E(M)$ need not be the intersection in E of the injective submodules containing M. (See Exercise (18.6).) In this connection prove that:

   (1) If E is injective, then every submodule of E has a unique injective envelope in E iff the intersection of every pair of injective submodules of E is injective.

   (2) If H, K, and $H \cap K$ are injective submodules of a module M, then so is $H + K$.

   (3) The converse of (2) fails in the sense that there exist injective submodules of the $\mathbb{Z}$-module $\mathbb{Q} \oplus \mathbb{Q}$ whose sum is injective and whose intersection is not injective.

8. Let P and E be left R-modules. Suppose P is E-projective and E is P-injective. Prove that every submodule of P is E-projective iff every factor module of E is P-injective. [Hint: Consider

$$
\begin{array}{ccc}
0 \to & L & \to P \\
 & \downarrow & \\
0 \leftarrow & E/K \leftarrow & E.]
\end{array}
$$

9. (1) Let $0 \to K \to P \to M \to 0$ and $0 \to K' \to P' \to M \to 0$ be exact with P and P' both projective. Prove

Schanuel's Lemma. $P \oplus K' \cong P' \oplus K$.

[Hint: Consider

$$
\begin{array}{ccccccccc}
 & & 0 & & 0 & & & & \\
 & & \downarrow & & \downarrow & & & & \\
 & & K' & = & K' & & & & \\
 & & \downarrow & & \downarrow & & & & \\
0 & \to & K & \to & Q & \xrightarrow{\pi'} & P' & \to & 0 \\
 & & \| & & {\scriptstyle \pi}\downarrow & & \downarrow{\scriptstyle g'} & & \\
0 & \to & K & \to & P & \xrightarrow{g} & M & \to & 0 \\
 & & & & \downarrow & & \downarrow & & \\
 & & & & 0 & & 0 & &
\end{array}
$$

where $Q = \{(p, p') \in P \times P' \mid g(p) = g'(p')\}$.]
(2) Let $0 \to K \to E \to M \to 0$ and $0 \to K \to E' \to M' \to 0$ be exact with E
and E' injective. Prove that:

$$E \oplus M' \cong E' \oplus M.$$

10. A ring R is (left) *hereditary* in case each of its left ideals is projective. For
example, every P.I.D. is left hereditary. Prove that:
(1) For a ring R the following are equivalent: (a) R is left hereditary;
(b) Every factor module of an injective left R-module is injective; (c) Every
submodule of a projective left R-module is projective. [Hint: Exercise
(18.8).]
(2) Let R be left artinian with $J = J(R)$ and let $e_1, \ldots, e_n$ be a complete
set of pairwise orthogonal primitive idempotents. (See Exercise (17.20).)
Prove that the following are equivalent: (a) R is left hereditary; (b) Every
maximal left ideal is projective; (c) Je_i is projective $(i = 1, \ldots, n)$; (d) $_R J$ is
projective. [Hint: (c) $\Rightarrow$ (a). Let $_R R = I > I_1 > \ldots > I_l = 0$ be a composition series. Consider the exact sequences $0 \to I_{k+1} \to I_k \to I_k/I_{k+1} \to 0$
and $0 \to Je \to Re \to I_k/I_{k+1} \to 0$. Now use Schanuel's Lemma (Exercise
(18.9)) and induction.]
(3) The ring R of $n \times n$ upper triangular matrices over a field K is both
left and right hereditary.

11. Let R be a commutative integral domain with field of quotients Q. For
each ideal $I \leq R$ define $I^{-1} = \{q \in Q \mid qI \subseteq R\}$. Then I is *invertible* in
case $I^{-1}I = R$ (i.e., in case there exist $q_1, \ldots, q_n \in I^{-1}$ and $a_1, \ldots, a_n \in I$
with $q_1 a_1 + \ldots + q_n a_n = 1$). We say that R is a *Dedekind domain* in case
every non-zero ideal of R is invertible. Prove that:
(1) For each non-zero ideal I, the multiplication $\lambda : q \mapsto \lambda(q)$ defines an
isomorphism $\lambda : I^{-1} \to Hom_R(I, R)$.
(2) A non-zero ideal I of R is invertible iff it is (finitely generated and)
projective. [Hint: Use (1) and The Dual Basis Lemma (Exercise (17.11)).]
(3) The following are equivalent: (a) R is a Dedekind domain; (b) R is
hereditary; (c) Every divisible R-module is injective.
(4) Every Dedekind domain is noetherian.

12. Prove that a ring R is left hereditary (and noetherian) iff the sum of every
pair (set) of injective submodules in any left R-module is injective.
Consequently, each module over a hereditary noetherian ring contains
(as a direct summand) a unique maximal injective submodule. [Hint: For
the sufficiency, let $E = M_1 = M_2$ in Exercise (5.10).]

13. Let R be a P.I.D. Two elements $a, b \in R$ are *equivalent* in case there
exists an invertible element $u \in R$ such that $a = ub$. The *primes* in R are
those non-invertible elements divisible only by elements equivalent to
themselves and invertible elements. Let P denote a set consisting of one
element from each equivalence class of primes in R. Prove that if Q is the
field of quotients of R, then $Q/R = \oplus_P R_{p^\infty}$ (where $R_{p^\infty} = \{(a/p^n) +$
$R \mid a \in R, n \in \mathbb{Z}\}$) is the minimal cogenerator in $_R M$. In particular,
$\mathbb{Q}/\mathbb{Z} \cong \oplus_\mathbb{P} \mathbb{Z}_{p^\infty}$ is the minimal cogenerator in $_\mathbb{Z} M$.

14. Let $R = \mathbb{LTM}_n(K)$, K a field, $E = \mathbb{M}_n(K)$ and $J = J(R)$. Show that E/J is the minimal left R-cogenerator.

15. Let E and Q be injective modules. Prove that if there exist mono-morphisms $f: E \to Q$ and $q: Q \to E$, then $E \cong Q$. [Hint: If $E = Q \oplus Q'$ and $H = E(\Sigma_{n \in \mathbb{N}} f^n(Q'))$, then $H \cong H \oplus Q'$.]

16. Let R be a ring such that for every triple $_RE$, $_RU$, $_RV$, if E is injective and U cogenerates V, then $Tr_E(V) \leq Tr_E(U)$. Prove that $R/J(R)$ is semisimple. Thus in general the assertion of Exercise (16.5.1) fails without the finiteness requirement.

17. Let U and M be left R-modules. Prove that U is M-injective iff $Im\,\gamma \leq U$ for each $\gamma: M \to E(U)$. In particular, U is quasi-injective iff it is stable under $End(_RE(U))$. [Hint: Exercise (16.13).]

18. Let U be a quasi-injective module. Prove that if $E(U) = E_1 \oplus E_2$ then $U = U_1 \oplus U_2$ where $E_i = E(U_i)(i = 1, 2)$. Conclude that U is indecom-posable iff $E(U)$ is indecomposable.

19. Let U_1 and U_2 be quasi-injective modules such that $E(U_1) \cong E(U_2)$. Prove that $U_1 \oplus U_2$ is quasi-injective iff $U_1 \cong U_2$.

20. Show that a finite abelian group is quasi-injective iff it is quasi-projective. [Hint: Exercises (17.19) and (18.19).]

21. Prove that every quasi-injective left module over a left artinian ring is injective modulo its annihilator.

22. Prove that for a left R-module U the following statements are equivalent: (a) U^A is quasi-injective for every set A; (b) U is injective modulo its annihilator; (c) $U = r_{E(U)}(l_R(U))$.

23. Recall that a module is co-semisimple in case each of its submodules is an intersection of maximal submodules and that a ring is left co-semisimple in case it has a semisimple left cogenerator. (See Exercises (9.14) and (13.10).) Prove that:

(1) For a left R-module M the following are equivalent: (a) M is co-semisimple; (b) Every finitely cogenerated factor module of M is semi-simple; (c) Every simple left R-module is M-injective.

(2) Submodules, factor modules and direct sums of co-semisimple modules are co-semisimple.

(3) For a ring R the following are equivalent: (a) $_RR$ is co-semisimple; (b) R is left co-semisimple; (c) Every left R-module is co-semisimple; (d) Every simple left R-module is injective; (e) Every short exact sequence $0 \to {}_RK \to {}_RM \to {}_RN \to 0$ with K finitely cogenerated splits. ((a) $\Leftrightarrow$ (d) is due to Villamayor.)

(4) If R is co-semisimple, then $I^2 = I$ for each left ideal $I \leq R$. [Hint: It will suffice to show $x \in (Rx)^2$ for each $x \in R$. If $x \notin (Rx)^2$, then for some maximal left ideal M of R, $x \notin M$ and $(Rx)^2 \leq M$. (See part (1).) But then $R = Rx + M$.]

(5) For a commutative ring R the following are equivalent: (a) R is co-semisimple; (b) $I^2 = I$ for each ideal I of R; (c) R is von Neumann regular. (This is due to Kaplansky.) [Hint: (c) $\Rightarrow$ (a). Let $I \leq R$ and $x \notin I$. By (c), $x = yx^2$ for some $y \in R$. Let M be maximal with $I \leq M$ and

$x \notin M$. Then M is a maximal ideal of R so $_R R$ is co-semisimple. See Exercise (15.13).]

(6) If R is the endomorphism ring of an infinite dimensional vector space M_D, then $R \times R^{op}$ is von Neumann regular but neither left nor right co-semisimple. [Hint: Exercise (18.4).] (Note: Cozzens [70] has shown that co-semisimple rings need not be von Neumann regular.)

24. A ring R is *left co-artinian* (*co-noetherian*) in case every submodule (factor module) of each finitely cogenerated left module is finitely generated (finitely cogenerated).

(1) Prove that R is left co-artinian (co-noetherian) iff $E(T)$ is noetherian (artinian) for every simple left R-module T.

(2) $\mathbb{Z}$ is co-noetherian but not co-artinian.

25. A module is faithful iff it cogenerates every projective module. (See Exercise (17.6).) A module is *co-faithful* in case it generates every injective module. Prove that:

(1) M is co-faithful iff M finitely cogenerates the regular module $_R R$. [Hint: For ($\Rightarrow$) observe that $1 \in \mathit{Tr}_{E(R)}M$.]

(2) Every faithful left R-module is co-faithful iff $_R R$ is finitely co-generated.

(3) Prove that every co-faithful quasi-injective module is injective.

26. A module is faithful iff it cogenerates a generator. (See Exercise (8.3).) A module is **-faithful* in case it generates a cogenerator. Prove that:

(1) $_R M$ is *-faithful iff M generates the minimal cogenerator in $_R \mathbf{M}$.

(2) co-faithful $\Rightarrow$ *-faithful $\Rightarrow$ faithful.

(3) If R is left artinian then all three are equivalent.

27. (1) **Theorem.** *If $_R E$ is a non-zero injective module and if $S = End(_R E)$, then $S/J(S)$ is von Neumann regular.* [Hint: Exercise (15.13). If $a \in S$, then $E = E(Ker\, a) \oplus E' = E'' \oplus E'a$ and $(a \mid E'): E' \to E'a$ is an isomorphism. Let $x = 0 \oplus (a \mid E')^{-1}$ and show that $Ker(axa - a) \trianglelefteq E$.]

(2) **Corollary.** *If R is left or right self-injective, then $R/J(R)$ is von Neumann regular.*

28. Let $_R U$ be quasi-injective. Prove that $End(_R U)$ is isomorphic to a factor ring of $End(_R E(U))$ and $End(_R U)/J(End(_R U))$ is von Neumann regular.

29. A Boolean ring R is *complete* in case for each $A \subseteq R$ there is an element $u \in R$ such that $l(A) = l(u)$. Prove that:

(1) A Boolean ring R is complete iff R is self-injective.

(2) Prove that the ring of all continuous functions from $\mathbb{Q}$ to $\mathbb{Z}_2$ (with the discrete topology) is von Neumann regular but not self-injective.

30. Let R be primitive with $Soc\, R \neq 0$. (See Exercise (17.5).) Then there is an idempotent $e \in R$ with Re and eR faithful and simple. Set

$$B = BiEnd(Re) = End(Re_{eRe}),$$

and identify $R = \lambda(R)$ with a dense subring of B. For each left (right), eRe-module U let U^* be the right (left) eRe-module $Hom_{eRe}(U, eRe)$. Prove that:

(1) $eB = Re^*$.

(2) $eR \trianglelefteq eB$ as right R-modules. [Hint: Show $Be = Re$.]

(3) If Re and eR are R-injective, then the eRe vector space Re is reflexive; i.e., the evaluation map $\sigma: Re \to ((Re)^*)^*$ defined via

$$[\sigma(x)](\gamma) = \gamma(x)$$

is an isomorphism. [Hint: Let $B' = BiEnd(eR_R)$. Then $((Re)^*)^* = (eB)^* = (eR)^* = B'e = Re$.]

(4) No infinite dimensional vector space is reflexive. [Hint: Let M_D have a basis $(x_\alpha)_{\alpha \in A}$. Define $f_\beta(x_\alpha) = \delta_{\alpha\beta} \in D$. If A is infinite, there exists $0 \neq g: M^* \to D$ such that $g(f_\alpha) = 0$ for all $\alpha \in A$. Then $g \neq \sigma(v)$ for $v \in M$.]

(5) **Theorem.** *If R has simple faithful injective projective left and right modules, then R is simple artinian.*

(6) Use (5) to prove that if M_D is an infinite dimensional vector space, then $End(M_D)$ is not both left and right self-injective. (See also Exercise (18.4.5).)

31. Let R be a left artinian ring. Prove that:

(1) If $Soc\ _RM = \oplus_A T_\alpha$ with each T_α simple, then $E(M) \cong \oplus_A E(T_\alpha)$.

(2) Every injective left R-module has a decomposition (whose terms are injective envelopes of simple modules) that complements direct summands.

(3) The numbers of isomorphism classes of simple, indecomposable projective and indecomposable injective left R-modules are all (finite and) equal. (See Exercise (17.20).)

32. Prove that every left cogenerator over a left artinian ring is balanced. [Hint: Let $_RE$ be an injective cogenerator. Apply (14.2) and Exercise (8.5) to $R(x_1, \ldots, x_n) \leq E^{(n)}$ to see that R operates densely on E. Then apply Exercise (16.19) to see that E is balanced. Now apply (14.1).]

§19. The Tensor Functors and Flat Modules

There is another important class of additive functors, in addition to the *Hom* functors, that ply their trade among module categories. This is the class of "tensor" functors that arises from the study of multilinear algebra. In a sense such functors serve to linearize multilinear functions.

Tensor Products of Modules

Given a right module M_R and a left module $_RN$ over a ring R and an abelian group A, a function

$$\beta: M \times N \to A$$

is said to be *R-balanced* in case for all $m, m_i \in M$, $n, n_i \in N$ and $r \in R$

(1) $\beta(m_1 + m_2, n) = \beta(m_1, n) + \beta(m_2, n)$;

(2) $\beta(m, n_1 + n_2) = \beta(m, n_1) + \beta(m, n_2)$;

(3) $\beta(mr, n) = \beta(m, rn)$.

The most familiar examples of such maps are the inner products of elementary linear algebra and ring multiplications $R \times R \to R$.

We shall not study R-balanced maps as such. For there is a natural way to trade each R-balanced map in for a linear map by using the concept of a tensor product. Let M_R and $_R N$ be modules. A pair (T, τ) consisting of an abelian group T and an R-balanced map $\tau: M \times N \to T$ is a *tensor product of* M_R *and* $_R N$ in case for every abelian group A and every R-balanced map $\beta: M \times N \to A$ there is a unique $\mathbb{Z}$-homomorphism $f: T \to A$ such that the diagram

commutes. If (T, τ) is a tensor product of M_R and $_R N$, then clearly, $f \circ \tau$ is R-balanced for each homomorphism $f: T \to A$. Thus, (T, τ) is a tensor product of M_R and $_R N$ if and only if for each abelian group A

$$f \leftrightarrow f \circ \tau$$

defines a one-to-one correspondence between $Hom_{\mathbb{Z}}(T, A)$ and the set of R-balanced maps $\beta: M \times N \to A$. Our first task is to show that not only do such tensor products exist, but that they are essentially unique. The uniqueness is particularly easy.

19.1. Proposition. *If* (T, τ) *and* (T', τ') *are two tensor products of* $(M_R, {}_R N)$ *then there is a* $\mathbb{Z}$-*isomorphism* $f: T \to T'$ *such that* $\tau' = f\tau$.

Proof. The hypotheses imply the existence of homomorphisms f and g such that

commute. Then the commuting of the diagrams

forces $gf = 1_T$. Similarly $fg = 1_{T'}$, whence f is an isomorphism.   □

Next we shall construct a tensor product of $(M_R, {}_R N)$ over R. For this let $F = \mathbb{Z}^{(M \times N)}$ be a free abelian group on $M \times N$. Then F has a free basis

$(x_\alpha)_{\alpha \in M \times N}$. For notational convenience let us simply write (m, n) for $x_{(m,n)}$. Then

$$F = \bigoplus_{M \times N} \mathbb{Z}(m, n).$$

Now let K be the subgroup of F generated by all elements of the form

$$(m_1 + m_2, n) - (m_1, n) - (m_2, n),$$

$$(m_1, n_1 + n_2) - (m, n_1) - (m, n_2),$$

$$(mr, n) - (m, rn),$$

and set $T = F/K$. Define $\tau : M \times N \to T$ via

$$\tau(m, n) = (m, n) + K.$$

19.2. Proposition. *With T and τ defined as above, (T, τ) is a tensor product of $(M_R, {}_R N)$ over R.*

Proof. Suppose $\beta : M \times N \to A$ is an R-balanced map. Since F is free on $M \times N$ there is a $\mathbb{Z}$-homomorphism $h : F \to A$ such that

commutes. Since β is R-balanced, $K \leq \operatorname{Ker} h$. Thus, there is a $\mathbb{Z}$-homomorphism $f : T \to A$ such that

commutes. Finally, since $\tau(M \times N)$ clearly spans T, f is uniquely determined by this diagram. $\qquad \square$

Given $(M_R, {}_R N)$, let (T, τ) be the tensor product constructed above. By Proposition 19.1 it is unique to within isomorphism. We write

$$T = M \otimes_R N$$

and for each $(m, n) \in M \times N$,

$$\tau(m, n) = m \otimes n.$$

We tend to be somewhat loose with our terminology and call $M \otimes_R N$ the tensor product of M and N. As we shall see, the notation $m \otimes n$ is ambiguous. That is, if $m \in M' \leq M$ and $n \in N' \leq N$, then $m \otimes n$ can have a vastly different meaning in $M' \otimes_R N'$ than in $M \otimes_R N$. Usually, however, the context removes this ambiguity. Now combining (19.1) and (19.2) we have that $M \otimes_R N$ is the unique (to within isomorphism) abelian group that contains a spanning set $\{m \otimes n \mid m \in M, n \in N\}$ satisfying

19.3. Proposition. *For each R-balanced map* $\beta : M \times N \to A$ *there exists a unique abelian group homomorphism*

$$f : M \otimes_R N \to A$$

such that for all $m \in M, n \in N$

$$f(m \otimes n) = \beta(m, n). \qquad \square$$

Also we have at once the following arithmetic properties of $M \otimes_R N$:

19.4. Proposition. *Each element of* $M \otimes_R N$ *can be expressed as a finite sum of the form*

$$\Sigma_i (m_i \otimes n_i) \qquad (m_i \in M, n_i \in N).$$

Moreover, for all $m, m_i \in M, n, n_i \in N$ *and* $r \in R.$
(1) $(m_1 + m_2) \otimes n = (m_1 \otimes n) + (m_2 \otimes n),$
(2) $m \otimes (n_1 + n_2) = (m \otimes n_1) + (m \otimes n_2),$
(3) $mr \otimes n = m \otimes rn. \qquad \square$

A few more words of caution are in order. Although

$$\tau(M \times N) = \{m \otimes n \mid m \in M, n \in N\}$$

spans $M \otimes_R N$, in general $\tau(M \times N) \neq M \otimes_R N$. Moreover, the representation of elements of $M \otimes_R N$ as finite sums $\Sigma_i(m_i \otimes n_i)$ need not be unique.

In general, the abelian group $M \otimes_R N$ is not an R-module. However, bimodule structures on M or N induce module structures on $M \otimes_R N$. Suppose, for example, that we have $({}_S M_R, {}_R N)$. Then for each $s \in S$ the mapping $\sigma_s : M \times N \to M \otimes_R N$ defined via

$$\sigma_s(m, n) = (sm) \otimes n$$

is R-balanced. Hence there is a unique $\mathbb{Z}$-homomorphism

$$v(s) : M \otimes_R N \to M \otimes_R N$$

such that

commutes. It is easy to see that $v : s \mapsto v(s)$ defines a unital ring homomorphism $S \to End^l(M \otimes_R N)$. Thus $M \otimes_R N$ is a left S-module with

$$s(m \otimes n) = (sm) \otimes n.$$

If $N = {}_R N_T$ is a bimodule, then a similar argument shows that $M \otimes_R N$ is a right T-module. In fact it is now easily checked that

19.5. Proposition. *If* ${}_S M_R$ *and* ${}_R N_T$ *are bimodules, then* $M \otimes_R N$ *is a left S-right T-bimodule with*

$$s(m \otimes n) = (sm) \otimes n, \qquad (m \otimes n)t = m \otimes (nt). \qquad \square$$

Since $_RR_R$, it follows from (19.5) that $M \otimes_R R$ is a right R-module and $R \otimes_R N$ is a left R-module.

19.6. Proposition. *For each right module M_R, there is an R-isomorphism $\eta: M \otimes_R R \to M$ such that*

$$\eta(m \otimes r) = mr, \qquad \eta^{-1}(m) = m \otimes 1$$

and for each left module $_RN$ there is an R-isomorphism $\mu: R \otimes_R N \to N$ such that

$$\mu(r \otimes n) = rn, \qquad \mu^{-1}(n) = 1 \otimes n.$$

Moreover, if $_SM_R$ $(_RN_S)$ is a bimodule, then η (μ) is a bimodule isomorphism.

Proof. We shall prove only the first assertion. Since $(m, r) \mapsto mr$ defines an R-balanced map $M \times R \to M$, there is an $\eta: M \otimes_R R \to M$ such that $\eta(m \otimes r) = mr$. Clearly η is an R-homomorphism. Also $\eta': M \to M \otimes_R R$ defined via $\eta'(m) = m \otimes 1$ is also an R-homomorphism by (19.4). Clearly $\eta \circ \eta' = 1_M$. Since $M \otimes_R R = \{m \otimes 1 \mid m \in M\}$, it is also clear that

$$\eta' \circ \eta = 1_{M \otimes_R R}. \qquad \square$$

Tensor Products of Homomorphisms

Enroute to the tensor functors we next develop a theory of a tensor product $f \otimes g$ of two R-homomorphisms.

Let M, M' be right R-modules, and let N, N' be left R-modules. Suppose further that $f: M \to M'$ and $g: N \to N'$ are R-homomorphisms. Define a map $(f, g): M \times N \to M' \otimes_R N'$ via

$$(f, g)(m, n) = f(m) \otimes g(n).$$

It is evident that (f, g) is R-balanced, so there is a unique $\mathbb{Z}$-homomorphism, which we shall denote by $f \otimes g$, from $M \otimes_R N$ to $M' \otimes_R N'$ such that the following diagram commutes:

Thus, in particular, $f \otimes g$ is characterized via

$$(f \otimes g)(m \otimes n) = f(m) \otimes g(n).$$

19.7. Lemma. *Consider M_R, M'_R, $_RN$, $_RN'$. For all $f_1, f_2, f \in Hom_R(M, M')$ and all $g_1, g_2, g \in Hom_R(N, N')$,*

(1) $(f_1 + f_2) \otimes g = (f_1 \otimes g) + (f_2 \otimes g)$,
(2) $f \otimes (g_1 + g_2) = (f \otimes g_1) + (f \otimes g_2)$,
(3) $f \otimes 0 = 0 \otimes g = 0$,
(4) $1_M \otimes 1_N = 1_{M \otimes_R N}$.

Proof. These identities clearly hold on the generators $m \otimes n$ of $M \otimes_R N$. ☐

19.8. Lemma. *Given R-homomorphisms* $f : M \to M', f' : M' \to M'', g : N \to N'$, *and* $g' : N' \to N''$,

$$(f' \otimes g')(f \otimes g) = (f'f) \otimes (g'g).$$

Proof. It works on all $m \otimes n$. ☐

19.9. Lemma. *Suppose that* $(M_R, (i_\alpha)_{\alpha \in A})$ *is a direct sum of* $(M_\alpha)_{\alpha \in A}$ *and that* $({}_R N, (j_\beta)_{\beta \in B})$ *is a direct sum of* $(N_\beta)_{\beta \in B}$. *Then* $(M \otimes_R N, (i_\alpha \otimes j_\beta)_{(\alpha, \beta) \in A \times B})$ *is a direct sum of* $(M_\alpha \otimes_R N_\beta)_{(\alpha, \beta) \in A \times B}$.

Proof. Let $(p_\alpha)_{\alpha \in A}$ and $(q_\beta)_{\beta \in B}$ be R-homomorphisms such that

$$M_\alpha \xrightarrow{i_\alpha} M \xrightarrow{p_\alpha} M_\alpha \qquad \text{and} \qquad N_\beta \xrightarrow{j_\beta} N \xrightarrow{q_\beta} N_\beta$$

satisfy

$$p_\alpha i_{\alpha'} = \delta_{\alpha \alpha'} 1_{M_\alpha} \qquad \text{and} \qquad q_\beta j_{\beta'} = \delta_{\beta \beta'} 1_{N_\beta}$$

and

$$\Sigma_A i_\alpha p_\alpha(m) = m \qquad \text{and} \qquad \Sigma_B j_\beta q_\beta(n) = n$$

for all $m \in M$ and $n \in N$. Then

$$M_\alpha \otimes_R N_\beta \xrightarrow{i_\alpha \otimes j_\beta} M \otimes_R N \xrightarrow{p_\alpha \otimes q_\beta} M_\alpha \otimes_R N_\beta$$

satisfy

$$(p_\alpha \otimes q_\beta)(i_{\alpha'} \otimes j_{\beta'}) = \delta_{(\alpha, \alpha')(\beta, \beta')} 1_{M_\alpha \otimes_R N_\beta},$$

$$(p_\alpha \otimes q_\beta)(m \otimes n) = 0 \text{ for almost all } (\alpha, \beta) \in A \times B,$$

$$\Sigma_{A \times B}(i_\alpha \otimes j_\beta)(p_\alpha \otimes q_\beta)(m \otimes n) = m \otimes n.$$

Now apply Proposition 6.21. ☐

The Tensor Functors

Let $U = {}_S U_R$ be a bimodule. Then it follows from (19.7) and (19.8) that there is an additive covariant functor

$$(U \otimes_R -) : {}_R\mathsf{M} \to {}_\mathbb{Z}\mathsf{M}$$

defined by

$$(U \otimes_R -) : M \mapsto U \otimes_R M$$

$$(U \otimes_R -) : f \mapsto 1_U \otimes f.$$

By (19.5) each $U \otimes_R M$ is a left S-module. We claim moreover that if $f : M \to M'$ is an R-homomorphism, then

$$U \otimes_R f : {}_S U \otimes_R M \to {}_S U \otimes_R M'$$

is an S-homomorphism. It suffices to check this on the generators $u \otimes m$ of $U \otimes_R M$. But for each $s \in S$, $u \in U$, and $m \in M$.

$$(U \otimes_R f)(su \otimes m) = (1_U \otimes f)(su \otimes m)$$
$$= su \otimes f(m) = s(u \otimes f(m))$$
$$= s((U \otimes_R f)(u \otimes m))$$

as claimed. Thus we may view this as an additive functor from $_R M$ to $_S M$, and write it

$$(_S U \otimes_R \_) : {}_R M \to {}_S M.$$

Similarly, there is an additive covariant functor

$$(\_ \otimes_S U_R) : M_S \to M_R$$

defined by

$$(\_ \otimes_S U_R) : N \mapsto N \otimes_S U_R$$
$$(\_ \otimes_S U_R) : g \mapsto g \otimes 1_U.$$

Finally, applying (19.9) we have

19.10. Theorem. *Let R and S be rings and let $_R U_S$ be a bimodule. Then*

$$(_S U \otimes_R \_) : {}_R M \to {}_S M$$

and

$$(\_ \otimes_S U_R) : M_S \to M_R$$

are both additive covariant functors that preserve (arbitrary) direct sums. □

There are other versions of tensor functors. Thus for example, a bimodule U_{R-S} gives rise to a functor

$$(U_{R-S} \otimes_R \_) : {}_R M \to M_S.$$

The properties of any one of these can be deduced from the others by simple translations via opposite rings. Indeed observe that

$$M \otimes_R N = N \otimes_{R^{op}} M.$$

Thus our usual practice shall be to state our results in terms of a functor $(_S U \otimes_R \_) : {}_R M \to {}_S M$, but we shall feel free to use the related versions for the other tensor functors.

Now let $_R M' \leq {}_R M$, let W be a left R-module, and let $i : M' \to M$ be the inclusion map. Informally, we often tend to view $Hom_R(W, M')$ as a submodule of $Hom_R(W, M)$. Although strictly speaking this identification is incorrect, it is justified by the fact that

$$Hom_R(W, i) : Hom_R(W, M') \to Hom_R(W, M)$$

is a monomorphism, or more generally, that the functor

$$Hom_R(W, \_):_R\mathsf{M} \to _Z\mathsf{M}$$

is left exact. Similarly, it is the left exactness of $Hom_R(\_, W)$ that keeps us from disaster when we view $Hom_R(M/M', W)$ as a submodule of $Hom(M, W)$.

Again let $_R M' \leq _R M$, but now let U be a right R-module. Then in general, $U \otimes_R M'$ cannot be identified with a submodule of $U \otimes_R M$. For example, as $\mathbb{Z}$-modules $\mathbb{Z} \leq \mathbb{Q}$, but for each $n > 1$,

$$\mathbb{Z}_n \otimes_{\mathbb{Z}} \mathbb{Z} \cong \mathbb{Z}_n \quad \text{and} \quad \mathbb{Z}_n \otimes_{\mathbb{Z}} \mathbb{Q} = 0.$$

To see the last assertion, note that if $x \in \mathbb{Z}_n$, then in $\mathbb{Z}_n \otimes_{\mathbb{Z}} \mathbb{Q}$,

$$x \otimes q = x \otimes n(n^{-1}q) = nx \otimes n^{-1}q = 0 \otimes n^{-1}q = 0.$$

This latter phenomenon is a consequence of the fact that in general the functors $(U \otimes_R \_)$ are not left exact. However, we shall soon see that each of these tensor functors is right exact. We begin with

19.11. Lemma. *Given modules* $_S U_R$ *and* $_S N$, *if* $f:_R M' \to _R M$ *is an R-homomorphism, then there exist* $\mathbb{Z}$-*isomorphisms* ϕ *and* ϕ' *such that the following diagram commutes:*

$$
\begin{array}{ccc}
Hom_R(M, Hom_S(U, N)) & \xrightarrow{\;Hom_R(f, Hom_S(U, N))\;} & Hom_R(M', Hom_S(U, N)) \\
\phi \downarrow & & \downarrow \phi' \\
Hom_S((U \otimes_R M), N) & \xrightarrow{\;Hom_S((U \otimes_R f), N)\;} & Hom_S((U \otimes_R M'), N)
\end{array}
$$

Proof. Let $\gamma \in Hom_R(M, Hom_S(U, N))$. Then it is easy to check that $(u, m) \mapsto \gamma(m)(u)$ is R-balanced. So there exists an S-homomorphism

$$\phi(\gamma):_S U \otimes_R M \to N$$

defined by $\phi(\gamma): u \otimes m \mapsto \gamma(m)(u)$. It is straightforward to check that the mapping ϕ defined by $\phi: \gamma \mapsto \phi(\gamma)$ is an isomorphism between the appropriate abelian groups with inverse $\phi^{-1}(\delta)(m): u \mapsto \delta(u \otimes m)$. Then with a parallel definition for ϕ', we have

$$\phi'(Hom_R(f, Hom_S(U, N))(\gamma))(u \otimes m') =$$
$$= \phi'(\gamma f)(u \otimes m') = \gamma(f(m'))(u)$$
$$= \phi(\gamma)(u \otimes f(m')) = \phi(\gamma) \circ (U \otimes_R f)(u \otimes m')$$
$$= Hom_S((U \otimes_R f), N)(\phi(\gamma))(u \otimes m'),$$

and the diagram commutes. $\qquad\qquad\square$

Note that this last result states that under certain circumstances a tensor functor can be traded for a *Hom* functor. Formally this lemma is the statement of an "adjoint" relationship. (See §21.)

Now let C be an injective cogenerator in the category $_Z\mathsf{M}$ of abelian groups. (See (18.19).) Let

$$(\quad)^* = Hom_{\mathbb{Z}}(\_, C).$$

Recall that if U is a right R-module, then U^* is a left R-module. Using this notation we can state the following "exactness test" whose proof involves the trade made possible by the last lemma.

19.12. Lemma. *Let* $f:M' \to M$ *and* $g:M \to M''$ *be R-homomorphisms in* $_R M$ *and let* U *be a right R-module. Then*

$$U \otimes_R M' \xrightarrow{U \otimes_R f} U \otimes_R M \xrightarrow{U \otimes_R g} U \otimes_R M''$$

is exact if and only if

$$Hom_R(M'', U^*) \xrightarrow{Hom_R(g, U^*)} Hom_R(M, U^*) \xrightarrow{Hom_R(f, U^*)} Hom_R(M', U^*)$$

is exact.

Proof. From (19.11) we have a commutative diagram

$$
\begin{array}{ccccc}
Hom_R(M'', U^*) & \xrightarrow{Hom_R(g, U^*)} & Hom_R(M, U^*) & \xrightarrow{Hom_R(f, U^*)} & Hom_R(M', U^*) \\
\downarrow{\scriptstyle \phi''} & & \downarrow{\scriptstyle \phi} & & \downarrow{\scriptstyle \phi'} \\
(U \otimes_R M'')^* & \xrightarrow{(U \otimes_R g)^*} & (U \otimes_R M)^* & \xrightarrow{(U \otimes_R f)^*} & (U \otimes_R M')^*
\end{array}
$$

where ϕ'', ϕ and ϕ' are isomorphisms. It follows that the top row is exact if and only if the bottom row is exact. But, since C is an injective cogenerator in $_Z M$, it follows from (18.1) and (18.14) that the bottom row is exact if and only if

$$U \otimes_R M' \xrightarrow{U \otimes_R f} U \otimes_R M \xrightarrow{U \otimes_R g} U \otimes_R M''$$

is exact.   ☐

19.13. Proposition. *The tensor functors are right exact. In particular, if*

$$0 \to M' \xrightarrow{f} M \xrightarrow{g} M'' \to 0$$

is exact in $_R M$*, then for every bimodule* $_S U_R$*,*

$$U \otimes_R M' \xrightarrow{U \otimes_R f} U \otimes_R M \xrightarrow{U \otimes_R g} U \otimes_R M'' \to 0$$

is exact in $_S M$*.*

Proof. Apply (19.12) to the sequence

$$0 \to Hom_R(M'', U^*) \xrightarrow{Hom_R(g, U^*)} Hom_R(M, U^*) \xrightarrow{Hom_R(f, U^*)} Hom_R(M', U^*)$$

which we know is exact by (16.6).   ☐

Flat Modules

We say that a module U_R is *flat relative to a module* $_R M$ (or that U is *M-flat*) in case the functor $(U \otimes_R \_)$ preserves the exactness of all short exact sequences with middle term M. Then, of course, U is M-flat if and only if for every submodule $K \leq M$ the sequence

$$0 \to U \otimes_R K \xrightarrow{U \otimes_R i_K} U \otimes_R M$$

is exact. A module V_R that is flat relative to every right R-module is called a *flat* right R-module. The theory of flat left modules is an obvious left-right symmetric version of that of flat right modules.

We continue our convention of letting ( )* denote the functor $Hom_{\mathbb{Z}}(\_, C)$ where C is a fixed injective cogenerator in $_{\mathbb{Z}}M$.

19.14. Lemma. *Let M be a left R-module. A right R-module V is M-flat if and only if V^* is M-injective. In particular V is flat if and only if V^* is injective.*

Proof. Apply (19.12) to the monomorphisms $0 \to K \to M$.   ☐

19.15. Proposition. *Let $(V_\alpha)_{\alpha \in A}$ be an indexed set of right R-modules. Then $\oplus_A V_\alpha$ is flat if and only if each V_α is flat..*

Proof. By (16.4), $(\oplus_A V_\alpha)^* \cong \Pi_A (V_\alpha)^*$ and by (16.11) the latter is injective if and only if each $(V_\alpha)^*$ is injective. Thus Lemma (19.14) applies.   ☐

19.16. Proposition. *Every projective module is flat.*

Proof. Since projective modules are isomorphic to direct summands of free modules (17.2), we need only show that the regular module R_R is flat. But if $f: M' \to M$ is a monomorphism in $_R\mathsf{M}$, then by (19.6) there are isomorphisms μ' and μ that make the diagram

$$
\begin{array}{ccc}
0 \to M' & \xrightarrow{f} & M \\
\mu' \uparrow & & \uparrow \mu \\
R \otimes_R M' & \xrightarrow{R \otimes_R f} & R \otimes_R M
\end{array}
$$

commute. Thus $R \otimes_R f$ is monic; and R_R is flat.   ☐

19.17. The Flat Test Lemma. *The following statements about a right R-module V are equivalent:*

(a) *V is flat;*

(b) *V is flot relative to $_R R$;*

(c) *For each (finitely generated) left ideal $I \le {}_R R$ the $\mathbb{Z}$-epimorphism $\mu_I : V \otimes_R I \to VI$ with $\mu_I(v \otimes a) = va$ is monic.*

Proof. (a) ⇔ (b). This is by the Injective Test Lemma and Lemma (19.14).

(b) ⇔ (c). The diagram

$$
\begin{array}{ccc}
V \otimes_R I & \xrightarrow{V \otimes_R i_I} & V \otimes_R R \\
\mu_I \downarrow & & \downarrow \mu \\
VI & \xrightarrow{i_{VI}} & V
\end{array}
$$

commutes, where μ is the isomorphism of (19.6). Thus since the inclusion map i_{VI} is a $\mathbb{Z}$-monomorphism, $V \otimes_R i_I$ is monic if and only if μ_I is monic.

Finally, the parenthetical version of (c) implies the non-parenthetical version. For let $v_i \in V$, $a_i \in R$ $(i = 1, \ldots, n)$ and suppose $\Sigma_i v_i \otimes a_i \in Ker \mu_I$; that is, $\Sigma_i v_i a_i = 0$. Then $\Sigma_i v_i \otimes a_i \in Ker \mu_K$ where $K = \Sigma_i Ra_i$. By hypothesis, $\Sigma_i v_i \otimes a_i = 0$ as an element of $V \otimes_R K$. So with $i_K : K \to I$ the inclusion map

$$0 = (V \otimes_R i_K)(\Sigma(v_i \otimes a_i)) = \Sigma(v_i \otimes a_i) \in V \otimes_R I. \qquad \square$$

From the Flat Test Lemma we obtain the following two additional tests for flatness.

19.18. Lemma. *Let V be a flat right R-module and suppose that the sequence*

$$0 \to K \xrightarrow{i_K} V \xrightarrow{f} V' \to 0$$

is exact in M_R. Then V' is flat if and only if for each (finitely generated) left ideal $I \leq {}_R R$

$$KI = K \cap VI.$$

Proof. The diagram of $\mathbb{Z}$-homomorphisms

$$
\begin{array}{ccccccc}
 & & 0 & & & & \\
 & & \downarrow & & & & \\
K \otimes_R I & \xrightarrow{i_K \otimes_R I} & V \otimes_R I & \xrightarrow{f \otimes_R I} & V' \otimes_R I & \to 0 \\
\mu \downarrow & & \mu_I \downarrow & & \mu'_I \downarrow & & \\
0 \to K \cap VI & \xrightarrow{\subseteq} & VI & \xrightarrow{(f|VI)} & V'I & \\
 & & \downarrow & & & & \\
 & & 0 & & & &
\end{array}
$$

is commutative and has exact rows and columns. Therefore by (3.14.3) and (3.14.4) μ'_I is monic if and only if μ is epic. But $Im \, \mu = KI \subseteq K \cap VI$, so that μ'_I is monic if and only if $KI = K \cap VI$. Thus the Flat Test Lemma applies. $\square$

19.19. Lemma. *A module V_R is flat if and only if for every relation*

$$\Sigma^n_{j=1} v_j a_j = 0 \qquad (v_j \in V, a_j \in R)$$

there exist elements $u_1, \ldots, u_m \in V$ and elements $c_{ij} \in R \, (i = 1, \ldots, m, j = 1, \ldots, n)$ such that

$$\Sigma^n_{j=1} c_{ij} a_j = 0 \qquad (i = 1, \ldots, m)$$

and

$$\Sigma^m_{i=1} u_i c_{ij} = v_j \qquad (j = 1, \ldots, n).$$

Proof. ($\Rightarrow$). Suppose that V_R is flat and $\Sigma^n_{j=1} v_j a_j = 0$. Let $I = \Sigma_j Ra_j$. Consider the free left R-module $F = \oplus^n_{j=1} Rx_j$ and the short exact sequence

$$0 \to K \xrightarrow{i_K} F \xrightarrow{f} I \to 0$$

where $f(x_j) = a_j$ for each $j = 1, \ldots, n$. By the Flat Test Lemma (19.17.c) $\Sigma_j(v_j \otimes f(x_j)) = \Sigma_j(v_j \otimes a_j) = 0$ as an element of $V \otimes_R I$. So in the exact sequence

$$0 \to V \otimes K \xrightarrow{V \otimes i_K} V \otimes F \xrightarrow{V \otimes f} V \otimes I \to 0$$

we have $\Sigma_j(v_j \otimes x_j) \in Ker(V \otimes f) = Im(V \otimes i_K)$. Thus there exist $u_i \in V$ and $k_i \in K$ with $\Sigma_j(v_j \otimes x_j) = \Sigma_i(u_i \otimes k_i)$. Now each $k_i \in F$, so $k_i = \Sigma_j c_{ij} x_j$ for each $i = 1, \ldots, m$ and some $c_{ij} \in R$. From this we get

$$\Sigma_j c_{ij} a_j = \Sigma_j c_{ij} f(x_j) = f(k_i) = 0 \qquad (i = 1, \ldots, n).$$

Moreover, this also gives

$$\Sigma_j (v_j \otimes x_j) = \Sigma_i (u_i \otimes k_i) = \Sigma_i (u_i \otimes (\Sigma_j c_{ij} x_j))$$
$$= \Sigma_j ((\Sigma_i u_i c_{ij}) \otimes x_j).$$

But $V \otimes F = \oplus_{j=1}^n Im(V \otimes i_{Rx_j})$ by (19.9), so $v_j = \Sigma_i u_i c_{ij}$ for each j.

($\Leftarrow$). Let $I \le {}_R R$ and suppose $a_j \in I$ and $v_j \in V$ with $\Sigma_j v_j a_j = 0$. Then by hypothesis there exist $u_j \in V$, $c_{ij} \in R$ such that in $V \otimes I$,

$$\Sigma_j (v_j \otimes a_j) = \Sigma_j ((\Sigma_i u_i c_{ij}) \otimes a_j)$$
$$= \Sigma_i (u_i \otimes \Sigma_j c_{ij} a_j) = 0$$

and the Flat Test Lemma (19.17) applies.   □

Products of Flat Modules

A finitely generated module ${}_R M$ is said to be *finitely presented* in case in every exact sequence

$$0 \to K \to F \to M \to 0$$

with F finitely generated and free the kernel K is also finitely generated. Observe that R is noetherian if and only if every finitely generated R-module is finitely presented. (See Proposition (10.19).) More generally, a ring R is *left coherent* if each of its finitely generated left ideals is finitely presented. These are the rings over which direct products of flat right modules are flat.

19.20. Theorem [S. U. Chase]. *For a ring R the following are equivalent:*
(a) *Every direct product of flat right R-modules is flat;*
(b) R_R^A *is flat for every set A;*
(c) R *is left coherent.*

Proof. (a) $\Rightarrow$ (b). By (19.16), R_R is flat.

(b) $\Rightarrow$ (c). Suppose $I \le {}_R R$ and F is a free module with free basis $x_1, \ldots, x_n$ that maps onto ${}_R I$

$$F \xrightarrow{f} I \to 0.$$

For each $j = 1, \ldots, n$ let $a_j = f(x_j)$ and let $K = Ker f$. To show that K is finitely generated define, in the direct product R^K of $card(K)$ copies of R, elements $v_j \in R^K$ via the equations

$$k = \pi_k(v_1) x_1 + \ldots + \pi_k(v_n) x_n \qquad (k \in K).$$

Then $0 = f(k) = \Sigma_j \pi_k(v_j) a_j$ or, equivalently,

$$\Sigma_{j=1}^n v_j a_j = 0 \in R^K.$$

By hypothesis R^K is flat so by (19.19) there exist $u_1, \ldots, u_m \in R^K$, $c_{ij} \in R$, with

$$\Sigma_{j=1}^n c_{ij} a_j = 0 \qquad \text{and} \qquad \Sigma_{i=1}^m u_i c_{ij} = v_j$$

for all i and j. Now let

$$k_i = \Sigma_{j=1}^n c_{ij} x_j \in F.$$

Then $f(k_i) = \Sigma_j c_{ij} a_j = 0$ so that $k_1, \ldots, k_n \in K$. But for each $k \in K$

$$k = \Sigma_j \pi_k(v_j) x_j = \Sigma_j \pi_k(\Sigma_i u_i c_{ij}) x_j$$
$$= \Sigma_i \pi_k(u_i)(\Sigma_j c_{ij} x_j) = \Sigma_i \pi_k(u_i) k_i;$$

hence K is spanned by $k_1, \ldots, k_n$.

(c) $\Rightarrow$ (a). Let R be a left coherent ring. The first step in this proof is to show, via (19.19), that the right R-module $(R^{(B)})^A$ is flat for all sets A and B. So suppose that

$$v_j \in (R^{(B)})^A \qquad \text{and} \qquad a_j \in R$$

with

$$\Sigma_{j=1}^n v_j a_j = 0.$$

Let F be the free left R-module with free basis $x_1, \ldots, x_n$ and let K be the kernel of the epimorphism

$$F \xrightarrow{f} \Sigma_{j=1}^n R a_j \to 0 \qquad (f(x_j) = a_j, j = 1, \ldots, n).$$

Then since R is left coherent we can write

$$K = \Sigma_{i=1}^m R k_i$$

where

$$k_i = \Sigma_{j=1}^n c_{ij} x_j \in K$$

and

$$\Sigma_{j=1}^n c_{ij} a_j = f(k_i) = 0 \qquad (i = 1, \ldots, m).$$

Now to find the u_i's observe that for all $\alpha \in A$, $\beta \in B$ we also have

$$\Sigma_{j=1}^n [v_j(\alpha)](\beta) x_j \in K.$$

Thus we may choose $b_{i\alpha\beta} \in R$ $(i = 1, \ldots, m)$ such that $b_{i\alpha\beta} = 0$ whenever $[v_j(\alpha)](\beta) = 0, (j = 1, \ldots, n)$ and

$$\Sigma_{j=1}^n [v_j(\alpha)](\beta) x_j = \Sigma_{i=1}^n b_{i\alpha\beta} k_i$$

to get $u_1, \ldots, u_m \in (R^{(B)})^A$ defined by

$$[u_i(\alpha)](\beta) = b_{i\alpha\beta} \qquad (\alpha \in A, \beta \in B, i = 1, \ldots, m);$$

so that

$$\Sigma_j [v_j(\alpha)](\beta) x_j = \Sigma_i [u_i(\alpha)](\beta) k_i$$
$$= \Sigma_i [u_i(\alpha)](\beta)(\Sigma_j c_{ij} x_j)$$
$$= \Sigma_j (\Sigma_i [u_i(\alpha)](\beta) c_{ij}) x_j$$

or equivalently

$$v_j = \Sigma_{i=1}^m u_i c_{ij} \qquad (j = 1, \dots, n).$$

Thus by (19.19) we see that $(R^{(B)})^A$ is always flat.

Now to complete the proof suppose that $(V_\alpha)_{\alpha \in A}$ are flat right R-modules and let B be a set such that the free right modules $F_\alpha \cong R^{(B)}$ map onto V_α. That is

$$0 \to K_\alpha \to F_\alpha \overset{g_\alpha}{\to} V_\alpha \to 0$$

is exact for all $\alpha \in A$. Then we have an exact sequence

$$0 \to \Pi_A K_\alpha \to \Pi_A F_\alpha \overset{\Pi g_\alpha}{\longrightarrow} \Pi_A V_\alpha \to 0$$

(see (6.25)) in which $\Pi_A F_\alpha \cong (R^{(B)})^A$ is flat. Now let I be a finitely generated left ideal in R. Then for any direct product of right R-modules $\Pi_A M_\alpha$ we have $(\Pi_A M_\alpha)I = \Pi_A (M_\alpha I)$ (see Exercise (15.3)). So applying (19.18) we have

$$(\Pi_A K_\alpha)I = \Pi_A(K_\alpha I) = \Pi_A(K_\alpha \cap F_\alpha I) = (\Pi_A K_\alpha) \cap (\Pi_A(F_\alpha I))$$
$$= (\Pi_A K_\alpha) \cap (\Pi_A F_\alpha)I$$

and the theorem is proved. $\square$

19. Exercises

1. Let R be a ring, $L \leq {}_R R$ and $I \leq R_R$. Prove that:
   (1) For each ${}_R M$ there is a $\mathbb{Z}$-isomorphism $f: R/I \otimes_R M \to M/IM$ such that $f: (r + I) \otimes m \mapsto rm + IM$. Deduce that as abelian groups

   $$R/I \otimes_R R/L \cong R/(I + L).$$

   (2) If $m, n \in \mathbb{N}$ and if $d = (m, n)$ is the greatest common divisor of m and n, then $\mathbb{Z}_m \otimes_{\mathbb{Z}} \mathbb{Z}_n \cong \mathbb{Z}_d$.
2. Let R be commutative. Let F and G be free R-modules with free bases $(x_\alpha)_{\alpha \in A}$ and $(y_\beta)_{\beta \in B}$, respectively. Prove that $F \otimes_R G$ is a free R-module with free basis $(x_\alpha \otimes y_\beta)_{(\alpha, \beta) \in A \times B}$.
3. Let L be an extension field of K and let V be a K-vector space with basis $(x_\alpha)_{\alpha \in A}$. Prove that $(1 \otimes x_\alpha)_{\alpha \in A}$ is a basis for the L-vector space $L \otimes_K V$.
4. Let R and S be rings and let $e \in R, f \in S$ be non-zero idempotents. Prove that:
   (1) For each ${}_R M_S$ there are isomorphisms η_M and μ_M

   $$\eta_M: eR \otimes_R M \to {}_{eRe} eM_S \qquad \text{and} \qquad \mu_M: M \otimes_S Sf \to {}_R Mf_{fSf}.$$

   such that $\eta_M(e \otimes m) = em$ and $\mu_M(m \otimes f) = mf$.
   (2) η_M is a natural transformation from the tensor functor $eR \otimes_R (\_)$ to T_e (see (0.13) and Exercise (4.17)).
5. Let R and S be algebras over a commutative ring K. Prove that:
   (1) $R \otimes_K S$ is a K-algebra with multiplications

$$(r \otimes s)(r' \otimes s') = rr' \otimes ss'$$

$$k(r \otimes s) = (kr) \otimes s = r \otimes (ks).$$

(2) $\alpha : r \mapsto r \otimes 1$ and $\beta : s \mapsto 1 \otimes s$ define algebra homomorphisms $\alpha : R \to R \otimes_K S$ and $\beta : S \to R \otimes_K S$ such that $\alpha(r)\beta(s) = \beta(s)\alpha(r)$ for all $r \in R, s \in S$.

6. Let R and S be rings. Consider the rings ($\mathbb{Z}$-algebras) $T = R \otimes_{\mathbb{Z}} S$, $U = R \otimes_{\mathbb{Z}} S^{op}$, and $V = R^{op} \otimes_{\mathbb{Z}} S$. (See Exercise (19.5).) Prove that:

   (1) Each bimodule $_R M_S$ induces modules $_U M$ where $(r \otimes s)m = rms$ and M_V where $m(r \otimes s) = rms$.

   (2) The maps α and β of Exercise (19.5.2) induce an $_R M_{S^{op}}$ structure on each module $_T M$.

7. For each left R-module M let M^* be the right R-module $Hom_R(_R M, _R R_R)$. Then by Exercise (19.6), $M \otimes_{\mathbb{Z}} M^*$ is a left $R \otimes_{\mathbb{Z}} R^{op}$-module. Prove that:

   (1) If $_R P$ is finitely generated projective, then $P \otimes_{\mathbb{Z}} P^*$ is finitely generated projective over $R \otimes_{\mathbb{Z}} R^{op}$.

   (2) If $_R G$ is a generator, then $G \otimes_{\mathbb{Z}} G^*$ is an $R \otimes_{\mathbb{Z}} R^{op}$-generator.

8. Let K be a field. Prove that as K-algebras $\mathbb{M}_m(K) \otimes_K \mathbb{M}_n(K) \cong \mathbb{M}_{mn}(K)$.

9. Recall that a module $_R U$ over a commutative integral domain R is torsion-free in case $l_R(u) = 0$ for all $0 \neq u \in U$. Prove that:

   (1) If R is a P.I.D., $_R U$ is flat iff $_R U$ is torsion-free. [Hint: $aR \otimes_R U = \{a \otimes u \mid u \in U\}$. Apply the Flat Test Lemma (19.17).]

   (2) If K is a field, then $RX + RY$ is a torsion-free $R = K[X, Y]$ ideal but is not R flat.

10. Let $\phi : R \to S$ be a ring homomorphism. Then $_S S_R$ and $_R S_S$ via ϕ. Consider the functors $T_\phi = (_S S \otimes_R \_)$ and $H_\phi = Hom_R(S_S, \_)$ from $_R M$ to $_S M$. Deduce from (19.11) that

    (1) If P is projective in $_R M$, then $T_\phi(P)$ is projective is $_S M$.

    (2) If E is injective in $_R M$, then $H_\phi(E)$ is injective in $_S M$.

11. A submodule $_R U \leq _R V$ is *pure* in V in case $IU = U \cap IV$ for each right ideal $I \leq R_R$. Thus if V is flat, then V/U is flat iff U is pure in V. (See (19.18).)

    (1) Prove that if $(U_\alpha)_{\alpha \in A}$ is a chain of pure submodules of V, then $\cup_A U_\alpha$ is pure in V.

    (2) Prove that if $K \leq V$, then there is a submodule $U \leq V$ maximal with respect to $U \leq K$ and U pure in V.

12. Prove that extensions of flat modules by flat modules are flat; i.e., if $0 \to V' \to V \to V'' \to 0$ is exact with V' and V'' flat, then V is flat. [Hint: (19.17) and (3.14.1).]

13. Prove that if V^A is flat for all sets A, then $(V^{(B)})^A$ is flat for all sets A and B.

14. Let $_S V_R$ be a bimodule and let $_S Q$ be an injective cogenerator in $_S M$. Prove that the following are equivalent: (a) V_R is flat; (b) $Hom_S(V_R, Q)$ is an injective left R-module; (c) There exists a left S-cogenerator C such that $Hom_S(V_R, C)$ is injective over R; (d) $Hom_S(V_R, E)$ is injective over R for each injective $_S E$.

15. Given modules M_R and $_R U$ let $\mathscr{F}(M)$ denote the class of left R-modules

that are M-flat and let $\mathcal{F}^{-1}(U)$ denote the class of right R-modules N such that U is N-flat. Prove that:

(1) $\mathcal{F}(M)$ is closed under direct sums and direct summands.

(2) $\mathcal{F}^{-1}(U)$ is closed under submodules, factor modules and direct sums.

(3) If V_R is flat, then $\mathcal{F}(V)$ is closed under extensions; i.e., if $0 \to U' \to U \to U'' \to 0$ is exact and U' and $U'' \in \mathcal{F}(V)$ then $U \in \mathcal{F}(V)$.

16. Prove that a ring is von Neumann regular iff each of its left modules is flat. [Hint: For ($\Leftarrow$) use (19.18) on $0 \to Ra \to R \to R/Ra \to 0$.]

17. Prove that if $0 \to K \to P \to M \to 0$ is exact with K and P finitely generated and P projective then M is finitely presented. [Hint: Schanuel's Lemma, Exercise (18.9).]

18. Let U_R and $_RM$ be modules. Define the *annihilator in M of U* to be

$$Ann_M(U) = \{m \in M \mid u \otimes m = 0 \text{ in } U \otimes_R M \text{ for all } u \in U\};$$

say that U_R is $_RM$-*faithful* in case $Ann_M(U) = 0$. Prove that:

(1) $Ann_M(U)$ is the unique smallest submodule K of M such that U is M/K-faithful.

(2) If $f: {}_RM \to {}_RN$, then $f(Ann_M(U)) \subseteq Ann_N(U)$. In particular, $Ann_M(U)$ is stable under endomorphisms of M.

(3) If $f: M \to N$ is epic and $Ker f \leq Ann_M(U)$, then

$$f(Ann_M(U)) = Ann_N(U).$$

(4) If $(U_\alpha)_{\alpha \in A}$ are right R-modules and M is a left R-module, then
$$Ann_M(\oplus_A U_\alpha) = \cap_A Ann_M(U_\alpha).$$

(5) If $(M_\alpha)_{\alpha \in A}$ are left R-modules and U is a right R-module,

$$Ann_{\oplus_A M_\alpha}(U) = \oplus_A Ann_{M_\alpha}(U).$$

(6) If U_R generates V_R, then $Ann_M(U) \leq Ann_M(V)$.

(7) U_R is $_RM$-faithful iff for every homomorphism $f: N \to M$, $U \otimes f = 0$ implies $f = 0$.

(8) $Ann_{_RR}(U) = r_R(U)$.

(9) If $I \leq {}_RR_R$, then $Ann_M(R/I) = IM$. [Hint: Exercise (19.1.1).]

19. A module W_R is said to be *completely faithful* in case $Ann_M(W) = 0$ for every left R-module M. Prove that:

(1) R_R is completely faithful;

(2) Every generator in M_R is completely faithful.

(3) The following statements about a module W_R are equivalent:

   (a) W_R is completely faithful;

   (b) For every homomorphism $f \in {}_RM$, if $W \otimes f = 0$, then $f = 0$;

   (c) For every homomorphism $f \in {}_RM$, if $W \otimes f$ is monic, then f is monic;

   (d) A sequence $M' \to M \to M''$ is exact in $_RM$ if the induced sequence $W \otimes M' \to W \otimes M \to W \otimes M''$ is exact.

20. Given modules $_SW_R$, $_RM$ and $_SC$, let $W^* = Hom_S(W, C) \in {}_RM$. Use the isomorphism ϕ of (19.11) to prove:

(1) $Ann_M(W) \leq Rej_M(W^*)$.

(2) If $_SC$ is a cogenerator, then $Ann_M(W) = Rej_M(W^*)$.

(3) If $_SC$ is a cogenerator, then W_R is completely faithful iff W^* is a cogenerator in $_R M$.

21. Prove that the following statements about a flat right R-module V are equivalent:

    (a) V is completely faithful;

    (b) $V \otimes M \neq 0$ whenever $_R M \neq 0$;

    (c) $V \otimes_R T \neq 0$ for every simple left R-module T;

    (d) $VI \neq V$ for every maximal left ideal I or R.

[Hint: Exercises (19.20) and (19.1.1).]

22. Let S be a direct sum of a set of representatives of the simple right modules over a ring R. For each left R-module M define $Trad\, M = Ann_M(S)$. Prove that:

(1) $Trad\,_R R = J(R)$. [Hint: Exercise (19.18.8).]

(2) If R is commutative, then $Trad\, M = Rad\, M$. [Hint: Exercises (19.18.9) and (15.5).]

(3) If $R/J(R)$ is semisimple, then $Trad\, M = Rad\, M$.

23. Prove that for a commutative ring R the following statements are equivalent: (a) R has a completely faithful semisimple module; (b) R is von Neumann regular; (c) R is co-semisimple. [Hint: See Exercise (18.23). For (b) $\Rightarrow$ (a) Let $S = \oplus_A T_\alpha$ with T_α simple. Then $S \otimes T_\alpha \neq 0$. Now apply Exercises (19.16) and (19.21).]

24. Suppose that R has faithful simple projective modules Re and eR. (See Exercise (17.5).) Prove that eR is injective iff $eR = Hom_{eRe}(Re, eRe)$. [Hint: Exercises (19.14) and (18.30).]

25. Let M_K be an infinite dimensional vector space over a field K and let R be the subring of $End(M_K)$ generated by the socle S and the scalar transformations $K1_M \subseteq End(M_K)$. [Note that S is just the set of elements of $End(M_K)$ of finite rank, so $\sigma \in R$ iff $\sigma - \alpha 1_M$ has finite rank for some $\alpha \in K$.] Prove that:

(1) R is von Neumann regular.

(2) To within isomorphism R has exactly two left and two right simple modules and that exactly three of them are injective.

(3) R is right but not left co-semisimple. (See Exercise (18.23).)

(4) R has completely faithful semisimple left and right modules.

26. Prove that R is left coherent iff a direct product of $card\, R$ copies of R_R is flat. [Hint: See the proof (b) $\Rightarrow$ (c) of (19.20).]

§20. Natural Transformations

At last we come to the central notion of categorical algebra, that of a natural transformation. It is by means of this that the intuitive idea of a "natural" homomorphism is made precise. Recall (0.13) that if $C = (\mathscr{C}, mor_C, \circ)$ and $D = (\mathscr{D}, mor_D, \circ)$ are categories and if F, G are covariant functors C to D, then a *natural transformation* from F to G is a map $\eta : M \mapsto \eta_M$ from $\mathscr{C}$ to mor_D.

such that for each $M \in \mathscr{C}$, $\eta_M : F(M) \to G(M)$ and for each $f : M \to N$ in C,

$$
\begin{array}{ccc}
F(M) & \xrightarrow{F(f)} & F(N) \\
{\scriptstyle \eta_M}\big\downarrow & & \big\downarrow{\scriptstyle \eta_N} \\
G(M) & \xrightarrow{G(f)} & G(N)
\end{array}
$$

commutes. (If F and G are contravariant, reverse the arrows $F(f)$ and $G(f)$ in this diagram.) We usually abbreviate this by $\eta : F \to G$. If each η_M is an isomorphism, then we call the natural transformation $\eta : F \to G$ a *natural isomorphism*.

Let F and G be functors (of the same variance) between two categories C and D. We say that F and G are *isomorphic* and write

$$
F \cong G
$$

in case there is a natural isomorphism $\eta : F \to G$. It is easy to check that this concept induces an equivalence relation on the class of functors from C to D.

Two Simple Examples

Several important isomorphisms in the preceding sections are natural isomorphisms of functors. Two of the most basic of these, given in (4.5) and (19.6), assert that for each left R-module M there are isomorphisms

$$
\operatorname{Hom}_R(R, M) \cong M \qquad \text{and} \qquad R \otimes_R M \cong M.
$$

These are actually natural isomorphisms of the functors $\operatorname{Hom}_R(R, \_)$ and $(R \otimes_R \_)$ with the identity functor on $_R$M. Specifically,

20.1. Proposition. *Let R be a ring. Then there are natural isomorphisms:*
(1) $\rho : 1_{_R\mathsf{M}} \to \operatorname{Hom}_R(_R R_R, \_)$ *where for each $_R M$, each $m \in M$, each $r \in R$, and each $\gamma \in \operatorname{Hom}_R(R, M)$*

$$
\rho_M(m) : r \mapsto rm \qquad \text{and} \qquad \rho_M^{-1}(\gamma) = \gamma(1);
$$

(2) $\mu : (_R R \otimes_R \_) \to 1_{_R\mathsf{M}}$ *where for each $_R M$, each $m \in M$, and each $r \in R$*

$$
\mu_M(r \otimes m) = rm \qquad \text{and} \qquad \mu_M^{-1}(m) = 1 \otimes m.
$$

Proof. We have already seen in (4.5) and (19.6) that ρ_M and μ_M are R-isomorphisms. Thus all that remains is to check their naturality. For (2) observe that

$$
\begin{array}{ccc}
R \otimes_R M & \xrightarrow{R \otimes_R f} & R \otimes_R M' \\
{\scriptstyle \mu_M}\big\downarrow & & \big\downarrow{\scriptstyle \mu_{M'}} \\
M & \xrightarrow{\quad f \quad} & M'
\end{array}
$$

commutes because

$$
\mu_{M'} \circ (R \otimes_R f)(r \otimes m) = \mu_{M'}(r \otimes f(m))
$$
$$
= rf(m) = f(rm) = f \circ \mu_M(r \otimes m).
$$

It is equally easy to prove the naturality of ρ.   ☐

Direct Sums and Products

Suppose that C and D are categories of modules and that all direct sums and products of members of D belong to D. If $(F_\alpha)_{\alpha \in A}$ is an indexed class of additive functors $F_\alpha : C \to D$ all of the same variance, then their *direct sum* and *direct product* are the additive functors

$$\oplus_A F_\alpha : C \to D \qquad \text{and} \qquad \Pi_A F_\alpha : C \to D$$

defined coordinatewise by

$$(\oplus_A F_\alpha)(f) = \oplus_A(F_\alpha(f)) \qquad \text{and} \qquad (\Pi_A F_\alpha)(f) = \Pi_A(F_\alpha(f)).$$

For instance if the F_α are covariant and $M \xrightarrow{f} M'$ in C, then

$$F_\alpha(M) \xrightarrow{F_\alpha(f)} F_\alpha(M') \qquad (\alpha \in A)$$

and (see Remark 6.25)

$$\Pi_A F_\alpha(M) \xrightarrow{\Pi_A F_\alpha(f)} \Pi_A F_\alpha(M').$$

The straightforward proof that these are additive functors is left to Exercise (20.4).

The first two assertions of the next proposition, simply rephrasings of (16.5), were used to prove that direct sums of projectives are projective and that direct products of injectives are injective. The final assertion of this proposition follows easily from (19.9).

20.2. Proposition. *Let $(U_\alpha)_{\alpha \in A}$ be an indexed set of left R-modules. Then, as functors from $_R M$ to $_{\mathbb{Z}} M$,*
  (1) $Hom_R(\oplus_A U_\alpha, \_) \cong \Pi_A Hom_R(U_\alpha, \_)$;
  (2) $Hom_R(\_, \Pi_A U_\alpha) \cong \Pi_A Hom_R(\_, U_\alpha)$;
  (3) $(\_ \otimes_R (\oplus_A U_\alpha)) \cong \oplus_A(\_ \otimes_R U_\alpha)$.   ☐

Endomorphism Rings and Bimodules

Recall that the Hom and tensor functors preserve certain module structure. For example, given modules $_R M_S$, $_R N$, K_R, then $Hom_R(M_S, N)$ and $K \otimes_R M_S$ are naturally a left S-module and a right S-module, respectively. (See (4.4) and (19.5).) This behavior is typical of additive functors and is a consequence of the following

20.3. Lemma. *Let $_R C$, $_S D$ and D_S be full subcategories of $_R M$, $_S M$ and M_S, respectively. Suppose that $F : {}_R C \to {}_S D$ and $H : {}_R C \to D_S$ are additive functors with F covariant and H contravariant. Then for each non-zero $_R M \in {}_R C$ these functors restrict to ring homomorphisms*

$$F : End(_R M) \to End(_S F(M)) \qquad \text{and} \qquad H : End(_R M) \to End(H(M)_S).$$

Proof. Let $f \in End(_RM)$. Then

$$_RM \xrightarrow{f} {_RM}$$

belongs to $_RM$. Hence

$$_sF(M) \xrightarrow{F(f)} {_sF(M)} \qquad \text{and} \qquad H(M)_S \xrightarrow{H(f)} H(M)_S$$

are endomorphisms of $F(M)$ and $H(M)$, respectively. Denoting composition in the categories by $\circ$ and multiplication in the endomorphism rings by juxtaposition we have, for $f, g \in End(_RM)$,

$$F(fg) = F(g \circ f) = F(g) \circ F(f) = F(f)F(g)$$

and

$$H(fg) = H(g \circ f) = H(f) \circ H(g) = H(f)H(g).$$

Thus, since F and H are additive and preserve identity maps, they restrict to the desired ring homomorphisms. □

Suppose now that we have $F : {_RC} \to {_sD}$ and $H : {_RC} \to D_S$ as in the lemma, and that $_RM_T$ is a bimodule such that $_RM \in {_RC}$. Then letting ρ denote scalar multiplication in M_T,

$$_RM \xrightarrow{\rho(t)} {_RM} \qquad (t \in T),$$

we have by Lemma (20.3) and (4.10) that

$$t \mapsto F(\rho(t)) \qquad \text{and} \qquad t \mapsto H(\rho(t)) \qquad (t \in T)$$

define ring homomorphisms

$$T \to End(_sF(M)) \qquad \text{and} \qquad T \to End(H(M)_S).$$

Thus (see 4.10) we obtain bimodules

$$_sF(M)_T \qquad \text{and} \qquad _TH(M)_S$$

where for $x \in F(M)$, $y \in H(M)$ and $t \in T$

$$xt = F(\rho(t))(x) \qquad \text{and} \qquad ty = H(\rho(t))(y).$$

These bimodules are called *the (canonical) bimodules* $_sF(M)_T$ *and* $_TH(M)_S$ *induced by* $_RM_T$.

The bimodule structures that were constructed earlier from the Hom and tensor functors are precisely the canonical ones for these functors. For example, consider the functor

$$Hom_R(M, \_) : {_RM} \to {_ZM}.$$

Let $_RU_T$. Then in §4 we made

$$Hom_R(_RM, _RU_T) = Hom_R(M, \_)(U)$$

into a right T-module with

$$(\gamma t)(m) = (\gamma(m))t = (\rho(t) \circ \gamma)(m),$$

i.e.,

$$\gamma t = Hom_R(M, \rho(t))(\gamma),$$

so that the right T-module is the canonical $(\mathbb{Z}, T)$-bimodule induced by $_RU_T$. Similarly $(_SV \otimes_R \_): {}_RM \to {}_SM$ and the bimodule

$$_SV \otimes_R U_T = (V \otimes_R \_)(U)$$

is the (S, T)-bimodule induced by $_RU_T$.

Let $_RC$ and $_SD$ be full subcategories of $_RM$ and $_SM$, respectively. If T is a ring, then the left R- right T-bimodules $_RU_T$ with $_RU$ in $_RC$ and the (R, T)-homomorphisms between them form a full subcategory $_RC_T$ of $_RM_T$. There is a similar subcategory $_SD_T$ of $_SM_T$. Let $F: {}_RC \to {}_SD$ be an additive covariant functor. Then equipping each $F(_RM_T)$ with its canonical bimodule structure, we have that F maps the objects of $_RC_T$ to objects in $_SD_T$. The next result implies that F, restricted to $_RC_T$, is a functor to $_SD_T$, and that natural transformations between pairs of such functors restrict to natural transformations of their restrictions.

20.4. Lemma. *Let $_RC$, $_SD$, and D_S be full subcategories of R- and S-modules. Let $F, F': {}_RC \to {}_SD$ be covariant and $H, H': {}_RC \to D_S$ be contravariant additive functors. Let $\eta: F \to F'$ and $v: H \to H'$ be natural transformations. Finally, let $_RU_T$ and $_RV_T$ be bimodules with $_RU, {}_RV \in {}_RC$ and let*

$$f: {}_RU_T \to {}_RV_T$$

be a bimodule homomorphism. Then with the canonical bimodule structure
  (1) $F(f): {}_SF(U)_T \to {}_SF(V)_T$ *and* $H(f): {}_TH(V)_S \to {}_TH(U)_S$;
  (2) $\eta_U: {}_SF(U)_T \to {}_SF'(U)_T$ *and* $v_U: {}_TH(U)_S \to {}_TH'(U)_S$;
are bimodule homomorphisms.

Proof. For (1) it will suffice to check that $F(f)$ and $H(f)$ are T-homomorphisms. Denote scalar multiplication in both U_T and V_T by ρ. If $t \in T$ then, since f is a right T-homomorphism, $f \circ \rho(t) = \rho(t) \circ f$. Thus, for all $x \in F(U)$, and $t \in T$,

$$F(f)(xt) = F(f) \circ F(\rho(t))(x) = F(f \circ \rho(t))(x)$$

$$= F(\rho(t) \circ f)(x) = F(\rho(t))(F(f)(x))$$

$$= (F(f)(x))t.$$

Similarly $H(f)$ is a left T-homomorphism.

For (2) simply observe that the diagrams

$$
\begin{array}{ccc}
F(U) & \xrightarrow{F(\rho(t))} & F(U) \\
{\scriptstyle \eta_U}\downarrow & & \downarrow{\scriptstyle \eta_U} \\
F'(U) & \xrightarrow{F'(\rho(t))} & F'(U)
\end{array}
\qquad \text{and} \qquad
\begin{array}{ccc}
H(U) & \xrightarrow{H(\rho(t))} & H(U) \\
{\scriptstyle v_U}\downarrow & & \downarrow{\scriptstyle v_U} \\
H'(U) & \xrightarrow{H'(\rho(t))} & H'(U)
\end{array}
$$

commute. $\square$

Applied to (20.2.1) for example, this Lemma implies that $Hom_R(\oplus_A U_\alpha, \_)$ and $\Pi_A Hom_R(U_\alpha, \_)$ restricted to $_R M_S$ are (still) isomorphic functors to M_S. As another application we have

20.5. Proposition. *Let* $\theta : _R U_S \to _R V_S$ *be a bimodule homomorphism. Then the following are natural transformations:*

(1) $\eta : Hom_R(V_S, \_) \to Hom_R(U_S, \_)$ *defined via* $\eta_M = Hom_R(\theta, M)$;

(2) $v : Hom_R(\_, U_S) \to Hom_R(\_, V_S)$ *defined via* $v_M = Hom_R(M, \theta)$;

(3) $\phi : (\_ \otimes_R U_S) \to (\_ \otimes_R V_S)$ *defined via* $\phi_N = N \otimes_R \theta$.

Moreover, if θ *is an isomorphism, then each of the above is a natural isomorphism.*

Proof. We'll do (1). Because $Hom_R(\_, M) : _R M \to _Z M$ is a contravariant additive functor, (20.4.1) implies that $\eta_M = Hom_R(\theta, M)$ is a left S-homomorphism. But, for each $_R M \xrightarrow{f} _R M'$

$$
\begin{array}{ccc}
Hom_R(V, M) & \xrightarrow{\ Hom_R(V, f)\ } & Hom_R(V, M') \\
\eta_M \downarrow & & \downarrow \eta_{M'} \\
Hom_R(U, M) & \xrightarrow{\ Hom_R(U, f)\ } & Hom_R(U, M')
\end{array}
$$

commutes because $Hom_R(\theta, M') \circ Hom_R(V, f) = Hom_R(U, f) \circ Hom_R(\theta, M)$. $\square$

Some Hom-tensor Relations

Given a triple $(_R M, _S W_R, _S N)$ of modules there is, by Proposition (19.11), an isomorphism

$$\phi = \phi_{MWN} : Hom_R(M, Hom_S(W, N)) \to Hom_S((W \otimes_R M), N)$$

defined via

$$[\phi(\gamma)](w \otimes m) = [\gamma(m)](w).$$

The importance of this isomorphism in our study of the tensor functors is that if W and N are fixed, then the indexed class (ϕ_{MWN}), indexed by the left R-modules M, is a natural isomorphism of contravariant functors

$$Hom_R(\_, Hom_S(W, N)) \cong Hom_S((W \otimes_R \_), N).$$

Speaking loosely we can view both

$$Hom_R(\_, Hom_S(\_, \_)) \quad \text{and} \quad Hom_S((\_ \otimes_R \_), \_)$$

as functors of "three variables" $(M_R, _S W_R, _S N)$ with mixed variance. Then because of (19.11) we shall say that the isomorphism ϕ_{MWN} is *natural in* M. It should be clear how to develop a theory of functors of several variables and of homomorphisms of such functors natural in various variables. We shall illustrate this by showing that the isomorphism ϕ_{MWN} is natural in each of the three variables. First suppose that $_S N \xrightarrow{g} _S N'$. Then the diagram of $\mathbb{Z}$-homomorphisms

$$Hom_R(M, Hom_S(W, N)) \xrightarrow{\ Hom_R(M, Hom_S(W, g))\ } Hom_R(M, Hom_S(W, N'))$$

$$\phi_{MWN} \downarrow \qquad\qquad\qquad\qquad\qquad\qquad\qquad \downarrow \phi_{MWN'}$$

$$Hom_S((W \otimes_R M), N) \xrightarrow{\ Hom_S(W \otimes_R M, g))\ } Hom_S(W \otimes_R M, N')$$

commutes, and so the isomorphism $\phi = \phi_{MWN}$ is *natural in N*. Finally, suppose $h : {}_S W_R \to {}_S W'_R$ is a bimodule homomorphism. Then by (20.4.1) $Hom_S(h, N)$ is a left R-homomorphism and $h \otimes_R M$ is a left S-homomorphism so that we can form $Hom_R(M, Hom_S(h, N))$ and $Hom_S((h \otimes_R M), N)$. It is easy to check that

$$Hom_R(M, Hom_S(W', N)) \xrightarrow{\ Hom_R(M, Hom_S(h, N))\ } Hom_R(M, Hom_S(W, N))$$

$$\phi_{MW'N} \downarrow \qquad\qquad\qquad\qquad\qquad\qquad\qquad \downarrow \phi_{MWN}$$

$$Hom_S((W' \otimes_R M), N) \xrightarrow{\ Hom_S((h \otimes_R M), N)\ } Hom_S((W \otimes_R M), N)$$

also commutes; hence the isomorphism $\phi = \phi_{MWN}$ is *natural in W*. Stated formally:

20.6. Proposition. *For every triple of modules* $({}_R M, {}_S W_R, {}_S N)$ *there is an isomorphism*

$$\phi : Hom_R(M, Hom_S(W, N)) \to Hom_S((W \otimes_R M), N),$$

defined via

$$\phi(\gamma)(w \otimes m) = [\gamma(m)](w),$$

that is natural in each of the three variables M, N, and W. ☐

For the Hom functors we have

20.7. Proposition. *For every triple of modules* $({}_R M, N_S, {}_R U_S)$ *there is an isomorphism*

$$\eta : Hom_R(M, Hom_S(N, U)) \to Hom_S(N, Hom_R(M, U)),$$

defined via

$$[\eta(\gamma)](n) : m \mapsto [\gamma(m)](n),$$

that is natural in each of the three variables M, N, and U.

Proof. The inverse of η is given by

$$[\eta^{-1}(\alpha)](m) : n \mapsto [\alpha(n)](m).$$

We leave the details as an exercise. ☐

Because of the following proposition it is often said that the formation of tensor products is associative.

20.8. Proposition. *For every triple of modules* $(M_R, {}_R W_S, {}_S N)$ *there is an isomorphism*

$$v : M \otimes_R (W \otimes_S N) \to (M \otimes_R W) \otimes_S N,$$

defined via

$$v: m \otimes_R (w \otimes_S n) \mapsto (m \otimes_R w) \otimes_S n,$$

that is natural in each of the three variables M, W, and N.

Proof. For each $m \in M$ the map $\beta_m: W \times N \to (M \otimes_R W) \otimes_S N$ defined by $\beta_m(w, n) = (m \otimes_R w) \otimes_S n$ is clearly S-balanced. Thus, for each $n \in M$ there is a unique homomorphism

$$v_m: W \otimes_R N \to (M \otimes_R W) \otimes_S N$$

such that

$$v_m(\Sigma_i w_i \otimes_S n_i) = \Sigma_i((m \otimes_R w_i) \otimes_S n_i).$$

The map $\gamma: M \times (W \otimes_S N) \to (M \otimes_R W) \otimes_S N$ defined by

$$\gamma: (m, \Sigma_i w_i \otimes_S n_i) = v_m(\Sigma_i w_i \otimes_S n_i)$$

is clearly R-balanced. So there is a homomorphism

$$v: M \otimes_R (W \otimes_S N) \to (M \otimes_R W) \otimes_S N$$

such that

$$v(m \otimes_R (w \otimes_S n)) = v_m(w \otimes_S n) = (m \otimes_R w) \otimes_S n.$$

With a similar argument it can be shown that there is a homomorphism

$$\mu_{MWN} = \mu: (M \otimes_R W) \otimes_S N \to M \otimes_R (W \otimes_S N)$$

such that

$$\mu((m \otimes_R w) \otimes_S n) = m \otimes_R (w \otimes_S n).$$

It is now easy to show that μ is an inverse of v, whence v is an isomorphism. Suppose next that $f: M_R \to M'_R$. Then it is evident that

$$\begin{array}{ccc} (M \otimes_R W) \otimes_S N & \xrightarrow{(f \otimes W) \otimes N} & (M' \otimes_R W) \otimes_S N \\ {\scriptstyle \mu_{MWN}} \downarrow & & \downarrow {\scriptstyle \mu_{M'WN}} \\ M \otimes_R (W \otimes_S N) & \xrightarrow{f \otimes (W \otimes N)} & M' \otimes_R (W \otimes_S N) \end{array}$$

commutes, whence μ is natural in M. To complete the proof it is simple to check the naturality of μ in the other two variables. □

If $_R U_T$ and $_S W_R$ are bimodules, then there are the two functors

$$Hom_R(U, Hom_S(W, \_)): {}_S\mathsf{M} \to {}_T\mathsf{M}$$

and

$$Hom_S((W \otimes_R U), \_): {}_S\mathsf{M} \to {}_T\mathsf{M}.$$

Viewed as functors to the category $_\mathbb{Z}\mathsf{M}$, they are naturally isomorphic (20.6). The fact is that they are isomorphic as functors to $_T\mathsf{M}$; to see this it will suffice to show that for each $N \in {}_S\mathsf{M}$

$$\phi = \phi_{UWN} : Hom_R(U, Hom_S(W, N)) \to Hom_S((W \otimes_R U), N)$$

is a T-homomorphism. But that ϕ is a T-homomorphism is a trivial consequence of the definition of the T-action on these modules. However, it is also a consequence of (20.4.2) and the fact that ϕ defines a natural isomorphism between the functors

$$Hom_R(\_, Hom_S(W, N)) :_R M \to M_{\mathbb{Z}}$$

$$Hom_S(W \otimes_R \_), N) :_R M \to M_{\mathbb{Z}}.$$

Of course similar observations apply in the other variables and to the natural transformations of (20.7) and (20.8).

Given a natural transformation $\eta : F \to G$ we often are concerned with those modules M for which η_M is an isomorphism (or merely monic or epic). The following lemma tells us that these classes of modules are closed under finite direct sums and direct summands.

20.9. Lemma. *Let* C *and* D *be full subcategories of the categories of left or right modules over rings R and S. Let F and G be additive functors from* C *to* D *and let $\eta : F \to G$ be a natural transformation. If*

$$0 \to M' -\oplus\to M -\oplus\to M'' \to 0$$

is split exact in C, *then η_M is monic (epic) if and only if both $\eta_{M'}$ and $\eta_{M''}$ are monic (epic).*

Proof. Consider the commutative diagrams with, by (16.2), split exact rows

$$0 \to F(M') -\oplus\to F(M) -\oplus\to F(M'') \to 0$$
$$\eta_{M'}\downarrow \qquad \eta_M\downarrow \qquad \eta_{M''}\downarrow$$
$$0 \to G(M') -\oplus\to G(M) -\oplus\to G(M'') \to 0$$

and

$$0 \to F(M'') -\oplus\to F(M) -\oplus\to F(M') \to 0$$
$$\eta_{M''}\downarrow \qquad \eta_M\downarrow \qquad \eta_{M'}\downarrow$$
$$0 \to G(M'') -\oplus\to G(M) -\oplus\to G(M') \to 0$$

obtained from

$$0 \to M' -\oplus\to M -\oplus\to M'' \to 0$$

and

$$0 \to M'' -\oplus\to M -\oplus\to M' \to 0. \qquad \square$$

Now (20.9) implies that if $\eta_{M_1}, ..., \eta_{M_n}$ are isomorphisms, then so is $\eta_{M_1\oplus...\oplus M_n}$. Therefore, if η_R is an isomorphism, then so is η_P for every finitely generated projective R-module P. Using this fact we derive the following two natural isomorphisms:

20.10. Proposition. *Given modules $_SP$, $_SU_T$ and $_TN$ there is a homomorphism, natural in P, U and N,*

$$\eta : Hom_S(P, U) \otimes_T N \to Hom_S(P, (U \otimes_T N))$$

defined via

$$\eta(\gamma \otimes_T n) : p \mapsto \gamma(p) \otimes_T n.$$

If $_SP$ is finitely generated and projective, then η is an isomorphism.

Proof. It is tedious but not difficult to check that η is a $\mathbb{Z}$-homomorphism that is natural in all three variables. Now for each $_SU_T$ and $_TN$ we have by (20.1.1) and (20.5.3)

$$Hom_S(S, U) \otimes_T N \cong U \otimes_T N \cong Hom_S(S, (U \otimes_T N))$$

via

$$\gamma \otimes_T n \mapsto \gamma(1) \otimes_T n \mapsto \rho(\gamma(1) \otimes_T n).$$

But for all $s \in S$

$$\eta(\gamma \otimes_T n)(s) = \gamma(s) \otimes_T n = s(\gamma(1) \otimes_T n)$$
$$= [\rho(\gamma(1) \otimes_T n)](s).$$

Thus

$$\eta : Hom_S(S, U) \otimes_T N \to Hom_S(S, (U \otimes_T N))$$

is the composite of these isomorphisms, and so is itself an isomorphism. So by (20.9) and (17.3)

$$\eta : Hom_S(P, U) \otimes_T N \to Hom_S(P, (U \otimes_T N))$$

is an isomorphism for every finitely generated projective $_SP$.   □

Similar arguments can be used to prove:

20.11. Proposition. *Given modules P_R, $_TU_R$ and $_TN$ there is a homomorphism, natural in P, U and N,*

$$v : P \otimes_R Hom_T(U, N) \to Hom_T(Hom_R(P, U), N)$$

defined via

$$v(p \otimes_R \gamma) : \delta \mapsto \gamma(\delta(p)).$$

If P_R is finitely generated and projective, then v is an isomorphism.   □

The U-dual Functors

Let $_RU_S$ be a bimodule. Then the pair of contravariant additive functors

$$Hom_R(-, {_RU_S}) : {_R}\mathsf{M} \to \mathsf{M}_S \qquad \text{and} \qquad Hom_S(-, {_RU_S}) : \mathsf{M}_S \to {_R}\mathsf{M}$$

is called the *U-dual*. For brevity we write

$$(\ )^* = Hom(-, {}_RU_S)$$

to denote either of these functors. So if $M_1 \xrightarrow{f} M_2$ in ${}_R\mathsf{M}$ then

$$M_2{}^* \xrightarrow{f^*} M_1{}^*$$

in M_S and

$$M_1{}^{**} \xrightarrow{f^{**}} M_2{}^{**}$$

in ${}_R\mathsf{M}$. The module M^* is said to be the *U-dual* of M and the map f^* is called the *U-dual* of f. Also M^{**} and f^{**} are called the *double dual* of M and f, respectively. For each M in ${}_R\mathsf{M}$ or M_S

$$[\sigma_M(m)](\gamma) = \gamma(m) \qquad (m \in M, \ \gamma \in M^*)$$

defines the *evaluation map*

$$\sigma_M : M \to M^{**}.$$

this evaluation map is easily seen to be an R-homomorphism if $M \in {}_R\mathsf{M}$ or an S-homomorphism if $M \in \mathsf{M}_S$. Moreover if $M_1 \xrightarrow{f} M_2$ then for all $m \in M_1$, $\gamma \in M_2{}^*$

$$\begin{aligned}
[f^{**}(\sigma_{M_1}(m))](\gamma) &= (\sigma_{M_1}(m) \circ f^*)(\gamma) \\
&= [\sigma_{M_1}(m)](\gamma \circ f) = \gamma(f(m)) \\
&= [\sigma_{M_2}(f(m))](\gamma)
\end{aligned}$$

so that the diagram

$$\begin{array}{ccc}
M_1 & \xrightarrow{\ f\ } & M_2 \\
{\scriptstyle \sigma_{M_1}}\downarrow & & \downarrow{\scriptstyle \sigma_{M_2}} \\
M_1{}^{**} & \xrightarrow{\ f^{**}\ } & M_2{}^{**}
\end{array}$$

commutes. Thus the evaluation maps yield natural transformations

$$\sigma : 1_{{}_R\mathsf{M}} \to ((\ )^*)^*$$

and

$$\sigma : 1_{\mathsf{M}_S} \to ((\ )^*)^*.$$

A module M is said to be *U-reflexive* in case σ_M is an isomorphism. If σ_M is monic, then M is *U-torsionless*.

A module M is *U*-torsionless if and only if U cogenerates M. In fact from the definition of σ_M we see that $m \in Ker\,\sigma_M$ if and only if $m \in Ker\,\gamma$ for all $\gamma : M \to U$. In other words

20.12. Proposition. *Let ${}_RU_S$ be a bimodule, and let M be a left R- or a right S-module. Then*

$$Ker\,\sigma_M = Rej_M(U). \qquad \qquad \square$$

We know of no such handy tests for reflexivity. However, by Lemma (20.9) the class of U-reflexive modules is closed under direct summands and finite direct sums.

20.13. Proposition. *Let* $M \cong M_1 \oplus \ldots \oplus M_n$. *Then* M *is* U-reflexive (tor-sionless) *if and only if each of* $M_1, \ldots, M_n$ *is* U-reflexive (torsionless).   □

20.14. Proposition. *Let* $_RU_S$ *be a bimodule and let* M *be a module in* $_R\mathsf{M}$ *or* M_S. *Then*

(1) $\sigma_M^* \circ \sigma_{M^*} = 1_{M^*}$.
(2) M^* *is* U-torsionless.
(3) *If* M *is* U-reflexive, *then* M^* *is* U-reflexive.

Proof. First observe that $\sigma_{M^*} : M^* \to M^{***}$ and $\sigma_M^* : M^{***} \to M^*$. If $\gamma \in M^*$, then for all $m \in M$

$$\sigma_M^*(\sigma_{M^*}(\gamma))(m) = (\sigma_{M^*}(\gamma) \circ \sigma_M)(m)$$
$$= [\sigma_M(m)](\gamma) = \gamma(m).$$

This proves (1). From (1) it follows that σ_{M^*} is a (split) monomorphism so (2) also holds. For (3), suppose that σ_M is an isomorphism. Then so is σ_M^*. But then (1) forces σ_{M^*} to be an isomorphism.   □

There is a useful test for the reflexivity of the regular modules $_RR$ and S_S.

20.15. Proposition. *Let* $_RU_S$ *be a bimodule and let*

$$\lambda : R \to End(U_S)$$

be left multiplication. Then $\sigma_{_RR}$ *is injective or surjective if and only if* λ *is.*

Proof. We know by (20.1) that there is an R-isomorphism

$$\rho_U : U \to Hom_R(R, U) = (_RR)^*.$$

By naturality (20.4.2) this is also an S-isomorphism. Thus since ( )* is a functor,

$$\rho_U^* : (_RR)^{**} \to (U_S)^*$$

is an isomorphism. But

$$\rho_U^*(\sigma_R(r))(u) = \sigma_R(r)(\rho_U(u)) = \rho_U(u)(r) = \lambda(r)(u)$$

so that

$$\rho_U^* \circ \sigma_R = \lambda.$$ □

20.16. Corollary. *Let* $_RU_S$ *be a bimodule. Then* $_RR$ *and* S_S *are* U-reflexive *if and only if* $_RU_S$ *is a faithfully balanced bimodule.*   □

One of the most important duals is the $_RR_R$ dual. We shall denote this dual by

$$( )^{\circledast} = Hom_R(-, R).$$

In the event that R is a field this is just our old friend from linear algebra, the vector-space dual.

If e is an idempotent in R then, as we saw in Proposition (4.6),

$$Re^{\circledast} \cong eR \qquad \text{and} \qquad Re^{\circledast\circledast} \cong Re.$$

Applying the preceding results we have, more generally,

20.17. Proposition. *Let P be a finitely generated projective R-module. Then*

(1) *P is R-reflexive.*

(2) *$P^{\circledast}$ is finitely generated and projective over R.*

Proof. Suppose P is a finitely generated projective left R-module. Then, by (17.3) there exist P' and n such that $P \oplus P' \cong {}_RR^{(n)}$. But by (4.11) $\lambda:R \to End(R_R)$ is an isomorphism. So we infer from (20.15) and (20.9) that P is R-reflexive. Moreover,

$$P^{\circledast} \oplus P'^{\circledast} \cong (P \oplus P')^{\circledast} \cong (R^{(n)})^{\circledast} \cong (R^{\circledast})^{(n)} \cong R^{(n)},$$

so that $P^{\circledast}$ is a finitely generated projective right R-module. □

20. Exercises

1. Let C, D, and E be categories, let $F, F', F'':\mathsf{C} \to \mathsf{D}$ and $G, G':\mathsf{D} \to \mathsf{E}$ be functors. Prove:

   (1) If $\eta:F \to F'$ and $\mu:F' \to F''$ are natural transformations (isomorphisms), then so is their "composite" $\mu \circ \eta:F \to F''$ defined by $(\mu \circ \eta)_M = \mu_M \circ \eta_M$.

   (2) If $\eta:F \to F'$ is a natural isomorphism, then its "inverse" $\eta^{-1}:F' \to F$ defined by $(\eta^{-1})_M = (\eta_M)^{-1}$ is a natural isomorphism.

   (3) $F \cong F$; $F \cong F' \Rightarrow F' \cong F$; and $F \cong F'$ and $F' \cong F'' \Rightarrow F \cong F''$.

   (4) $F \cong F'$ and $G \cong G' \Rightarrow G \circ F \cong G' \circ F'$.

2. Let R be commutative. Prove that for each $M \in {}_R\mathsf{M}$ the functors $(M \otimes_R \_)$ and $(\_ \otimes_R M)$ from ${}_R\mathsf{M}$ to ${}_R\mathsf{M}$ are isomorphic.

3. Let $\phi:R \to S$ be a ring homomorphism so that, via ϕ, we have bimodules ${}_RS_S$ and ${}_SS_R$. Let $F_{\phi}:{}_S\mathsf{M} \to {}_R\mathsf{M}$ be the change of rings functor (of Exercise (4.15)) induced by ϕ. Consider the functors

   $$T_{\phi} = ({}_SS \otimes_R \_) \qquad H_{\phi} = Hom_R({}_RS_S, \_)$$

   from ${}_R\mathsf{M}$ to ${}_S\mathsf{M}$. Prove that

   (1) $Hom_S({}_SS_R, \_) \cong F_{\phi} \cong ({}_RS \otimes_S \_)$.

   (2) $T_{\phi} \circ F_{\phi} \cong 1_{{}_S\mathsf{M}} \cong H_{\phi} \circ F_{\phi}$ if ϕ is surjective;

4. Let C and D be categories of modules with D closed under the formation of direct sums and products. Let $(F_{\alpha})_{\alpha \in A}$ be an indexed set of functors of the same variance from C to D. Define $\oplus_A F_{\alpha}$ and $\Pi_A F_{\alpha}$ from C to D by

$$\oplus_A F_\alpha : M \mapsto \oplus_A F_\alpha(M) \qquad \oplus_A F_\alpha : f \mapsto \oplus_A F_\alpha(f)$$

$$\Pi_A F_\alpha : M \mapsto \Pi_A F_\alpha(M) \qquad \Pi_A F_\alpha : f \mapsto \Pi_A F_\alpha(f).$$

Prove that these are additive functors and that $\eta : \oplus_A F_\alpha \to \Pi_A F_\alpha$ is a natural transformation where $\eta_M : \oplus_A F_\alpha(M) \to \Pi_A F_\alpha(M)$ is the inclusion map.

5. Let $\mathbf{C}$ be a full subcategory of $_R\mathbf{M}$ and let $F : \mathbf{C} \to \mathbf{C}$ be isomorphic to the identity functor $F \cong 1_\mathbf{C}$ via $\eta : 1_\mathbf{C} \to F$. Prove that for all $M, N \in \mathbf{C}$ the restriction $F : Hom_R(M, N) \to Hom_R(F(M), F(N))$ of F is an isomorphism whose inverse is given by $g \mapsto \eta_N^{-1} \circ g \circ \eta_M$ for each

$$g \in Hom_R(F(M), F(N)).$$

6. Let $\mathbf{C}$ be a full subcategory of $_R\mathbf{M}$ and let $_R U_S$ and $_R V_S$ be bimodules with $_R U$ and $_R V$ both in $\mathbf{C}$. Consider the functors $Hom_R(U_S, \_)$ and $Hom_R(V_S, \_)$ from $\mathbf{C}$ to $_S\mathbf{M}$ and $Hom_R(\_, U_S)$ and $Hom_R(\_, V_S)$ from $\mathbf{C}$ to $\mathbf{M}_S$. Prove that:

(1) If $\phi : Hom_R(U_S, \_) \to Hom_R(V_S, \_)$ is a natural isomorphism, then $\phi_U(1_U) : V \to U$ and $\phi_V^{-1}(1_V) : U \to V$ are inverse (R, S)-isomorphisms.

(2) If $\phi : Hom_R(\_, U_S) \to Hom_R(\_, V_S)$ is a natural isomorphism, then $_R U_S \cong {}_R V_S$.

7. Let $\mathbf{C}$ and $\mathbf{D}$ be full subcategories of $_R\mathbf{M}$ and $_S\mathbf{M}$, respectively. Let $T : \mathbf{C} \to \mathbf{D}$ and $H : \mathbf{D} \to \mathbf{C}$ be additive covariant functors. Then T is a *left adjoint* of H and H is a *right adjoint* of T, or simply (T, H) is an *adjoint pair* in case for each $M \in \mathbf{C}$ and $N \in \mathbf{D}$ there is a $\mathbb{Z}$-isomorphism

$$\phi = \phi_{MN} : Hom_R(M, H(N)) \to Hom_S(T(M), N)$$

that is natural in both M and N. Prove the following version of Kan's Theorem:

(1) If T and T' are both left adjoints of H, then $T \cong T'$. [Hint: Exercise (20.6).]

(2) Let $\mathbf{C} = {}_R\mathbf{M}$ and $\mathbf{D} = {}_S\mathbf{M}$ and let (T, H) be an adjoint pair. Let $_S U_R$ be the canonical bimodule $_S U_R = T(_R R_R)$. Then $H \cong Hom_S(U_R, \_)$ and $T \cong (_S U \otimes_R \_)$. In particular, if H' is a right adjoint of T, then $H \cong H'$.

8. An additive functor $T : {}_R\mathbf{M} \to {}_S\mathbf{M}$ is *faithful* in case $T(f) = 0$ implies $f = 0$. So for example, if $_R C$ is a cogenerator, then $Hom_R(\_, C)$ is a faithful functor. Let $H : {}_S\mathbf{M} \to {}_R\mathbf{M}$ be a right adjoint of $T : {}_R\mathbf{M} \to {}_S\mathbf{M}$ (see Exercise (20.7)). Prove that:

(1) If $_S N$ is injective and T is exact, then $_R H(N)$ is injective. [Hint: The functor $Hom_R(\_, H(N))$ is exact.]

(2) If $_S N$ is a cogenerator and $_R H(N)$ is injective, then T is exact. [Hint: (18.14).]

(3) If $_S N$ is a cogenerator and T is faithful, then $_R H(N)$ is a cogenerator.

(4) If $_R H(N)$ is a cogenerator for some $_S N$, then T is faithful.

(5) If $_R M$ is projective and H is exact, then $_S T(M)$ is projective.

(6) If $_R M$ is a generator and $_S T(M)$ is projective, then H is exact.

(7) If $_RM$ is a generator and H is faithful, then $_ST(M)$ is a generator.

(8) If $_ST(M)$ is a generator for some $_RM$, then H is faithful.

9. Given modules N_S and $_RU_S$, use the isomorphism

$$Hom_R(\_, Hom_S(N, {}_RU_S)) \cong Hom_S(N, Hom_R(\_, {}_RU_S))$$

of Proposition (20.7) to prove that:

(1) N_S projective and $_RU$ injective implies $Hom_S(N, {}_RU_S)$ injective.

(2) N_S a generator and $Hom_S(N, {}_RU_S)$ injective implies $_RU$ injective.

(3) N_S a generator and $_RU$ a cogenerator implies $Hom_S(N, {}_RU_S)$ a cogenerator.

(4) $Hom_S(N, {}_RU_S)$ a cogenerator implies $_RU$ a cogenerator.

10. Given modules $_SN$ and $_RW_S$, use the isomorphism

$$(\_ \otimes_R (W \otimes_S N)) \cong ((\_ \otimes_R W) \otimes_S N)$$

of Proposition (20.8) to prove:

(1) $_SN$ flat and $_RW$ flat implies $_RW \otimes {}_SN$ flat.

(2) $_SN$ completely faithful and $_RW \otimes {}_SN$ flat implies $_RW$ flat.

(3) $_SN$ and $_RW$ completely faithful implies $_RW \otimes {}_SN$ completely faithful.

(4) $_RW \otimes {}_SN$ completely faithful implies $_RW$ completely faithful.

11. Let $x_1, \ldots, x_n \in P_R$ and $f_1, \ldots, f_n \in P^* = Hom_R(P, R)$. Prove that $(x_1, \ldots, x_n)$, $(f_1, \ldots, f_n)$ is a dual basis for P_R iff $(f_1, \ldots, f_n)$, $(\sigma(x_1), \ldots, \sigma(x_n))$ is a dual basis for $_RP^*$. (Exercise (17.11).)

12. Let $(\ )^{\circledast}$ denote the $_RR_R$ dual so that $M^{\circledast} = Hom_R(M, R)$. For each pair N_R, M_R of right R-modules there is a map

$$\theta : N \otimes_R M^{\circledast} \to Hom_R(M, N)$$

such that $\theta(n \otimes \gamma) : m \mapsto n\gamma(m)$. Prove that:

(1) θ is a $\mathbb{Z}$-homomorphism natural in M and N.

(2) The following are equivalent:

    (a) P_R is finitely generated projective;

    (b) $\theta : (P \otimes_R (\_)^{\circledast}) \to Hom_R(\_, P)$ is a natural isomorphism;

    (c) $\theta : (\_ \otimes_R P^{\circledast}) \to Hom_R(P, \_)$ is a natural isomorphism;

    (d) $\theta : P \otimes_R P^{\circledast} \to Hom_R(P, P)$ is a $\mathbb{Z}$-isomorphism.

[Hint: (a) $\Rightarrow$ (b). Let $(x_i), (f_i)$ be a dual basis for P_R. Let $g \in Hom_R(M, P)$. Then $g = \theta(\Sigma_i (x_i \otimes f_i g)$. If $\theta(\Sigma_j y_j \otimes \gamma_j) = 0$, then write y_j in terms of the dual basis to infer

$$\Sigma_j y_j \otimes \gamma_j = \Sigma_i (x_i \otimes \Sigma_j f_i(y_j)\gamma_j) = \Sigma_i (x_i \otimes 0) = 0.$$

(d) $\Rightarrow$ (a). Consider $\theta^{-1}(1_P)$ and the Dual Basis Lemma. (Exercise (17.11).).]

13. Prove that every finitely presented flat module is projective. In particular, every finitely generated flat module over a noetherian ring is projective. [Hint: Let $P_2 \to P_1 \to V_R \to 0$ be exact with the P_i finitely generated projective. Apply the functors $V \otimes_R (\_)^*$ and $Hom_R(\_, V)$ and Exercises (16.4) and (20.12).]

14. Prove that a ring R is von Neumann regular iff every finitely presented left R-module is projective. [Hint: Exercise (19.16).]

15. (1) Prove that if $_RE$ is injective and I is an ideal of R, then $Hom_R(I_R, E) \cong E/\mathbf{r}_E(I)$.
    (2) Prove that if R is right artinian and left hereditary, then it is right hereditary. [Hint: Exercises (20.13), (19.14), and (18.10).]

16. Let $_RU_S$ be a bimodule. Let $_R\mathscr{R}(U)$ denote the class of U-reflexive left R-modules. Then $_R\mathscr{R}(U)$ is closed under the formation of finite direct sums and direct summands (20.13). Let

$$0 \to {}_RK \to {}_RM \to {}_RN \to 0$$

be exact. Prove that
(1) If $_RU$ and U_S are injective and N is U-torsionless, then M is reflexive iff K and N are reflexive.
(2) If $_RU$ and U_S are injective and $_RU$ is a cogenerator, then $_R\mathscr{R}(U)$ is closed under submodules, epimorphic images, and extensions (of modules in $_R\mathscr{R}(U)$ by modules in $_R\mathscr{R}(U)$).

17. Let C and D be full subcategories of $_RM$ and $_SM$, say. Let F and G be additive functors from C to D that preserve direct sums and let $\eta : F \to G$ be a natural transformation. Let $(M_\alpha)_{\alpha \in A}$ be an indexed class in C with a direct sum $(M, (\iota_\alpha)_{\alpha \in A})$ in C. Prove that η_M is an isomorphism iff each η_{M_α} ($\alpha \in A$) is an isomorphism.

18. Let $_RP$ be finitely generated and projective with $S = End(_RP)$, and let $C(P)$ denote the full subcategory of $_RM$ whose objects are modules M such that there are sets X and Y and an exact sequence

$$P^{(Y)} \to P^{(X)} \to M \to 0.$$

Prove that if $H = Hom_R(P, \underline{\ \ })$ and $T = (P \otimes_S \underline{\ \ })$, then

$$H : C(P) \to {}_SM \qquad \text{and} \qquad T : {}_SM \to C(P)$$

and these functors define an equivalence of categories in the sense that $H \circ T \cong 1_{{}_SM}$ and $T \circ H \cong 1_{C(P)}$. [Hint: The first isomorphism of functors follows from (20.10); the second uses Exercise 16.3 and the Five Lemma, among other things.]

Chapter 6

Equivalence and Duality for Module Categories

So far our emphasis has been on studying rings in terms of the module categories they admit—that is, in terms of the representations of the rings as endomorphism rings of abelian groups. As we shall see the Wedderburn Theorem for simple artinian rings can be interpreted as asserting that a ring R is simple artinian if and only if the category $_R\mathsf{M}$ is "the same" as the category $_D\mathsf{M}$ for some division ring D. On the other hand, if D is a division ring, then the theory of duality from elementary linear algebra asserts that the categories $_D\mathsf{FM}$ and FM_D of finitely generated left D-vector spaces and right D-vector spaces are "duals" of one another.

These are examples of two related general theories that are of truly fundamental importance to the study of rings and modules. Although historically they were not studied in the context of categories and functors, it is in that context that their significance and their simplicity are clear. The principal work on the subject was done by Morita [58a]. In this chapter we treat the basic theory including what are sometimes known as the "Morita Theorems".

§21. Equivalent Rings

Let C and D be arbitrary categories. Then a covariant functor

$$F : \mathsf{C} \to \mathsf{D}$$

is a *category equivalence* in case there is a functor (necessarily covariant)

$$G : \mathsf{D} \to \mathsf{C}$$

and natural isomorphisms

$$GF \cong 1_\mathsf{C} \quad \text{and} \quad FG \cong 1_\mathsf{D}.$$

A functor G with this property (also a category equivalence) is called an *inverse equivalence* of F. Two categories are *equivalent* in case there exists a category equivalence from one to the other. We write

$$\mathsf{C} \approx \mathsf{D}$$

in case C and D are equivalent. It is easy to check that this defines an equivalence relation on the class of all categories. (See Exercise (21.2).)

For the remainder of this section (excluding the exercises) our interest will be restricted to module categories. Thus, we shall revert to our earlier convention and assume that all functors between such categories are additive. Thus for two such categories to be equivalent there must be an additive equivalence from one to the other.

Definitions and Notation

Two rings R and S are (*Morita*) *equivalent*, abbreviated

$$R \approx S$$

in case

$$_R\mathsf{M} \approx {}_S\mathsf{M},$$

i.e., in case there are additive equivalences between these categories of modules. As we shall see in §22, the categories $_R\mathsf{M}$ and $_S\mathsf{M}$ are equivalent if and only if M_R and M_S are equivalent.

Since the study of the properties of equivalent pairs of rings entails a fair amount of notation, it will be especially useful to pause to assemble most of it in one place.

21.1. Let R and S be a pair of equivalent rings. Specifically, assume that

(1) $\qquad\qquad F:{}_R\mathsf{M} \to {}_S\mathsf{M} \qquad$ and $\qquad G:{}_S\mathsf{M} \to {}_R\mathsf{M}$

are inverse (additive) equivalences. In particular,

$$GF \cong 1_{_R\mathsf{M}} \qquad \text{and} \qquad FG \cong 1_{_S\mathsf{M}} \; ;$$

that is, there exist natural isomorphisms

(2) $\qquad\qquad \eta:GF \to 1_{_R\mathsf{M}} \qquad$ and $\qquad \zeta:FG \to 1_{_S\mathsf{M}} \; .$

This means (see §20), in the case of η, that for each $_R M$ there is an isomorphism $\eta_M:GF(M) \to M$ in $_R\mathsf{M}$ such that for each M, M' in $_R\mathsf{M}$ and each $f:M \to M'$ in $_R\mathsf{M}$, the diagram

(3)

$$
\begin{array}{ccc}
M & \xrightarrow{\;f\;} & M' \\
{\scriptstyle \eta_M}\big\uparrow & & \big\uparrow{\scriptstyle \eta_{M'}} \\
GF(M) & \xrightarrow{GF(f)} & GF(M')
\end{array}
$$

commutes. (Of course, parallel remarks apply to ζ.) Now for each $_R M$ in $_R\mathsf{M}$ and each $_S N$ in $_S\mathsf{M}$, there are $\mathbb{Z}$-homomorphisms

(4)
$$\phi = \phi_{MN}:Hom_S(N, F(M)) \to Hom_R(G(N), M)$$
$$\theta = \theta_{MN}:Hom_S(F(M), N) \to Hom_R(M, G(N))$$

defined via

$$\phi_{MN}:\gamma \mapsto \eta_M \circ G(\gamma)$$
$$\theta_{MN}:\delta \mapsto G(\delta) \circ \eta_M^{-1}.$$

The natural isomorphism ζ determines a pair of homomorphisms similar to ϕ and θ; however, we have no need to introduce special notation for these here. In practice, there is almost never any real ambiguity about the domain of the η_M, ζ_N, ϕ_{MN}, and θ_{MN}. Thus, for the most part we shall clean up our notation by omitting these subscripts.

It should now be plausible from say (3) above, that if $R \approx S$, then the behaviour of $_R M$ and $_S M$ is the same "to within isomorphism". To expand on this we first prove:

21.2. Proposition. *Let* $F : {}_R M \to {}_S M$ *be a category equivalence. Then for each* M, M' *in* $_R M$ *the restriction of* F *to* $Hom_R(M, M')$ *is an abelian group isomorphism*

$$F : Hom_R(M, M') \to Hom_S(F(M), F(M'))$$

such that $F(f)$ *is an epimorphism (monomorphism) in* $_S M$ *if and only if* f *is an epimorphism (monomorphism) in* $_R M$. *Moreover, if* $M \neq 0$, *then this restriction*

$$F : End(_R M) \to End(_S F(M))$$

is a ring isomorphism.

Proof. Since F is additive, these restrictions are abelian group homomorphisms. The latter is a ring homomorphism by (20.3). To finish the proof we shall adopt the notation of (21.1). Then clearly for each M and M' in $_R \mathcal{M}$

$$H : Hom_S(F(M), F(M')) \to Hom_R(M, M')$$

defined by

$$H : g \to \eta_{M'} G(g) \eta_M^{-1}$$

is a $\mathbb{Z}$-homomorphism. Moreover, it is monic, for if $H(g) = 0$, then $G(g) = 0$, so

$$g = \zeta_{F(M')} FG(g) \zeta_{F(M)}^{-1} = 0.$$

But now, for all $f \in Hom_R(M, M')$

$$HF(f) = \eta_{M'} GF(f) \eta_M^{-1} = f.$$

It follows that H is an epimorphism. Thus H is an isomorphism with inverse F. Therefore F is an isomorphism. Now it is clear from (21.1.3) that f is monic (epic) if and only if $GF(f)$ is monic (epic). So suppose f is monic and that for some h in $_S M$

$$F(f)h = 0.$$

Then since G is an additive functor and $GF(f)$ is monic, $GF(f)G(h) = 0$, and hence $G(h) = 0$. But then $FG(h) = 0$, so from the version of (21.1.3) for ζ, it is clear that $h = 0$, whence $F(f)$ is monic (3.4.d). The remainder of the proof is entirely similar and will be omitted. □

The Fundamental Lemma

One of the most important facts concerning equivalent rings is that the homomorphisms ϕ and θ of (21.1) are natural isomorphisms. That is, the pairs (G, F) and (F, G) of (21.1) are *adjoint* pairs of functors (see Exercise (20.7)). This adjoint relationship of equivalences F and G provides a very powerful bit of machinery. It is described in the following lemma.

21.3. Lemma. *Let R and S be equivalent rings. Then, in the notation of (21.1). the homomorphisms*

$$\phi : Hom_S(N, F(M)) \to Hom_R(G(N), M)$$

$$\theta : Hom_S(F(M), N) \to Hom_R(M, G(N))$$

are isomorphisms natural in each variable. In particular, for each

$$\gamma \in Hom_S(N_1, F(M_1)), \qquad \delta \in Hom_S(F(M_2), N_2)$$

$$\bar{\gamma} \in Hom_R(G(N_1), M_1), \qquad \bar{\delta} \in Hom_S(M_2, G(N_2)),$$

and for each

$$h : M_1 \to M_2, \qquad k : N_2 \to N_1$$

we have

(1) $\phi(F(h)\gamma k) = h\phi(\gamma)G(k)$,
(2) $\theta(k\delta F(h)) = G(k)\theta(\delta)h$,
(3) $\phi^{-1}(h\bar{\gamma}G(k)) = F(h)\phi^{-1}(\bar{\gamma})k$,
(4) $\theta^{-1}(G(k)\bar{\delta}h) = k\theta^{-1}(\bar{\delta})F(h)$.

Finally, $\phi(\gamma)$ is a monomorphism (epimorphism) if and only if γ is a monomorphism (epimorphism), and $\theta(\delta)$ is a monomorphism (epimorphism) if and only if δ is a monomorphism (epimorphism).

Proof. The $\mathbb{Z}$-homomorphism induced by G

$$G : Hom_S(N, F(M)) \to Hom_R(G(N), GF(M))$$

is an isomorphism by (21.2). Since $\eta_M : GF(M) \to M$ is an isomorphism, so is

$$Hom_R(G(N), \eta_M) : Hom_R(G(N), GF(M)) \to Hom_R(G(N), M)$$

(see (16.2)). Thus, since it is the composite of these two maps,

$$\phi : Hom_S(N, F(M)) \to Hom_R(G(N), M)$$

is a $\mathbb{Z}$-isomorphism. Also, with h, k, and γ as given in the hypothesis,

$$\phi(F(h)\gamma k) = \eta_{M_2} GF(h)G(\gamma)G(k)$$

$$= \eta_{M_2} GF(h)\eta_{M_1}^{-1} \eta_{M_1} G(\gamma)G(k)$$

$$= h\phi(\gamma)G(k).$$

That θ is an isomorphism and that the identities (2), (3), and (4) hold are proved similarly and therefore will be omitted. The equations (1) and (2) mean that ϕ and θ are natural in both M and N. For instance, taking $k = 1_N$ we see from (1) that for each $M_1 \xrightarrow{h} M_2$ in $_R\mathsf{M}$ the diagram

$$
\begin{array}{ccc}
Hom_S(N, F(M_1)) & \xrightarrow{\ Hom_S(N, F(h))\ } & Hom_S(N, F(M_2)) \\
{\scriptstyle \phi_{M,N}} \downarrow & & \downarrow {\scriptstyle \phi_{M_2N}} \\
Hom_R(G(N), M_1) & \xrightarrow{\ Hom_R(G(N), h)\ } & Hom_R(G(N), M_2)
\end{array}
$$

commutes.

For the final assertion, let $\gamma \in Hom_S(N, F(M))$. Then $\phi(\gamma) = \eta_M \circ G(\gamma)$. So since η_M is an isomorphism, $G(\gamma)$ is a monomorphism (epimorphism) if and only if $\phi(\gamma)$ is a monomorphism (epimorphism). But by (21.2) $G(\gamma)$ is monic (epic) if and only if γ is.   □

Remark: It should be observed that we can use ϕ and θ to "transform" certain diagrams in $_R\mathsf{M}$ (respectively, $_S\mathsf{M}$) to corresponding diagrams in $_S\mathsf{M}$ (respectively, $_R\mathsf{M}$). For example, (21.3.1) asserts that the composite

$$
\begin{array}{ccc}
N_1 & \xrightarrow{\ \gamma\ } & F(M_1) \\
{\scriptstyle k} \uparrow & & \downarrow {\scriptstyle F(h)} \\
N_2 & \xrightarrow{\ F(h)\gamma k\ } & F(M_2)
\end{array}
$$

is transformed by ϕ to

$$
\begin{array}{ccc}
G(N_1) & \xrightarrow{\ \phi(\gamma)\ } & M_1 \\
{\scriptstyle G(k)} \uparrow & & \downarrow {\scriptstyle h} \\
G(N_2) & \xrightarrow{\ \phi(F(h)\gamma k)\ } & M_2.
\end{array}
$$

Properties Preserved by Equivalence

Now we have the wherewithal to prove the basic properties of categorical equivalences. The first of these is that such equivalences "preserve exactness".

21.4. Proposition. *Let* $F : _R\mathsf{M} \to _S\mathsf{M}$ *be a category equivalence. Then a sequence*

$$0 \to M' \xrightarrow{f} M \xrightarrow{g} M'' \to 0$$

is (split) exact in $_R\mathsf{M}$ *if and only if the sequence*

$$0 \to F(M') \xrightarrow{F(f)} F(M) \xrightarrow{F(g)} F(M') \to 0$$

is (split) exact in $_S\mathsf{M}$.

Proof. We shall use the notation of (21.1). Then since η is a natural isomorphism, the diagram

$$0 \to \quad M' \quad \xrightarrow{\ f\ } \quad M \quad \xrightarrow{\ g\ } \quad M'' \quad \longrightarrow \quad 0$$

$$0 \to GF(M') \xrightarrow{\ GF(f)\ } GF(M) \xrightarrow{\ GF(g)\ } GF(M'') \to 0$$

commutes, and it follows that either row is (split) exact if and only if the other is. So to prove both implications of the proposition it will suffice to prove that F preserves (split) short exact sequences. The "split" part follows because F is additive (16.2). Let

$$0 \to M' \xrightarrow{f} M \xrightarrow{g} M'' \to 0$$

be exact in $_R\mathsf{M}$. Then by (21.2), $F(f)$ is monic, $F(g)$ is epic, and also $F(g)F(f) = F(gf) = 0$. Thus all that remains to prove is that $Ker\, F(g) \subseteq Im\, F(f)$. To this end let $K = Ker\, F(g)$ and let $i_K : K \to F(M)$ be the inclusion map. Then $\phi(i_K) : G(K) \to M$ and by (21.3.1) $g\phi(i_K) = \phi(F(g)i_K) = 0$. Thus $Im\, \phi(i_K) \subseteq Ker\, g = Im f$; and by The Factor Theorem (3.6.2) there is a $\bar{\gamma} \in Hom_R(G(K), M')$ such that $f\bar{\gamma} = \phi(i_K)$. Now applying (21.3.3) we have

$$i_K = \phi^{-1}(f\bar{\gamma}) = F(f)\phi^{-1}(\bar{\gamma}),$$

so that $Ker\, F(g) = Im\, i_K \subseteq Im\, F(f)$. $\qquad\qquad\qquad\qquad\qquad\qquad\square$

21.5. Proposition. *Let $F : _R\mathsf{M} \to _S\mathsf{M}$ be a category equivalence. Then*
(1) *A pair $(M, (p_\alpha)_{\alpha\in A})$ is a direct product of $(M_\alpha)_{\alpha\in A}$ if and only if $(F(M), (F(p_\alpha))_{\alpha\in A})$ is a direct product of $(F(M_\alpha))_{\alpha\in A}$;*
(2) *A pair $(M, (i_\alpha)_{\alpha\in A})$ is a direct sum of $(M_\alpha)_{\alpha\in A}$ if and only if $(F(M), (F(i_\alpha))_{\alpha\in A})$ is a direct sum of $(F(M_\alpha))_{\alpha\in A}$.*

Proof. We shall do (1); the proof of the other part is dual. Suppose then that $(M, (p_\alpha)_{\alpha\in A})$ is a product of $(M_\alpha)_{\alpha\in A}$, and suppose that in $_S\mathsf{M}$ there are homomorphisms $g_\alpha : N \to F(M_\alpha)$. Then in $_R\mathsf{M}$, these induce $\phi(g_\alpha) : G(N) \to M_\alpha$, so there exists a unique $f : G(N) \to M$ such that $\phi(g_\alpha) = p_\alpha f$ for each $\alpha \in A$. So by (21.3) $\phi^{-1}(f)$ is unique with the property that

$$g_\alpha = \phi^{-1}(p_\alpha f) = F(p_\alpha)\phi^{-1}(f) \qquad (\alpha \in A).$$

Conversely, suppose that $(F(M), (F(p_\alpha))_{\alpha\in A})$ is a product of $(F(M_\alpha))_{\alpha\in A}$, and suppose that in $_R\mathsf{M}$ there are homomorphisms $g_\alpha : K \to M_\alpha$. Then in $_S\mathsf{M}$ these induce $F(g_\alpha) : F(K) \to F(M_\alpha)$, so there is a unique homomorphism $g : F(K) \to F(M)$ such that

$$F(q_\alpha) = F(p_\alpha)g \qquad (\alpha \in A).$$

Finally, by (21.2) there is a unique $g' \in Hom_R(K, M)$ with $F(g') = g$, and we are done. $\qquad\qquad\qquad\qquad\qquad\qquad\qquad\qquad\qquad\qquad\qquad\square$

21.6. Proposition. *Let R and S be equivalent rings via an equivalence $F : _R\mathsf{M} \to _S\mathsf{M}$. Let M, M', and U be left R-modules. Then*
(1) *U is M-projective (M-injective) if and only if $F(U)$ is $F(M)$-projective ($F(M)$-injective);*
(2) *U is projective (injective) if and only if $F(U)$ is projective (injective);*

(3) *U generates (cogenerates) M if and only if $F(U)$ generates (cogenerates) $F(M)$;*

(4) *U is a generator (a cogenerator) (faithful) if and only if $F(U)$ is a generator (a cogenerator) (faithful);*

(5) *A monomorphism (epimorphism) $f: M \to M'$ is essential (superfluous) if and only if $F(f): F(M) \to F(M')$ is essential (superfluous);*

(6) *$f: M \to M'$ is an injective envelope (projective cover) if and only if $F(f): F(M) \to F(M')$ is an injective envelope (projective cover).*

Proof. We again adopt the notation of (21.1). Then for (1) suppose that U is M-projective, and that in $_S\mathsf{M}$ there is a diagram

$$
\begin{array}{c}
F(U) \\
\downarrow {\scriptstyle g} \\
F(M) \xrightarrow{\ f\ } N \longrightarrow 0
\end{array}
$$

with f an epimorphism. Then $\theta(f)$ is epic in $_R\mathsf{M}$, so there is an h such that

$$
\begin{array}{c}
U \\
{\scriptstyle h}\nearrow \quad \downarrow {\scriptstyle \theta(g)} \\
M \xrightarrow[\theta(f)]{} G(N) \longrightarrow 0
\end{array}
$$

commutes. Now, by (21.3.4), $g = \theta^{-1}(\theta(g)) = \theta^{-1}(\theta(f)h) = fF(h)$, whence $F(U)$ is $F(M)$-projective. We omit the rest of this proof.

(2) This is immediate from (1).

(3) This is an easy consequence of (21.4) and (21.5).

(4) This is by (3). (Note that U is faithful iff U cogenerates a generator (Exercise (8.3).)

(5) Suppose $g: F(M') \to N$ is a homomorphism in $_S\mathsf{M}$ such that $gF(f)$ is monic. Then by (21.3)

$$\phi(gF(f)) = \phi(g)f$$

is monic. So if f is an essential monomorphism, $\phi(g)$ is monic (5.13). Thus, again by (21.3), g is monic. Applying (5.19) we have therefore that $F(f)$ is essential. We omit the rest of this proof.

(6) This is immediate from (1) and (5).   □

Of considerable importance in the present study is the fact that submodule lattices are "preserved" by equivalences. To state this formally, we add to our list of notation. Thus, if $K \leq M$, we let

$$i_{K \leq M}: K \to M$$

denote the inclusion monomorphism.

21.7. Proposition. *Let R and S be equivalent rings via an equivalence $F: {}_R\mathsf{M} \to {}_S\mathsf{M}$. Then for each left R-module M, the mapping defined by*

$$\Lambda_M: K \mapsto \operatorname{Im} F(i_{K \leq M})$$

is a lattice isomorphism from the lattice of submodules of M onto the lattice of submodules of F(M).

Proof. Since F is a functor, it is easy to see that Λ is order preserving (see Exercise (16.2)). On the other hand, adopting the notation of (21.1), for each $N \leq F(M)$ define

$$\Gamma_M(N) = Im\,\phi(i_{N \leq F(M)}).$$

Then Γ_M is a function from the submodules of $F(M)$ to those of M. It also is order preserving by (21.3.1). Now for $K \leq M$, let

$$N = \Lambda_M(K).$$

Then since $F(i_{K \leq M})$ is monic (21.2), there is an isomorphism $h:F(K) \to N$ making

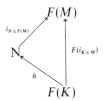

commute. But then by (21.3)

$$\phi(i_{N \leq F(M)})G(h) = \phi(i_{N \leq F(M)}h) = \phi(F(i_{K \leq M}))$$

$$= i_{K \leq M}\phi(1_{F(K)}),$$

and so since $G(h)$ and $\phi(1_{F(K)})$ are isomorphisms ((21.2) and (21.3)),

$$\Gamma_M\Lambda_M(K) = Im\,\phi(i_{N \leq F(M)}) = Im\,i_{K \leq M} = K.$$

Next, let $N \leq F(M)$, and let

$$K = \Gamma_M(N).$$

Then there is an isomorphism γ making

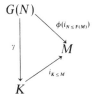

commute. Applying ϕ^{-1} and (21.3),

$$i_{N \leq F(M)} = \phi^{-1}(i_{K \leq M}\gamma) = F(i_{K \leq M})\phi^{-1}(\gamma)$$

whence, since $\phi^{-1}(\gamma)$ is an isomorphism (21.3), we have

$$\Lambda_M\Gamma_M(N) = Im\,F(i_{K \leq M}) = Im\,i_{N \leq F(M)} = N$$

and the proof is complete. $\qquad\qquad\qquad\qquad\qquad\qquad\qquad\qquad\qquad\qquad\quad\square$

21.8. Proposition. *Let R and S be equivalent rings, via an equivalence $F : {}_R M \to {}_S M$, and let M and M' be left R-modules. Then*

(1) *M is simple (semisimple) if and only if $F(M)$ is simple (semisimple);*

(2) *M is finitely generated (finitely cogenerated) if and only if $F(M)$ is finitely generated (finitely cogenerated);*

(3) *M is artinian (noetherian) if and only if $F(M)$ is artinian (noetherian);*

(4) *$c(M) = c(F(M))$; that is, M and $F(M)$ have the same composition length;*

(5) *M is indecomposable if and only if $F(M)$ is indecomposable.*

Proof. Each of these is simply an assertion about the lattices of submodules of M and $F(M)$. ☐

21.9. Corollary. *Let R and S be equivalent rings. Then R is semisimple, left artinian, left noetherian, primitive, or a ring with zero radical, respectively, if and only if so is S.*

Proof. The semisimple case is by (21.8) and (13.9). The artinian and noetherian cases follow from (21.8) because by (10.19) and (10.20) a ring is left artinian (noetherian) if and only if each of its finitely generated left modules is artinian (noetherian). For the remaining cases, recall that R is primitive $(J(R) = 0)$ iff R has a faithful simple (semisimple) module. ☐

It should now be abundantly clear that the categories of left modules over equivalent rings do have essentially the same structure. Also it is easy to show that much of the "two sided" structure in the rings themselves is the same.

21.10. Proposition. *If R and S are equivalent rings, then $Cen\, R \cong Cen\, S$.*

Proof. Adopt the notation of (21.2). Then by (21.2) $R \cong End({}_R R) \cong End({}_S F(R))$. But since ${}_R R$ is a generator, ${}_S F(R)$ is a generator (21.6). So by (17.9) $S \cong BiEnd({}_S F(R))$. Now apply Exercise (4.6) to get

$$Cen\, R \cong Cen(End({}_S F(R)))$$

$$= Cen(BiEnd({}_S F(R))) \cong Cen\, S. \qquad ☐$$

21.11. Proposition. *Let R and S be equivalent rings via an equivalence $F : {}_R M \to {}_S M$. For each (two sided) ideal I of R, set*

$$\Phi(I) = l_S(F(R/I)).$$

Then the mapping defined by

$$\Phi : I \mapsto \Phi(I)$$

is an isomorphism of the lattice of ideals of R and the lattice of ideals of S. Moreover, for each ideal I of R, there is an equivalence

$$R/I \approx S/\Phi(I).$$

Proof. As usual we adopt the notation of (21.1). If I is an ideal of R, then $F(R/I)$ is a left S-module, so its left annihilator $\Phi(I)$ is clearly an ideal of S.

Similarly, if K is an ideal of S, then

$$\Gamma(K) = I_R(G(S/K))$$

is an ideal of R. We claim that Φ and Γ define inverse mappings of the ideal lattices. By viewing $F(R/I)$ first as a faithful $S/\Phi(I)$ module, it is clear that as S-modules, $F(R/I)$ cogenerates $S/\Phi(I)$ (see 8.22) and $S/\Phi(I)$ generates $F(R/I)$. Thus, by (21.6), R/I cogenerates $G(S/\Phi(I))$ and $G(S/\Phi(I))$ generates R/I as R-modules. In particular these two R-modules must have the same annihilator (Exercise (8.2)). That is, $\Gamma\Phi(I) = I$. Similarly for each ideal K of S, $\Phi\Gamma(K) = K$. Also note that if I and I' are ideals of R with $I \subseteq I'$, then there is a natural R-epimorphism $R/I \to R/I' \to 0$. Thus, there is an S-epimorphism (21.2) $F(R/I) \to F(R/I') \to 0$ whence $\Phi(I) \subseteq \Phi(I')$, and Φ is order preserving. Similarly, so is Γ. Thus Φ is a lattice isomorphism.

For the final assertion let I be an ideal of R and let $v : R \to R/I$ be the natural ring homomorphism. Then the change of rings functor

$$T_v : {}_{R/I}\mathsf{M} \to {}_R\mathsf{M}$$

is an equivalence from ${}_{R/I}\mathsf{M}$ to the full subcategory $\mathsf{G}_R(R/I)$ of ${}_R\mathsf{M}$ whose objects are generated by R/I. Thus by (21.6.3) the functor F, restricted to $\mathsf{G}_R(R/I)$ defines an equivalence with the full subcategory $\mathsf{G}_S(F(R/I))$ of ${}_S\mathsf{M}$ whose objects are generated by $F(R/I)$. But as we have seen above, $S/\Phi(I)$ generates $F(R/I)$. Also R/I generates $G(S/\Phi(I))$, whence $F(R/I)$ generates $S/\Phi(I)$. Thus $\mathsf{G}_S(F(R/I)) = \mathsf{G}_S(S/\Phi(I))$ and we have ${}_{R/I}\mathsf{M} \approx {}_{S/\Phi(I)}\mathsf{M}$.   □

21.12. Corollary. *If $R \approx S$ and if R is simple, then S is simple.*

21.13. Corollary. *If $F : {}_R\mathsf{M} \to {}_S\mathsf{M}$ is an equivalence, then $R/J(R) \approx S/J(S)$ and $J(S) = I_S(F(R/J(R)))$.*

Proof. According to (21.9) and (21.11) an ideal P of R is primitive if and only if the corresponding ideal $\Phi(P)$ is a primitive ideal of S. Thus by (21.11)

$$J(S) = \cap \{\Phi(P) \,|\, P \text{ is a primitive ideal of } R\}$$

$$= \Phi(J(R)) = I_S(F(R/J(R))).$$

21. Exercises

1. Let C and D be categories. Let $F, F' : \mathsf{C} \to \mathsf{D}$ and $G, G' : \mathsf{D} \to \mathsf{C}$ be covariant functors with $F \cong F'$ and $G \cong G'$. Prove that if F and G are inverse equivalences, then F' and G' are inverse equivalences. [Hint: First show F and G' are inverse equivalences. Say $\phi : G' \to G$, $\eta : FG \to 1_\mathsf{D}$ and $v : GF \to 1_\mathsf{C}$ are isomorphisms. Consider $\eta' : FG' \to 1_\mathsf{D}$ and $v' : G'F \to 1_\mathsf{C}$ defined via $\eta'_N = \eta_N \circ F(\phi_N)$ and $v'_M = v \circ \phi_{F(M)}$.]
2. Prove that the relation $\approx$ of equivalence is an equivalence relation on the class of all categories. [Hint: Let $\mathsf{C}, \mathsf{D}, \mathsf{E}$ be categories. Let $F : \mathsf{C} \to \mathsf{D}$, $G : \mathsf{D} \to \mathsf{E}$, $H : \mathsf{D} \to \mathsf{C}$ and $K : \mathsf{E} \to \mathsf{D}$ be covariant and let

$\eta : HF \to 1_C$ and $v : KG \to 1_D$ be natural isomorphisms. Then show that $\eta_M \circ H(v_{F(M)}) : HKGF(M) \to M$ defines a natural isomorphism.]

3. Let C and D be categories. Prove that a covariant functor $F : C \to D$ is an equivalence iff it is both faithful and full, and each $D \in D$ is isomorphic to some $F(C)$ with $C \in C$. [Note: F is *full* in case $F(mor_C(C, C'))$ $= mor_D(F(C), F(C'))$.]

4. A covariant functor $F : C \to D$ is an *isomorphism* in case there is a covariant functor $G : D \to C$ with $FG = 1_D$ and $GF = 1_C$. If such a pair of functors exist, then C and D are *isomorphic*. Clearly isomorphic categories are equivalent, but the converse fails. For example, let $_R T$ be a simple module and let C be the full subcategory of $_R M$ whose object class is $\{T\}$. Let $\mathscr{D}$ be a non-empty set of R-modules isomorphic to T, and let D be the full subcategory of $_R M$ with object class $\mathscr{D}$. Prove that $C \approx D$ but that C and D are isomorphic iff $\mathscr{D}$ is a singleton.

5. Let C_1 and C_2 be categories. Their *product* is the category $C_1 \times C_2$ with objects $\mathscr{C}_1 \times \mathscr{C}_2$, morphisms $mor_{C_1} \times mor_{C_2}$ applied coordinate-wise, and composition the product composition. Prove that if R_1 and R_2 are rings then

$$_{R_1} M \times {}_{R_2} M \approx {}_{R_1 \times R_2} M.$$

6. This exercise culminates in a direct proof of the important fact that a ring R is equivalent to each of its matrix rings $M_n(R)$ $(n = 1, 2, \dots)$. This fact is also an immediate corollary of the general characterization of equivalence in the next section. (See (22.6).) Let R be a ring and let $e \in R$ be a non-zero idempotent. Set $S = eRe$. Prove that:

(1) There is a natural homomorphism $\eta : (Re \otimes_S eR \otimes_R \_) \to 1_{_R M}$ such that for each $_R M$,

$$\eta_M : ae \otimes eb \otimes m \mapsto aebm.$$

(2) η is a natural isomorphism iff $ReR = R$. [Hint: ($\Leftarrow$). Suppose $\Sigma_j s_j e t_j m_j = 0$. Let $1 = \Sigma_i a_i e b_i$. Then

$$\Sigma_j (s_j e \otimes e t_j \otimes m_j) = \Sigma_i (a_i e \otimes e b_i \otimes \Sigma_j s_j e t_j m_j) = 0,$$

so η_M is monic.]

(3) If $ReR = R$, then $R \approx S = eRe$. [Hint: Consider the functors $F = (eR \otimes_R \_) : {}_R M \to {}_S M$ and $G = (Re \otimes_S \_) : {}_S M \to {}_R M$. Apply (1) and (2).]

(4) For each $n \in \mathbb{N}$, $R \approx M_n(R)$.

7. Let $F : {}_R M \to {}_S M$ be an additive covariant functor. Then F is a *projector* (*injector*) in case for each projective module $_R P$ (injective module $_R Q$), its image $_S F(P)$ is projective ($_S F(Q)$ is injective). For example, see Exercise (19.10). Let $G : {}_S M \to {}_R M$ be a left adjoint of F. Prove that F is an injector (G is a projector) iff G is exact (F is exact). [Hint: See Exercise (20.8).]

8. Let $_S P_R$ be a bimodule and let $_R Q_S = Hom_R({}_S P_R, {}_R R_R)$. Consider the three functors

$$F_P = ({}_SP \otimes_R \quad): {}_RM \to {}_SM$$
$$G_P = ({}_RQ \otimes_S \quad): {}_SM \to {}_RM$$
$$H_P = Hom_S({}_SP_R, \quad): {}_SM \to {}_RM.$$

(1) By (20.6), (F_P, H_P) is an adjoint pair. Prove that F_P is a projector iff ${}_SP$ is projective. [Hint: Exercise (21.7).]

(2) Prove that if P_R is finitely generated projective, then (G_P, F_P) is an adjoint pair. [Hint: Exercise (20.7(2)) and Proposition 20.11.]

(3) Prove that if P_R is finitely generated projective, then F_P is an injector iff Q_S is flat.

(4) Let both P_R and ${}_SP$ be finitely generated projective. Then ${}_SP_R$ is *Frobenius* in case as (R, S)-bimodules

$$Hom_R({}_SP_R, {}_RR_R) \cong Hom_S({}_SP_R, {}_SS_S).$$

Prove that ${}_SP_R$ is Frobenius iff $G_P \cong H_P$. [Hint: (20.10).]

(5) Let ${}_SP_R$ be Frobenius. Prove that if ${}_RR$ is injective (${}_SS$ is injective), then ${}_SP$ (P_R) is injective.

9. Let R be a finite dimensional algebra over a field S. Then R is called a Frobenius algebra if ${}_SR_R$ is a Frobenius bimodule. Prove that:

(1) If R is a Frobenius algebra over S, then R is both left and right self injective.

(2) The following are equivalent: (a) R is Frobenius; (b) ${}_RR_S \cong Hom_S({}_SR_R, S)$; (c) There is an R-balanced map $\phi: R_R \times {}_RR \to S$ such that if $0 \neq x \in R$, then $\phi(x, R) \neq 0$ and $\phi(R, x) \neq 0$. [Hint: (c) $\Rightarrow$ (a). Define $\theta: Hom_R({}_SR_R, R) \to Hom_S({}_SR_R, S)$ by $\theta(f)(x) = \phi(x, f(1))$. A dimension argument implies that θ is epic.]

(3) If R is simple, then R is Frobenius. [Hint: Part (2.b) and Exercise (20.17).]

(4) If G is a finite group, then the group algebra $R = SG$ is Frobenius. [Hint: Let $e \in G$ be the identity. Define $\lambda: R \to S$ by $\lambda(f) = f(e)$ and $\phi: R \times R \to S$ by $\phi(f, f') = \lambda(ff')$. Now use (2.c).]

10. Let $\phi: S \to R$ be a ring homomorphism. Then via ϕ, R has the module structure ${}_SR_R$ and ${}_RR_S \cong Hom_R({}_SR_R, {}_RR_R)$ with R_R and ${}_RR$ projective. So the change of ring functor $F = ({}_SR \otimes_R -): {}_RM \to {}_SM$ has left adjoint $G = ({}_RR \otimes_S -): {}_SM \to {}_RM$ and right adjoint $H = Hom_S({}_SR_R, -): {}_SM \to {}_RM$. (See Exercise (21.9).) If ${}_RM$, then we simply write ${}_SM$ for $F({}_RM)$. Prove that:

(1) If ${}_SR$ is (finitely generated) projective, and if ${}_RM$ is (finitely generated) projective, then ${}_SM$ is (finitely generated) projective. [Hint: Exercise (20.8). Also see Exercise (19.16).]

(2) If R_S is flat and if ${}_RM$ is injective, then ${}_SM$ is injective.

(3) If $_SR$ is a generator and if $_RM$ is a generator, then $_SM$ is a generator. [Hint: Exercise (20.8).]

(4) F and G are inverse equivalences iff ϕ is an isomorphism.

11. Let R and S be equivalent rings via an equivalence $F:_R M \to _S M$ with inverse $G:_S M \to _R M$. Prove each of the following:

(1) $_RM$ is finitely presented iff $_SF(M)$ is finitely presented.

(2) $_RM$ has a decomposition that complements (maximal) direct summands iff $F(M)$ has such a decomposition.

12. Let R and S be rings with $R \approx S$. Prove that:

(1) R is von Neumann regular iff S is von Neumann regular. [Hint: Exercise (20.14).]

(2) R is left hereditary iff S is left hereditary.

(3) R is left self-injective iff S is left self-injective.

(4) R is co-semisimple iff S is co-semisimple.

§22. The Morita Characterizations of Equivalence

The prototype of (Morita) equivalence is provided by a ring R and the ring $\mathbb{M}_n(R)$ of $n \times n$ matrices over R. Indeed, the Wedderburn characterization of simple artinian rings (13.3) may be viewed as one of the earliest treatments of the theory of equivalence of rings. In this section we shall give the complete characterizations of equivalence, generalizing the Wedderburn-Artin theory, that are due to Morita [58a]. We begin with various necessary conditions.

A left R-module $_RP$ is a *progenerator* (or a *left R-progenerator*) in case it is a finitely generated projective generator. In particular, $_RR$ is a progenerator. Indeed, $_RP$ is a progenerator if and only if there are integers m and n and modules P' and R' with

$$R^{(m)} \cong P \oplus P' \qquad \text{and} \qquad P^{(n)} \cong R \oplus R'.$$

22.1. Theorem. *Let R and S be equivalent rings via inverse equivalences $F:_R M \to _S M$ and $G:_S M \to _R M$. Set*

$$P = F(R) \qquad \text{and} \qquad Q = G(S).$$

Then P and Q are naturally bimodules $_SP_R$ and $_RQ_S$ such that

(1) $_SP_R$ *and* $_RQ_S$ *are faithfully balanced;*

(2) $P_R, _SP, _RQ$ *and* Q_S *are all progenerators;*

(3) $_SP_R \cong Hom_S(Q, S) \cong Hom_R(Q, R)$ *and* $_RQ_S \cong Hom_R(P, R) \cong Hom_S(P, S)$;

(4) $F \cong Hom_R(Q, \_)$ *and* $G \cong Hom_S(P, \_)$;

(5) $F \cong (P \otimes_R \_)$ *and* $G \cong (Q \otimes_S \_)$.

Proof. We shall assume all of the notation of (21.1). From (20.3) we see that the bimodule structure $_RR_R$ and the additivity of the functor F induce a canonical bimodule structure $_SP_R = F(R)$ where the right R-scalar multiplication is given by the ring homomorphism $r \mapsto F(\rho(r))$ of R into $End(_SP) = End(_SF(R))$. Observe that this is just the composition of the two isomorphisms

$$R \cong End(_RR) \quad \text{and} \quad End(_RR) \cong End(_SF(R))$$

of (4.11) and (21.2).

Now the left S-module $_SP = F(_RR)$ is a progenerator since $_RR$ is clearly a progenerator and F is an equivalence ((21.6) and (21.8)). In particular, since $_SP$ is a generator, it is balanced (17.8.1). But as we have just observed R, with its natural action on $_SP$, is isomorphic to $End(_SP)$. Thus, $_SP_R$ is a faithfully balanced bimodule. Therefore, by (17.7), P_R is also a progenerator. Similarly, $_RQ = G(_SS)$ has, induced by $_SS_S$, a natural faithfully balanced bimodule structure $_RQ_S$ with $_RQ$ and Q_S both progenerators. This gives (1) and (2).

Next, let $_RM$ be a left R-module. Then since the $\mathbb{Z}$-isomorphism ϕ of (21.1) is natural in the first variable,

$$\phi : Hom_S(S, F(M)) \to Hom_R(G(S), M) = Hom_R(Q, M)$$

is a left S-isomorphism (20.4). But then the S-isomorphisms

$$F(M) \cong Hom_S(S, F(M)) \cong Hom_R(Q, M)$$

are natural in M whence $F \cong Hom_R(Q, \_)$. Similarly, $G \cong Hom_S(P, \_)$. Then at $_RR_R$ and at $_SS_S$ these natural isomorphisms are (see (20.4)) bimodule isomorphisms

$$_SP_R = {_SF(R)_R} \cong Hom_R(Q, R)$$

and

$$_RQ_S = {_RG(S)_S} \cong Hom_S(P, S).$$

But then, applying (1) and (20.7)

$$_SP_R \cong Hom_R(Q, R) \cong Hom_R(Q, Hom_S(Q, Q))$$
$$\cong Hom_S(Q, Hom_R(Q, Q)) \cong Hom_S(Q, S)$$

and

$$_RQ_S \cong Hom_S(P, S) \cong Hom_S(P, Hom_R(P, P))$$
$$\cong Hom_R(P, Hom_S(P, P)) \cong Hom_R(P, R).$$

Thus, we have (3) and (4).

Finally, (5) follows from (4) and (3) because by (20.5), (20.11) and (20.1), there are natural isomorphisms

$$Hom_R(Q, \_) \cong Hom_R(Hom_R(P, R), \_)$$
$$\cong P \otimes_R Hom_R(R, \_)$$
$$\cong (P \otimes_R \_)$$

and similarly

$$Hom_S(P, \_) \cong (Q \otimes_S \_).$$ □

Now it is an easy matter to prove the basic characterization of equivalent rings.

22.2. Theorem [Morita]. *Let R and S be rings and let*

$$F : {}_R\mathsf{M} \to {}_S\mathsf{M} \qquad and \qquad G : {}_S\mathsf{M} \to {}_R\mathsf{M}$$

be additive functors. Then F and G are inverse equivalences if and only if there exists a bimodule ${}_SP_R$ such that:

(1) *${}_SP$ and P_R are progenerators;*
(2) *${}_SP_R$ is balanced;*
(3) *$F \cong (P \otimes_R \_)$ and $G \cong Hom_S(P, \_)$.*

Moreover, if there is a bimodule ${}_SP_R$ satisfying these conditions, then with

$$Q = Hom_R(P, R),$$

we have ${}_RQ_S$ with ${}_RQ$ and Q_S progenerators and

$$F \cong Hom_R(Q, \_) \qquad and \qquad G \cong (Q \otimes_S \_).$$

Proof. The final assertion as well as the necessity of the conditions (1), (2), and (3) are immediate from (21.1). So, conversely, suppose that ${}_SP_R$ is a bimodule satisfying (1) and (2). Then for each ${}_RM$ and each ${}_SN$ there are natural isomorphisms:

$$Hom_S(P, P \otimes_R M) \cong Hom_S(P, P) \otimes_R M \qquad (20.10)$$

$$\cong R \otimes_R M \qquad\qquad \text{(by (2))}$$

$$\cong M \qquad\qquad (20.1)$$

and

$$P \otimes_R Hom_S(P, N) \cong Hom_S(Hom_R(P, P), N) \qquad (21.11)$$

$$\cong Hom_S(S, N) \qquad\qquad \text{(by (2))}$$

$$\cong N \qquad\qquad (21.1).$$

Thus $F = (P \otimes_R \_)$ and $G = Hom_S(P, \_)$ are inverse equivalences.   □

22.3. Corollary. *If R and S are rings, then ${}_R\mathsf{M} \approx {}_S\mathsf{M}$ if and only if $\mathsf{M}_R \approx \mathsf{M}_S$.*

Proof. Let ${}_R\mathsf{M} \approx {}_S\mathsf{M}$. Then by (22.2) there is a balanced ${}_SP_R$ with ${}_SP$ and P_R progenerators and with

$$(P \otimes_R \_) \qquad and \qquad Hom_S(P, \_)$$

inverse equivalences. Then by (22.1) $Q = Hom_S(P, S)$ is a balanced bimodule ${}_RQ_S$ with ${}_RQ$ and Q_S progenerators. Thus, we may view Q as a bimodule ${}_{S^{op}}Q_{R^{op}}$ that is balanced and with $Q_{R^{op}}$ and ${}_{S^{op}}Q$ both progenerators. Then (22.2) asserts that ${}_{R^{op}}\mathsf{M} \approx {}_{S^{op}}\mathsf{M}$ or equivalently that $\mathsf{M}_R \approx \mathsf{M}_S$. The converse now follows by using opposite rings again.   □

In particular, we see from this corollary that, as we have claimed earlier, there is no concern with just "left" and "right" equivalence for rings. Indeed, that left semisimple rings are the same as right semisimple rings, is also a corollary of (22.3). (See Corollary 21.9.)

Among the most useful tests for equivalence are those given in the next corollary.

22.4. Corollary. *For two rings R and S the following assertions are equivalent:*

(a) $R \approx S$;
(b) *There is a progenerator P_R with $S \cong End(P_R)$;*
(c) *There is a progenerator $_RQ$ with $S \cong End(_RQ)$.*

Proof. (a) $\Rightarrow$ (b). This is immediate from (22.2).

(b) $\Rightarrow$ (a). We may assume that $S = End(P_R)$. By (17.8) we know that since P_R is a generator, it is balanced and finitely generated projective over $S = End(P_R)$. Then also since $_SP_R$ is balanced and P_R is finitely generated projective, an application of (17.7) gives us that $_SP$ is a generator. Now (22.2) applies and R and S are equivalent via $F = (P \otimes_R \_)$ and $G = Hom_S(P, \_)$. $\square$

22.5. Corollary. *Let R be a ring. If P_R is a progenerator, then R and $S = End(P_R)$ are equivalent. In fact, if*

$$P^{\circledast} = Hom_R(P, R),$$

then $_SP_R$ and $_RP^{\circledast}_S$ are bimodules and

$$(P \otimes_R \_) : {_R}M \to {_S}M$$

$$(P^{\circledast} \otimes_S \_) : {_S}M \to {_R}M$$

are inverse equivalences. $\square$

Now we can prove what is perhaps the most important special case. (See also Exercise (21.6).)

22.6. Corollary. *Let R be a ring and let $n > 0$ be a natural number. Then R and $\mathbb{M}_n(R)$ are equivalent rings.*

Proof. The matrix ring $\mathbb{M}_n(R)$ is isomorphic to the endomorphism ring of the free right R-module $R^{(n)}$. (See (13.2).) But $R^{(n)}$ is clearly a right R-progenerator. $\square$

22.7. Corollary. *If R and S are equivalent rings, then there is a positive integer n and an idempotent matrix $e \in \mathbb{M}_n(R)$ such that*

$$S \cong e\mathbb{M}_n(R)e.$$

Proof. We may assume that $R^{(n)} = P \oplus P'$. Then $S \cong End(P_R) \cong e\,End(R^{(n)})e$ where e is the idempotent for P in the given decomposition of $R^{(n)}$. $\square$

22. Exercises

1. Let P_R be a finitely generated projective right R-module, let $P^\circledast$ be its R-dual $Hom_R(P, R)$. Then $_RP^\circledast$ is finitely generated and projective (20.17). The ideal $T = Tr_R(P)$ is simply called the *trace* of P. (See (8.21).) Since R is a generator, P_R is a progenerator iff $T = R$. Prove that:
   (1) $T^2 = T$; $PT = P$; and $T = Tr_R(P^\circledast)$.
   (2) The natural maps $P \otimes_R T \to P_R$ and $T \otimes_R P^\circledast \to {}_RP^\circledast$ are isomorphisms. [Hint: If $(x_i), (f_i)$ is a dual basis for P, then $x \mapsto \Sigma\, x_i \otimes f_i(x)$ gives one inverse.]
   (3) If $e \in R$ is an idempotent and $P \cong eR$, then $T = ReR$.
2. Let P and $P^\circledast$ be as in Exercise (22.1). Assume that R is a prime ring (Exercise (14.10)). Prove the following generalization of Exercise (17.4.1): P_R is faithful iff $_RP^\circledast$ is faithful.
3. Every generator is faithful (Exercise (8.3)). Concerning a converse, prove that:
   (1) A projective module P_R is a generator iff for each ideal I of R, if $PI = P$, then $I = R$. [Hint: Exercise (22.1).]
   (2) Faithful finitely generated projective modules need not be generators.
   (3) If R is commutative and P_R is finitely generated projective, then P_R is a progenerator iff it is faithful. [Hint: Let T be the trace of P. Then T is a finitely generated ideal with $T^2 = T$. Use Exercise (7.12) for $T = Re$. If P_R is faithful, $e = 1$.]
4. The category $_R$FM of finitely generated left R-modules characterizes the entire category $_R$M to within category equivalence. Indeed, let R and S be rings; prove that:
   (1) If $_SN$ and $_SN'$ are finitely generated, then an S-homomorphism $f : N \to N'$ is epic iff for each $h \in {}_S$FM, $hf = 0$ implies $h = 0$. [Hint: (3.3) and Exercise (10.1)]
   (2) If $_RM$, $_RM'$ are in $_R$FM and if $F : {}_R$FM $\to {}_S$FM is a category equivalence, then $f : M \to M'$ is a (split) epimorphism iff $F(f) : F(M) \to F(M')$ is a (split) epimorphism.
   (3) $_R$FM $\approx {}_S$FM iff $_R$M $\approx {}_S$M. [Hint: ($\Rightarrow$). Let $F : {}_R$FM $\to {}_S$FM be an equivalence with inverse equivalence G. If $R^{(n)} \to G(S)$ is a split epimorphism, then so is $F(R)^{(n)} \to FG(S) \cong S$.]
5. Let $_SP_R$ and $_RQ_S$ be bimodules, and let (θ, ϕ) be a pair of bimodule homomorphisms

$$\theta : P \otimes_R Q \to {}_SS_S \qquad \text{and} \qquad \phi : Q \otimes_S P \to {}_RR_R$$

such that for all $x, y \in P$ and $f, g \in Q$,

$$\theta(x \otimes f)y = x\phi(f \otimes y) \qquad \text{and} \qquad f\theta(x \otimes g) = \phi(f \otimes x)g.$$

Then (θ, ϕ) is a *Morita pair* for (P, Q). Clearly then (ϕ, θ) is a Morita pair for (Q, P). For each $x \in P$ and each $f \in Q$ define $\hat{\phi}(x) : Q \to R$ and $\bar{\phi}(f) : P \to R$ by

$$\hat{\phi}(x):f \mapsto \phi(f \otimes x) \quad \text{and} \quad \bar{\phi}(f):x \mapsto \phi(f \otimes x).$$

Prove that:

(1) $\hat{\phi}$ is an (S, R)-homomorphism $_SP_R \to Hom_R(_RQ_S, _RR_R)$ and $\bar{\phi}$ is an (R, S)-homomorphism $_RQ_S \to Hom_R(_SP_R, _RR_R)$.

(2) If θ is epic, then

   (i) θ is an isomorphism;

   (ii) P_R and $_RQ$ are finitely generated projective;

   (iii) $_SP$ and Q_S are generators;

   (iv) $\hat{\phi}$ and $\bar{\phi}$ are isomorphisms;

   (v) $\lambda:S \to End(P_R)$ and $\rho:S \to End(_RQ)$ are ring isomorphisms.

[Hint: For (iii) observe that $Tr_S(P) \geq \hat{\theta}(Q)(P) = \theta(P \otimes Q)$. Next suppose that $\Sigma_i \theta(x_i \otimes f_i) = 1_S \in S$. Then

$$\Sigma_j(y_j \otimes g_j) = \Sigma_i(x_i \otimes f_i(\Sigma_j \theta(y_j \otimes g_j))),$$

and it follows that θ is monic. Also (x_i) and $(\bar{\phi}(f_i))$ form a dual basis for P_R. (See Exercise (17.11).) If $\bar{\phi}(f) = 0$, then show that $f = f1_S = 0$. If $g \in Hom_R(P, R)$, then $g = \bar{\phi}(\Sigma_i g(x_i)f_i)$. If $\lambda(s) = 0$, then $s = s1_S = 0$. Finally, if $u \in End(P_R)$, then $u = \lambda(\Sigma_i \theta(ux_i \otimes f_i))$.]

6. Let P_R be a right R-module and let $S = End(P_R)$. Then $_SP_R$. Let $P^\circledast = Hom_R(_SP_R, _RR_R)$. Then $_RP^\circledast_S$. Prove that:

(1) There is a Morita pair (θ_P, ϕ_P) for $(P, P^\circledast)$ via

$$\theta_P(x \otimes f):y \mapsto xf(y) \quad \text{and} \quad \phi_P(f \otimes x) = f(x).$$

(2) P_R is finitely generated projective iff θ_P is an isomorphism. [Hint: The Dual Basis Lemma (Exercise (17.11)).]

(3) P_R is a generator iff ϕ_P is an isomorphism.

(4) P_R is a progenerator iff θ_P and ϕ_P are both epic.

7. Prove the following version of Morita's Theorem (22.2): Two rings R and S are equivalent iff there exist bimodules $_SP_R$ and $_RQ_S$ and a Morita pairing (θ, ϕ) of (P, Q) with both θ and ϕ epic. [Hint: Exercises (22.5) and (22.6).]

8. Let $_SP_R$ and $_RQ_S$ be bimodules, let (θ, ϕ) be a Morita pair for (P, Q) such that θ, ϕ are epic. Thus, by Exercise (22.7), $R \approx S$. For each ideal I of R and each ideal I' of S define

$$\Theta(I) = \theta(PI \otimes Q) \quad \text{and} \quad \Phi(I') = \phi(QI' \otimes P).$$

Prove that:

(1) Θ and Φ define inverse lattice isomorphisms between the ideal lattices of R and S.

(2) If H and I are ideals of R, then $\Theta(IH) = \Theta(I)\Theta(H)$.

(3) Prove that I is a prime (nilpotent) ideal of R iff $\Theta(I)$ is a prime (nilpotent) ideal of S.

(4) Prove that N is the lower nilradical of R iff $\Theta(N)$ is the lower nilradical of S.

9. Let K be a commutative ring and let R, S, T be K-algebras. Consider the K-algebras

$$R^T = T \otimes_K R \qquad \text{and} \qquad S^T = T \otimes_K S.$$

(See Exercise (19.5).) Prove that if $R \approx S$, then $R^T \approx S^T$. [Hint: Exercise (22.7).]

10. Two modules M and N are *similar*, abbreviated $M \sim N$, in case there exist natural numbers m and n and modules M' and N' with

$$M \oplus M' \cong N^{(m)} \qquad \text{and} \qquad N \oplus N' \cong M^{(n)}.$$

(1) Prove that $\sim$ defines an equivalence relation on the class $_R\mathcal{M}$.

(2) Prove that $_RP$ is similar to $_RR$ iff P is a progenerator.

11. Let M_K and N_K be non-zero modules over a ring K. Let $R = End(M_K)$ and $S = End(N_K)$. Consider the bimodules

$$_SP_R = Hom_K(_RM_K, {}_SN_K) \qquad \text{and} \qquad _RQ_S = Hom_K(_SN_K, {}_RM_K).$$

(1) Prove that there is a Morita pair (θ, ϕ) for (P, Q) such that for each $f \in P$ and $g \in Q$,

$$\theta(f \otimes g) = fg \qquad \text{and} \qquad \phi(g \otimes f) = gf.$$

(2) Prove that both θ and ϕ (of part (1)) are epic iff M_K and N_K are similar. (See Exercise (22.10).)

(3) Infer that if M_K and N_K are similar, then $End(M_K) \approx End(N_K)$. [Hint: Exercise (22.7).]

12. Let $F : {}_R\mathsf{M} \to {}_S\mathsf{M}$ be a category equivalence. Prove that $_RM$ is flat iff $_SF(M)$ is flat. Deduce that R is coherent iff S is coherent.

13. Let P_R be finitely generated and projective, and let $S = End(P_R)$. Then P_R is a *projector* (*injector*) in case the associated functor

$$F_P = (P \otimes_R -) : {}_R\mathsf{M} \to {}_S\mathsf{M}$$

is a projector (injector), (See Exercises (21.7), (21.8).) Let $T = Tr_R(P)$ be the trace of P_R. Now F_P is an equivalence iff $T = R$. Prove that:

(1) If $_RT$ is projective, then $_SP_R$ is a projector. [Hint: By Exercise (22.1), $P^\circledast = Hom_R(P, R)$ generates the projective module $_RT$, so for some A there is a split epimorphism $P \otimes_R (P^\circledast)^{(A)} \to P \otimes_R T$ in $_S\mathsf{M}$. But $P \otimes_R P^\circledast \cong S$ and $_SP \otimes_R T \cong {}_SP$ is projective. (Exercises (22.6) and (22.1).)]

(2) If T_R is flat, then $_SP_R$ is an injector. [Hint: It will suffice to show that $P^\circledast_S \cong T \otimes_R P^\circledast_S$ is flat. But $P \otimes_S P^\circledast = R$, so if $_SI \leq {}_SS$, and $L = Ker(P^\circledast \otimes I \to S)$, then $P \otimes L = 0$, so $T \otimes L = 0$.]

14. Prove that if R is hereditary (von Neumann regular), then every finitely generated projective module P_R is a projector (injector).

15. Let K be a field and $T = K[X]/X^2K[X]$. Let R be the subring of $\mathbb{M}_3(T)$ consisting of all matrices of the form

$$\begin{bmatrix} \alpha & \mu X & \nu X \\ 0 & \beta & \eta X \\ 0 & 0 & \gamma \end{bmatrix}$$

where $\alpha, \beta, \gamma, \mu, \eta, \nu \in K$. Let

$$
e = \begin{bmatrix} 0 & 0 & 0 \\ 0 & 1 & 0 \\ 0 & 0 & 0 \end{bmatrix}, \quad
f = \begin{bmatrix} 1 & 0 & 0 \\ 0 & 1 & 0 \\ 0 & 0 & 0 \end{bmatrix}, \quad
g = \begin{bmatrix} 0 & 0 & 0 \\ 0 & 1 & 0 \\ 0 & 0 & 1 \end{bmatrix}.
$$

Prove that:

(1) eR is both an (R, eRe)-projector and an (R, eRe)-injector.

(2) ReR is neither left R-projective nor right R-flat.

(3) fR is an (R, fRf)-injector, but not an (R, fRf)-projector.

(4) gR is an (R, gRg)-projector, but not an (R, gRg)-injector.

16. Let R be von Neumann regular and left self-injective. Prove that if P_R is finitely generated projective, then $End(P_R)$ is von Neumann regular and left self-injective. In particular, if $e \in R$ is a non-zero idempotent, then eRe is von Neumann regular and left self-injective. [Hint: By Morita it is true if P_R is free. Now use (5.9) and Exercise (22.14).]

(Note: The conclusion of this exercise is false if P_R is not finitely generated (Exercise (18.4)) or if R is not von Neumann regular. (See Rosenberg and Zelinsky [61].))

17. Let R be hereditary. Prove that if P_R is finitely generated projective, then $End(P_R)$ is hereditary. In particular, if $e \in R$ is a non-zero idempotent, eRe is hereditary.

§23. Dualities

Let C and D be two categories. Then a pair (H', H'') of contravariant functors

$$H' : C \to D \quad \text{and} \quad H'' : D \to C$$

is a *duality* between C and D in case there are natural isomorphisms

$$H'' H' \cong 1_C \quad \text{and} \quad H' H'' \cong 1_D.$$

The general theories of dualities and of equivalences really are dual. That is, if $op: C \to C^{op}$ denotes the canonical contravariant functor from a category to its opposite category (Exercise (23.1)), then it is easy to see that the pair (H', H'') is a duality between C and D if and only if

$$(op) \circ H' : C \to D^{op} \quad \text{and} \quad H'' \circ (op) : D^{op} \to C$$

are inverse equivalences. (See Exercise (23.2).)

Because of this last observation there is no need to develop anew the general theory of duality. However, when interpreted specifically in categories of modules, things are not so simple; the theory of duality in module categories is markedly different from that of equivalences. Of course, non-trivial dualities between certain module categories do exist—for example, the well-known dualities of finite dimensional vector spaces. But for no two rings R and S is there a duality between $_R$M and $_S$M (or M_S); the difficulty here is that

$(_R\mathsf{M})^{op}$ is *not* equivalent to $_S\mathsf{M}$ or M_S for any ring S. (See Exercise (24.1).) Thus, although a unified treatment of parts of the two theories is possible, we shall treat them separately. We shall economize some, however, by omitting proofs that are patently dual to previous ones.

Our interest is in dualities (H', H'') between categories of modules; and in this context we require that H' and H'' be additive functors.

Reflexive Modules and Duality

We first check that dualities between module categories do exist. Thus let R and S be rings, let $_R U_S$ be a non-zero bimodule, and set

$$H' = Hom_R(\_, U) : {}_R\mathsf{M} \to \mathsf{M}_S$$

$$H'' = Hom_S(\_, U) : \mathsf{M}_S \to {}_R\mathsf{M}$$

Then for each $_R M$ the evaluation map $\sigma_M : M \to H''H'(M)$, defined via

$$\sigma_M(x)(f) = f(x),$$

and for each N_S the evaluation map $\sigma_N : N \to H'H''(N)$, defined via

$$\sigma_N(y)(g) = g(y),$$

are natural homomorphisms (see §20). That is, they define natural transformations

$$\sigma : 1_{_R\mathsf{M}} \to H''H' \qquad \text{and} \qquad \sigma : 1_{\mathsf{M}_S} \to H'H''.$$

Now let $_R\mathsf{R}[U]$ and $\mathsf{R}_S[U]$ be the full subcategories of $_R\mathsf{M}$ and M_S whose objects are the U-reflexive modules. Then (20.14) since H' and H'' are functors between these categories of reflexive modules, we have from §20 the following result:

23.1. Proposition. *Let R and S be rings and let $_R U_S$ be a bimodule. Then the functors*

$$H' = Hom_R(\_, U) \qquad \text{and} \qquad H'' = Hom_S(\_, U)$$

define a duality between the categories $_R\mathsf{R}[U]$ and $\mathsf{R}_S[U]$ of U-reflexive modules. Indeed, for each $M \in {}_R\mathsf{R}[U]$ and each $N \in \mathsf{R}_S[U]$, the evaluation maps

$$\sigma_M : M \to H'' H'(M) \qquad \text{and} \qquad \sigma_N : N \to H' H''(N)$$

are natural isomorphisms.   □

Thus there do exist non-trivial dualities. For example, every finitely generated projective module is $_R R_R$ reflexive. Indeed, from (20.17) we infer that $H' = Hom_R(\_, {}_R R)$ and $H'' = Hom_R(\_, R_R)$ define a duality between the full subcategories $_R\mathsf{FP}$ and FP_R of $_R\mathsf{M}$ and M_R whose object classes are the finitely generated projective modules.

The Fundamental Lemma

The main purpose of the rest of this section is to prove that, with very modest restrictions, every duality between module categories is of the form described in the last proposition. In our discussion of dualities it will be convenient (but not at all necessary) to deal with dualities between categories of left R-modules and of right S-modules.

We proceed next to assemble suitable notation, dual to that of (21.1), and to establish certain basic isomorphisms dual to those of (21.2) and (21.3) for equivalences.

23.2. Let $_R\mathsf{C}$ and D_S be full subcategories of left R-modules and right S-modules, respectively. That is, $_R\mathsf{C}$ is a full subcategory of $_R\mathsf{M}$ and D_S is a full subcategory of M_S. Let the pair of functors

(1) $H' : {_R\mathsf{C}} \to \mathsf{D}_S$ and $H'' : \mathsf{D}_S \to {_R\mathsf{C}}$

be a duality between $_R\mathsf{C}$ and D_S. Then

$$H''H' \cong 1_{_R\mathsf{C}}   \text{and}   H'H'' \cong 1_{\mathsf{D}_S},$$

that is, there exist natural isomorphisms

(2) $\eta : H''H' \to 1_{_R\mathsf{C}}$ and $\zeta : H'H'' \to 1_{\mathsf{D}_S}$.

In particular, for η this means that for each $_R M$ in $_R\mathsf{C}$ there is an isomorphism $\eta_M : H''H'(M) \to M$ such that for each M_1, M_2 in $_R\mathsf{C}$ and each R-homomorphism $f : M_1 \to M_2$ the diagram

(3)
$$
\begin{array}{ccc}
M_1 & \xrightarrow{\;\;f\;\;} & M_2 \\
{\scriptstyle \eta_{M_1}}\uparrow & & \uparrow{\scriptstyle \eta_{M_2}} \\
H''H'(M_1) & \xrightarrow{H''H'(f)} & H''H'(M_2)
\end{array}
$$

commutes. Again similar remarks apply to ζ. For each $_R M$ in $_R\mathsf{C}$ and each N_S in D_S there are $\mathbb{Z}$-homomorphisms

(4)
$$\mu = \mu_{MN} : Hom_S(N, H'(M)) \to Hom_R(M, H''(N))$$
$$v = v_{MN} : Hom_S(H'(M), N) \to Hom_R(H''(N), M)$$

defined via

$$\mu_{MN} : \gamma \mapsto H''(\gamma) \circ \eta_M^{-1}$$
$$v_{MN} : \delta \mapsto \eta_M \circ H''(\delta)$$

for each $\gamma \in Hom_S(N, H'(M))$ and $\delta \in Hom_S(H'(M), N)$.

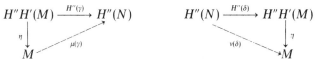

As in the case of equivalences we shall frequently omit the subscripts on η, ζ, μ and v.

Now we have two results whose proofs are dual to those of (21.2) and (21.3) and will be omitted.

23.3. Proposition. *Let (H', H'') be a duality between full subcategories $_R\mathsf{C}$ and D_S of $_R\mathsf{M}$ and M_S. Then for each M_1, M_2 in $_R\mathsf{C}$ and each N_1, N_2 in D_S the restrictions of H' to $Hom_R(M_1, M_2)$ and of H'' to $Hom_S(N_1, N_2)$ are abelian group isomorphisms*

$$H' : Hom_R(M_1, M_2) \to Hom_S(H'(M_2), H'(M_1))$$

$$H'' : Hom_S(N_1, N_2) \to Hom_R(H''(N_2), H''(N_1)).$$

If M in $_R\mathsf{C}$ and N in D_S are non-zero, then these maps are ring isomorphisms

$$End(_RM) \to End(H'(M)_S)$$

$$End(N_S) \to End(_RH''(N)). \qquad \square$$

Observe that here if $f, g \in Hom_R(M, M)$, then since the functor H' is contravariant

$$H'(fg) = H'(g)H'(f).$$

But (see (20.3)) if we view f and g as right operators in $End(_RM)$ and $H'(f)$ and $H'(g)$ as left operators in $End(H'(M)_S)$, according to our usual convention (§4), then the order is straightened out to give an isomorphism

$$End(_RM) \to End(H'(M)_S).$$

Note also that our statement (23.3) is not the complete dual of (21.2). Indeed, in (21.2) we have that under equivalences, monomorphisms and epimorphisms are preserved. Now it is true that for dualities, "monomorphisms" and "epimorphisms" are reversed—but here we mean monomorphisms and epimorphisms *in the categories* $_R\mathsf{C}$ and D_S, and these need not be monomorphisms and epimorphisms in the categories $_R\mathsf{M}$ and M_S. (See Exercise (4.2).)

23.4. Lemma. *Let (H', H'') be a duality between full subcategories $_R\mathsf{C}$ and D_S of $_R\mathsf{M}$ and M_S. Then in the notation of (23.2) the homomorphisms*

$$\mu : Hom_S(N, H'(M)) \to Hom_R(M, H''(N))$$

$$v : Hom_S(H'(M), N) \to Hom_R(H''(N), M)$$

are isomorphisms natural in each variable. In particular, for each

$$\gamma \in Hom_S(N_1, H'(M_1)), \qquad \delta \in Hom_S(H'(M_2), N_2),$$

$$\bar{\gamma} \in Hom_R(M_1, H''(N_1)), \qquad \bar{\delta} \in Hom_R(H''(N_2), M_2),$$

and for each

$$h : M_2 \to M_1, \qquad k : N_2 \to N_1,$$

we have

(1) $\mu(H'(h)\gamma k) = H''(k)\mu(\gamma)h$

(2) $v(k\delta H'(h)) = hv(\delta)H''(k)$
(3) $\mu^{-1}(H''(k)\bar{\gamma}h) = H'(h)\mu^{-1}(\bar{\gamma})k$
(4) $v^{-1}(h\bar{\delta}H''(k)) = kv^{-1}(\bar{\delta})H'(h).$ ☐

For example, the assertion (1) can be illustrated by the diagrams

$$
\begin{array}{ccc}
N_1 & \xrightarrow{\gamma} & H'(M_1) \\
{\scriptstyle k}\uparrow & & \downarrow{\scriptstyle H'(h)} \\
N_2 & \xrightarrow{H'(h)\gamma k} & H'(M_2)
\end{array}
\qquad
\begin{array}{ccc}
H''(N_1) & \xleftarrow{\mu(\gamma)} & M_1 \\
{\scriptstyle H''(k)}\downarrow & & \uparrow{\scriptstyle h} \\
H''(N_2) & \xleftarrow{\mu(H'(h)\gamma k)} & M_2
\end{array}
$$

A Characterization of Dualities

The main result of this section, due to Morita [58a], characterizes dualities between full subcategories of modules as "U-dualities" for some bimodule U provided only that the categories be "closed under isomorphic images" and contain the appropriate regular modules. Thus,

23.5. Theorem [Morita]. *Let R and S be rings and let $_R\mathsf{C}$ and D_S be full subcategories of $_R\mathsf{M}$ and M_S such that*

$$_RR \in {}_R\mathsf{C} \qquad and \qquad S_S \in \mathsf{D}_S,$$

and such that every module in $_R\mathscr{M}$ (respectively, $\mathscr{M}_S$) isomorphic to one in $_R\mathsf{C}$ (D_S) is in $_R\mathsf{C}$ (C_S). If (H', H'') is a duality between the categories $_R\mathsf{C}$ and D_S,

$$H' : {}_R\mathsf{C} \to \mathsf{D}_S \qquad and \qquad H'' : \mathsf{D}_S \to {}_R\mathsf{C},$$

then there is a bimodule $_RU_S$ such that
(1) $_RU \cong H''(S)$ and $U_S \cong H'(R)$;
(2) There are natural isomorphisms

$$H' \cong Hom_R(\_, U) \qquad and \qquad H'' \cong Hom_S(\_, U);$$

(3) All $M \in {}_R\mathsf{C}$ and all $N \in \mathsf{D}_S$ are U-reflexive.

Proof. We adopt the notation of (23.2). By hypothesis $_RR \in {}_R\mathsf{C}$ and $S_S \in \mathsf{D}_S$. Thus

$$U = H'(R) \in \mathsf{D}_S \qquad and \qquad V = H''(S) \in {}_R\mathsf{C};$$

and (see §20) the regular bimodules $_RR_R$ and $_SS_S$ induce canonical bimodule structures

$$_RU_S = H'(R) \qquad and \qquad _RV_S = H''(S).$$

Since μ is natural in both variables, for each $M \in {}_R\mathsf{C}$ and each $N \in \mathsf{D}_S$,

$$\mu_{RN} : Hom_S(N, H'(R)) \to Hom_R(R, H''(N))$$

is a left R-isomorphism and

$$\mu_{MS} : Hom_S(S, H'(M)) \to Hom_R(M, H''(S))$$

is a right S-isomorphism. (See (20.4).) Moreover, since the first of these isomorphisms is natural in N,

$$\mu_{RS}: Hom_S(S, H'(R)) \to Hom_R(R, H''(S))$$

is an (R, S)-isomorphism. Whence (4.5), as (R, S)-bimodules

$$U \cong Hom_S(S, H'(R)) \cong Hom_R(S, H''(S)) \cong V.$$

Now since μ_{MS} is natural in M, it induces a natural S-isomorphism

$$Hom_S(S, H'(\ )) \cong Hom_R(\_, H''(S)).$$

Similarly μ_{RN} induces a natural R-isomorphism

$$Hom_R(R, H''(\ )) \cong Hom_S(\_, H'(R)).$$

So by (20.1.1) and (20.5.2), there are natural isomorphisms:

$$H' \cong Hom_S(S, H'(\ )) \cong Hom_R(\_, H''(S)) \cong Hom_R(\_, U),$$

and

$$H'' \cong Hom_R(R, H''(\ )) \cong Hom_S(\_, H'(R)) \cong Hom_S(\_, U).$$

This gives both (1) and (2).

Now to prove (3) we may assume that there is a bimodule $_RU_S$ and that

$$H' = Hom_R(\_, U) \qquad \text{and} \qquad H'' = Hom_S(\_, U)$$

gives a duality between $_RC$ and D_S. Let $N \in D_S$. In order to prove that N is U-reflexive we first show that the natural S-isomorphism

$$\zeta_N: H'H''(N) \to N$$

determines an R-automorphism $\alpha: H''(N) \to H''(N)$. Indeed, define α via

$$(\alpha(g))(n) = (\zeta_N^{-1}(n))(g)$$

for $g \in H''(N) = Hom_S(N, U)$ and $n \in N$. It is easy to check that α is a (monic) R-endomorphism of $H''(N)$. Now since $_RU_S$ is a bimodule

$$H'(U) = Hom_R(U, U)$$

is an (S, S)-bimodule and, by the naturality of v (23.4.2),

$$v = v_{UN}: Hom_S(H'(U), N) \to Hom_R(H''(N), U) = H'H''(N)$$

is a right S-isomorphism. Thus,

$$\gamma_N = v_{UN}^{-1} \circ \zeta_N^{-1}: N \to Hom_S(H'(U), N)$$

is a natural isomorphism. That is, these γ_N induce a natural isomorphism of functors

$$\gamma: 1_{D_S} \to Hom_S(H'(U), \_).$$

Thus, for each $N \in D_S$,

(1) $\psi: Hom_S(N, U) \to Hom_S(Hom_S(H'(U), N), H''H'(U))$

defined via

$$\psi : g \mapsto Hom_S(H'(U), g)$$

is a $\mathbb{Z}$-isomorphism. (See Exercise (20.5).) Also since $_RU_S$ is a bimodule, $H''H'(U)$ is a left R- right S-bimodule via H'' and H' (§20). Then, since γ_N is an isomorphism,

(2) $Hom_S(\gamma_N, H''H'(U)) : Hom_S(Hom_S(H'(U), N), H''H'(U)) \to$
$$Hom_S(N, H''H'(U))$$

is also a $\mathbb{Z}$-isomorphism. Since η is natural, $\eta_U : H''H'(U) \to U$ is an S-isomorphism and

(3) $$Hom_S(N, \eta_U) : Hom_S(N, H''H'(U)) \to Hom_S(N, U)$$

is a $\mathbb{Z}$-isomorphism. Now throwing these together, we have for each $g \in Hom_S(N, U)$ and each $n \in N$,

$$((Hom_S(N, \eta_U) \circ Hom_S(\gamma_N, H''H'(U))) \circ \psi)(g))(n)$$
$$= (\eta_U \circ Hom_S(H'(U), g) \circ \gamma_N)(n)$$
$$= \eta_U(g \circ \gamma_N(n)) = \eta_U(Hom_S(\gamma_N(n), U)(g))$$
$$= (\eta_U \circ H''(\gamma_N(n)))(g) = (v_{UN}(\gamma_N(n)))(g)$$
$$= (\zeta_N^{-1}(n))(g) = (\alpha(g))(n).$$

Thus α is just the composite of the three isomorphisms (1), (2), and (3); hence α is an R-automorphism of $H''(N)$ as claimed.

Now given $N \in D_S$, we know that $N \cong H'H''(N)$, via ζ_N, so in particular, N is isomorphic to the U-dual of a module and thus, N is U-torsionless (20.14). So to see that N is U-reflexive, we need only show that if $\xi \in H'H''(N)$, then there is an $n \in N$ such that $\xi(g) = g(n)$ for all $g \in H''(N)$. But let $\xi \in H'H''(N)$. Then since α is an endomorphism of $H''(N)$, we have

$$\xi \circ \alpha \in H'H''(N) = Im\, \zeta_N^{-1}.$$

Thus there does exist an $n \in N$ such that

$$\xi \circ \alpha = \zeta_N^{-1}(n).$$

Then for all $g \in H''(N)$,

$$\xi(g) = (\zeta_N^{-1}(n))(\alpha^{-1}(g)) = \alpha(\alpha^{-1}(g))(n) = g(n),$$

whence N is U-reflexive. Similarly, each $M \in {}_RC$ is U-reflexive and the proof is complete. □

23. Exercises

1. Let $C = (\mathscr{C}, mor, \circ)$ be a category and let $C^{op} = (\mathscr{C}^{op}, mor^{op}, *)$ where $\mathscr{C}^{op} = \mathscr{C}$, where $mor^{op}(A, B) = mor(B, A)$ for each pair A, B of $\mathscr{C}^{op}$, and

where $f * g = g \circ f$. For each $A \in \mathscr{C}$ and each morphism f in C, let $op(A) = A$ and $op(f) = f$. Prove that:

(1) C^{op} is a category and $\mathsf{C}^{op\,op} = \mathsf{C}$.

(2) $op : \mathsf{C} \to \mathsf{C}^{op}$ is a contravariant functor.

(3) (op, op) is a duality between C and C^{op}.

2. Let C and D be categories and $H' : \mathsf{C} \to \mathsf{D}$ and $H'' : \mathsf{D} \to \mathsf{C}$ be contravariant functors. Prove that (H', H'') is a duality between C and D iff

$$(op) \circ H' : \mathsf{C} \to \mathsf{D}^{op} \qquad \text{and} \qquad H'' \circ (op) : \mathsf{D}^{op} \to \mathsf{C}$$

are inverse equivalences. (See Exercise (23.1).)

3. Let C be a category. Recall (Exercise (3.4)) that a morphism $f : A \to B$ in C is a *monomorphism* (an *epimorphism*) iff it is cancellable on the left (on the right). Suppose (H', H'') is a duality between categories C and D. Prove that a morphism $f : A \to B$ in C is a monomorphism (an epimorphism) iff $H'(f) : H'(B) \to H'(A)$ is an epimorphism (a monomorphism) in D. Thus in particular, the statements "f is a monomorphism" and "f is an epimorphism" are *dual*.

4. Each of the following terms has been defined in module categories $_R\mathsf{M}$. Extend these definitions to arbitrary categories C in such a way that in each pair the terms are dual to one another (i.e., so that, under a duality between C and D, the first corresponds to the second).

(1) Projective object; injective object.

(2) Superfluous epimorphism; essential monomorphism.

(3) Projective cover; injective envelope.

(4) Generator; cogenerator.

(5) Direct sum; direct product.

(6) Simple object; simple object.

5. Let R be the ring of upper triangular 2×2 matrices over a field S. Let $_R U_S$ be the set of column vectors $S^{(2)}$.

(1) Show that $_R U$ is the unique (to within isomorphism) simple object in the category $_R\mathsf{R}[U]$. (Note that $_R U$ is not simple in $_R\mathsf{M}$.)

(2) Describe $_R\mathsf{R}[U]$ and $\mathsf{R}[U]_S$.

6. Let R be a finite dimensional algebra over a field K. Then an R-module is finitely generated iff it is finite dimensional as a K-vector space iff it has a composition series. Recall that a finite dimensional K-space M and its dual

$$M^* = Hom_K(M, K)$$

have the same dimension. Assuming that all modules below are finitely generated, prove that:

(1) $(\ )^* : {}_R\mathsf{FM} \to \mathsf{FM}_R$ and $(\ )^* : \mathsf{FM}_R \to {}_R\mathsf{FM}$ define a duality. [Hint: The usual evaluation map

$$\sigma_M : M \to M^{**}$$

with $[\sigma_M(m)](\gamma) = \gamma(m)$ is an R-isomorphism.]

(2) $0 \to M_1 \overset{f}{\to} M_2 \overset{g}{\to} M_3 \to 0$ is (split) exact iff its dual $0 \to M_3^* \overset{g^*}{\to} M_2^* \overset{f^*}{\to} M_1^* \to 0$ is (split) exact.

(3) M is simple, semisimple, indecomposable (respectively) iff M^* is.

(4) $c(M) = c(M^*)$.

(5) $Soc(M^*) \cong (M/Rad\ M)^*$ and $M^*/Rad\ M^* \cong (Soc\ M)^*$.

(6) M is injective (projective) iff M^* is projective (injective).

(7) $f : M_1 \to M_2$ is an injective envelope (projective cover) iff $f^* : M_2^* \to M_1^*$ is a projective cover (injective envelope). (Recall (Exercise (17.20)) that projective covers exist over artinian rings.)

(8) M is a generator (cogenerator) iff M^* is a cogenerator (generator).

(9) $l_R(M) = r_R(M^*)$.

(10) If e is a primitive idempotent in R, then $Re/Je \cong (eR/eJ)^*$ $(J = J(R))$ and $E(Re/Je) \cong (eR)^*$.

7. Let R and S be rings and let $_R U_S$ be a bimodule with both $_R U$ and U_S injective. Let $_R\mathsf{FLM}$ and FLM_S be the full subcategories of $_R\mathsf{M}$ and M_S of modules of finite length. Prove that $H' = Hom_R(\_, U)$ and $H'' = Hom_S(\_, U)$ define a duality (H', H'') between $_R\mathsf{FLM}$ and FLM_S iff for each simple R-module $_R T$ and each simple S-module V_S, the modules $H'(T)$ and $H''(V)$ are simple. [Hint: Induct on the length to show that each $M \in {}_R\mathsf{FLM}$ and each $N \in \mathsf{FLM}_S$ is U-reflexive.]

8. Prove that if R is commutative, then there exists a duality (H', H'') between $_R\mathsf{FLM}$ and itself. [Hint: Let $_R U$ be the minimal injective cogenerator of R. Apply Exercise (23.7).]

9. Let $(P, \leq)$ be a poset (see (0.5)). Define a category $\mathsf{C}(P, \leq) = (P, mor, \circ)$ where for each $a, b \in P$,

$$mor(a, b) = \begin{cases} \{(a, b)\} & \text{if} \quad a \leq b \\ \varnothing & \text{if} \quad a \nleq b \end{cases}$$

and where $(a, b) \circ (b, c) = (a, c)$ if $a \leq b$ and $b \leq c$. A category in which every isomorphism is an identity is a *hull category*. Prove that:

(1) $(P, \leq) \mapsto \mathsf{C}(P, \leq)$ defines a one-to-one correspondence between the class of all posets and the class of all hull categories $\mathsf{C} = (\mathscr{C}, mor, \circ)$ such that $\mathscr{C}$ is a set and for each $A, B \in \mathscr{C}$, $mor(A, B)$ is at most a singleton.

(2) $(P, \leq)$ is a lattice iff $\mathsf{C}(P, \leq)$ is closed under finite direct sums and direct products. (See Exercise (23.4.5).)

(3) If $(P, \leq)$ and $(P', \leq')$ are posets, then a mapping $f : P \to P'$ is order-preserving (order-reversing) iff the induced map $\mathsf{C}(P, \leq) \to \mathsf{C}(P', \leq')$ is a covariant (contravariant) functor.

(4) Two posets $(P, \leq)$ and $(P', \leq')$ are isomorphic (anti-isomorphic) iff there is an equivalence (a duality) between $\mathsf{C}(P, \leq)$ and $\mathsf{C}(P', \leq')$. [Hint: In $\mathsf{C}(P, \leq)$, there is an isomorphism $a \to b$ iff $a = b$.]

(5) If $(P, \leq)$ is a poset (lattice), then $(P, \geq)$ is a poset (lattice) and $\mathsf{C}(P, \geq) = \mathsf{C}(P, \leq)^{op}$. Thus, the class of posets (lattices) is self dual. In particular, duality in posets is a special case of categorical duality.

§24. Morita Dualities

The dualities that are of interest in practice are between subcategories of $_R\mathsf{M}$ and M_S that do contain $_R R$ and S_S, and for these, Theorem 23.5 permits a substantial simplification. Therefore, throughout this section we shall let R and S be rings and let $_R U_S$ be a fixed bimodule. Also for each module $_R M$ and each module N_S we set

$$M^* = H'(M) = Hom_R(M, U)$$
$$N^* = H''(N) = Hom_S(N, U),$$

and we use σ for both natural transformations

$$\sigma_M : M \to M^{**} \quad \text{and} \quad \sigma_N : N \to N^{**}.$$

We say that $_R U_S$ defines a *Morita duality* or that the duality given by

$$Hom_R(\_, U) \quad \text{and} \quad Hom_S(\_, U)$$

is a *Morita duality* in case
   (1) $_R R$ and S_S are U-reflexive;
   (2) Every submodule and every factor module of a U-reflexive module is U-reflexive.

A Characterization of Morita Dualities

A familiar example of a Morita duality is afforded by the case in which $R = S$ is a division ring and $_R U_R = {}_R R_R$. Then the reflexive modules are the finite dimensional R-vector spaces and U defines a Morita duality. In general, however, Morita dualities appear only infrequently.

24.1. Theorem. *Let R and S be rings. Then for a bimodule $_R U_S$ the following statements are equivalent:*
   (a) $_R U_S$ *defines a Morita duality;*
   (b) *Every factor module of $_R R$, S_S, $_R U$ and U_S is U-reflexive;*
   (c) $_R U_S$ *is a balanced bimodule such that $_R U$ and U_S are injective cogenerators.*

Proof. (a) $\Rightarrow$ (b). Assume (a). Since by (4.5) $U_S \cong (_R R)^*$, since by hypothesis R is reflexive, and since by (20.14) duals of reflexives are reflexive, we know that U_S is reflexive. Similarly, $_R U$ is reflexive. Now (b) follows from (a) and the reflexivity of $_R R$, S_S, $_R U$ and U_S.

(b) $\Rightarrow$ (c). Assume (b). Then by (20.16) $_R U_S$ is a balanced bimodule, so it will suffice to prove that $_R U$ is an injective cogenerator. To see that $_R U$ is injective, let I be a left ideal of R and consider the natural short exact sequence

$$0 \to I \xrightarrow{f} R \xrightarrow{g} R/I \to 0.$$

Then there is a factor module T of $U_S \cong R^*$ and right S-homomorphisms h and k so that

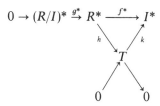

is commutative with the horizontal and diagonal sequences exact. Now T, a factor of U, is U-reflexive by hypothesis. Thus by (20.14) its dual T^* is U-reflexive. Since $\sigma_R f = f^{**}\sigma_I = h^*k^*\sigma_I$, we have a commutative diagram

$$
\begin{array}{ccccccccc}
0 & \to & T^* & \xrightarrow{h^*} & R^{**} & \xrightarrow{g^{**}} & (R/I)^{**} & \to & 0 \\
& & \uparrow{\scriptstyle k^* \circ \sigma_I} & & \uparrow{\scriptstyle \sigma_R} & & \uparrow{\scriptstyle \sigma_{R/I}} & & \\
0 & \to & I & \xrightarrow{f} & R & \xrightarrow{g} & R/I & \longrightarrow & 0.
\end{array}
$$

Since R and R/I are reflexive and $(\ )^* = Hom(\_, U)$ is left exact, the two right-hand vertical maps are isomorphisms and the rows are exact. Thus, since the diagram commutes, the map $k^* \circ \sigma_I : I \to T^*$ is an isomorphism. But then I, isomorphic to T^*, is U-reflexive, so $k^* : I^{**} \to T^*$ is an isomorphism. But then (see (16.2)) $k^{**} : T^{**} \to I^{***}$ is also an isomorphism; so since I^*, isomorphic to the dual of a U-reflexive module, and T are U-reflexive, we see from the commutative diagram

$$
\begin{array}{ccc}
T^{**} & \xrightarrow{k^{**}} & I^{***} \\
\uparrow{\scriptstyle \sigma_T} & & \uparrow{\scriptstyle \sigma_{I^*}} \\
T & \xrightarrow{k} & I^*
\end{array}
$$

that k is an isomorphism. Whence, since $Im f^* = Im k$,

$$ 0 \to (R/I)^* \to R^* \to I^* \to 0 $$

is exact. Thus, by (18.3), $_R U$ is injective. Also, any simple left R-module is a factor of R whence by hypothesis every simple left R-module is U-reflexive. Therefore, every simple left R-module is U-torsionless and by (18.15) $_R U$ is a cogenerator. Similarly, U_S is an injective cogenerator.

(c) $\Rightarrow$ (a). Assume (c). Since cogenerators are faithful (8.22), $_R U_S$ is a faithfully balanced bimodule. But then, by (20.16), $_R R$ and S_S are U-reflexive. Let M be U-reflexive, let $K \le M$ and consider the commutative diagram

$$
\begin{array}{ccccccccc}
0 & \to & K & \longrightarrow & M & \longrightarrow & M/K & \longrightarrow & 0 \\
& & \downarrow{\scriptstyle \sigma_K} & & \downarrow{\scriptstyle \sigma_M} & & \downarrow{\scriptstyle \sigma_{M/K}} & & \\
0 & \to & K^{**} & \to & M^{**} & \to & (M/K)^{**} & \to & 0.
\end{array}
$$

Since $_R U$ and U_S are injective, the bottom row is exact. Since U cogenerates

M/K, $\sigma_{M/K}$ is monic. But now, since σ_M is an isomorphism, it easily follows that $\sigma_{M/K}$ and σ_K are also isomorphisms (see Exercise (3.11)). □

24.2. Corollary. *Let $_RU_S$ define a Morita duality. Then a sequence of R-homomorphisms*

$$M_1 \xrightarrow{f} M_2 \xrightarrow{g} M_3$$

is exact if and only if

$$M_3^* \xrightarrow{g^*} M_2^* \xrightarrow{f^*} M_1^*$$

is exact. In particular, f is epic (monic) iff f^* is monic (epic).

Proof. Since $_RU_S$ defines a Morita duality, $_RU$ is an injective cogenerator.
□

Of course, if $_RU_S$ defines a Morita duality, then the obvious version of (24.2) for S-homomorphisms also holds.

Annihilators

A module $_RM$ together with its $_RU_S$-dual M^* determine a "pairing" of the two modules M, M^* in U, an (R, S)-bilinear map $M \times M^* \to U$ defined via

$$(x, f) \mapsto f(x).$$

In general, such pairings of modules form the basis of the general theory of annihilators.

Suppose that $_RM$ and N_S are modules. If $_RU_S$ is a bimodule then a function

$$\mu : M \times N \to U$$

is (R, S)-*bilinear* in case for all $m, m' \in M$, $n, n' \in N$, $r \in R$, and $s \in S$

$$\mu(m + m', n) = \mu(m, n) + \mu(m', n)$$

$$\mu(m, n + n') = \mu(m, n) + \mu(m, n')$$

$$\mu(rm, ns) = r\mu(m, n)s.$$

Let $\mu : M \times N \to U$ be (R, S)-bilinear. For each subset $A \subseteq M$ the *right annihilator of A in N* (with respect to μ) is defined to be

$$r_N(A) = \{n \in N \,|\, \mu(a, n) = 0 \quad (a \in A)\}.$$

For each subset $B \subseteq N$, the *left annihilator of B in M* (with respect to μ) is

$$l_M(B) = \{m \in M \,|\, \mu(m, b) = 0 \quad (b \in B)\}.$$

Clearly, $r_N(A)$ and $l_M(B)$ are submodules of N and M, respectively. Thus μ induces mappings

$$M' \mapsto r_N(M') \qquad (M' \leq M)$$

$$N' \mapsto l_M(N') \qquad (N' \leq N)$$

between the lattices of submodules of $_RM$ and of N_S.

Note that if Q is a third ring and if $_RM_Q$ and $_QN_S$ are bimodules, then the tensor product

$$\otimes : M \times N \to M \otimes_Q N$$

is a Q-balanced (R, S)-bilinear map.

Suppose that M is a left R-module. Then R-scalar multiplication determines an $(R, \mathbb{Z})$-bilinear map

$$\mu : R \times M \to {}_RM_{\mathbb{Z}}.$$

Thus it is clear that the annihilators with respect to this map μ are simply the annihilators of subsets of M and R as defined in §2. That is, the present general definition of annihilators is consistent with the earlier special one. Indeed the proof of the following important result involves just trivial modifications of those of (2.15) and (2.16), and so will be omitted.

24.3. Proposition. *Let R and S be rings, let $_RM$, N_S and $_RU_S$ be modules and let $\mu : M \times N \to U$ be (R, S)-bilinear. Then for all submodules M', M'', and M_α $(\alpha \in A)$ of M:*
(1) $M' \leq M''$ *implies* $r_N(M') \geq r_N(M'')$;
(2) $M' \leq l_M r_N(M')$;
(3) $r_N(M') = r_N l_M r_N(M')$;
(4) $r_N(\Sigma_A M_\alpha) = \cap_A r_N(M_\alpha)$;
(5) $r_N(\cap_A M_\alpha) \geq \Sigma_A r_N(M_\alpha)$.
Moreover, analogous statements hold for the S-submodules of N. □

As we mentioned in §2 the inequalities of (2) and (5) of (24.3) cannot be strengthened to equalities. (See Exercise (2.15).) However, in certain important cases arising from dualities, they are equalities.

For the present then we shall assume that $_RM$, N_S, and $_RU_S$ are modules. Then there are (R, S)-bilinear maps

$$M \times M^* \to U \qquad \text{and} \qquad N^* \times N \to U$$

defined by

$$(m, f) \mapsto f(m) \qquad \text{and} \qquad (g, n) \mapsto g(n).$$

respectively. Thus, if $A \subseteq M$ and $B \subseteq M^*$,

$$r_{M^*}(A) = \{f : M \to U \,|\, A \subseteq \mathrm{Ker}\, f\}$$

$$l_M(B) = \cap \{\mathrm{Ker}\, f \,|\, f \in B\}.$$

Until further notice we shall compute annihilators with respect to these bilinear maps induced by the U-duals.

24.4. Lemma. *Let $_RM$ and $_RU_S$ be modules. Then for each submodule $K \leq M$*

$$Rej_{M/K}(U) = l_M(r_{M^*}(K))/K.$$

In particular,

    (1) $l_M(M^*) = Rej_M(U) = Ker\,\sigma_M;$

    (2) $l_M(r_{M^*}(K)) = K$ *iff* U *cogenerates* M/K.

Proof. First recall (24.3.2) that $K \leq l_M(r_{M^*}(K))$. If $x \in l_M(r_{M^*}(K))$ and $f: M/K \to U$, then the natural epimorphism $n_K: M \to M/K$ composed with f gives $f \circ n_K \in r_{M^*}(K)$. So $f \circ n_K$ is annihilated by x. That is, $f(x + K) = f \circ n_K(x) = 0$, and $x + K \in Ker\,f$. Thus,

$$Rej_{M/K}(U) \supseteq l_M(r_{M^*}(K))/K.$$

But every $g \in r_{M^*}(K)$ (so $K \leq Ker\,g$) factors through n_K (3.6.1). That is, $g = f \circ n_K$ where $f: M/K \to U$. Thus if $x + K \in Rej_{M/K}(U)$, then for each such g we have $g(x) = f \circ n_K(x) = f(x + K) = 0$; so $x \in l_M(r_{M^*}(K))$. Now for part (1) of the last statement take $K = 0$ and recall (20.12). For part (2) apply (8.13).   □

Properties of Morita Dualities

Each bimodule $_RU_S$ determines a duality (H', H'') between the categories $_R\mathsf{R}[U]$ and $\mathsf{R}_S[U]$ of U-reflexive modules (23.1). The duality (H', H'') dualizes properties in the categories $_R\mathsf{R}[U]$ and $\mathsf{R}_S[U]$, but in general it need not dualize them to properties in $_R\mathsf{M}$ and M_S. (See Exercises (23.4) and (23.5).) For Morita dualities, however, we can claim more. (See also (24.2) and Exercises (24.4) through (24.6).)

We have already proved (in (20.14)) the first statement of the following theorem.

24.5. Theorem. *Let R and S be rings, let $_RU_S$ define a Morita duality, and let $_RM$ and N_S be U-reflexive. Then M^* and N^* are U-reflexive, and with respect to the canonical pairings induced by U-duality,*

    (1) *For each $K \leq M$ and each $L \leq M^*$,*

$$l_M r_{M^*}(K) = K \qquad and \qquad r_{M^*} l_M(L) = L;$$

    (2) *For each $L \leq N$ and each $K \leq N^*$,*

$$r_N l_{N^*}(L) = L \qquad and \qquad l_{N^*} r_N(K) = K;$$

    (3) *The lattices of submodules of M and M^* are anti-isomorphic via the mapping $K \mapsto r_{M^*}(K)$;*

    (4) *The lattices of submodules of N and N^* are anti-isomorphic via the mapping $L \mapsto l_{N^*}(L)$.*

Proof. Since $_RU$ and U_S are cogenerators the first assertion of (1) and (2) follow from Lemma (24.4). So letting $M^* = N$, we have, by the first part of (2),

$$r_{M^*} l_{M^{**}}(L) = L \qquad (L \leq M^*).$$

But since σ_M is an isomorphism, we have for $L \leq M^*$

$$l_{M^{**}}(L) = \{\sigma_M(x) \mid \sigma_M(x)(h) = 0 \text{ for all } h \in L\}$$
$$= \{\sigma_M(x) \mid h(x) = 0 \text{ for all } h \in L\}$$
$$= \sigma_M(l_M(L))$$

and hence

$$r_{M^*}(l_{M^{**}}(L)) = \{f \in M^* \mid \sigma_M(y)(f) = 0 \text{ for all } y \in l_M(L)\}$$
$$= r_{M^*}(l_M(L)).$$

Thus the second assertion of (1) follows from the first of (2) and, by symmetry, we have the second of (2). Now (3) and (4) follow from (1) and (2) and (24.3). $\square$

24.6. Theorem. *Let R and S be rings. If there exists a bimodule $_RU_S$ that defines a Morita duality, then,*

(1) *The lattices of two-sided ideals of R and S are isomorphic;*

(2) *The centers of R and S are isomorphic;*

(3) *Every finitely generated or finitely cogenerated left R- (right S-) module is reflexive;*

(4) *A left R- (right S-) module is finitely generated projective if and only if its U-dual is finitely cogenerated injective.*

Proof. For (1), recall that $R^* = \operatorname{Hom}_R(R, U)$ is isomorphic to $_RU_S$ as an (R, S)-bimodule (4.5). Also for each ideal I of R, its annihilator $r_{R^*}(I)$ is clearly an (R, S)-submodule of R^*. On the other hand, for any (R, S)-submodule V of R^*, its left annihilator $l_R(V)$ is clearly an ideal of R. So it follows from (24.5) that the lattices of ideals of R and of (R, S)-submodules of $_RU_S$ are anti-isomorphic. Similarly the lattices of ideals of S and of (R, S)-submodules of $_RU_S$ are anti-isomorphic. Thus the ideal lattices of R and of S are isomorphic.

From the fact (24.1) that $_RU_S$ is faithfully balanced it follows easily (see Exercise (4.5)) that $Cen\,R \cong Cen\,S$.

It follows from (20.13) that the finite direct sums of reflexive modules are reflexive. Thus, since $_RR$ is a generator and $_RU$ is a cogenerator (24.1) and since both are reflexive, (3) follows from the definition of a Morita duality.

Finally, for (4), since R is reflexive and $R^* \cong U$, a module $_RM$ is finitely generated projective if and only if for some integer n there is a split epimorphism $R^{(n)} - \oplus \to M \to 0$ if and only if there is a split epimorphism $0 \to M^* - \oplus \to U^{(n)}$. However, by (24.5) and (24.1), U is a finitely cogenerated injective cogenerator, so this last condition is equivalent to M^* being finitely cogenerated injective. $\square$

Dualities of Finitely Generated Modules

As we have noted before, there is a duality between the finitely generated left and right modules over a division ring. We conclude this section by showing that in order to have a duality between $_R$FM and FM$_S$ it is necessary that R be left artinian and S be right artinian.

24.7. Lemma [Osofsky]. *If a bimodule $_RU_S$ defines a Morita duality, then no infinite direct sum of non-zero left R-modules is U-reflexive.*

Proof. Suppose that $_RU_S$ defines a Morita duality and that $(M_\alpha)_{\alpha \in A}$ is an indexed set of non-zero submodules of $_RM$ such that $M = \oplus_A M_\alpha$ is U-reflexive. Let $(p_\alpha)_{\alpha \in A}$ be the projections for this direct sum. Since each M_α is U-torsionless, we see at once that there is an $f \in M^*$ such that $(f \mid M_\alpha) \neq 0$ for each $\alpha \in A$. Since $\cap_A Ker\, p_\alpha = 0$, we have by (24.5) that

$$\Sigma_A \, \mathbf{r}_{M^*}(Ker\, p_\alpha) = M^*.$$

Thus

$$f = g_{\alpha_1} + \dots + g_{\alpha_n}$$

with

$$g_{\alpha_i} \in \mathbf{r}_{M^*}(Ker\, p_{\alpha_i}) \qquad (i = 1, \dots, n).$$

If $\alpha \in A \backslash \{\alpha_1, \dots, \alpha_n\}$, then $M_\alpha \subseteq Ker\, p_{\alpha_i}\, (i = 1, \dots, n)$ and

$$f(M_\alpha) \subseteq g_{\alpha_1}(M_\alpha) + \dots + g_{\alpha_n}(M_\alpha) = 0.$$

Thus we have $A = \{\alpha_1, \dots, \alpha_n\}$. $\qquad\qquad\qquad\qquad\qquad\qquad\qquad\qquad\quad \square$

24.8. Theorem. *Let R and S be rings. Then the following are equivalent:*
(a) *There exists a duality between the category $_R$FM of finitely generated left R-modules and the category FM$_S$ of finitely generated right S-modules;*
(b) *R is left artinian and some bimodule $_RU_S$ defines a Morita duality;*
(c) *S is right artinian and some bimodule $_RU_S$ defines a Morita duality.*
Moreover, if R, S and U satisfy either of the last two conditions, then a left R- (right S-) module is U-reflexive iff it is finitely generated iff it is finitely cogenerated.

Proof. (a) $\Rightarrow$ (b). Assume (a). Then by (23.5) there is a bimodule $_RU_S$ that defines a duality between $_R$FM and FM$_S$; and all members of $_R$FM and FM$_S$ are U-reflexive. Now since each factor module of $_RR$, S_S, $_RU \cong (S)^*$ and $U_S \cong (R)^*$ is finitely generated we have by (24.1) that U defines a Morita duality. Hence it follows from (24.5) that every finitely generated left R-module M, being isomorphic to the U-dual of the finitely generated S-module M^*, is finitely cogenerated. Thus R is left artinian by (10.18).

(b) $\Rightarrow$ (a). Suppose that R is left artinian and $_RU_S$ defines a Morita duality. By (24.6.3) every finitely generated left R- (right S-) module is U-reflexive. Moreover, if $M \in {}_R$FM, then M is finitely cogenerated (10.18) and by (24.5) $M^* \in$ FM$_S$. Let $N \in$ FM$_S$. Then N^* is U-reflexive. So, since U defines a Morita duality, the semisimple module $N^*/Rad\, N^*$ cannot be an infinite direct sum of simples (24.7). Thus $N^*/Rad\, N^*$ is finitely generated, and hence by (15.21), $N^* \in {}_R$FM. We have now shown that $_RU_S$ defines a duality between $_R$FM and FM$_S$.

(a) ⟺ (c). By symmetry.
The last statement follows from (24.7), (15.21), and (24.5).   □

24.9. Corollary [Azumaya, Morita]. *Let R and S be rings and let $_RU_S$ be a bimodule. Then*

$$Hom_R(\_, U) \quad and \quad Hom_S(\_, U)$$

is a duality between the categories $_R$FM and FM$_S$ iff $_RR$ and S_S are artinian, $_RU$ and U_S are finitely generated injective cogenerators, and $_RU_S$ is balanced. □

(The conditions of this corollary are not the weakest possible. See Morita [58a] and Azumaya [59].)

24. Exercises

1. Prove that for no rings R and S is there a duality between $_R$M and M$_S$.
2. Prove that for no ring R and bimodule $_ZU_R$ does $_ZU_R$ define a Morita duality.
3. Let $_RM$, N_S, and $_RU_S$ be modules, and let $\mu: M \times N \to U$ be (R, S)-bilinear. Prove that if for each $M' \leq M$ and each $N' \leq N$

$$l_M r_N(M') = M' \quad and \quad r_N l_M(N') = N',$$

then for each indexed set $(M_\alpha)_{\alpha \in A}$ of submodules of M,

$$r_N(\cap_A M_\alpha) = \Sigma_A r_N(M_\alpha)$$

and for each indexed set $(N_\alpha)_{\alpha \in A}$ of submodules of N,

$$l_M(\cap_A N_\alpha) = \Sigma_A l_M(N_\alpha).$$

4. Let $_RU_S$ define a Morita duality. Let M be U-reflexive. Prove that:
   (1) M is simple, semisimple, indecomposable, of finite length n, respectively, iff $M^* = Hom(M, U)$ is.
   (2) M is finitely generated (noetherian) iff M^* is finitely cogenerated (artinian).
5. Let $_RU_S$ define a Morita duality. Let M_1, M_2 be U-reflexive, and let $f: M_1 \to M_2$. Prove that:
   (1) f is a superfluous epimorphism (essential monomorphism) iff for each h in the category $_RR[U]$, fh is epic implies h is epic. (hf is monic implies h is monic.) [Hint: See (5.13) and (5.15).]
   (2) $f: M_1 \to M_2$ is a superfluous epimorphism iff $f^*: M_2^* \to M_1^*$ is an essential monomorphism.
6. Again let $_RU_S$ define a Morita duality and let M be U-reflexive. Prove that:
   (1) If M is finitely generated, then $E(M^*)^*$ is a projective cover for M.
   (2) If M is finitely cogenerated and $P(M^*)$ is a projective cover for M^*, then $(P(M^*))^*$ is an injective envelope of M.
   (3) $Soc\, M^* \cong (M/Rad\, M)^*$ and $M^*/Rad\, M^* \cong (Soc\, M)^*$.
   (4) If R is left artinian, then M is faithful over R iff M^* is faithful over S.

7. Let $_RU_S$ define a Morita duality. Since U_S is isomorphic to $(_RR)^*$ and $_RU$ is isomorphic to $(S_S)^*$, there exist lattice anti-isomorphisms between the submodule lattices of $_RR$ and U_S and between those of S_S and $_RU$. (See (24.5).) Prove that:

   (1) For each $_RI \leq {}_RR$ and each $V_S \leq U_S$, $l_R(r_U(I)) = I$ and $r_U(l_R(V)) = V$.

   (2) For each $K_S \leq S_S$ and each $_RW \leq {}_RU$, $r_S(l_U(K)) = K$ and $l_U(r_S(W)) = W$.

8. Prove that if $_RU_S$ defines a Morita duality, then:

   (1) $R/J(R)$ and $S/J(S)$ are semisimple. [Hint: The sum of all minimal submodules in U_S is the sum of a finite number of them.]

   (2) $Soc\,_RU = Soc\,U_S$. [Hint: By (1) the intersection of the maximal left ideals in R equals the intersection of the maximal two-sided ideals in R.]

9. (1) Let R be a ring with radical $J = J(R)$ and let $_RT$ be simple. Prove that if R/J is semisimple and $J^2 = 0$, then $E(T)/T \cong Hom_R(J, T)$. [Hint: Exercise (20.15).]

   (2) P. M. Cohn [61] has shown that there is a division ring D and a division subring C of D such that D_C is finite dimensional and $_CD$ is not. Let R be the ring of all matrices

   $$r = \begin{bmatrix} d_1 & d_2 \\ 0 & c \end{bmatrix}$$

   with $d_1, d_2 \in D$ and $c \in C$. Prove that R is both left and right artinian but that $_R\mathsf{FM}$ has a duality with FM_S for no ring S. [Hint: Let $T \leq R$ consist of those r with $d_2 = c = 0$. Then $_RT$ is simple and $E(T)$ is not finitely generated (use (1)). Apply (24.9).]

10. A ring R is a *cogenerator ring* in case both $_RR$ and R_R are cogenerators. Prove that for a ring R the following are equivalent: (a) R is a cogenerator ring; (b) $_RR$ and R_R are injective cogenerators; (c) $_RR_R$ defines a Morita duality. [Hint: (a) $\Rightarrow$ (b). By (a) there is a set A with $_RR \leq E = E(R) \leq {}_RR^A$. If $\pi_\alpha(1) = e_\alpha$ for each $\alpha \in A$, let $I = \Sigma_A e_\alpha R \leq R_R$. Then $l_R(I) = 0$ so $I = R$. Say $1 = \Sigma_A e_\alpha r_\alpha$ with almost all r_α zero. Then $x \mapsto \Sigma_A \pi_\alpha(x)r_\alpha$ is a split epimorphism $E \to R$.]

11. A ring R has the *double annihilator property* in case

   $$r_R l_R(I) = I \qquad \text{and} \qquad l_R r_R(I') = I'$$

   for each right ideal I and each left ideal I'. Suppose that R has the double annihilator property. Then prove that the following are equivalent for R: (a) left artinian; (b) right artinian; (c) left noetherian; (d) right noetherian.

12. Prove that a ring R is a cogenerator ring iff $_RR$ and R_R are injective and R has the double annihilator property. [Hint: (18.15) and Exercise (24.7).]

13. A ring R is *quasi-Frobenius* in case it is an artinian cogenerator ring. Prove that if R is artinian, then the following are equivalent: (a) R is quasi-Frobenius; (b) $_RR$ and R_R are injective; (c) R has the double annihilator property; (d) the $_RR_R$-dual ( )* defines a duality between

$_R$FM and FM$_R$. [Hint: (b) $\Rightarrow$ (a). Exercise (18.31). (c) $\Rightarrow$ (b). Let $I \leq {}_RR$ and let $f : I \rightarrow R$. Induct on the length of a minimal generating set for I. If $I = Rx$, then $f(x) \in r_R l_R(x) = xR$, so $f(x) = xa$ for some $a \in R$. Suppose $I = I_1 + I_2$ and $f(x_i) = x_i a_i$ for all $x_i \in I_i \ (i = 1, 2)$. Then

$$a_1 - a_2 \in r_R(I_1 \cap I_2) = r_R(I_1) + r_R(I_2)$$

by Exercise (24.3), say $a_1 - a_2 = b_1 + b_2$. Then $f(x) = x(a_1 + b_1)$ for all $x \in I$.]

14. (1) Prove that every Frobenius algebra is quasi-Frobenius. (See Exercise (21.9).)

    (2) Prove that every semisimple ring is quasi-Frobenius.

    (3) Prove that if $R = R_1 \times \ldots \times R_n$, then R is a cogenerator ring (quasi-Frobenius ring) iff each R_i is.

    (4) Prove that if $R \approx S$, then R is a cogenerator ring (quasi-Frobenius ring) iff S is.

15. Quasi-Frobenius rings (introduced by Nakayama in 1939) have motivated much of the study of duality via bimodules. Each of the following conditions serves to characterize them. Although there is not room for the converses here, prove that a quasi-Frobenius ring R satisfies each of:

    (1) R is left noetherian and left or right self-injective.

    (2) R is left noetherian and $_RR$ or R_R is a cogenerator.

    (3) R is left or right noetherian and every faithful left R-module is a generator.

    (4) R is left (right) artinian and the $_RR_R$-dual of each simple module is simple.

    (5) Every projective left R-module is injective.

    (6) Every injective left R-module is projective.

Chapter 7
Injective Modules, Projective Modules, and Their Decompositions

In this chapter we return to the study of decompositions of modules —specifically of injective and projective modules. First we examine characterizations of noetherian rings in terms of the structure of injective modules. Then, after considering the decomposition theory of direct sums of countably generated modules, we proceed to the study of semiperfect and perfect rings (those over which all finitely generated modules and, respectively, all modules have projective covers). In the final section we show that the structure of the endomorphism ring of a finitely generated module determines whether direct sums of copies of that module have decompositions that complement direct summands.

§25. Injective Modules and Noetherian Rings—The Faith–Walker Theorems

Recall (18.13) that a ring R is left noetherian if and only if the class of injective left R-modules is closed under the formation of direct sums. In this section we pursue this further and study the relation between the decomposition theory of the injective modules in $_R\mathsf{M}$ and the finiteness conditions of R.

The annihilator $l_R(X)$ in R of a subset X of a left R-module M is a left ideal of R; such left ideals will be called M-*annihilator left ideals*. The set $\mathscr{L}_R(M)$ of all M-annihilator left ideals, ordered by set inclusion, is a complete lattice where for each $\mathscr{A} \subseteq \mathscr{L}_R(M)$ the meet and join in $\mathscr{L}_R(M)$ of $\mathscr{A}$ are

$$\cap \mathscr{A} \qquad \text{and} \qquad l_R(r_M(\Sigma\mathscr{A})),$$

respectively. (See Exercise (2.16).) In general, $\mathscr{L}_R(M)$ is not a sublattice of the lattice of left ideals of R since not every sum of M-annihilator left ideals need be an M-annihilator left ideal. (See Exercise (2.15).) The lattice $\mathscr{L}_R(M)$ is anti-isomorphic to the lattice $\mathscr{R}_M(R)$ of annihilators in M of subsets in R via the inverse maps

$$I \mapsto r_M(I) \qquad \text{and} \qquad A \mapsto l_R(A)$$

for each $I \in \mathscr{L}_R(M)$ and each $A \in \mathscr{R}_M(R)$. (See Exercise (2.16).) Now if $_R E$ is an injective module, the lattice $\mathscr{L}_R(E)$ determines whether $E^{(A)}$ is injective for all sets A.

25.1. Theorem [Faith]. *The following statements about an injective left R-module E are equivalent:*

(a) $E^{(A)}$ *is injective for all sets A;*

(b) *The E-annihilator left ideals in R satisfy the ascending chain condition;*

(c) $E^{(\mathbb{N})}$ *is injective.*

Proof. (a) $\Rightarrow$ (c). This implication is immediate. (Also note that Exercise (25.4) yields its converse directly.)

(c) $\Rightarrow$ (b). Assume that (c) holds but that (b) does not. Then there is a strictly increasing sequence

$$I_1 \subset I_2 \subset \ldots$$

in $\mathscr{L}_R(E)$. The right annihilators of this sequence

$$r_E(I_1) \supset r_E(I_2) \supset \ldots$$

are strictly decreasing. Choose $x_n \in r_E(I_n) \backslash r_E(I_{n+1})$ and let

$$I = \cup_{n=1}^{\infty} I_n.$$

Then for each $a \in I$ there exists an $n > 0$ such that $ax_{n+k} = 0$ for all $k = 1, 2, \ldots$. Thus, the map

$$f: a \mapsto (ax_1, ax_2, \ldots) \qquad (a \in I)$$

is an R-homomorphism $f: I \to E^{(\mathbb{N})}$. But now by The Injective Test Lemma there exists a

$$y = (y_1, \ldots, y_n, 0, \ldots) \in E^{(\mathbb{N})}$$

such that, for all $a \in I$

$$(ax_1, ax_2, \ldots) = f(a) = ay$$
$$= (ay_1, \ldots, ay_n, 0, \ldots).$$

But this is contrary to our choice of $x_{n+1} \notin r_E(I_{n+2})$.

(b) $\Rightarrow$ (a). Assume (b). Then every non-empty collection of E-annihilator left ideals contains a maximal element. (See Exercise (10.9).) Let $I \leq {}_R R$ and consider an R-homomorphism

$$f: I \to E^{(A)}.$$

Since E^A is injective and since $E^{(A)} \leq E^A$, there exists an $x \in E^A$ such that $f(a) = ax$ for all $a \in I$. For each subset $B \subseteq A$ let $x_B = \iota_B \pi_B(x)$, i.e.,

$$\pi_\alpha(x_B) = \begin{cases} \pi_\alpha(x), & \text{if } \alpha \in B \\ 0, & \text{otherwise.} \end{cases}$$

If we let F range over the finite subsets of A, then by hypothesis, the set of E-annihilator left ideals of the form

$$l_R(x_{A \backslash F}) = l_R(\{\pi_\alpha(x) \mid \alpha \in A \backslash F\})$$

contains a maximal element $l_R(x_{A \setminus F_0})$. By maximality, if F is a finite subset of A, then

$$F \supseteq F_0 \quad \text{implies} \quad l_R(x_{A \setminus F}) = l_R(x_{A \setminus F_0}).$$

Now for each $a \in I$, since $f(a) \in E^{(A)}$, there is a finite subset $F_a \supseteq F_0$ such that

$$a\pi_\alpha(x) = \pi_\alpha(ax) = \pi_\alpha(f(a)) = 0 \quad (\alpha \in A \setminus F_a).$$

Thus, for every $a \in I$ we have $a \in l_R(x_{A \setminus F_a}) = l_R(x_{A \setminus F_0})$, whence

$$f(a) = ax - ax_{A \setminus F_0} = ax_{F_0} \quad (a \in I).$$

But since $x_{F_0} \in E^{(A)}$ it follows from the Injective Test Lemma that $E^{(A)}$ is injective. □

We may view Theorem (18.13) as a test to determine whether a ring is noetherian. The preceding theorem together with the next lemma gives rise to a much more economical test.

25.2. Lemma. *Let I be a left ideal of R and let M be a left R-module. Then M cogenerates R/I if and only if I is the annihilator of a subset of M.*

Proof. It is easy to check that

$$l_R(r_M(I)) = \cap \{ Ker f \mid f : {}_R R \to M \text{ with } I \le Ker f \}.$$

Hence $l_R(r_M(I))/I = Rej_{R/I}(M)$. □

25.3. Theorem. *Let R be a ring with minimal left R-cogenerator C_0. Then the following are equivalent:*
(a) *R is left noetherian;*
(b) *An infinite direct sum of copies of some left R-cogenerator is injective;*
(c) *$C_0^{(\mathbb{N})}$ is injective.*

Proof. (a) $\Rightarrow$ (c). Recall that the minimal cogenerator is a direct sum of injective (envelopes of simple) modules (18.16). Thus $C_0^{(\mathbb{N})}$ is a direct sum of injective modules. So this implication follows from (18.13).

(c) $\Rightarrow$ (b). This is immediate.

(b) $\Rightarrow$ (a). By (25.1) and (25.2). □

Decompositions of Injective Modules

25.4. Lemma. *The endomorphism ring of every indecomposable injective module is local.*

Proof. Let E be an indecomposable injective left R-module. Let $t \in End({}_R E)$. Then since

$$Ker t \cap Ker(1 - t) = 0,$$

it follows from (18.12) that either t or $1 - t$ is monic. But the image of a monomorphism $E \to E$ is a direct summand of E (18.7), so either t or $1 - t$ is invertible. Thus $End({}_R E)$ is local (see (15.15)). □

25.5. Proposition. *If an injective module E has an indecomposable decomposition $E = \oplus_A E_\alpha$, then that decomposition complements direct summands.*

Proof. By the above lemma an indecomposable decomposition of an injective module $E = \oplus_A E_\alpha$ satisfies the hypothesis of Azumaya's Theorem (12.6). Let K be a direct summand of E and choose a subset $B \subseteq A$ maximal with respect to

$$(\oplus_B E_\beta) \cap K = 0.$$

Then the submodule $(\oplus_B E_\beta) + K = (\oplus_B E_\beta) \oplus K$ is injective (18.2); thus for some $E' \leq E$,

$$E = E' \oplus (\oplus_B E_\beta) \oplus K.$$

We claim that $E' = 0$. For if $E' \neq 0$, then by (12.6) E' contains an indecomposable direct summand and there is a $\gamma \in A$ and a direct summand E'' of E' such that

$$E = E_\gamma \oplus (E'' \oplus (\oplus_B E_\beta) \oplus K)$$

contrary to the maximality of B. Thus $E = (\oplus_B E_\beta) \oplus K$, and the given decomposition complements direct summands.   □

The injective modules over R need not have indecomposable decompositions. Indeed, we now prove that they all do if and only if R is noetherian.

25.6. Theorem. *For a ring R the following are equivalent:*
(a) *R is left noetherian;*
(b) *Every injective left R-module is a direct sum of indecomposable modules;*
(c) *Every injective left R-module has a decomposition that complements direct summands;*
(d) *Every injective left R-module has a decomposition that complements maximal direct summands.*

Proof. (a) $\Rightarrow$ (b). Let E be an injective left module over a left noetherian ring R. If $0 \neq x \in E$, then by (18.12.3) we may assume $E = E(Rx) \oplus E'$ for some $E' \leq E$. If $(M_\alpha)_{\alpha \in A}$ is a set of independent submodules of $E(Rx)$, then $(Rx \cap M_\alpha)_{\alpha \in A}$ is an independent set of submodules of Rx. Thus, since Rx is noetherian, all but finitely many of the $Rx \cap M_\alpha$ are zero. But $Rx \trianglelefteq E(Rx)$, so it follows that $E(Rx)$ contains no infinite independent set of non-zero submodules. Thus it contains no infinite ascending chain of direct summands and, by (10.14), $E(Rx) = E_1 \oplus \ldots \oplus E_n$ where each E_i is indecomposable. This shows that every non-zero injective E contains an indecomposable direct summand. Now, by The Maximal Principle there exists a maximal independent set $\{E_\alpha \mid \alpha \in A\}$ of indecomposable direct summands of E. Let $E' = \oplus_A E_\alpha$. Then, since R is left noetherian, E' is injective and $E = E' \oplus E''$. But E'' must be zero, for otherwise, since it is injective, it contains an indecomposable direct summand. Therefore $E = \oplus_A E_\alpha$ is an indecomposable decomposition of E.

(b) $\Rightarrow$ (c). This follows from (25.5).

(c) $\Rightarrow$ (d). This is trivial.

(d) $\Rightarrow$ (a). Let $\mathscr{S}$ be an irredundant set of representatives of the simple left R-modules (see (18.16)); then

$$C_0 = \oplus_{T\in\mathscr{S}} E(T),$$

is the minimal cogenerator for $_R$M. Let

$$E = E(C_0^{(\mathbb{N})}).$$

Then by hypothesis there is a decomposition

$$E = \oplus_A E_\alpha$$

that complements maximal direct summands. For each of the simple modules $T\in\mathscr{S}$, let

$$A(T) = \{\alpha \in A \mid E_\alpha \cong E(T)\}.$$

Now for each $n > 0$, the injective submodule $E(T)^{(n)}$ is isomorphic to a direct summand of E. So by (12.2)

$$card(A(T)) \geq n,$$

and hence $A(T)$ is infinite. Let

$$B = \cup_{T\in\mathscr{S}} A(T).$$

Then it is clear that $C_0^{(\mathbb{N})}$ is isomorphic to a direct summand of the direct summand $\oplus_B E_\beta$ of E. Therefore $C_0^{(\mathbb{N})}$ is injective, and by (25.3) R is left noetherian. $\qquad\square$

The Main Faith–Walker Theorem

A module M is c-*generated* in case there is an indexed set $(x_\gamma)_{\gamma\in C}$ that spans M with $c = card\, C$. Thus (see (8.1)) if C is a set, then M is $card\, C$ generated if and only if M is an epimorphic image of the free module $R^{(C)}$. It is clear therefore that every epimorphic image of a c-generated module is c-generated. In general, of course, submodules of c-generated modules need not be c-generated.

25.7. Lemma. *Let c be an infinite cardinal. Let $M = \oplus_A M_\alpha$ and let $N \leq M$. If N is c-generated, then there is a subset $B \subseteq A$ with $card\, B \leq c$ and $N \leq \oplus_B M_\beta$.*

Proof. Let $(x_\gamma)_{\gamma\in C}$ span N and let $card\, C = c$. Then each x_γ is in a finite sum of the M_α, so there is a function $F : \gamma \mapsto F(\gamma)$ from C to the finite subsets of A with $x_\gamma \in \oplus_{F(\gamma)} M_\alpha$ for each $\gamma \in C$. Set $B = \cup_C F(\gamma)$. Then since C is infinite, $card\, C \geq card\, B$. $\qquad\square$

Observe that if $\mathscr{A}$ is a set of modules, then each module in the set is c-generated where

$$c = card\,(\cup\mathscr{A}).$$

Now an indecomposable injective left R-module must be the injective envelope of each of its non-zero submodules. (See (18.12.3).) It follows that the set

$$\{E(R/I) \mid I \leq R\}$$

contains an isomorphic copy of each indecomposable injective. Since this is a set, we have that there is a cardinal number c such that every indecomposable injective left R-module is c-generated. Then it follows from Theorem (25.6) that if R is left noetherian, there is a cardinal number c such that every injective left R-module is a direct sum of c-generated modules. In fact, the converse is true.

25.8. Theorem [Faith and Walker]. *A ring R is left noetherian if and only if there exists a cardinal number c such that every injective left R-module is a direct sum of c-generated modules.*

Proof. As we have just observed, the condition is necessary. Conversely, assume the existence of a cardinal number satisfying the stated condition. It is clear that any larger cardinal will then also satisfy this condition. Now to prove that R is left noetherian it will suffice to show that if $_RE$ is injective, then $E^{(\mathbb{N})}$ is also injective (25.3). So let $_RE$ be injective. Now any module spanned by a set C has at most $(card\,R) \cdot (card\,C)$ elements. Thus, our assumption implies that there is an infinite cardinal number c that is greater than both $card\,E$ and $card\,R$ and such that every injective module is a direct sum of modules of cardinality at most c. Let B be a set with

$$card\,B > 2^c.$$

The direct product E^B is injective (18.2), so by hypothesis

$$E^B = \oplus_A E_\alpha$$

where each E_α has cardinality at most c. We claim that there is a partition $\{A_0, A_1, A_2, \ldots\}$ of A such that

$$card\,A_n \leq c \qquad (n = 1, 2, \ldots)$$

and

$$\oplus_{A_n} E_\alpha = Q_n \oplus Q'_n \qquad (n = 1, 2, \ldots)$$

with $Q_n \cong E$. Once this claim is established, we will be done, for assuming our claim,

$$E^B = (\oplus_{n=1}^\infty Q_n) \oplus ((\oplus_{n=1}^\infty Q'_n) \oplus (\oplus_{A_0} E_\alpha))$$

and so $E^{(\mathbb{N})} \cong \oplus_{n=1}^\infty Q_n$, a direct summand of an injective, is injective. Now, to establish the claim suppose $A_1, \ldots, A_n$ are disjoint subsets of A such that $card\,A_i \leq c$ $(i = 1, \ldots, n)$ and

$$\oplus_{A_i} E_\alpha = Q_i \oplus Q'_i \qquad (i = 1, \ldots, n)$$

with $Q_i \cong E$. Set

$$D = A_1 \cup \ldots \cup A_n$$

and observe that $card(\oplus_D E_\alpha) \leq n \cdot c^2 = c$. For each $\beta \in B$, let $\iota_\beta : E \to E^B$ be the natural injection. Since $\{(\oplus_D E_\alpha) \cap \iota_\beta(E) \mid \beta \in B\}$ is a set of independent submodules of $\oplus_D E_\alpha$ and since $\oplus_D E_\alpha$ has at most $2^c (< card\, B)$ subsets, there exists a $\beta \in B$ with $\oplus_D E_\alpha \cap \iota_\beta(E) = 0$. Thus the projection of E^B on $\oplus_{A \setminus D} E_\alpha$ is monic on $\iota_\beta(E)$. In particular,

$$\oplus_{A \setminus D} E_\alpha = Q \oplus V$$

for some $Q \cong {}_R E$. So by (25.7) there is a subset $A_{n+1} \subseteq A \setminus D$ with $card\, A_{n+1} \leq c$ with $Q \leq \oplus_{A_{n+1}} E_\alpha$. Now a standard induction argument establishes the existence of $A_1, A_2, \ldots$. Finally, set $A_0 = A \setminus \cup_{n=1}^\infty A_n$. $\qquad \square$

25. Exercises

1. A non-zero module H is *uniform* in case each of its non-zero submodules is essential in H. It is *co-uniform* in case each of its proper submodules is superfluous in H.
   (1) Prove that a non-zero module H is uniform iff $E(H)$ is indecomposable.
   (2) Suppose that $p : P \to H$ is a projective cover. Prove that H is co-uniform iff P is $End({}_R P)$ is local.
2. The *Goldie dimension* $G.dim(M)$ of a module M is the infimum of those cardinal numbers c such that $card\, A \leq c$ for every independent set $(M_\alpha)_{\alpha \in A}$ of non-zero submodules of M. Thus, $G.dim(0) = 0$ and $G.dim(M) = 1$ iff M is uniform. Prove that:
   (1) For a module M the following are equivalent: (a) $G.dim(M) = n$; (b) there exists an independent sequence $H_1, \ldots, H_n$ of uniform submodules of M with $H_1 \oplus \ldots \oplus H_n \trianglelefteq M$; (c) $E(M) = E_1 \oplus \ldots \oplus E_n$ with each E_i indecomposable.
   (2) A module M has finite Goldie dimension iff M contains no infinite independent set of non-zero submodules.
   (3) If M_1 and M_2 are R-modules, then

   $$G.dim(M_1) + G.dim(M_2) = G.dim(M_1 \oplus M_2).$$

   (4) If R is left noetherian and c is a cardinal number, then the following are equivalent: (a) $G.dim(M) = c$; (b) there exists an independent set of uniform submodules $(H_\gamma)_{\gamma \in C}$ with $card\, C = c$ and $\oplus_C H_\gamma \trianglelefteq M$; (c) $E(M) = \oplus_C E_\gamma$ with $card\, C = c$ and each E_γ indecomposable.
3. Prove that the following are equivalent: (a) R is left noetherian; (b) $U^{(A)}$ is quasi-injective whenever ${}_R U$ is; (c) $E^{(A)}$ is quasi-injective whenever ${}_R E$ is injective.
4. Prove that $\oplus_A E_\alpha$ is injective if $\oplus_C E_\gamma$ is injective for each countable

subset $C \subseteq A$. [Hint: If $\oplus_A E_\alpha$ is not injective then there exists $I \leq {}_R R$ and $f : I \to \oplus_A E_\alpha$ such that $\pi_\alpha(Im f) \neq 0$ for a countably infinite number of $\alpha \in A$.]

5. Prove that R is left noetherian iff $\oplus_{n=1}^{\infty} E(T_n)$ is injective for each sequence $T_1, T_2, \ldots$ of simple left R-modules.

§26. Direct Sums of Countably Generated Modules—With Local Endomorphism Rings

The results of §12 show the importance of those decompositions $M = \oplus_A M_\alpha$ in which each M_α has a local endomorphism ring. It is apparently an open question whether the direct summands of such a module M also have decompositions whose terms have local endomorphism rings. In this section we show that if in addition each M_α has a countable spanning set, then every direct summand of M does have such a decomposition.

A Theorem of Kaplansky

Recall that a module N is c-generated in case N has a spanning set $(x_\gamma)_{\gamma \in C}$ such that $card\ C = c$. In particular, a module with a countable spanning set is *countably generated*. The following important theorem was first proved by Kaplansky in the countably generated case, and later extended to its present form by C. Walker.

26.1. Theorem. *Let c be an infinite cardinal. If a module M is a direct sum of c-generated submodules then so is every direct summand of M.*

Proof. Let $M = \oplus_A M_\alpha$ and suppose that each M_α is c-generated. Suppose also that $M = K \oplus L$ and let

$$(K_\beta)_{\beta \in B} \qquad \text{and} \qquad (L_\gamma)_{\gamma \in C}$$

denote the c-generated submodules of K and L, respectively. Let $\mathscr{P}$ denote the set of ordered triples

$$(A', B', C')$$

such that

(i) $A' \subseteq A$, $B' \subseteq B$, $C' \subseteq C$;

(ii) $\oplus_{A'} M_\alpha = (\oplus_{B'} K_\beta) \oplus (\oplus_{C'} L_\gamma)$.

Define a partial ordering $\leq$ on $\mathscr{P}$ by $(A', B', C') \leq (A'', B'', C'')$ in case $A' \subseteq A''$, $B' \subseteq B''$, $C' \subseteq C''$. It is easy to check that $(\mathscr{P}, \leq)$ is inductive; so there is a maximal element

$$(A', B', C') \in \mathscr{P}.$$

To prove the theorem we shall show that $A' = A$. Let e and f be the idempotents in $End({}_R M)$ for K and L, respectively, in the decomposition

$M = K \oplus L$. Suppose $A' \neq A$, and let $\alpha \in A \backslash A'$. By Lemma (25.7) each c-generated submodule of M is contained in a sum of at most c of the M_α. In particular, if $D \subseteq A$ is of cardinality at most c, then both

$$(\oplus_D M_\delta)e \qquad \text{and} \qquad (\oplus_D M_\delta)f$$

are c-generated submodules of M, so their sum, also c-generated, is contained in a sum of at most c of the M_α. So by a standard induction argument it follows that there exists an increasing sequence

$$D_1 \subseteq D_2 \subseteq \ldots$$

of subsets of A each of cardinality at most c such that

$$M_\alpha \le M_\alpha e + M_\alpha f \le \oplus_{D_1} M_\delta$$

$$\oplus_{D_1} M_\delta \le (\oplus_{D_1} M_\delta)e + (\oplus_{D_1} M_\delta)f \le \oplus_{D_2} M_\delta$$

$$\vdots$$

$$\oplus_{D_n} M_\delta \le (\oplus_{D_n} M_\delta)e + (\oplus_{D_n} M_\delta)f \le \oplus_{D_{n+1}} M_\delta$$

$$\vdots$$

Set $D = \cup_{n=1}^\infty D_n$. Since the M_δ are independent and $\alpha \notin A'$, it is clear that $D \nsubseteq A'$. Note also that

$$(\oplus_D M_\delta)e \le \oplus_D M_\delta \qquad \text{and} \qquad (\oplus_D M_\delta)f \le \oplus_D M_\delta$$

and that $\oplus_D M_\delta$ is c^2-generated $= c$-generated. Now set

$$M' = \oplus_{A'} M_\alpha, \quad K' = \oplus_{B'} K_\beta, \quad L' = \oplus_{C'} L_\gamma.$$

Then by hypothesis $M' = K' \oplus L'$. Also set

$$M'' = \oplus_{A' \cup D} M_\nu, \quad K'' = M''e, \quad L'' = M''f.$$

Then

$$K'' = (K' + L' + \oplus_D M_\delta)e \le K' + (\oplus_D M_\delta) \le M'',$$

and

$$L'' = (K' + L' + \oplus_D M_\delta)f \le L' + (\oplus_D M_\delta) \le M''.$$

So since $M'' \le K'' \oplus L''$, we infer that $M'' = K'' \oplus L''$. Now K' and L' are direct summands of M contained in K'' and L'', respectively. So for some submodules $K_1' \le K$ and $L_1' \le L$, we have $K'' = K' \oplus K_1'$ and $L'' = L' \oplus L_1'$. Thus

$$M'' = K'' \oplus L'' = M' \oplus (K_1' \oplus L_1').$$

Now

$$K_1' \oplus L_1' \cong M''/M' \cong \oplus_{D \backslash A'} M_\delta$$

is non-zero and c-generated. This contradicts the maximality of (A', B', C') in $\mathscr{P}$. $\qquad \square$

Every free module is a direct sum of countably generated (indeed cyclic) modules. Since every projective module is a direct summand of a free module, it follows at once that

26.2. Corollary. *Every projective module is a direct sum of countably generated modules.*

Also, Theorem (26.1) combines with (25.8) to yield

26.3. Corollary. *A ring R is left noetherian if and only if there exists a left R-module H such that every left R-module can be embedded in a direct sum of copies of H.* ☐

Proof. Since a module can be embedded in an injective module, the necessity follows from (25.8) and the fact that every c-generated module is isomorphic to a submodule of the direct sum of the modules in the set $\{R^{(C)}/K \mid K \leq R^{(C)}\}$ when $card\, C = c$.

Conversely, if H satisfies the stated condition then every injective left R-module is isomorphic to a direct summand of a direct sum of copies of H. Hence this implication follows from (26.1) and (25.8). ☐

Countably Generated Modules with Local Endomorphism Rings

Before we state the main result of the section, we pause to insert a lemma of some interest in its own right. Recall that if a module has decompositions

$$M = \oplus_A M_\alpha = K \oplus L$$

with $K \neq 0$ and each $End(M_\alpha)$ a local ring, then Azumaya's Theorem (12.6) implies that K has a direct summand isomorphic to one of the M_α. The next lemma implies that on the other hand each M_α is isomorphic to a direct summand of either K or L.

26.4. Lemma. *Let M be a module with a decomposition $M = K \oplus L$. Let N be a direct summand of M such that $N = N_1 \oplus \dots \oplus N_n$ with each $End(N_i)$ a local ring. Then there exist direct summands $K' \leq K$ and $L' \leq L$ such that*

$$M = N \oplus K' \oplus L'.$$

Proof. Suppose that N, H, H', K and L are submodules of $_R M$, that $End(_R N)$ is a local ring, and that

$$M = N \oplus H \oplus H' = H \oplus K \oplus L.$$

We shall prove that there are direct summands $K' \leq K$, $L' \leq L$ such that

$$M = N \oplus H \oplus K' \oplus L'.$$

Then with an obvious induction argument

$$(N_{n+1} = N \quad \text{and} \quad N_1 \oplus \dots \oplus N_n = H)$$

the proof of the Lemma follows. Now let e, e', f be idempotents in $End(_R M)$ with e and e' orthogonal,

$$K = Me, \quad L = Me', \quad H = M(1 - e - e')$$

and

$$N = Mf, \quad H \oplus H' = M(1 - f).$$

Since $f \operatorname{End}(_R M) f \cong \operatorname{End}(_R N)$ is a local ring with identity f, and since $f = fef + fe'f$ (for $f(1 - e - e')f = 0$), we may assume that fef is invertible in $f \operatorname{End}(_R M) f$. So let $s \in \operatorname{End}(_R M)$ with

$$s = fsf \quad \text{and} \quad sfef = fefs = f.$$

Then in $\operatorname{End}(_R M)$, $(ese)^2 = e(sfef)se = efse = ese$ so that

$$M = (\operatorname{Im} ese) \oplus (\operatorname{Ker} ese).$$

But $L \oplus H = M(1 - e) \le \operatorname{Ker} ese$, so

$$\operatorname{Ker} ese = (\operatorname{Ker} ese) \cap (K \oplus L \oplus H)$$
$$= ((\operatorname{Ker} ese) \cap K) \oplus L \oplus H.$$

Set

$$K' = (\operatorname{Ker} ese) \cap K.$$

Then

$$M = Mese \oplus K' \oplus L \oplus H$$

and

$$p : m \mapsto mese$$

defines the projection of M on $Mese$ along $K' \oplus L \oplus H$. Consider

$$(p \mid N) : N \to Mese.$$

Since

$$Mese = (Mes)ese \subseteq (Mf)ese = Nese \subseteq Mese,$$

the image of $(p \mid N)$ is $Mese$. And since

$$fese = (fef)s(fe)$$

$(p \mid N)$ is a composite of monomorphisms. Thus $(p \mid N)$ is an isomorphism and (see 5.5) $M = N \oplus K' \oplus L \oplus H$. This establishes the initial claim and hence also the lemma. □

Now it is not difficult to prove the main result of this section

26.5. Theorem [Crawley–Jónsson–Warfield]. *If a module M is a direct sum of countably generated modules, each with local endomorphism ring, then so is every direct summand of M.*

Proof. By hypothesis $M = \oplus_A M_\alpha$ with each M_α countably generated and with each $\operatorname{End}(M_\alpha)$ local. By Theorem (26.1) it follows that each direct summand of M is a direct sum of countably generated submodules (necessarily direct summands) of M. So it will suffice to prove the result for a countably generated direct summand of M. Thus, let

$$M = K \oplus L$$

with K countably generated. Let $x \in K$. Then there is a finite set $G \subseteq A$ with $x \in \oplus_G M_\alpha$. Set $N = \oplus_G M_\alpha$. By (26.4) there are direct summands $K' \leq K$ and $L' \leq L$ such that

$$M = N \oplus K' \oplus L'.$$

Let

$$H = K \cap (N \oplus L').$$

Then $x \in K \cap N \subseteq H$, and by modularity

$$K = K \cap M = K \cap ((N \oplus L') \oplus K')$$

$$= H \oplus K'.$$

Since L' is a direct summand of L, there is a submodule $I \leq L$ with $L = I \oplus L'$. Thus

$$N \cong M/(K' \oplus L') \cong H \oplus I.$$

But by (12.7) the decomposition $N = \oplus_G M_\alpha$ complements direct summands. Hence, in particular, H is isomorphic to $\oplus_F M_\alpha$ for some finite subset $F \subseteq A$. Now let $x_1, x_2, \ldots$ be a spanning set for K. Assume that there exists a direct decomposition

$$K = H_1 \oplus \ldots \oplus H_n \oplus K_n$$

of K and finite subsets $F_1, \ldots, F_n$ of A such that

$$x_1, \ldots, x_n \in H_1 \oplus \ldots \oplus H_n,$$

$$H_i \cong \oplus_{F_i} M_\alpha \qquad (i = 1, \ldots, n).$$

Then there exist $h_n \in H_1 \oplus \ldots \oplus H_n$ and $k_n \in K_n$ with

$$x_{n+1} = h_n + k_n.$$

Since $k_n \in K_n$ the above argument assures us that there exist submodules H_{n+1}, K_{n+1} of K_n and a finite set $F_{n+1} \subseteq A$ with

$$K_n = H_{n+1} \oplus K_{n+1},$$

$$k_n \in H_{n+1} \cong \oplus_{F_{n+1}} M_\alpha.$$

Thus a straightforward induction argument shows that there exists a sequence $H_1, H_2, \ldots$ of submodules of K and a sequence $F_1, F_2, \ldots$ of finite subsets of A with

$$K = \oplus_{n=1}^{\infty} H_n \quad \text{and} \quad H_n \cong \oplus_{F_n} M_\alpha \qquad (n = 1, 2, \ldots).$$

This completes the proof of the theorem. $\qquad\qquad\qquad\qquad\qquad\qquad\square$

26.6. Corollary. *Let* $M = \oplus_A M_\alpha$ *where each* M_α *is countably generated and has a local endomorphism ring. If* $M = \oplus_B N_\beta$ *is any other decomposition of* M, *then there is a partition* $(A_\beta)_{\beta \in B}$ *of* A *such that*

$$N_\beta \cong \oplus_{A_\beta} M_\alpha \qquad (\beta \in B).$$

Proof. This is an immediate consequence of Theorem (26.5) and Asumaya's Theorem (12.6).   □

Since $R \cong End(_R R)$ (see (4.11)), Theorem (26.5) also tells us that projective modules over local rings are free. We record this as the following corollary.

26.7. Corollary. *Every projective module over a local ring is free.*

This result, originally proved by Kaplansky, is very likely the inspiration for a major portion of the results in the next section.

26. Exercises

1. Let $\mathscr{C}$ be a class of left R-modules and let $_R M$ be countably generated. Suppose that, for every direct summand K of M, each element of K belongs to a direct summand of K that is isomorphic to a member of $\mathscr{C}$. Prove that M is isomorphic to a direct sum of members of $\mathscr{C}$.

2. Let R be a von Neumann regular ring. Prove that:
   (1) Every finitely generated submodule of a projective R-module is a direct summand. [Hint: Let K be a finitely generated submodule of $R^{(n)}$. Let $F = Hom_R(R^{(n)}, \_)$. Then $F: _R M \to _{End(R^{(n)})} M$ is a category equivalence and $0 \to F(K) \to F(R^{(n)})$ splits.]
   (2) If $_R P$ is projective then there exist idempotents $(e_\alpha)_{\alpha \in A}$ in R such that $P \cong \oplus_A Re_\alpha$.

3. Prove that the following are equivalent: (a) R is a local ring; (b) Every projective left R-module is free and has an indecomposable decomposition that complements maximal direct summands; (c) Every finitely generated projective left R-module is free and has a decomposition that complements direct summands; (d) The decomposition $R^{(2)} = \iota_1(R) \oplus \iota_2(R)$ complements direct summands.

4. Indecomposable injective modules have local endomorphism rings (25.4) and by (25.6) every direct sum of indecomposable injective modules over a noetherian ring satisfies the conclusion of the Crawley–Jønsson–Warfield Theorem (26.5). Show that the field of fractions of $\mathbb{R}[X]$ (i.e., the rational functions) is an indecomposable injective module over a noetherian ring that is not countably generated. [Hint: Exercise (18.13).]   □

§27. Semiperfect Rings

Let R be a local ring and let $_RP$ be finitely generated projective. Then P/JP is a vector space over the division ring $R/J(R)$. A remarkable fact suggested by Exercise (26.3) is that the very civilized decomposition theory of the vector space P/JP "lifts" to that of the module P. The ability to perform such lifting of decompositions is not restricted to local rings, but it clearly does depend on the regular module $_RR$. Since the decompositions of $_RR$ are determined by the idempotents of R we turn now to a study of a phenomenon known as

Lifting Idempotents

The idempotents in a ring R represent idempotents in every factor ring of R (see (7.10)). However, idempotent cosets in a factor ring of R need not have idempotent representatives in R. For example, $\mathbb{Z}$ has but two idempotents, while $\mathbb{Z}_6$ has four.

Let I be an ideal in a ring R and let $g + I$ be an idempotent element of R/I. We say that this idempotent can be *lifted* (*to e*) *modulo* I in case there is an idempotent $e \in R$ such that $g + I = e + I$. We say that *idempotents lift modulo* I in case every idempotent in R/I can be lifted to an idempotent in R.

Intuitively, the smaller the ideal I, the more likely that idempotents lift modulo I. A nil ideal is small enough.

27.1. Proposition. *If I is a nil ideal in a ring R then idempotents lift modulo I.*

Proof. Suppose that I is a nil ideal in R and $g \in R$ satisfies $g + I = g^2 + I$. Then letting n be the nilpotency index of $g - g^2$ we can use the binomial formula as follows

$$0 = (g - g^2)^n = \Sigma_{k=0}^n \binom{n}{k} g^{n-k} (-g^2)^k = \Sigma_{k=0}^n (-1)^k \binom{n}{k} g^{n+k}$$

$$= g^n - g^{n+1} (\Sigma_{k=1}^n (-1)^{k-1} \binom{n}{k} g^{k-1})$$

to obtain $t = \Sigma_{k=1}^n (-1)^{k-1} \binom{n}{k} g^{k-1} \in R$ such that

$$g^n = g^{n+1} t \qquad \text{and} \qquad gt = tg.$$

Now

$$e = g^n t^n = (g^{n+1} t) t^n = g^{n+1} t^{n+1}$$

$$= g^{n+2} t^{n+2} = \ldots = g^{2n} t^{2n} = e^2,$$

so $e = g^n t^n$ is idempotent; and also

$$g + I = g^n + I = g^{n+1} t + I$$

$$= (g^{n+1} + I)((t + I) = (g + I)(t + I) = gt + I$$

so that $g + I = (g + I)^n = (gt + I)^n = e + I.$ $\qquad \square$

Orthogonality Relations

In general, if a pair of orthogonal idempotents $g_1 + I$ and $g_2 + I$ in R/I lifts to idempotents e_1 and e_2 in R, there is no guarantee that e_1 and e_2 will be orthogonal. (For example, consider an upper triangular matrix ring.) We propose to show, however, that for each ideal $I \subseteq J(R)$ (e.g., if I is nil) orthogonality can be preserved. This fact hinges on properties of projective covers established in the following two lemmas.

27.2. Lemma. *Let $_RM$ have a decomposition $M = M_1 \oplus \ldots \oplus M_n$ such that each term M_i has a projective cover. Then an R-homomorphism*

$$p : P \to M$$

is a projective cover if and only if P has a decomposition $P = P_1 \oplus \ldots \oplus P_n$ such that for each $i = 1, \ldots, n$

$$(p|P_i) : P_i \to M_i$$

is a projective cover.

Proof. Let $q_i : Q_i \to M_i$ $(i = 1, \ldots, n)$ be projective covers. Then it follows inductively from (16.11) and (5.20) that

$$\oplus_{i=1}^{n} q_i : Q_1 \oplus \ldots \oplus Q_n \to M_1 \oplus \ldots \oplus M_n$$

is a projective cover. Thus, letting $q_i = (p \mid P_i)$ we see that the condition is sufficient. The necessity follows from the last statement of (17.17). $\square$

27.3. Lemma. *A cyclic module $_RM$ has a projective cover if and only if $M \cong Re/Ie$ for some idempotent $e \in R$ and some left ideal $I \subseteq J(R)$. For e and I satisfying this condition the natural map*

$$Re \to Re/Ie \to 0$$

is a projective cover.

Proof. The natural map $Re \to Re/Ie$ has kernel Ie. So if $I \subseteq J(R)$, then $Ie \subseteq J(R)e \ll Re$. Conversely, suppose $_RM$ has a projective cover $p : P \to M$. If M is cyclic, then there is an epimorphism $f : R \to M$. So by (17.17) we may assume $R = P \oplus P'$ with $p = (f|P)$. Thus for some idempotent $e \in R$, $P = Re$ and $Ie = \operatorname{Ker} p \ll Re$. Whence $Ie \subseteq J(R)e \subseteq J(R)$ and $M \cong Re/Ie$. $\square$

Now we return to the problem of lifting orthogonal sets of idempotents.

27.4. Proposition. *Let R be a ring and let I be an ideal of R with $I \subseteq J(R)$. Then the following are equivalent:*

(a) *Idempotents lift modulo I;*

(b) *Every direct summand of the left R-module R/I has a projective cover;*

(c) *Every (complete) finite orthogonal set of idempotents in R/I lifts to a (complete) orthogonal set of idempotents in R.*

Proof. (a) $\Rightarrow$ (b). A direct summand of $_RR/I$ is also one of $_{R/I}R/I$ and so is generated by an idempotent of R/I. Assuming (a), we can lift any such idempotent, so it will suffice to prove that if $e \in R$ is idempotent, then $(Re + I)/I$ has a projective cover in $_RM$. But

$$(Re + I)/I \cong Re/(I \cap Re) = Re/Ie$$

and so (27.3) applies.

(b) $\Rightarrow$ (c). Let $g_1, \ldots, g_n \in R$ be a complete orthogonal set of idempotents modulo I. (This will suffice since any finite orthogonal set can be expanded to a complete orthogonal set.) Since $I \leq J(R) \ll R$, the natural map $n_I : R \to R/I$ is a projective cover. By hypothesis each term in

$$R/I = (R/I)(g_1 + I) \oplus \ldots \oplus (R/I)(g_n + I)$$

has a projective cover, so by (27.2) and (7.2) there is a complete orthogonal set of idempotents $e_1, \ldots, e_n \in R$ such that

$$(R/I)(e_i + I) = n_I(Re_i) = (R/I)(g_i + I) \qquad (i = 1, \ldots, n).$$

But then applying the uniqueness part of (7.2.c) we have

$$e_i + I = g_i + I \qquad (i = 1, \ldots, n).$$

(c) $\Rightarrow$ (a). This is clear. $\square$

The Basic Characterizations

A ring R is called *semiperfect* in case $R/J(R)$ is semisimple and idempotents lift modulo $J(R)$. So for example, local rings are semiperfect. From (15.16), (15.19) and (27.1) it follows that a left (or right) artinian ring is semiperfect. It is worthy of note that in a semiperfect ring the radical is the unique largest ideal containing no non-zero idempotents. (See (15.12).)

The next theorem, which gives the basic characterizations of semiperfect rings, depends on the following lemma. Note that this lemma is a dual of the fact (18.12.2) that if $M \trianglelefteq N$, then the injective envelopes of M and N are the same.

27.5. Lemma. *Let $f : M \to N$ be a superfluous epimorphism and let $p : P \to M$ be an R-homomorphism. Then $p : P \to M$ is a projective cover if and only if $fp : P \to N$ is a projective cover.*

Proof. Clearly it will suffice to prove that p is a superfluous epimorphism if and only if fp is. So suppose p, as well as f, is a superfluous epimorphism; then certainly fp is epic. To see that fp is superfluous we use (5.15): If h is a homomorphism with fph epic, then ph is epic, whence h is epic. Thus fp is superfluous. Conversely, if fp is a superfluous epimorphism, then p is epic by (5.15) and p is superfluous because $Ker\, p \leq Ker\, fp \ll P$. $\square$

Observe that, since the definition of semiperfect rings is left-right symmetric, the right-hand versions of conditions (c) and (d) below also serve to characterize semiperfect rings.

27.6. Theorem. *For a ring R the following statements are equivalent:*

(a) *R is semiperfect;*

(b) *R has a complete orthogonal set $e_1, \dots, e_n$ of idempotents with each $e_i R e_i$ a local ring;*

(c) *Every simple left R-module has a projective cover;*

(d) *Every finitely generated left R-module has a projective cover.*

Proof. Throughout this proof let $J = J(R)$ be the radical of R.

(a) $\Rightarrow$ (b). If R is semiperfect, then we can, by (27.4), lift the idempotents (7.2) for a semisimple decomposition of R/J to obtain a complete orthogonal set $e_1, \dots, e_n$ of idempotents in R with each

$$Re_i/Je_i \cong (R/J)(e_i + J)$$

simple. Then by (17.20) each $e_i R e_i$ is local.

(b) $\Rightarrow$ (c). Given (b), each Re_i/Je_i is simple by (17.20), and has a projective cover by (27.3). But each simple left R-module is isomorphic to a factor of $R/J \cong Re_1/Je_1 \oplus \dots \oplus Re_n/Je_n$, and so is isomorphic to one of the Re_i/Je_i. (See (9.4).)

(c) $\Rightarrow$ (d). Assume (c) and let $\mathscr{P}$ be a complete set of projective covers of simple left R-modules. Then by (17.9) and (8.9), $\mathscr{P}$ generates every left R-module. Let $_RM$ be finitely generated. Then there is a sequence $P_1, \dots, P_n$ in $\mathscr{P}$ and an epimorphism

$$P = P_1 \oplus \dots \oplus P_n \overset{f}{\to} M \to 0.$$

Since $f(JP) = JM$ we infer that there is an epimorphism

$$P_1/JP_1 \oplus \dots \oplus P_n/JP_n \cong P/JP \to M/JM \to 0.$$

But each P_i/JP_i is simple (17.19), so M/JM is a finite direct sum of simple modules (9.4). Therefore, by (27.2), M/JM has a projective cover. But $JM \ll M$ by Nakayama's Lemma (15.13), so $M \to M/JM$ is a superfluous epimorphism. Now apply (27.5).

(d) $\Rightarrow$ (a). Assume (d). Since this implies in particular that every direct summand of R/J has a projective cover, idempotents lift modulo J by (27.4). To see that R/J is semisimple, let $J \leq K \leq {}_RR$. Then, since the cyclic R-module R/K has a projective cover, we have by (27.3)

$$R/K \cong Re/Ie$$

for some left ideal $Ie \subseteq Je$. But then $J \cdot Re/Ie \cong J \cdot R/K = 0$ so that $Je = JRe \subseteq Ie$. Thus $Ie = Je$ and

$$R/K \cong Re/Je \cong (R/J)(e + J)$$

is projective over R/J. Hence K/J is a direct summand of R/J. Thus R/J is semisimple. $\quad\square$

27.7. Corollary. *Let $e_1, \dots, e_n \in R$ be non-zero orthogonal idempotents with $1 = e_1 + \dots + e_n$. Then R is semiperfect if and only if each $e_i R e_i$ is semiperfect.*

Proof. ($\Rightarrow$). By (4.15), (7.2) and (27.6.b) R is semiperfect iff $_RR$ is a direct sum of modules with local endomorphism rings. But by (12.7) then so is every direct summand of $_RR$. Hence e_i, the identity of $e_iRe_i \cong End(_RRe_i)$, must satisfy (27.6.b).

($\Leftarrow$). If each e_i is a sum of idempotents satisfying (27.6.b), then so is $1 = e_1 + \dots + e_n$.   □

27.8. Corollary. *Let R be a semiperfect ring. If $_RP$ is a non-zero finitely generated projective module, then $End(_RP)$ is semiperfect. In particular, every ring Morita equivalent to R is semiperfect.*

Proof. Since property (c) of (27.6) is clearly categorical, (21.6), (21.8), we have the final assertion. On the other hand, $End(_RP)$ is of the form eSe for some S Morita equivalent to R and some idempotent $e \in S$. Thus (27.7) applies.   □

27.9. Corollary. *If R is semiperfect, then so is every factor ring of R.*

Proof. It is easy to show that condition (b) of (27.6) is preserved under surjective ring homomorphisms.   □

Projective Modules

Let R be semisimple. Then $_RM$ has a minimal projective generator of the form

$$Re = Re_1 \oplus \dots \oplus Re_m$$

with $e = e_1 + \dots + e_m$ an idempotent and $e_1, \dots, e_m$ pairwise orthogonal idempotents that generate a complete set of pairwise non-isomorphic simple R-modules. Now $eRe \cong End(_RRe)$ is a ring direct sum of the division rings $e_1Re_1, \dots, e_mRe_m$ and is Morita equivalent to R. Of course this is just one formulation of the Wedderburn–Artin structure of semisimple rings. Moreover, in this analysis, if M is a left R-module, then there exist sets $A_1, \dots, A_m$, unique to within cardinality, with

$$M \cong Re_1^{(A_1)} \oplus \dots \oplus Re_m^{(A_m)}.$$

Thus, R and the category $_RM$ are effectively characterized by the simple modules $Re_1, \dots, Re_m$ and the division rings $e_1Re_1, \dots, e_mRe_m$, or equivalently by the ring eRe. If R is semiperfect, then in view of the last theorem it is hardly surprising that the above reduction of the semisimple ring $R/J(R)$ "lifts", at least in part, to a reduction of R itself.

Let R be semiperfect. A module $_RM$ is *primitive* in case there is a primitive idempotent $e \in R$ with

$$M \cong Re.$$

Thus, for a semisimple ring the primitive modules are just the simple ones. A set $e_1, \dots, e_m$ of idempotents of R is *basic* in case it is pairwise orthogonal, and

$$Re_1, \dots, Re_m$$

is a complete irredundant set of representatives of the primitive left R-modules. Now the semisimple ring $\overline{R} = R/J(R)$ clearly has a basic set

$\bar{e}_1, \ldots, \bar{e}_m$ of idempotents. By (27.4.c) and (17.18), this set lifts to a basic set $e_1, \ldots, e_m$ for R.

27.10. Proposition. *Let R be semiperfect with $J = J(R)$. Then every complete set of orthogonal primitive idempotents for R contains a basic set. Moreover, for pairwise orthogonal primitive idempotents $e_1, \ldots, e_m \in R$ the following are equivalent:*

(a) *$e_1, \ldots, e_m$ is a basic set of primitive idempotents for R;*

(b) *$Re_1, \ldots, Re_m$ is an irredundant set of representatives of the indecomposable projective left R-modules;*

(c) *$Re_1/Je_1, \ldots, Re_m/Je_m$ is an irredundant set of representatives of the simple left R-modules;*

(d) *$e_1 + J, \ldots, e_m + J$ generate the simple blocks in the block decomposition of the semisimple ring R/J.*

Finally, these are also equivalent to the corresponding versions of (b) *and* (c) *for right R-modules.*

Proof. As we noted above, the equivalence of (a) and (c) follows from (27.4) and (17.18). Also (c) and (d) are equivalent by (13.7). Every primitive left R-module is indecomposable and projective. So to prove the equivalence of (a) and (b) it will suffice to prove that every indecomposable projective R-module is primitive. Suppose then that $_RP$ is a non-zero projective. Then for some $_RP'$ and some set A

$$P \oplus P' \cong R^{(A)}.$$

By (27.6.b) and (4.15), $R^{(A)}$ is a direct sum of primitive modules each with a local endomorphism ring. Thus by (12.6) P has a primitive direct summand, so P is indecomposable iff it is primitive. The first statement is now easy. For if $f_1, \ldots, f_n$ form a complete set of orthogonal primitive idempotents for R, then they are primitive modulo J (17.20), and hence they form such a set for the semisimple ring R/J. But clearly such a complete set for R/J contains a basic set for R/J. Now apply the equivalence of (a) and (d) to lift to a basic set for R. The final assertion follows from the left right symmetry of (d). □

A basic set of idempotents for a semisimple ring can be used to characterize all modules. A basic set for a semiperfect ring characterizes all projective modules.

27.11. Theorem. *Let R be a semiperfect ring and let $e_1, \ldots, e_m$ be a basic set of primitive idempotents for R. If $_RP$ is projective, then there exist sets $A_1, \ldots, A_m$ (unique to within cardinality and possibly empty) such that*

$$P \cong Re_1^{(A_1)} \oplus \ldots \oplus Re_m^{(A_m)}.$$

Proof. By (27.6) the regular module $_RR$ is a direct sum of primitive left R-modules. So there exist sets $C_1, \ldots, C_m$ with

$$_RR \cong Re_1^{(C_1)} \oplus \ldots \oplus Re_m^{(C_m)}.$$

Let P be projective. Then there is a set A and a module P' such that

$$P \oplus P' \cong R^{(A)} \cong Re_1^{(C_1 \times A)} \oplus \ldots \oplus Re_m^{(C_m \times A)}.$$

Now (27.6) each $e_i Re_i \cong End(_R Re_i)$ is a local ring. So the existence assertion of the Theorem follows from the Crawley-Jønsson-Warfield result (26.5). The uniqueness assertion follows from (12.6).   □

This structure of projective modules over semiperfect rings and Azumaya's extension of the Krull–Schmidt Theorem (12.6) readily yield part of the following characterization of semiperfect rings in terms of the decomposition theory of their projective modules.

27.12. Theorem. *The following statements about a ring R are equivalent:*

(a) *R is semiperfect;*

(b) *Every projective (left) R-module has an indecomposable decomposition that complements maximal direct summands;*

(c) *Every finitely generated projective (left) R-module has a decomposition that complements direct summands;*

(d) *The free (left) R-module $R^{(2)}$ has a decomposition that complements direct summands.*

Proof. (a) ⇒ (b). This follows from (27.11) and Azumaya's Theorem (12.6) since in (27.6.b) each $e_i Re_i \cong End(_R Re_i)$ is local.

(b) ⇒ (c). Assume (b), and let P be a finitely generated projective R-module. Then every direct summand of P is a finite direct sum of indecomposable modules. Thus a decomposition of P that complements maximal direct summands must complement all direct summands. (See (12.2).)

(c) ⇒ (d). This is clear.

(d) ⇒ (a). Assume (d). Then by (12.3) each direct summand of $R^{(2)}$ has a decomposition that complements direct summands. In particular, $_R R$ has such a decomposition, say

$$R = Re_1 \oplus \ldots \oplus Re_n.$$

But then

$$R \oplus R = Re_1 \oplus Re_1 \oplus \ldots \oplus Re_n \oplus Re_n$$

is an indecomposable decomposition and so (12.5) complements direct summands. Since each Re_i appears at least twice in this latter decomposition, its endomorphism ring

$$e_i Re_i \cong End(_R Re_i)$$

is local (12.10). Thus, by (27.6) R is semiperfect.   □

Projective Covers of Finitely Generated Modules

Let R be semiperfect with basic set $e_1, \ldots, e_m$ of idempotents. It follows from (27.11) that a finitely generated projective module P is then characterized by a set $k_1, \ldots, k_m$ of non-negative integers, namely, those for which

$$P \cong Re_1^{(k_1)} \oplus \ldots \oplus Re_m^{(k_m)}.$$

Moreover, the induced decomposition of P is very well behaved in the sense (27.12) that it complements direct summands. For finitely generated but not necessarily projective modules the situation is not so easy. However such a module $_RM$ does have a projective cover $P \to M \to 0$ and, as we shall see, the projective module P must be finitely generated. In fact, the projective module P is completely characterized by the semisimple module M/JM. Thus, although the module M and its decomposition theory may defy a complete analysis, the existence and accessibility of its projective cover give us some solid information about M.

27.13. Characterization of Projective Covers. Let R be semiperfect with basic set of idempotents $e_1, \ldots, e_m$ and let $_RM$ be finitely generated. Then M/JM is finitely generated and semisimple, so (see (27.10.c)) there exist unique non-negative integers $k_1, \ldots, k_m$ such that

$$M/JM \cong (Re_1/Je_1)^{(k_1)} \oplus \ldots \oplus (Re_m/Je_m)^{(k_m)}.$$

Set

$$P = Re_1^{(k_1)} \oplus \ldots \oplus Re_m^{(k_m)}.$$

Thus P is finitely generated projective and $P/JP \cong M/JM$. But by (15.13), $JP \ll P$, so the natural epimorphism $P \to P/JP \to 0$ is a projective cover. Also by (15.13), $JM \ll M$, so we deduce from (27.5) that there is a projective cover $P \to M \to 0$. Now projective covers, when they exist, are unique to within isomorphism (17.17), so to summarize:

If $M/JM \cong (Re_1/Je_1)^{(k_1)} \oplus \ldots \oplus (Re_m/Je_m)^{(k_m)}$, then there is a projective cover $P \to M \to 0$ if and only if

$$P \cong Re_1^{(k_1)} \oplus \ldots \oplus Re_m^{(k_m)}.$$

Basic Rings

An idempotent e of a semiperfect ring R is called a *basic idempotent* of R in case e is the sum

$$e = e_1 + \ldots + e_m$$

of a basic set $e_1, \ldots, e_m$ of primitive idempotents of R. Since a basic set of primitive idempotents is pairwise orthogonal, every basic set sums to a basic idempotent. If

$$e = e_1 + \ldots + e_m \qquad \text{and} \qquad f = f_1 + \ldots + f_m$$

are basic idempotents for R, then clearly

$$Re = Re_1 \oplus \ldots \oplus Re_n \cong Rf_1 \oplus \ldots \oplus Rf_n = Rf,$$

so as rings

$$eRe \cong End(_RRe) \cong End(_RRf) \cong fRf.$$

A ring S is a *basic ring* for R in case S is isomorphic to eRe for some basic idempotent $e \in R$. Thus for each semiperfect ring R a basic ring exists and is uniquely defined to within isomorphism. We shall feel free therefore to speak of "the" basic ring eRe of R.

As we suggested earlier the Wedderburn–Artin Theorem can be viewed as saying, at least in part, that a semisimple ring and its category of modules is completely determined by its basic ring. This extends to semiperfect rings.

27.14. Proposition. *A semiperfect ring is Morita equivalent to its basic ring. Moreover, two semiperfect rings are Morita equivalent if and only if their basic rings are isomorphic.*

Proof. Let R be semiperfect with basic idempotent e. By (27.10) and (17.9) Re is a progenerator for $_R\mathrm{M}$. Thus by (22.4) R and $eRe \cong End(_RRe)$ are Morita equivalent; that is, R and its basic ring are Morita equivalent. From this we deduce at once that if R and S are semiperfect with isomorphic, hence equivalent, basic rings, then R and S are equivalent.

Conversely, suppose R and S are equivalent semiperfect rings via an equivalence

$$F : _R\mathrm{M} \to {}_S\mathrm{M}.$$

If $e = e_1 + \ldots + e_m$ is a basic idempotent for R, then it follows from (27.10) and the results of §21 that

$$F(Re) \cong F(Re_1) \oplus \ldots \oplus F(Re_m)$$

where $F(Re_1), \ldots, F(Re_m)$ is an irredundant set of representatives of the indecomposable projective left S-modules. Thus by (27.10), if f is a basic idempotent for S, then $F(Re) \cong Sf$ and we have (see (21.2))

$$eRe \cong End(_RRe) \cong End(_SF(Re)) \cong fSf.$$

So the basic rings of R and S are isomorphic.   □

A semiperfect ring R is a *basic ring* in case 1 is a basic idempotent for R. Now if R is semiperfect with basic idempotent $e = e_1 + \ldots + e_m$, then eRe is semiperfect (27.6.b) with exactly m isomorphism classes of indecomposable projective left modules ((27.14), (21.6) and (21.8)) which must be represented by $eRe_1, \ldots, eRe_m$. Thus e is a basic idempotent for eRe and so fortunately the basic ring of a semiperfect ring is a basic ring. The importance of all this is that the basic rings form canonical representatives of the semiperfect rings with respect to Morita equivalence. Concerning basic rings we note

27.15. Proposition. *Let R be a semiperfect ring. Then R is a basic ring if and only if $R/J(R)$ is a direct sum of division rings.*

Proof. Observe simply that by (27.10), (a) and (c), an idempotent e of R is basic iff $e + J(R)$ is basic in $R/J(R)$.   □

Central Idempotents

Let R be semiperfect. The general structure of the regular modules $_RR$ and

R_R have been pretty well nailed down. There exists a complete set $e_1, \ldots, e_n$ of pairwise orthogonal primitive idempotents, so

$$R = Re_1 \oplus \ldots \oplus Re_n$$

and of course, each Re_i/Je_i is simple and each e_iRe_i is a local ring. Now we look briefly at the two-sided decomposition theory of R. Because 1 is a sum of primitive idempotents, R does have a "block decomposition" (7.9). That is, there exist unique orthogonal central idempotents $u_1, \ldots, u_t$ in R such that $1 = u_1 + \ldots + u_t$ and each u_jRu_j is an indecomposable ring. In other words, R is the ring direct sum

$$R = u_1Ru_1 \dot{+} \ldots \dot{+} u_tRu_t$$

with each u_iRu_i indecomposable. We call these rings (ideals of R) the *blocks* of R. Given any primitive idempotent $e \in R$ it is clear (see §7) that $eu_j \neq 0$ for exactly one j and $eu_j \neq 0$ iff $eu_j = e$. Thus a primitive idempotent e of R belongs to exactly one block. Moreover, it is easy to see (Exercise (27.9)) that the members of a pair of left modules M_1, M_2 over a semiperfect ring R have a common composition factor (i.e., contain submodules $L_i < K_i < M_i$ such that K_i/L_i ($i = 1, 2$) are isomorphic simple modules) iff there exists a primitive idempotent $e \in R$ that annihilates neither M_1 nor M_2. Thus, considering the equivalence relation $\approx$ of §7 we have

27.16. Theorem. *Every semiperfect ring R has a block decomposition. If $e_1, \ldots, e_m$ is basic set of idempotents for R, then two primitive idempotents e and f of R belong to the same block if and only if there exist idempotents $e_{i_1}, \ldots, e_{i_l}$ in that basic set such that $Re \cong Re_{i_1}$, $Rf \cong Re_{i_l}$ and the members of each consecutive pair Re_{i_j}, $Re_{i_{j+1}}$ ($j = 1, \ldots, l-1$) have a common composition factor.* □

Again let R be semiperfect. Then as we just noted, R has a block decomposition. If I is any ideal of R (in particular, if $I = J(R)$) then the factor ring R/I is semiperfect and so has a block decomposition. Since $u + I$ is a central idempotent of R/I whenever u is a central idempotent of R, it is clear that the factor of each block of R is a sum of blocks of R/I. In general, however, the blocks of R/I (i.e., in effect, the central idempotents of R/I) do not "lift" to ones of R. For example, consider the ring R of $n \times n$ upper triangular matrices over a field with $J = J(R)$. Then R is indecomposable, but R/J is a direct sum of n copies of the field. (See Exercise (7.15.3).) Note, on the other hand, that in this example, R/J^2 is indecomposable, so central idempotents lift modulo J^2. More generally,

27.17. Proposition. *Let I be an ideal of a ring R such that*

$$\cap_{n=1}^{\infty} I^n = 0.$$

Let $e \in R$ be idempotent. Then $e \in \text{Cen } R$ if and only if $e + I^2 \in \text{Cen } R/I^2$. Thus, if idempotents lift modulo I^2, then central idempotents lift to central idempotents.

Proof. Clearly it will suffice to show that if $e = e^2 \in R$ and $e + I^2$ is central in R/I^2, then e is central in R. So let e be central modulo I^2. Then

$$eR(1 - e) \subseteq I^2 \qquad \text{and} \qquad eI \subseteq Ie + I^2.$$

Suppose that $eR(1 - e) \subseteq I^n$. Then

$$eR(1 - e) \subseteq eI^n(1 - e)$$
$$\subseteq (Ie + I^2)I^{n-1}(1 - e)$$
$$\subseteq IeR(1 - e) + I^{n+1}(1 - e)$$
$$\subseteq I^{n+1}.$$

So by induction $eR(1 - e) \subseteq I^n$ for all $n = 1, 2, \dots$. Similarly $(1 - e)Re \subseteq I^n$ for all n. Our hypothesis then forces $eR(1 - e) = (1 - e)Re = 0$. So for each $x \in R$

$$ex = exe = xe$$

and indeed, e is central in R. $\qquad \qquad \square$

27.18. Corollary. *Let R be semiperfect with $J = J(R)$ and $\cap_{n=1}^{\infty} J^n = 0$ (e.g., let R be left artinian). Then R is indecomposable if and only if R/J^2 is indecomposable.*

27. Exercises

1. Let I be an ideal of R with $I \subseteq J(R)$ such that idempotents lift modulo I. Prove:
   (1) If $0 \neq f = f^2 \in R$, then idempotents lift modulo fIf in the ring fRf. [Hint: Suppose $g = fgf$ and $g - g^2 \in fIf$. Then show that

   $$Rf/If = (Rf/If)(g + fIf) \oplus (Rf/If)((f - g) + fIf),$$

   and consider the proof (b) $\Rightarrow$ (c) of (27.4).]
   (2) Countable orthogonal sets of idempotents can be lifted modulo I. [Hint: Suppose $g_i g_j \in \delta_{ij} g_i + I$ $(i, j \in \mathbb{N})$ and $e_1, \dots, e_n$ are orthogonal idempotents with $e_i - g_i \in I$ $(i = 1, \dots, n)$. Let $f_n = 1 - (e_1 + \dots + e_n)$ and show that $f_n g_{n+1} f_n - g_{n+1} \in I$.]
2. Prove that for a ring R with $J = J(R)$ the following are equivalent:
   (a) For each $n = 1, 2, \dots$, if K is a direct summand of $(R/J)^{(n)}$, there exists a direct summand P of $R^{(n)}$ such that $(P + J^{(n)})/J^{(n)} = K$; (b) For each $n = 1, 2, \dots$, every direct summand of $R^{(n)}/J^{(n)}$ has a projective cover over R; (c) Idempotents lift modulo $\mathbb{M}_{(n)}(J)$ in each matrix ring $\mathbb{M}_n(R)$ $(n = 1, 2, \dots)$; (d) Idempotents lift modulo the radical of every ring that is Morita equivalent to R. (Question: Is "idempotents lift modulo $J(R)$" categorical?)
3. From Lemma (25.4) and Theorem (27.6) we see that an injective module has a finite indecomposable decomposition iff its endomorphism ring is

semiperfect. Prove:

(1) If $_R U_S$ defines a Morita duality, then R and S are semiperfect.

(2) If R is any ring and $_R E$ is injective with $S = End(_R E)$, then idempotents lift mod $J(S)$. [Hint: If $K = Ker(g - g^2) \trianglelefteq E$ (18.20) then $E = E(K) \oplus E(K(1 - g))$. Let e and $1 - e$ be the corresponding idempotents.)

4. Let $R = \{m/n \in \mathbb{Q} \mid 2 \nmid n \text{ and } 3 \nmid n \ (m/n \text{ in lowest terms})\}$.

   (1) Prove that R is a ring and that $2R$ and $3R$ are the only maximal ideals in R.

   (2) Show that $R/J(R)$ is semisimple but idempotents do not lift modulo $J(R)$.

5. Show that every commutative semiperfect ring is a basic ring and isomorphic to a finite direct product of local rings.

6. Prove that the following are equivalent: (a) R is semiperfect with local basic ring; (b) R is semiperfect and has (to within isomorphism) only one simple left module; (c) R is Morita equivalent to a local ring; (d) R is isomorphic to $\mathbb{M}_n(S)$ for some local ring S.

7. Calculate the basic ring of the ring of matrices of the form

$$\begin{bmatrix} A_{11} & A_{12} & \cdots & A_{1n} \\ & A_{22} & \cdots & A_{2n} \\ & & \ddots & \vdots \\ 0 & & & A_{nn} \end{bmatrix}$$

where $A_{ij} \in \mathbb{M}_{m_i \times m_j}(D)$, D a division ring. Also show that this ring is indecomposable.

8. Let e be a basic idempotent in a semiperfect ring R. Prove that Re is isomorphic to a direct summand of every generator in $_R\mathbb{M}$. Conclude that Re is the unique (to within isomorphism) minimal generator in $_R\mathbb{M}$.

9. Let e be an idempotent in R such that $Re/J(R)e$ is simple (e.g., let e be primitive in a semiperfect ring). Prove that $_R M$ has a composition factor isomorphic to $Re/J(R)e$ if and only if $eM \neq 0$.

§28. Perfect Rings

Two properties that are characteristic of a semiperfect ring R are (1) that every finitely generated module has a projective cover and (2) that every finitely generated projective module has a decomposition that complements direct summands. A "defect" in both of these is the restriction to finitely generated modules. Interestingly enough, however, without this restriction the two resulting conditions are still equivalent and characterize the class of so-called *perfect* rings, the object of study in this section. Unlike semiperfect rings there is a loss of symmetry and we are forced to distinguish between left perfect and right perfect rings. As we shall see, the perfect rings (left or right) are replete with very strong properties; indeed, many classical results about artinian rings, themselves perfect, extend easily to left and right perfect rings.

T-nilpotence

Certain fundamental properties of perfect rings depend on a generalization of the concept of nilpotence. We first encounter this in the study of changes of basis in a free module.

28.1. Lemma. *Let $a_1, a_2, \ldots$ be a sequence in the ring R. Let F be the free left R-module with free basis $x_1, x_2, \ldots$, let*

$$y_n = x_n - a_n x_{n+1} \qquad (n \in \mathbb{N}),$$

and finally, let G be the submodule of F spanned by $y_1, y_2, \ldots$. Then
  (1) *G is free with free basis $y_1, y_2, \ldots$;*
  (2) *$G = F$ iff for each $k \in \mathbb{N}$ there is an $n \geq k$ such that $a_k \ldots a_n = 0$.*

Proof. Let $n \geq k$ and let $r_k, \ldots, r_n \in R$. Then a routine computation gives

$$r_k y_k + \ldots + r_n y_n = r_k x_k + (r_{k+1} - r_k a_k) x_{k+1} + \ldots + \\ (r_n - r_{n-1} a_{n-1}) x_n - r_n a_n x_{n+1}.$$

Thus if $r_k y_k + \ldots + r_n y_n = 0$, the independence of the x's clearly forces $r_k = r_{k+1} = \ldots = r_n = 0$, so the y's are independent. This gives (1). Suppose next that $x_k \in G$, say $x_k = r_1 y_1 + \ldots + r_n y_n$. Then clearly $r_1 = \ldots = r_{k-1} = 0$. Comparing the coefficients of $x_k, \ldots, x_n$ in this equation we see that $r_k = 1$, $r_{k+1} = r_k a_k$, $r_{k+2} = r_{k+1} a_{k+1}, \ldots, r_n = r_{n-1} a_{n-1}$, and $r_n a_n = 0$. So $a_k a_{k+1} \ldots a_n = 0$. This gives the necessity in (2). For the converse, let $k \leq n$; then

$$x_k = y_k + a_k y_{k+1} + \ldots + (a_k \ldots a_{n-1}) y_n + (a_k \ldots a_n) x_{n+1}.$$

So if $a_k \ldots a_n = 0$, then $x_k \in G$.   □

28.2. Lemma. *With the hypotheses of Lemma (28.1) if G is a direct summand of F, then the chain $a_1 R \geq a_1 a_2 R \geq \ldots$ of principal right ideals terminates.*

Proof. By (28.1.1) there is an isomorphism $F \to G$ via $x_n \mapsto y_n$. Suppose the inclusion map $G \to F$ splits. Then there is an endomorphism $s \in End(_R F)$ such that $y_n s = x_n (n \in \mathbb{N})$. For each $m \in \mathbb{N}$ write

$$x_m s = \sum_k c_{mk} x_k$$

as a linear combination of $x_1, x_2, \ldots$. Then

$$x_n = y_n s = (x_n - a_n x_{n+1})s = \sum_k (c_{nk} - a_n c_{n+1 k}) x_k,$$

and so

$$c_{nk} - a_n c_{n+1 k} = \delta_{nk}.$$

Now for some k, $c_{1n} = 0$ for all $n \geq k$. So for each $n \geq k$,

$$
\begin{aligned}
-a_1 \ldots a_n c_{n+1\,n} &= a_1 \ldots a_{n-1}(1 - c_{nn}) \\
&= a_1 \ldots a_{n-1} - a_1 \ldots a_{n-1} c_{nn} \\
&= a_1 \ldots a_{n-1} - a_1 \ldots a_{n-2} c_{n-1\,n} \\
&\qquad\vdots \\
&= a_1 \ldots a_{n-1} - a_1 c_{1n} \\
&= a_1 \ldots a_{n-1}.
\end{aligned}
$$

That is, for each $n \geq k$, $a_1 \ldots a_{n-1} \in a_1 \ldots a_n R$. □

A subset I of a ring R is *left T-nilpotent* ("T" for "transfinite") in case for every sequence $a_1, a_2, \ldots$ in I there is an n such that

$$a_1 \ldots a_n = 0.$$

The subset I is *right T-nilpotent* in case for each $a_1, a_2, \ldots$ in I

$$a_n \ldots a_1 = 0$$

for some n.

Observe that if I is left or right T-nilpotent, then it is nil because $a, a, a, \ldots$ is a sequence in I whenever $a \in I$. On the other hand, even for ideals I left T-nilpotence does not imply right T-nilpotence (see Exercise (15.8)), so in particular, nil ideals need not be right (or left) T-nilpotent. Also note that (28.1.2) may be rephrased to assert that $y_1, y_2, \ldots$ is a free basis for the free module F in that Lemma iff for each $k \in \mathbb{N}$ the sequence $a_k, a_{k+1}, \ldots$ is left T-nilpotent.

The importance of the concept of T-nilpotence is due to the fact that the radical $J = J(R)$ of a ring R is left T-nilpotent precisely when a "Nakayama's Lemma" (15.13) holds for all left modules, finitely generated or not.

28.3. Lemma. *Let J be a left ideal in a ring R. Then the following are equivalent:*

(a) *J is left T-nilpotent;*
(b) *$JM \neq M$ for every non-zero left R-module M;*
(c) *$JM \ll M$ for every non-zero left R-module M;*
(d) *$JF \ll F$ for the countably generated free module $F = R^{(\mathbb{N})}$.*

Proof. (a) ⇒ (b). Suppose that $JM = M \neq 0$. Let $\mathscr{S}$ be the set of finite sequences $a_1, \ldots, a_n$ in J such that

$$a_1 \ldots a_n \in J \backslash l_R(M).$$

Then, since $JM = M \neq 0$, $J \nsubseteq l_R(M)$ so $\mathscr{S}$ contains sequences of length one. But also if $a_1 \ldots, a_n$ belongs to $\mathscr{S}$, then

$$0 \neq a_1 \ldots a_n M = a_1 \ldots a_n JM$$

so there exists a sequence $a_1, ..., a_n, a_{n+1}$ in $\mathcal{S}$. Since $0 \notin J \backslash I_R(M)$, induction guarantees a sequence $a_1, a_2, ...$ such that $a_1 ... a_n \neq 0$ for all $n = 1, 2, ...$

(b) $\Rightarrow$ (c). Assume (b). Suppose M is a left R-module and $K < M$ is a proper submodule. Then by (b), $J \cdot (M/K) \neq M/K$. But $(JM + K)/K = J \cdot (M/K)$ whence $JM + K \neq M$. In other words, $JM \ll M$.

(c) $\Rightarrow$ (d). This is clear.

(d) $\Rightarrow$ (a). Let $F \cong R^{(\mathbb{N})}$ have free basis $x_1, x_2, ...$, let $a_1, a_2, ...$ be a sequence in J, and let $G = \Sigma_{i=1}^{\infty} R(x_i - a_i x_{i+1})$ as in Lemma (28.1). Then clearly $G + JF = F$. But, assuming (d), this implies that $G = F$. So by (28.1.2), $a_1 ... a_n = 0$ for some n. $\qquad \square$

Bass's Theorem P

A ring R is *left perfect* (*right perfect*) in case each of its left (right) modules has a projective cover. It follows from (27.6) that left perfect rings and right perfect rings are both semiperfect. However (see Exercise (28.2)) right perfect rings need not be left perfect. The pioneering work on perfect rings was done by H. Bass [60] and most of the principal characterizations of left perfect rings are contained in the following version of Theorem P from that paper.

28.4. Theorem [Bass]. *Let R be a ring with radical $J = J(R)$. Then the following statements are equivalent:*

(a) *R is left perfect;*

(b) *R/J is semisimple and J is left T-nilpotent;*

(c) *R/J is semisimple and every non-zero left R-module contains a maximal submodule;*

(d) *Every flat left R-module is projective;*

(e) *R satisfies the minimum condition for principal right ideals;*

(f) *R contains no infinite orthogonal set of idempotents and every non-zero right R-module contains a minimal submodule.*

Proof. (a) $\Rightarrow$ (c). Suppose R is left perfect. Then R/J is semisimple by (27.6). Moreover, if $_R M \neq 0$, then there is a projective module P with superfluous submodule $K \ll P$ such that $M \cong P/K$. Since P is projective, P has a maximal submodule L (17.14). Since $K \ll P$, $K \subseteq L$. Thus L/K is a maximal submodule in $P/K \cong M$.

(c) $\Rightarrow$ (b). Since J annihilates every simple module, if (c) holds, then $JM \neq M$ whenever $_R M \neq 0$. Thus this implication follows from (28.3).

(b) $\Rightarrow$ (a). Assume (b). Then R is semiperfect by (27.1). Let M be a non-zero left R-module. Then M/JM is semisimple, so (27.10.c) there exists an indexed set $(e_\alpha)_{\alpha \in A}$ of primitive idempotents in R with

$$\oplus_A Re_\alpha / Je_\alpha \cong M/JM.$$

Let $P = \oplus_A Re_\alpha$. Since J is left T-nilpotent, by (28.3) both $JP \ll P$ and $JM \ll M$. Thus M has a projective cover by (27.5)

$$\begin{array}{c} P \\ {}^{p}\swarrow \quad \downarrow {}^{p'} \\ M \xrightarrow{f} M/JM \to 0. \end{array}$$

(a) $\Rightarrow$ (d). Assume (a). Let ${}_R U$ be flat and let $f : P \to U$ be a projective cover. Then $K = Ker f \ll P$, so by (9.13) and (17.10), $K \le JP$. Since ${}_R P$ and ${}_R U$ are flat, the maps

$$\mu_1 : J \otimes_R P \to JP \qquad \text{and} \qquad \mu_2 : J \otimes_R U \to JU$$

with $\mu_1(j \otimes p) = jp$ and $\mu_2(j \otimes u) = ju$ are isomorphisms (19.17). Now it is easy to check that the diagram

$$\begin{array}{ccc} JP & \xrightarrow{(f \,|\, JP)} & JU \\ {}_{\mu_1}\big\uparrow & & \big\uparrow{}_{\mu_2} \\ J \otimes K \xrightarrow{J \otimes i_K} J \otimes_R P & \xrightarrow{J \otimes f} & J \otimes_R U \end{array}$$

commutes (where $i_K : K \to P$ is the inclusion map). Therefore, since the bottom row is exact and μ_1 and μ_2 are isomorphisms, we have

$$K = Ker f = Ker(f \,|\, JP)$$
$$= \mu_1(Ker(J \otimes f))$$
$$= \mu_1(Im(J \otimes i_K)) = JK.$$

Thus $JK = K$; but then, since (a) $\Leftrightarrow$ (b), $K = 0$ (28.3) and $U \cong P$ is projective.

(d) $\Rightarrow$ (e). Clearly every descending chain of principal right ideals is of the form $a_1 R \ge a_1 a_2 R \ge \dots$ for some sequence $a_1, a_2, \dots$ of elements of R. Given such a sequence, let G and F be the modules of Lemma (28.1). Then by (28.2) we need only show that F/G is flat. By (28.1), $y_1, \dots, y_n, x_{n+1}, x_{n+2}, \dots$ is a free basis for F, so each submodule $G_n = \Sigma_{i=1}^n R y_i$ is a direct summand of F and each factor module F/G_n is free, hence flat. Now G is the union of these G_n, so (see Exercise (19.11)) we see that F/G is flat by applying the right-left symmetric version of (19.18)

$$G \cap IF = (\textstyle\bigcup_\mathbb{N} G_n) \cap IF = \textstyle\bigcup_\mathbb{N}(G_n \cap IF)$$
$$= \textstyle\bigcup_\mathbb{N} IG_n = IG.$$

(e) $\Rightarrow$ (f). Assume (e). Then R contains no infinite orthogonal set of idempotents because if $e_1, e_2, \dots$ are non-zero orthogonal idempotents, then $(1 - e_1)R > (1 - e_1 - e_2)R > \dots$. Suppose $0 \ne x \in M$ and xR contains no simple submodule. Then, since xR itself is not simple, there is an $a_1 \in R$ with $xR > xa_1 R > 0$ such that $xa_1 R$ contains no simple submodule. Thus proceeding inductively we can obtain a sequence $a_1, a_2, \dots$ in R such that

$$xa_1 R > xa_1 a_2 R > \dots.$$

Therefore $a_1 R > a_1 a_2 R > \dots$ contrary to (e).

(f) $\Rightarrow$ (b). Assume (f). Let $a_1, a_2, \dots$ be a sequence in J and suppose that

$a_1 \ldots a_n \neq 0$ for all n. Then by the Maximal Principal there exists a right ideal $I \leq R_R$ maximal with respect to

$$a_1 \ldots a_n \notin I \qquad (n = 1, 2, \ldots).$$

Now R/I is a non-zero right R-module, so by (f) there exists a right ideal K with $I < K \leq R_R$ and K/I simple. By maximality, there exists an n such that $a_1 \ldots a_n \in K$. But then also $a_1 \ldots a_n a_{n+1} \in K \backslash I$. So, since K/I is simple, there exists an $r \in R$ with

$$(a_1 \ldots a_n)(1 - a_{n+1}r) \in I.$$

But $a_{n+1} \in J$, so $1 - a_{n+1}r$ is invertible. This clearly contradicts $a_1 \ldots a_n \notin I$. Thus J is left T-nilpotent. In particular, J is nil so idempotents lift modulo J (27.1). Now, using the hypothesis that R contains no infinite set of orthogonal idempotents we have that R contains a complete orthogonal set $e_1, \ldots, e_n$ of primitive idempotents. (See 10.14) and Exercise (10.11).) Since idempotents lift modulo J, these must also be primitive modulo J. (See (27.4) and (17.18).) But by (f) each $(e_i R + J)/J$ contains a minimal right ideal of R/J. So, since minimal right ideals of a ring with zero radical are direct summands ((15.10) and Exercise (13.8)), $(e_i R + J)/J$ must be simple. Thus R/J is semisimple and the proof is complete. □

28.5. Remark. It is worthwhile to observe that (28.3) and the proof (f) $\Rightarrow$ (b) of (28.4) show that for a ring R:

(1) *If every left R-module has a maximal submodule, then $J(R)$ is left T-nilpotent;*

(2) *If every right R-module has a minimal submodule, then $J(R)$ is left T-nilpotent.*

The $\mathbb{Z}$-modules $\mathbb{Z}_{p^\infty}$ and $\mathbb{Z}$ show that the converses of both (1) and (2) are false. On the other hand, it is easy to see from (28.4) that:

(3) *If R is left perfect with $J = J(R)$, then for all modules $_R M$ and N_R*

$$Rad\, M = JM \ll {_R}M \qquad and \qquad Soc\, N = l_N(J) \trianglelefteq N_R.$$

There exist rings (e.g., cosemisimple rings) whose modules all satisfy $Rad\, M = JM \ll {_R}M$ that are not left perfect. However, if $Soc\, N = l_N(J) \trianglelefteq N_R$ for all right modules N_R, then R is left perfect (see (15.17) and part (2) above); but there do exist non-perfect rings all of whose non-zero right modules have minimal submodules (i.e., $Soc\, N \trianglelefteq N_R$ for all N_R). (See Exercise (28.5).)

28.6. Corollary. *If R is left perfect, then the endomorphism ring of every finitely generated projective left R-module is left perfect. In particular any ring Morita equivalent to R is left perfect.*

Proof. Clearly "left perfect" is categorical (21.6). Moreover if $_R P$ is finitely generated projective, then there is a ring S and an idempotent $e \in S$ such that $R \approx S$ and $End(_R P) \cong eSe$. But by (28.4.b) eSe must be left perfect. □

28.7. Corollary. *If R is perfect, then so is every factor ring of R.*

Proof. Let R be left perfect. Since $R/J(R)$ is semisimple, if I is an ideal of R then (see Exercise (9.9)) $J(R/I) = (J(R) + I)/I$ is left T-nilpotent. ☐

There is an especially important class of (two-sided) perfect rings. A ring R is *semiprimary* in case R/J is semisimple and J is nilpotent. The semiprimary rings form a class of rings that contains both the left and the right artinian rings. However, the ring R of 2×2 upper triangular real matrices with all diagonal entries rational,

$$R = \begin{bmatrix} \mathbb{Q} & \mathbb{R} \\ 0 & \mathbb{Q} \end{bmatrix}$$

is a semiprimary ring that is neither left nor right artinian.

28.8. Corollary. *Every semiprimary ring, hence every left or right artinian ring, is perfect.* ☐

Recall (18.13) that direct sums of injective left R-modules are injective if (and only if) R is left noetherian. Right artinian rings are left perfect and right coherent (the latter because they are right noetherian (15.20)). Thus since their flat modules and projective modules coincide (28.4.d), Chase's Theorem (19.20) implies the following partial dual to (18.13).

28.9. Corollary. *If R is right artinian, then every direct product of projective left R-modules is projective.*

Note. Chase actually proved that every direct product of projective left R-modules is projective if and only if R is left perfect and right coherent.

There are analogues of (27.7) for both perfect and semiprimary rings. We shall obtain them via the following lemma.

28.10. Lemma. *Let $e_1, \dots, e_n$ be a complete orthogonal set of idempotents for R and let I be an ideal of R. Then I is left T-nilpotent (nilpotent) if and only if each $e_i I e_i$ is left T-nilpotent (nilpotent).*

Proof. One implication of each version is clear.

Conversely, assume that I is not left T-nilpotent. Then there is a sequence $a_1, a_2, \dots$ in I such that

$$a_1 a_2 \dots a_n \neq 0 \qquad (n = 1, 2, \dots).$$

Let $\mathscr{S}$ be the set of finite sequences $x_1, \dots, x_m$ in $\{e_1, \dots, e_n\}$ such that

$$x_1 a_1 x_2 a_2 \dots x_m a_m a_{m+1} \dots a_{m+k} \neq 0$$

for each $k = 1, 2, \dots$. Then since $1 = e_1 + \dots + e_n$ it is easy to see that $\mathscr{S}$ is not empty and that every sequence $x_1, \dots, x_n$ belonging to $\mathscr{S}$ has a proper extension $x_1, \dots, x_n, x_{n+1}$ that belongs to $\mathscr{S}$. Thus there exists an infinite sequence $x_1, x_2, \dots$ in $\{e_1, \dots, e_n\}$ such that each product

$$x_1 a_1 x_2 a_2 \dots x_m a_m x_{m+1} \neq 0.$$

But then there is an e_i such that $x_k = e_i$ for infinitely many k, whence $e_i I e_i$ is not left T-nilpotent. For the nilpotent version, let $\mathscr{S}$ be the set of sequences $x_1, \ldots, x_m$ in $\{e_1, \ldots, e_n\}$ such that

$$x_1 I x_2 I \ldots x_m I^k \neq 0$$

for each $k = 1, 2, \ldots$. $\square$

Now, recalling (27.7), we have at once

28.11. Proposition. *Let $e_1, \ldots, e_n$ be a complete orthogonal set of idempotents in a ring R. Then R is left perfect (right perfect) (semiprimary) if and only if each $e_i R e_i$ is left perfect (right perfect) (semiprimary).* $\square$

28.12. Example. The semiprimary non-artinian ring

$$R = \begin{bmatrix} \mathbb{Q} & \mathbb{R} \\ 0 & \mathbb{Q} \end{bmatrix}$$

provides an example that is relevant to these last results. It is hereditary, hence right coherent, so the converse of (28.9) is false. Moreover, it has a complete orthogonal set of idempotents e_1, e_2 such that $e_1 R e_1 \cong e_2 R e_2 \cong \mathbb{Q}$, so there is no artinian analogue of (28.11).

Projective Modules

Let R be left perfect. Then, since R is semiperfect, it has a basic set of idempotents $e_1, \ldots, e_m$ such that the sequence $Re_1/Je_1, \ldots, Re_m/Je_m$ includes exactly one copy of each simple left R-module (see (27.10).) Let M be a left R-module. Then M/JM is semisimple, so there exist sets $A_1, \ldots, A_m$, unique to within cardinality, such that

$$M/JM \cong (Re_1/Je_1)^{(A_1)} \oplus \ldots \oplus (Re_m/Je_m)^{(A_m)}.$$

Set

$$P = Re_1^{(A_1)} \oplus \ldots \oplus Re_m^{(A_m)}.$$

Then $P/JP \cong M/JM$, P is projective, $JP \ll P$ and $JM \ll M$ (28.3), so by (27.5) there is a projective cover $P \xrightarrow{p} M \to 0$. Thus, by uniqueness of projective covers (17.17), we conclude

28.13. Proposition. *Let R be left perfect with basic set $e_1, \ldots, e_m$ of primitive idempotents. Let $_R M$ be a left R-module. Let $A_1, \ldots, A_m$ be sets and let*

$$P = Re_1^{(A_1)} \oplus \ldots \oplus Re_m^{(A_m)}.$$

Then there is a projective cover $P \to M \to 0$ if and only if

$$M/JM \cong (Re_1/Je_1)^{(A_1)} \oplus \ldots \oplus (Re_m/Je_m)^{(A_m)}.$$ $\square$

Since a left perfect ring R is semiperfect, if $e_1, \ldots, e_m$ is a basic set of idempotents for R, the projective left and right R-modules are those modules of the form

$$_R P \cong Re_1^{(A_1)} \oplus \ldots \oplus Re_m^{(A_m)} \qquad \text{and} \qquad Q_R \cong e_1 R^{(A_1)} \oplus \ldots \oplus e_m R^{(A_m)}.$$

Our next result implies that if R is left perfect, the induced decomposition of $_RP$ complements direct summands, and hence is as well behaved as could be desired. Moreover, the existence of such decompositions for all projective left modules is characteristic of left perfect rings.

28.14. Theorem. *The following statements about a ring R are equivalent:*

(a) *R is left perfect;*

(b) *Every projective left R-module has a decomposition that complements direct summands;*

(c) *The countably generated free module $F = R^{(\mathbb{N})}$ has a decomposition that complements direct summands.*

Proof. Let $J = J(R)$.

(a) $\Rightarrow$ (b). Assume (a) and let $_RP$ be projective. Since R is semiperfect, P has a decomposition $P = \oplus_A P_\alpha$ as a direct sum of primitive submodules. For each $H \leq P$, set

$$\bar{H} = (H + JP)/JP.$$

Then, each $\bar{P}_\alpha \cong P_\alpha/JP_\alpha$ is simple (see (27.10)). Suppose $P = U \oplus V$. Since $\bar{P} = \oplus_A \bar{P}_\alpha$ is semisimple and $\bar{P} = \bar{U} \oplus \bar{V}$, there exists, by (9.2), a set $B \subseteq A$ such that

$$\bar{P} = \bar{U} \oplus (\oplus_B \bar{P}_\beta).$$

We claim that this decomposition "lifts". Indeed, $P = U + (\Sigma_B P_\beta) + JP$ and $JP \ll P$ (28.3) imply

$$P = U + (\oplus_B P_\beta).$$

Set $L = \Sigma_B P_\beta$. So to complete the proof of this implication we need only show $U \cap L = 0$. But since U and L are direct summands of P, we have $JU = U \cap JP$ and $JL = L \cap JL$. Therefore, since $\bar{P} = \bar{U} \oplus \bar{L}$,

$$U \cap L \subseteq JP \cap U \cap L = JU \cap JL.$$

Now the natural map $U \times L \to U + L = P$ is epic, hence split, and it has kernel $K = \{(u, -u) \mid u \in U \cap L\} \leq J(U \times L)$. Since $J(U \times L) \ll U \times L$, we infer $K = 0$ and $U \cap L = 0$.

(b) $\Rightarrow$ (c). This is clear.

(c) $\Rightarrow$ (a). Let F be the free left R-module with free basis $x_1, x_2, \ldots$ and suppose that $F \cong R^{(\mathbb{N})}$ has a decomposition that complements direct summands. Then so does $R^{(2)}$, by (12.3). Thus R is semiperfect by (27.12). Let $e_1, \ldots, e_n$ be a complete orthogonal set of primitive idempotents of R. Let $a_1, a_2, \ldots$ be a sequence in J, and let $y_k = x_k - a_k x_{k+1}$ and $G = \Sigma_{k=1}^{\infty} R y_k \leq F$ as in Lemmas (28.1) and (28.2). Since both sequences $a_1, 0, a_3, 0, \ldots$ and $0, a_2, 0, a_4, \ldots$ satisfy the condition of (28.1.2), both $y_1, x_2, y_3, x_4, \ldots$ and $x_1, y_2, x_3, y_4, \ldots$ are free bases for F. Thus F has an indecomposable decomposition

$$F = (\oplus_{k=1}^{\infty}(\oplus_{i=1}^{n} Re_i y_{2k-1})) \oplus (\oplus_{k=1}^{\infty}(\oplus_{i=1}^{n} Re_i x_{2k}))$$

which, by (12.5), must complement direct summands; and $\oplus_{k=1}^{\infty} Ry_{2k}$ is one of those direct summands. Therefore there exist left ideals $I_1, I_2, \ldots$ (each of the form $R(e_{i_1} + \ldots + e_{i_t})$ or 0) such that

$$F = (\oplus_{k=1}^{\infty} R y_{2k}) \oplus (\oplus_{k=1}^{\infty} I_{2k-1} y_{2k-1}) \oplus (\oplus_{k=1}^{\infty} I_{2k} x_{2k}).$$

We claim that the sum of the first two terms is $G = \oplus_{k=1}^{\infty} R y_k$. For if we apply the projections p_k for $F = \oplus_{k=1}^{\infty} R x_k$ to the above decomposition of F, we obtain

$$R x_1 = p_1(F) = I_1 x_1$$

and for $l = 1, 2, \ldots,$

$$R x_{2l+1} = p_{2l+1}(F) = p_{2l+1}(R y_{2l}) + p_{2l+1}(I_{2l+1} y_{2l+1})$$

$$= (R a_{2l} + I_{2l+1}) x_{2l+1}.$$

Then since $R a_{2l} \leq J \ll R$, we must have

$$R = I_1 = I_3 \ldots.$$

Therefore G is a direct summand of F. Now by (28.2) there is an $n \in \mathbb{N}$ and an $r \in R$ such that $a_1 \ldots a_n = a_1 \ldots a_n a_{n+1} r$ so, since $1 - a_{n+1} r$ is invertible, $a_1 \ldots a_n = 0$ as desired.   □

By Corollary (15.23), a ring that is left or right perfect ring and left noetherian must be a left artinian ring. Thus from (25.6) and (28.14) we have

28.15. Corollary. *A ring R is left artinian if and only if each of its injective left modules and each of its projective left modules has a decomposition that complements direct summands.*   □

28. Exercises

1. Let $_R F$ be a free module with free basis $x_1, x_2, \ldots$. This basis determines a ring isomorphism $\varphi : End\,(_R F) \to \mathbb{RFM}_{\mathbb{N}}(R)$. (See Exercise (8.12).) Consider the two $\mathbb{N}$ square matrices over R:

$$A = \begin{bmatrix} 1 & -a_1 & 0 & \cdots \\ 0 & 1 & -a_2 & \cdots \\ 0 & 0 & 1 & \cdots \\ \cdot & \cdot & \cdot & \cdots \\ \cdot & \cdot & \cdot & \cdots \\ \cdot & \cdot & \cdot & \cdots \end{bmatrix} \qquad B = \begin{bmatrix} 1 & a_1 & a_1 a_2 & a_1 a_2 a_3 & \cdots \\ 0 & 1 & a_2 & a_2 a_3 & \cdots \\ 0 & 0 & 1 & a_3 & \cdots \\ \cdot & \cdot & \cdot & \cdot & \cdots \\ \cdot & \cdot & \cdot & \cdot & \cdots \\ \cdot & \cdot & \cdot & \cdot & \cdots \end{bmatrix}.$$

Then $A \in \mathbb{RFM}_{\mathbb{N}}(R)$ and $A, B \in \mathbb{CFM}_{\mathbb{N}}(R)$. Let $\varphi(s) = A$.

(1) Use the fact that $AB = BA = 1$, the $\mathbb{N} \times \mathbb{N}$ identity matrix, to prove Lemma 28.1.

(2) Prove that $Im\,s$ is a direct summand of F iff A has a right inverse in $\mathbb{RFM}_{\mathbb{N}}(R)$.

(3) Use (2) to prove Lemma 28.2.

2. Let R be the ring of all the $\mathbb{N}$-square lower triangular matrices over a field Q that are constant on the diagonal and have only finitely many non zero entries off the diagonal. Prove that R is left perfect but not right perfect. [Hint: See Exercise (15.8).]

3. Let $J = J(R)$ and let $_RP$ be projective. Prove that $JP \ll P$ iff P/JP has a projective cover.

4. Prove that R is left perfect iff every semisimple left R-module has a projective cover. [Hint: Apply (27.6), (28.3), and Exercise (28.3).]

5. Verify the assertions of Remark (28.5). For the last one try $R = K1_V + Soc(End(V_K))$ where V_K is an infinite dimensional vector space.

6. Prove that R is left perfect iff R/J is semisimple and every factor ring of R has a non-trivial right socle.

7. Prove that R is left artinian iff every factor ring of R is left finitely cogenerated. [Hint: By Exercise (28.6) R is right perfect. In such a ring $J^n = J^{n+1}$ implies $J^n = 0$.]

8. Prove that if $_RU_S$ defines a Morita duality and R is left or right perfect, then R is left artinian. [Hint: If R is left perfect, let $I \leq S_S$ and show, via (24.7), that $(S/I)^*$ is finitely generated, and then apply (24.5). If R is right perfect and $I \leq {}_RR$ then R/I is finitely cogenerated.]

9. Prove that, for a left perfect ring R with $J = J(R)$, the following are equivalent: (a) R/J^2 is right artinian; (b) J/J^2 is a finitely generated right R-module; (c) R is right artinian. [Hint: For (b) $\Rightarrow$ (c). Suppose that $J = j_1R + \dots + j_kR + J^2$ and that $J^l \nsubseteq J^{l+1}$ $(l = 1,2,\dots)$. Show that J is not left T-nilpotent by considering the set $\mathscr{S}$ of finite sequences $x_1,\dots,x_n$ in $\{j_1,\dots,j_k\}$ such that $x_1 \dots x_n J^l \nsubseteq J^{l+n+1}$ $(l = 1, 2,\dots)$.]

10. Construct a right perfect ring with radical J such that $\cap_{n=1}^{\infty} J^n \neq 0$. [Hint: Modify Exercise (15.8).]

11. Prove the assertions of Example (28.12).

§29. Modules with Perfect Endomorphism Rings

We have seen earlier that in rings satisfying certain finiteness conditions nil one sided ideals are nilpotent. In this final section we begin by considering some more such conditions. As one consequence we shall show that every module of finite length has a semiprimary endomorphism ring. Then using the results of §28 we characterize those finitely generated modules M for which each direct sum $M^{(A)}$ has a decomposition that complements direct summands. Finally, these combine to show that every direct sum of copies of a module of finite length has such a decomposition.

A ring R has the *maximum condition for right annihilators* in case every non empty set of right annihilators $r_R(A)(A \subseteq R)$ in R has a maximal element; an equivalent formulation is that every increasing chain of right annihilators

$$r_R(A_1) \leq r_R(A_2) \leq \dots$$

has finite length. See Exercise (10.9).) Moreover, this is also equivalent to the *minimum condition for left annihilators* (see (24.5)).

29.1. Proposition. *Let R have the maximum condition for right annihilators. If I is a right T-nilpotent one sided ideal, then I is nilpotent.*

Proof. Let I be a non nilpotent one sided ideal. The assumption on right annihilators implies that for some $n \in \mathbb{N}$

$$r_R(I^n) = r_R(I^{n+1}) = \ldots.$$

Since I is not nilpotent, $I^{n+1} \neq 0$, so $I^n x_1 \neq 0$ for some $x_1 \in I \backslash r_R(I^n)$. But then $x_1 \notin r_R(I^{n+1})$, so $I^{n+1} x_1 \neq 0$. Clearly then there is sequence $x_1, x_2, \ldots$ in I such that for each k,

$$x_k \ldots x_1 \in I \backslash r_R(I^n).$$

In particular, I is not right T-nilpotent. $\qquad\qquad\qquad\square$

The following important "nil implies nilpotence" theorem is due to Small and Fisher.

29.2. Theorem. *Let $_R M$ be either artinian or noetherian. Then every nil one sided ideal of $\operatorname{End}(_R M)$ is nilpotent.*

Proof. Suppose that $_R M$ is artinian. Let $S = \operatorname{End}(_R M)$. Since M_S is faithful, each right annihilator of S is of the form $r_S(K)$ for some $_R K \leq\, _R M$. So since $_R M$ is artinian, S has the maximum condition on right annihilators (see (24.3)). Now suppose that I is a nil one sided ideal of S. Let C be the set of all first terms s_1 of sequences $s_1, s_2, \ldots$ in I satisfying

$$s_n \ldots s_1 \neq 0 \qquad (n = 1, 2, \ldots).$$

In view of (29.1) it will suffice to prove that $C = \emptyset$. On the contrary let us assume that $C \neq \emptyset$. Since $_R M$ is artinian, there is an $s_1 \in C$ for which $M s_1$ is minimal in $\{Ms \mid s \in C\}$. Clearly, $Cs_1 \cap C \neq \emptyset$, so there is an $s_2 \in C$ for which Ms_2 is minimal in $\{Ms \mid ss_1 \in C\}$. Notice that $Ms_2 s_1 = Ms_1$. An obvious induction argument now shows that there exists a sequence $s_1, s_2, \ldots$ in C such that for each n, Ms_n is minimal in $\{Ms \mid ss_{n-1} \ldots s_1 \in C\}$. In particular, if $n \geq m$, then

$$Ms_n \ldots s_m = Ms_m.$$

For each n, set

$$p_n = s_n \ldots s_1.$$

Then $Mp_n = Ms_1 = Mp_m$ for all m and n. Therefore given n, if $x \in M$, there is an $x' \in M$ with $x'p_{n+1} = xp_n$, so $x = x - x's_{n+1} + x's_{n+1}$ and

(1) $$M = \operatorname{Ker} p_n + \operatorname{Im} s_{n+1}.$$

Now suppose that $n \geq m$ and $s_m p_n \neq 0$. If $k \geq m$, then $Ms_k \ldots s_m p_n = Ms_m p_n$, so $s_k \ldots s_m s_n \ldots s_m \ldots s_1 = s_k \ldots s_m p_n \neq 0$. The minimality of Ms_m then implies

$Ms_m s_n \dots s_m = Ms_m \neq 0$, clearly contrary to the nilpotence of $s_n \dots s_m$. Therefore $s_m p_n = 0$ whenever $n \geq m$; that is,

(2) $$Im\ s_m \leq \bigcap_{n \geq m} Ker\ p_n.$$

Using modularity (2.5) and (1) and (2) an easy induction shows that

(3) $$\bigcap_{k=1}^{n} Ker\ p_k + \sum_{j=2}^{n+1} Im\ s_j = M.$$

Since $_R M$ is artinian, there is an n such that $\bigcap_{k=1}^{n} Ker\ p_k \leq Ker\ p_{n+1}$. But by (2), $Ker\ p_{n+1} \geq \sum_{j=2}^{n+1} Im\ s_j$, so by (3) we have the desired contradiction, $Ker\ p_{n+1} = M$.

The dual proof for the case in which $_R M$ is noetherian will be omitted. $\square$

29.3. Corollary. *Every module of finite length has a semiprimary endomorphism ring.*

Proof. Let $_R M$ have finite length n and let $S = End\ (_R M)$. By (12.8), (12.9), and (27.6) S is semiperfect. So in view of (29.2) it will suffice to prove that $J(S)$ is nil. But if $a \in J(S)$, then by Fitting's Lemma (11.7)

$$M = Im\ a^n \oplus Ker\ a^n.$$

Thus $(a | Im\ a^n)$ is an automorphism of $Im\ a^n$ and there is an $s \in S$ such that $(sa | Im\ a^n) = 1_{Im\ a^n}$. Since $1_M - sa$ is invertible in S (15.3), we have $Im\ a^n \leq Ker\ (1_M - sa) = 0$. $\square$

In order to study the decompositions of direct sums of copies of a finitely generated module we show next that such direct sums "look like" free modules over its endomorphism ring. Specifically we prove

29.4. Lemma. *Let M be a finitely generated left R-module with endomorphism ring $S = End(_R M)$. Let $_S P$ denote the category of projective left S-modules and let $_M S$ denote the category of direct summands of direct sums of copies of M. Then the functors*

$$Hom_R(M_S, -) : _M S \to _S P$$

and

$$(_R M \otimes_S -) : _S P \to _M S$$

are inverse category equivalences.

Proof. Since $(M \otimes_S -)$ and $Hom_R(M, -)$ preserve direct sums (Theorem (19.10) and Exercise (16.3)), we see at once that they are functors between the desired categories. For instance, a split exact sequence

$$0 \to N - \oplus \to M^{(A)} - \oplus \to N' \to 0$$

goes to a split exact sequence

$$0 \to Hom_R(M, N) - \oplus \to Hom_R(M, M^{(A)}) - \oplus \to Hom_R(M, N') \to 0$$

(see (16.2)) in which the middle term is isomorphic to $_S S^{(A)}$. Now for each module N in $_M S$ define $\eta_N : M \otimes_S Hom_R(M, N) \to N$ via

$$\eta_N(m \otimes \gamma) = \gamma(m).$$

Then it is routine to check that this yields a natural transformation

$$\eta : (M \otimes_S Hom_R(M, \_)) \to 1_{_m S}$$

Moreover, since $_R M$ is finitely generated, the functor $(M \otimes_S Hom_R(M, \_))$ preserves direct sums (Theorem (19.10) and Exercise (16.3)). So we need only check that η_M is an isomorphism to see that η is actually a natural isomorphism (Exercise (20.17)). But this is the case by (19.6) because $Hom_R(M_S, M) = {}_S S$. On the other hand, for each projective module P in $_S P$, let $\gamma : P \to Hom_R(M, (M \oplus_S P))$ via

$$v_p(p) : m \to m \otimes p$$

to obtain, in a similar manner, a natural isomorphism

$$v : 1_{_s P} \to Hom_R(M, (M \otimes_S \_)).$$

(Observe here that $v_{_S S} : {}_S S \to Hom_R(M, (M \otimes_S S))$ is the canonical isomorphism.) □

29.5. Theorem. *Let M be a finitely generated left R-module. Then the following are equivalent:*

(a) *$M^{(A)}$ has a decomposition that complements direct summands for every set A;*

(b) *$M^{(\mathbb{N})}$ has a decomposition that complements direct summands;*

(c) *$End(_R M)$ is left perfect.*

Proof. Under the equivalence of (29.4) $M^{(A)}$ corresponds to the free left S-module $S^{(A)}$, and it is not hard to show that if N corresponds to P under such an equivalence then N has a decomposition that complements direct summands iff P does. (See Exercise (21.11.2).) Thus (28.14) applies. □

29.6. Corollary. *Let $M = \oplus_A M_\alpha$ be an indecomposable decomposition such that each term has finite length and the modules $(M_\alpha)_{\alpha \in A}$ represent only finitely many isomorphism classes. Then this decomposition complements direct summands.*

Proof. If each M_α is isomorphic to one of $M_{\alpha_1}, ..., M_{\alpha_n}$, then M is isomorphic to a direct summand of

$$(M_{\alpha_1} \oplus ... \oplus M_{\alpha_n})^{(A)}$$

and the module $M_{\alpha_1} \oplus ... \oplus M_{\alpha_n}$ has a perfect (indeed, semiprimary) endomorphism ring by (29.3). Thus the corollary follows from (29.5) and (12.3). □

Every module over a semisimple ring has a decomposition that complements direct summands. On the other hand, in this chapter we have seen that every injective and every projective left module over a ring R has a decomposition that complements direct summands iff R is left artinian (28.15). A ring R is a ring of *finite module type* in case there exist R-modules $M_1, \ldots, M_n$ of finite length such that every R-module is isomorphic to a direct sum of copies of the M_i. The literature on these rings is too extensive for us to cover here. However, we do note that by (29.6) every module over a ring of finite module type does have a decomposition that complements direct summands; and that such a ring need not be semisimple. (See Exercise (29.5).)

29. Exercises

1. Prove that if $_RE$ is injective and noetherian, then $S = End(_RE)$ is semi-primary. [Hint: By Exercise (27.3), S is semiperfect. Now use (18.20) and (29.2).]

2. Let R satisfy the maximum condition for left annihilators and the maximum condition for right annihilators. Prove that every nil one sided ideal of R is nilpotent. [Hint: There is an n such that every chain of annihilators has length $\le n$.]

3. Prove that the left perfect ring of Exercise (28.2) has the maximum condition for right annihilators but that not every left T-nilpotent ideal is nilpotent.

4. Let R be the ring of polynomials over $\mathbb{Z}_2$ in countably many indeterminants $X_1, X_2, \ldots$. Let I be the ideal generated by $\{X_1^2, X_2^2, \ldots\}$. Prove that $S = R/I$ has a non nilpotent ideal J each element of which is nilpotent of nilpotency index 2.

5. Let R be the ring of upper triangular 2×2 matrices over a field. Let $e = \begin{bmatrix} 0 & 0 \\ 0 & 1 \end{bmatrix} \in R$. Prove that every left R-module M has a decomposition $M = E \oplus S$ such that $E \cong Re^{(A)}$ and S is semisimple. Conclude that R is a non-semisimple ring of finite module type. [Hint: Consider a maximal indepent subset of $\{Rex \mid x \in M \text{ and } J(R)ex \ne 0\}$.]

Chapter 8

Classical Artinian Rings

In our concluding chapter we present basic results on several types of artinian rings that have come to be regarded as classical due to their natural origins and the influence they have had on the literature of ring and module theory. These include artinian rings with duality, quasi-Frobenius (or QF) rings, QF-3 rings, and serial rings.

§30. Artinian Rings with Duality

In this section we present a theorem of Azumaya and Morita which yields several necessary and sufficient conditions that the category $_R$FM of finitely generated left modules over a ring R have a duality with the category FM$_S$ of finitely generated right modules over a ring S. From §23 and §24 we see that in order for such a duality to exist the ring R must be left artinian and possess a finitely generated injective cogenerator. This pair of conditions is the basic characterization. Also from §23 and §24 it follows that in order to have a duality between $_R$FM and FM$_S$, S must be right artinian; and the duality is isomorphic to the $_RU_S$-dual for some bimodule $_RU_S$. This bimodule is both left and right faithful (indeed, a cogenerator) and the U-dual takes simples to simples in the sense that $T^* = Hom(T, U)$ is a simple right S- (left R-) module whenever T is a simple left R- (right S-) module. These conditions also serve to characterize the existence of a duality between $_R$FM and FM$_S$. They are intimately connected to the annihilator condition of Theorem 24.5 by the first theorem of this section.

Duality Theorems

A bilinear mapping (see §24) $\mu: {}_RM \times N_S \to {}_RU_S$ is called *non-degenerate* in case $l_M(N) = 0$ and $r_N(M) = 0$. Note that scalar multiplication yields a bilinear mapping $_RR \times {}_RU_S \to {}_RU_S$ that is non-degenerate iff $_RU$ is faithful; and that the usual bilinear map $M \times M^* \to {}_RU_S$ is non-degenerate iff $_RM$ is U-torsionless.

30.1. Theorem. *Let $_RU_S$ be a bimodule such that the $_RU_S$-dual $(\ )^*$ takes simples to simples. Let*

$$\mu: {}_RM \times N_S \to {}_RU_S$$

be a non-degenerate bilinear map. If either $_R M$ or N_S has a composition series, then:

(1) For each $K \leq M$ and each $L \leq N$,

$$l_M(r_N(K)) = K \quad and \quad r_N(l_M(L)) = L;$$

(2) The induced mappings

$$\lambda :_R M \to N^* \quad and \quad \rho : N_S \to M^*$$

defined by $\lambda(m) : n \mapsto \mu(m, n)$ and $\rho(n) : m \mapsto \mu(m, n)$ are isomorphisms;

(3) All submodules and factor modules of M and N are U-reflexive;

(4) $_R U$ is M-injective and U_S is N-injective.

Proof. Suppose that the $_R U_S$-dual does take simples to simples. If W is a left R- or right S-module of finite length and K is a maximal submodule of W, then, from the exact sequence

$$0 \to (W/K)^* \xrightarrow{n_K^*} W^* \xrightarrow{i_K^*} K^*,$$

we see that $c(W^*) \leq c(K^*) + 1$. Thus, arguing inductively on composition length, we have

$$c(W^*) \leq c(W).$$

Moreover, the inequalities

$$c(W) \geq c(W^*) \geq c(W^{**})$$

imply that W is U-reflexive iff W is U-torsionless.

Now to prove the theorem, let

$$\mu :_R M \times N_S \to {}_R U_S$$

be non-degenerate and suppose it is $_R M$ that has a composition series. For each $K \leq M$, define

$$\rho^K : N/r_N(K) \to K^*$$

via

$$\rho^K(n + r_N K) : k \mapsto \mu(k, n) \quad (n \in N, k \in K);$$

and for each $L \leq N$, define

$$\lambda_L : l_M(L) \to (N/L)^*$$

via

$$\lambda_L(x) : n + L \mapsto \mu(x, n) \quad (x \in l_M(L), n \in N).$$

Then it is easy to check that ρ^K and λ_L are monomorphisms (ρ^K is monic for any μ and λ_L is monic because μ is non-degenerate). These monomorphisms and our discussion in the first paragraph of this proof yield the following inequalities for each $K \leq M$:

$$c(K) \leq c(l_M(r_N(K))) \tag{24.3.2}$$

$$\leq c((N/r_N(K))^*) \quad (\text{using } \lambda_{r_M(K)})$$

$$\leq c(N/r_N(K)) \quad (\text{first paragraph})$$

$$\leq c(K^*) \leq c(K) \quad (\text{using } \rho^K).$$

These must all be equalities. Since they are, we see that

$$l_M(r_N(K)) = K$$

and that ρ^K is an isomorphism for all $K \leq M$. Moreover, since μ is non-degenerate,

$$\rho = \rho^M : N \to M^*$$

is an isomorphism. Noting that $N \cong M^*$ also has a composition series, it follows by symmetry that (1) and (2) hold, and that M and N (being isomorphic to U-duals of one another), and all of their submodules, are U-torsionless (20.14). But factor modules of N and, by symmetry, of M are also U-torsionless because of the isomorphisms $\rho^{l_M(L)}$ which yield

$$N/L = N/r_N(l_M(L)) \cong l_M(L)^*.$$

Thus, since U-torsionless modules of finite length are U-reflexive, (3) also holds. Finally, since ρ^K is an isomorphism, if $K \leq M$, then every map $h: {}_R K \to {}_R U$ is of the form

$$h: k \mapsto \mu(k, n_h) \tag{$k \in K$}$$

for some $n_h \in N$; such a map can be extended to $\bar{h}: M \to {}_R U$ via

$$\bar{h}: m \mapsto \mu(m, n_h). \tag{$m \in M$}$$

Thus, ${}_R U$ is M-injective, by symmetry U_S is N-injective, and the proof is complete. $\square$

Let R be a semisimple ring. Then, since all R-modules are semisimple, projective, and injective, and R contains a copy of each of its simple modules, an R-module U is faithful if and only if it is a generator if and only if it is a cogenerator (see (17.9) and (18.15)). Suppose that U is a finitely generated faithful left module over a semisimple ring R and let $S = End({}_R U)$. Then ${}_R U$ is a progenerator, so (see (22.4) and (21.9)) S is semisimple, and (see (17.9)) ${}_R U_S$ is a faithfully balanced module. But as we have just noted, ${}_R U$ and U_S, being faithful modules over semisimple rings, are injective cogenerators, so by (24.1) the ${}_R U_S$-dual defines a Morita duality. In particular (see (24.5)), the ${}_R U_S$-dual takes simples to simples—a fact which we now extend to any finitely cogenerated injective cogenerator over a ring that is semisimple modulo its radical.

30.2. Lemma. *Let $R/J(R)$ be semisimple and let $_R U$ be a finitely cogenerated injective cogenerator with $End(_R U) = S$. Then*

(1) *$Soc \, _R U = Soc \, U_S$;*

(2) *The $_R U_S$-dual $(\ )^*$ takes simples to simples.*

Proof. (1) Since $_R U$ is a finite direct sum of indecomposable injectives (18.18), $S = End(_R U)$ must be semiperfect by (25.4) and (27.6); and, letting $V = Soc(_R U) \le U$, it follows from (18.21) that $J(S) = r_S(V)$. Now we have

$$Soc(U_S) = l_U(J(S)) \quad (15.17)$$

$$= l_U(r_S(V)) \quad (18.21)$$

$$= l_U(r_{U^*}(V)) \quad (U^* = End(_R U))$$

$$= V \quad (24.4.2).$$

(2) Using (1), write $Soc_R U = V = Soc \, U_S$. Then, since $_R V$ is finitely generated and contains a copy of every simple left R-module (18.15) and $S/J(S)$ is naturally isomorphic to $End(_R V) = End(_{R/J(R)} V)$ by (18.21), we see from the discussion preceding the lemma that the $_{R/J(R)} V_{S/J(S)}$-dual takes simples to simples. But if $_R T$ and T'_S are simple, then $Hom_R(T, U) \cong Hom_{R/J(R)}(T, V)$ and $Hom_S(T', U) \cong Hom_{S/J(S)}(T', V)$, so the $_R U_S$-dual takes simples to simples. $\square$

30.3. Lemma. *If R is a left artinian ring, then every cogenerator $_R U$ is balanced.*

Proof. If R is left artinian and $_R U$ is a cogenerator, then, since R has only finitely many isomorphism classes of simple modules, $_R U = U_0 \oplus U'$, where U_0 is a finitely cogenerated injective cogenerator (18.16). By (14.1.1), since U_0 cogenerates U', $BiEnd(_R U)$ embeds in $BiEnd(_R U_0)$. Thus, we may assume that U is a finitely cogenerated injective cogenerator. Then letting $S = End(_R U)$, the $_R U_S$ duals of simples are simples by (30.2); and since $_R U$ is faithful, the bilinear map

$$\mu: {}_R R \times U_S \to {}_R U_S$$

given by scalar multiplication is non-degenerate. Thus, since $_R R$ has a composition series, we see by (30.1.2) that $_R U$ is balanced, i.e., $\lambda: R \to (U_S)^* = BiEnd(_R U)$ is an isomorphism. $\square$

From Theorem 23.5 it follows that to have a duality

$$H': {}_R FM \to FM_S, \qquad H'': FM_S \to {}_R FM,$$

there must exist a bimodule $_R U_S$ such that

$$H' \cong Hom_R(\_, U) = (\ )^*, \qquad H'' \cong Hom_S(\_, U) = (\ )^*,$$

and every finitely generated left R- and right S-module is U-reflexive. If $_R U_S$ is such a bimodule, then $_R U \cong (S_S)^*$ and $U_S \cong (_R R)^*$ are finitely generated and are, by Theorem 24.1, injective cogenerators. Moreover, by Theorem 24.8, R is a left artinian ring, S is a right artinian ring, and a left R- or right

S-module is U-reflexive iff it has a composition series. We now are ready to prove the main theorem of this section which gives several necessary and sufficient conditions on R, S, and $_RU_S$ to ensure that these phenomena occur.

30.4. Theorem [Azumaya, Morita]. *Let R be a left artinian ring and let $_RU_S$ be a bimodule. Then the following are equivalent:*

(a) *The $_RU_S$-dual ( )* defines a duality between $_R$FM and FM$_S$;*

(b) *R, U, and S satisfy*
   (i) *S is right artinian,*
   (ii) *all finitely generated left R-modules and right S-modules are U-reflexive;*

(c) *R, U, and S satisfy*
   (i) *S is right artinian ($_RU$ is finitely generated),*
   (ii) *$_RU$ and U_S are faithful,*
   (iii) *all simple left R-modules and right S-modules are U-reflexive;*

(d) *R, U, and S satisfy*
   (i) *$_RU$ is finitely generated (S is right artinian),*
   (ii) *$_RU$ and U_S are faithful,*
   (iii) *the $_RU_S$-dual ( )* takes simples to simples;*

(e) *R, U, and S satisfy*
   (i) *S_S is U-reflexive (i.e., right multiplication $\rho: S \to End(_RU)$ is an isomorphism),*
   (ii) *$_RU$ is a finitely generated injective cogenerator;*

(f) *R, U, and S satisfy*
   (i) *S_S is U-reflexive,*
   (ii) *$_RU$ and U_S are injective cogenerators;*

(g) *R, U, and S satisfy*
   (i) *$_RU$ is finitely generated (S is right artinian),*
   (ii) *for each $I \leq {}_RR$ and each $V \leq U_S$, $l_R(r_U(I)) = I$ and $r_U(l_R(V)) = V$,*
   (iii) *for each $K \leq S_S$ and each $W \leq {}_RU$, $r_S(l_U(K)) = K$ and $l_U(r_S(W)) = W$.*

Proof. (a) $\Rightarrow$ (b). This implication follows from Theorem 24.8.

(b) $\Rightarrow$ (c). The regular modules $_RR$ and S_S are U-torsionless iff they are faithful (e.g., Ker $\sigma_R = Rej_R(U) = l_R(U)$ by (20.12) and (8.22)). The remaining parts of this implication are clear.

(c) $\Rightarrow$ (d). Assume (c) and let T be a simple left R-module. Then since S is semiprimary (29.3) and $T^{**} \neq 0$ implies $T^* \neq 0$, T^* contains a maximal submodule M. Taking the dual of the natural exact sequence

$$0 \longrightarrow M \xrightarrow{\; i_M \;} T^* \xrightarrow{\; n_M \;} T^*/M \to 0,$$

we obtain an exact sequence

$$0 \longrightarrow (T^*/M)^* \xrightarrow{\; n_M^* \;} T^{**} \xrightarrow{\; i_M^* \;} M^*.$$

Being the U-dual of a simple module, $(T^*/M)^* \neq 0$. Thus, since T^{**} is simple, n_M^* is epic and, consequently, $i_M^* = 0$. But T^* is U-torsionless (see (20.14)) or, equivalently, U cogenerates T^*. Hence, by (8.11.2), $i_M^* = 0$ implies $i_M = 0$, so that $M = 0$. Thus, T^* is simple for every simple left R- (and, similarly, every simple right S-) module T. Now we see that either version of (c) implies both the parenthetical and the non-parenthetical version of (d) by applying (30.1) to the non-degenerate bilinear map

$$_RU \times S_S \to {_RU_S}$$

given by scalar multiplication $(u, s) \mapsto us$.

(d) $\Rightarrow$ (e). This implication follows from Theorem 30.1 applied to the non-degenerate bilinear maps

$$_RU \times S_S \to {_RU_S} \quad \text{and} \quad _RR \times U_S \to {_RU_S}$$

given by scalar multiplication. The first application shows that S_S is U-reflexive because, by (20.15), S_S is U-reflexive iff $\rho : S \to End(_RU)$ is an isomorphism. The second shows that $_RU$ is R-injective, so since every simple left R-module is U-torsionless, $_RU$ is an injective cogenerator (see (18.3) and (18.15)).

(e) $\Rightarrow$ (f). By (30.2), Theorem 30.1 applied to scalar multiplication

$$_RU \times S_S \to {_RU_S}$$

gives the proof of this implication.

(f) $\Rightarrow$ (a). In view of (30.3) the hypotheses of (f) imply that $_RU_S$ is a balanced bimodule. Thus, this implication follows from (24.1) and (24.8).

(d) $\Rightarrow$ (g). By (30.1).

(g) $\Rightarrow$ (d). Assume (g). Then

$$l_R(U) = l_R(r_U(0)) = 0$$

and

$$r_S(U) = r_S(l_U(0)) = 0,$$

so that $_RU$ and U_S are faithful. If I is a maximal left ideal in R, then clearly $r_U(I)$ is a minimal submodule of U_S. But it is easy to see that

$$(R/I)^* \cong r_{R^*}(I) \cong r_U(I)$$

for any $I \le {_RR}$ (see (4.5)). This and a symmetric argument show that the $_RU_S$-dual takes simples to simples.   □

Over a left artinian ring a finitely cogenerated injective cogenerator is always of the form

$$_RU \cong E(T_1)^{n_1} \oplus \cdots \oplus E(T_k)^{n_k},$$

where $T_1, \ldots, T_k$ represent all simple left R-modules. Thus, we have

30.5. Corollary. *Let R be a ring. Then there is a duality between $_RFM$ and FM_S for some ring S if and only if R is a left artinian ring over which the injective envelope of each simple left R-module is finitely generated.*   □

Next we consider some artinian rings that have *self-duality*, i.e., a duality between their categories finitely generated left and right modules. It remains an open problem to determine which artinian rings have self-duality.

Artin Algebras

Let K be a subring of the center of a ring R, so that R is a K-algebra. Then a left module M is an R-K-bimodule with $mk = km$, and if C is a K-module, then $Hom_K(_RM, C) \in M_R$ by (4.4) and, similarly, given N_R we also have $Hom_K(N_R, C) \in {}_RM$. If K is artinian and $_KR$ is finitely generated, then R is called an *artin algebra* over K. In this case, R is artinian and an R-module is finitely generated if and only if it is finitely generated as a K-module. Moreover, R is an artin algebra if and only if *Cen R* is artinian and R is finitely generated as a *Cen R*-module (Exercise 30.2). Of course, any commutative artinian ring is an artin algebra.

30.6. Proposition. *Let R be an artin algebra over K with $C = E(K/J(K))$, the minimal cogenerator over K. Then $D = Hom_K(\_, C)$ defines a duality between $_R$FM and FM$_R$.*

Proof. Since $K/J(K)$ is a ring direct sum of fields,

$$Soc\ C = S_1 \oplus \cdots \oplus S_n,$$

where the S_i are the distinct simple K-modules, and $S_i \cong K/I_i$ for some maximal ideal I_i $(i = 1, \ldots, n)$. But then

$$Hom_K(S_i, C) \cong Hom_K(K/I_i, K/I_i) \cong S_i$$

for $i = 1, \ldots, n$, so by (30.4 (d) and (b)) every finitely generated K-module is C-reflexive. In particular, the evaluation K-maps σ_M and σ_N are isomorphisms for all finitely generated $_RM$ and N_R, so, since they are in fact R-maps, we have $DD \cong 1_{_R\text{FM}}$ and $DD \cong 1_{\text{FM}_R}$. $\square$

QF Rings

We turn our attention to those rings for which the $_RR_R$-dual $(\ )^*$ defines a duality between the category of finitely generated left and right modules over R. From the results of §24, we see at once that such a ring must be left and right self-injective and left and right artinian. These conditions are, in fact, both necessary and sufficient. A ring satisfying them is called a *quasi-Frobenius (or QF) ring*. Nakayama introduced QF rings in 1938. They are, in a sense, the minimal categorical generalization of group algebras. Their basic characterizations are presented in

30.7. Theorem. *The following statements about a left artinian ring R are equivalent:*

(a) *R is QF;*

(b) *R is left or right self-injective;*

(c) *$_RR$ or R_R is a cogenerator;*

(d) *For each left ideal $I \leq {}_RR$ and each right ideal $K \leq R_R$,*

$$l_R(r_R(I)) = I \quad and \quad r_R l_R(K) = K;$$

(e) *The $_RR_R$-dual $(\ )^*$ defines a duality between $_R$FM and FM$_R$.*

Proof. (a) $\Leftrightarrow$ (d) $\Leftrightarrow$ (e). These are by (30.4).

(a) $\Rightarrow$ (b). This is obvious.

(b) $\Leftrightarrow$ (c). Let $e_1, \ldots, e_n$ be a basic set of primitive idempotents for R. If R_R is injective, then $e_1 R, \ldots, e_n R$ must be pairwise non-isomorphic indecomposable injectives in M_R. Since their socles are essential (see (28.8) and (28.5)), they must be the injective envelopes of the n distinct simple right R-modules. Thus, every simple right R-module embeds in R_R, and R_R is a cogenerator by (18.15). Conversely, if R_R is a cogenerator, then, by (18.16), R_R must have direct summands isomorphic to the n indecomposable injective right R-modules; they must be $e_1 R, \ldots, e_n R$, so R_R is injective.

(c) $\Rightarrow$ (e). Since (b) $\Leftrightarrow$ (c), if $_R R$ is a cogenerator then $_R R_R$ satisfies (30.4(e)) and (e) follows. But if R_R is a cogenerator then $r_R(l_R(K)) = K$ for every $K \leq R_R$ by (25.2). Since R is left noetherian, it follows that R is right artinian, so the version of (30.4) for right artinian rings applies in this case. $\qquad\square$

Several other characterizations of QF rings now follow from (30.4).

30.8. Corollary. *Let R be a left artinian ring. Then the following are equivalent:*

(a) *R is QF;*

(b) *Every finitely generated left and right R-module is $_R R_R$-reflexive;*

(c) *Every simple left and right R-module is $_R R_R$-reflexive;*

(d) *The $_R R_R$-dual ( )* takes simples to simples;*

(e) *Every left and every right cyclic R-module is R-torsionless.*

Proof. The equivalence of (a), (b), (c), and (d) is immediate from (30.7) and (30.4); and considering (30.7(d)) and (25.2), we see that (e) is equivalent to (a). $\qquad\square$

A left noethenian ring that is either left or right self-injective is also QF, as we shall see using the following lemma.

30.9. Lemma. *If R is left self-injective, then*

(1) *$r_R(I_1 \cap I_2) = r_R(I_1) + r_R(I_2)$ for every pair of left ideals $I_1, I_2 \leq {}_R R$;*

(2) *$r_R(l_R(K)) = K$ for every finitely generated right ideal $K \leq R_R$.*

Proof. (1) As noted in (2.16) we always have, for $I_1, I_2 \leq {}_R R$,

$$r_R(I_1) + r_R(I_2) \subseteq r_R(I_1 \cap I_2).$$

Let $x \in r_R(I_1 \cap I_2)$. Then, checking that

$$\varphi : a_1 + a_2 \mapsto a_2 x \quad (a_i \in I_i)$$

defines an R-homomorphism

$$\varphi : I_1 + I_2 \to R,$$

we see that there is, by the Injective Test Lemma, an element $y \in R$ such that φ is right multiplication by y. But then

$$a_1 y = \varphi(a_1 + 0) = 0x = 0$$

for all $a_1 \in I_1$ and

$$a_2(x - y) = \varphi(a_2) - \varphi(a_2) = 0$$

for all $a_2 \in I_2$ so that

$$x = y + x - y \in r(I_1) + r(I_2).$$

(2) First we prove that $r_R(l_R(x)) = xR$ whenever $x \in R$ and $_RR$ is injective. The inclusion $xR \subseteq r_R(l_R(x))$ always holds (see (2.15)). If $a \in r_R(l_R(x))$, then since $rx = 0$ implies $ra = 0$,

$$\theta : rx \mapsto ra \quad (r \in R)$$

defines an R-homomorphism

$$\theta : Rx \to Ra.$$

Using injectivity, we see that θ is right multiplication by some $y \in R$. Hence,

$$a = \theta(x) = xy \in xR$$

and the reverse inclusion holds. Now using (1) and (2.16), if

$$K = x_1 R + \cdots + x_n R$$

is a finitely generated right ideal, we have

$$r_R(l_R(K)) = r_R(l_R(x_1 R) \cap \cdots \cap l_R(x_n R))$$
$$= r_R(l_R(x_1)) + \cdots + r_R(l_R(x_n))$$
$$= x_1 R + \cdots + x_n R = K. \qquad \square$$

30.10. Theorem. *Every left self-injective left or right noetherian ring is QF.*

Proof. Suppose the $_RR$ is injective and that R is either left or right noetherian. Then the ascending chain condition ensures that R has a complete set of primitive idempotents $e_1, \ldots, e_n$ ((10.14) and (7.5)); each Re_i is an indecomposable injective module so its endomorphism ring, isomorphic to $e_i Re_i$, is local (25.4). Thus, R is semiperfect by (27.6). Letting $J = J(R)$, we see from the ascending chain of ideals

$$l_R(J) \leq l_R(J^2) \leq \cdots$$

that, for some $n > 0$,

$$l_R(J^n) = l_R(J^{n+1}).$$

By (15.17(e)), the right socle of $R/l_R(J^n)$ is $l_R(J^{n+1})/l_R(J^n)$. If R is left noetherian, then, by Lemma 30.9, R has the descending chain condition on principal right ideals, so the preceding equality implies $R/l_R(J^n) = 0$ (see (28.8)). Thus, in this case, $J^n = RJ^n = 0$. If, on the other hand, R is right noetherian, then (30.9) yields

$$J^n = r_R(l_R(J^n)) = r_R(l_R(J^{n+1})) = J^{n+1},$$

so, by Lemma 15.13, $J^n = 0$. Thus, R is a semiprimary and noetherian; hence artinian, on one side or the other, so R is QF by (30.7). $\qquad \square$

30. Exercises

1. Prove that if R is a QF ring with basic idempotent e, then the following are equivalent:
   (a) $_RM$ is faithful;
   (b) $_RM$ is a cogenerator;
   (c) $_RM$ is a generator;
   (d) Re is isomorphic to a direct summand of $_RM$. In particular, every faithful R-module is balanced.

2. Let R be an artin algebra over $K \subseteq Cen\ R$. Prove that:
   (1) $_RM$ is finitely generated iff $_KM$ is finitely generated.
   (2) $Cen\ R$ is artinian and R is finitely generated over $Cen\ R$.
   (3) R is indecomposable iff $Cen\ R$ is local.
   (4) Exercise (23.6) is valid with ( )* replaced by the artin algebra duality D.
   (5) If $L = Cen\ R$, $C_1 = E(K/J(K))$, and $C_2 = E(L/J(L))$, then the functors $Hom_K(-, C_1)$ and $Hom_L(-, C_2)$ are isomorphic on $_LFM$, and hence on $_RFM$. [Hint: L is an artin K-algebra; show that $C_2 \cong Hom_K(L, C_1)$.]

3. Let $I \le {}_RR_R$ and $M \in {}_RM$.
   (1) Prove that if $IM = 0$, then $E(_{R/I}M) = \mathbf{r}_{E(M)}(I)$.
   (2) Conclude that if R is left artinian with Morita duality, then so is R/I.

4. Schofield [85, pp. 215–218] has shown that there is a division ring E with division subring F such that dim $_FE = 2$, dim $E_F = 3$, and $\dim(Hom_F(E_F, F_F)_E) = 1$. Prove that the ring

$$R = \begin{bmatrix} E & E \\ 0 & F \end{bmatrix}$$

is artinian with all indecomposable left and right injective modules finitely generated, but R does not have self-duality. [Hint: See Exercise (24.9).]

§31. Injective Projective Modules

In this section we present a characterization of injective projective modules over artinian rings, and examine the structure of QF, QF-3, and QF-2 rings.

Projective and Injective Modules

We begin by recalling results from §25, §27, and §28 that specify the structure of the projective and injective modules over a left artinian ring R with radical $J = J(R)$. Since R is semiperfect, according to (27.10) it has a basic set of primitive idempotents $e_1, \ldots, e_n$ such that

$$Re_1, \ldots, Re_n \quad \text{and} \quad e_1R, \ldots, e_nR$$

are complete irredundant sets of the indecomposable projective left and right R-modules, and the simple left and right R-modules are similarly represented by

$$Re_1/Je_1, \ldots, Re_n/Je_n \quad \text{and} \quad e_1R/e_1J, \ldots, e_nR/e_nJ.$$

Since R is perfect, indeed semiprimary (28.8), every R-module has an essential socle (28.4(f)), so the indecomposable injective R-modules are

$$E(Re_1/Je_1), \ldots, E(Re_n/Je_n) \quad \text{and} \quad E(e_1R/e_1J), \ldots, E(e_nR/e_nJ).$$

Moreover, by (27.11), the projective left (right) R-modules are the direct sums of copies of the $Re_i(e_iR)$, and since R is left noetherian, according to (25.6) the injective left R-modules are the direct sums of copies of the $E(Re_i/Je_i)$. These direct sum decompositions of injective and projective modules over R are unique in a very strong sense (see (25.6), (28.14) and §12).

Suppose now that R is a QF ring. Then $Re_1, \ldots, Re_n$ must be $E(Re_1/Je_1), \ldots, E(Re_n/Je_n)$ in some order, so we see that

31.1. Proposition. *A module over a QF ring is injective if and only if it is projective.*   □

Injective Projective Modules

Our next objective is to determine just when an indecomposable projective Re_i is injective. To do so we shall employ

31.2. Lemma. *Let E be an injective left module over a ring R with $S = End(_RE)$, and let f be an idempotent in R such that $r_E(fR) = 0$. Then the natural homomorphism*

$$\theta : Hom_R(-, {}_RE_S) \to Hom_{fRf}(fR \otimes -, fE)$$

is an isomorphism. Thus,

$$fE \cong fR \otimes E$$

is fRf-injective and there is a natural isomorphism

$$S \cong End(_{fRf}fE).$$

Proof. First, for each $\gamma \in Hom_R(R/RfR, E)$, we have $fR\, Im\, \gamma = 0$; so, by hypothesis, $\gamma = 0$. Then, from

$$0 \to RfR \to R \to R/RfR \to 0$$

and the injectivity of E,

$$0 \to 0 \to Hom_R(R, E) \to Hom_R(RfR, E) \to 0,$$

is exact so that $E \cong Hom_R(R, E) \cong Hom_R(RfR, E)$. Next consider the epimorphism $Rf \otimes fR \to RfR$ given by multiplication, with kernel K,

$$0 \to K \to Rf \otimes_{fRf} fR \to RfR \to 0.$$

Then for every $\sum_i (a_i f \otimes f b_i) \in K$ and every $\gamma \in Hom_R(K, E)$, we have

$$fR\gamma\left(\sum_i (a_i f \otimes f b_i)\right) = \gamma\left(f \otimes fR \sum_i a_i f b_i\right) = 0.$$

So since $r_E(fR) = 0$, we have $Hom_R(K, E) = 0$ and

$$Hom_R(RfR, E) \cong Hom_R((Rf \otimes_{fRf} fR), E).$$

Thus,

$$E \cong Hom_R(R, E)$$
$$\cong Hom_R((Rf \otimes fR), E)$$
$$\cong Hom_{fRf}(fR, Hom_R(Rf, E)) \quad \text{(by (20.6))}$$
$$\cong Hom_{fRf}(fR, fE).$$

Then for every $_RM$, we have

$$Hom_R(M, E) \cong Hom_R(M, Hom_{fRf}(fR, fE))$$
$$\cong Hom_{fRf}(fR \otimes M), fE) \quad \text{(by (20.6))}$$
$$\cong Hom_{fRf}(fM, fE).$$

Now $Hom_R(\_, E)$ is exact and, for each module $_{fRf}N$, the isomorphism

$$N \cong fR \otimes Hom_{fRf}(fR, N)$$

is natural, so $Hom_{fRf}(\_, fE)$ is exact on $_{fRf}M$, and fE is fRf-injective. Finally, applying the natural isomorphism θ to $_RE$ we deduce that

$$S \cong End(_RE) \cong End(_{fRf}fE). \qquad \square$$

We note that as a consequence of the next theorem there is a one-to-one correspondence between the indecomposable injective projective left and right modules over any one-sided artinian ring.

31.3. Theorem. *Let R be a left or right artinian ring with $J = J(R)$, and let e be a primitive idempotent in R. Then Re is injective if and only if there is a primitive idempotent $f \in R$ such that*

$$Soc\ Re \cong Rf/Jf \quad and \quad Soc\ fR \cong eR/eJ.$$

Proof. $(\Rightarrow)$. Assume that Re is injective. Then there is a primitive idempotent $f \in R$ such that $Re = E(T)$ with $T \cong Rf/Jf$. We claim that

$$l_{fR}(Re) = r_{Re}(fR) = 0.$$

Suppose that $fr \neq 0$. Then $Rfr/Jfr \cong Rf/Jf \cong T$, so the Injective Test Lemma yields $frRe \neq 0$. On the other hand, $fT \neq 0$, where $T = Soc(Re)$, so $r_{Re}(fR) = 0$. Now since Re is injective and $r_{Re}(fR) = 0$, we have by (31.2) that fRe is fRf-injective and that

$$eRe \cong End(Re) \cong End(_{fRf}fRe).$$

We claim next that $fT \cong f(Soc\ Re)$ is simple. Indeed, let $0 \neq L \leq {}_{fRf}fRe$. Then $T \leq RL$, so $fT \leq fRL = L$. Then $fT = Soc({}_{fRf}fRe)$. But $fT \cong fRf/fJf$, so that

$$fRe = E(fRf/fJf)$$

is a finitely cogenerated injective cogenerator over fRf. Thus, from (30.2.1) we have

$$Soc(fRe_{eRe}) = Soc({}_{fRf}fRe)$$
$$\cong Hom_{fRf}(fRf, Soc\ fRe)$$
$$\cong Hom_{fRf}(fRf/fJf, fRe)$$

which is simple by (30.2.2). Now since $l_{fR}(Re) = 0$, no minimal right ideal in fR is annihilated by e; so from $(Soc\ fR)e \leq Soc(fRe_{eRe})$ we see that $Soc\ fR \cong eR/eJ$.

($\Leftarrow$). Conversely, suppose e and f are primitive idempotents with

$$T = Soc\ Re \cong Rf/Jf \quad \text{and} \quad S = Soc\ fR \cong eR/eJ.$$

Then, since $S \nsubseteq l_{fR}(Re)$ and $T \nsubseteq r_{Re}(fR)$,

$$l_{fR}(Re) = 0 = r_{Re}(fR).$$

Also one easily checks that fT and Se are the unique minimal submodules of ${}_{fRf}fRe$ and fRe_{eRe}, respectively, so

$$Soc({}_{fRf}fRe) = fTe = fSe = Soc(fRe_{eRe})$$

and both are simple. Thus, from

$$Hom_{fRf}(fRf/fJf, fRe) \cong Hom_{fRf}(fRf, fSe),$$

we see that the ${}_{fRf}fRe_{eRe}$-duals of simples are simple. Now, since R is left or right artinian, either ${}_{fRf}fR$ or Re_{eRe} has a composition series so we can apply (30.1) to the non-degenerate bilinear map

$$fR \times Re \to {}_{fRf}fRe_{eRe}$$

via multiplication in R to see that ${}_{fRf}fRe$ is injective (see also (16.13) and (18.3)) and that

$$Re \cong Hom_{fRf}(fR, fRe)$$

over R (as well as over eRe) via $\rho(re): fx \mapsto fxre$. But then using (20.6), we have natural isomorphisms of functors

$$Hom_R(\_, Re) \cong Hom_R(\_, Hom_{fRf}(fR, fRe))$$
$$\cong Hom_{fRf}((fR \otimes \_), fRe).$$

However, fRe is fRf-injective and fR is R-projective, so $Hom_R(\_, Re)$ is exact and Re is injective (as in Exercise (20.8.1)). $\square$

The condition of the following immediate corollary was Nakayama's defining condition for QF rings. The permutation σ therein is known as the *Nakayama permutation*.

31.4. Corollary. *A left artinian ring R with basic set of idempotents $e_1, \ldots, e_n$ and $J = J(R)$ is QF if and only if there is a permutation σ of $\{1, \ldots, n\}$ such that*

$$\operatorname{Soc} Re_i \cong Re_{\sigma(i)}/Je_{\sigma(i)} \quad \text{and} \quad \operatorname{Soc} e_{\sigma(i)}R \cong e_i R/e_i J$$

for $i = 1, \ldots, n$.

QF-3 Rings

A faithful left (or right) R-module U is said to be a *minimal faithful module* in case it is isomorphic to a direct summand of each faithful left (respectively, right) R-module. This rather unusual usage of the adjective "minimal" has become accepted in this particular context. For example, if e is a basic idempotent for a QF ring R, then $Re = E(Re/Je)$ is such a minimal faithful module; that is, it is faithful and appears as a direct summand of every faithful left R-module (Exercise (30.1)). Finite-dimensional algebras having such minimal faithful modules were first studied by R. M. Thrall. An excellent account of more general cases appears in [Tachikawa, 73] which includes the following fundamental characterization of minimal faithful modules by Colby and Rutter.

31.5. Lemma. *Let $_R U$ be left R-module. If $_R U$ is minimal faithful, then $_R U$ is both injective and projective, and there is a sum $e = e_1 + \cdots + e_k$ of orthogonal primitive idempotents in R with*

$$U \cong Re_1 \oplus \cdots \oplus Re_k = Re,$$

such that

$$Re_i \cong E(T_i), \qquad\qquad (i = 1, \ldots, k)$$

where $T_1, \ldots, T_k$ is an irredundant set of representatives of the minimal left ideals in R. Conversely, if $T_1, \ldots, T_k$ are pairwise non-isomorphic simple modules with

$$U = E(T_1 \oplus \cdots \oplus T_k) = E(T_1) \oplus \cdots \oplus E(T_k)$$

faithful and projective, then $_R U$ is a minimal faithful left R-module.

Proof. Suppose that $_R U$ is a minimal faithful left R-module. Then U must isomorphic to a direct summand of the regular module $_R R$, so there is an idempotent $e \in R$ such that $U \cong Re$. But U must also be isomorphic to a direct summand of the (faithful) minimal cogenerator $C_0 = \bigoplus_{T \in S_0} E(T)$ (see (18.16)). Since $U \cong Re$ is cyclic, the image of any embedding of U in C_0 must be contained in a finite direct sum of the $E(T)$ ($T \in S_0$). Thus, there is a finite irredundant set of simple left R-modules $(T_\varphi)_{\varphi \in F}$ such that U is isomorphic to a direct summand of $\oplus_F E(T_\varphi)$. But then (see (25.5)) among the T_φ there must exist $T_1, \ldots, T_k$ with

$$U \cong E(T_1) \oplus \cdots \oplus E(T_k) \cong Re,$$

and we can write e as a sum of primitive orthogonal idempotents $e = e_1 + \cdots + e_k$ with $Re_i \cong E(T_i)$ $(i = 1, \ldots, k)$ where, since every minimal left ideal must embed in a faithful module, $T_1, \ldots, T_k$ is an irredundant set of representatives of the minimal left ideals in R.

Conversely, if $T_1, \ldots, T_k$ are pairwise non-isomorphic simple modules such that $E(T_i)$ is projective for $i = 1, \ldots, k$, then, since faithful modules cogenerate all projective modules (see Exercise (17.6)), if $_R M$ is faithful, there exist monomorphisms (i.e., maps whose kernels *do not* contain T_i)

$$0 \longrightarrow E(T_i) \xrightarrow{\gamma_i} M \quad (i = 1, \ldots, k).$$

Since the T_i are pairwise non-isomorphic simple modules $\gamma_1(T_1), \ldots, \gamma_k(T_k)$ must be independent. But then (see (6.24)) so are their essential extensions $Im\ \gamma_1, \ldots, Im\ \gamma_k$. Hence, the injective module $E(T_1 \oplus \cdots \oplus T_k) \cong Im\ \gamma_1 + \cdots + Im\ \gamma_k$ is isomorphic to a direct summand of M. $\qquad\square$

A ring R is said to be a *left (right) QF-3 ring* in case it has a minimal faithful left (right) R-module. A ring is a *QF-3 ring* in case it is both left and right QF-3. Thus, every QF-ring is QF-3.

Next we give several characterizations of one-sided artinian QF-3 rings. In particular, we shall see that in the presence of either minimum condition "left QF-3" and "right QF-3" are equivalent. It is to be noted, however, that unlike the QF case, there do exist left artinian QF-3 rings that are not right artinian. (See Exercise (31.2).)

31.6. Theorem. *The following statements about a left artinian ring R are equivalent:*

(a) *R is left (right) QF-3;*

(b) *R has a faithful injective left (right) ideal;*

(c) *R has a faithful injective projective left (right) module;*

(d) *$E(_R R)$ $(E(R_R))$ is projective;*

Moreover, if R is QF-3, then the minimal faithful R-modules are of the form

$$Re = Re_1 \oplus \cdots \oplus Re_k \quad and \quad fR = f_1 R \oplus \cdots \oplus f_k R$$

with $e = e_1 + \cdots + e_k$ and $f = f_1 + \cdots + f_k$ sums of orthogonal primitive idempotents such that

$$Soc\ Re_i \cong Rf_i / Jf_i \quad and \quad Soc\ f_i R \cong e_i R / e_i J \quad (j = 1, \ldots, k)$$

and these simple modules are irredundant sets of representatives of the minimal left and right ideals.

Proof. (a) $\Rightarrow$ (b). These are consequences of Lemma 31.5.

(b) $\Rightarrow$ (c). Trivial.

(c) $\Rightarrow$ (d). The hypothesis (c) implies that R is cogenerated by an injective projective left (right) module Q (see (8.22)), i.e., there is a left (right)

R-monomorphism

$$0 \to R \to Q^A$$

for some set A. In the left-hand case, since $_R R$ is finitely cogenerated, we may take A to be finite. In the right-hand case, direct products of projective right R-modules are projective (28.9). Thus, in either case, the regular module R embeds in an injective projective module, so its injective envelope is projective.

(d) $\Rightarrow$ (a). If (d) holds, then $E(_R R)$ is a direct sum of injective envelopes of simple modules; so choosing one from each isomorphism class, we can find pairwise non-isomorphic simple modules $T_1, \ldots, T_k$ such that $E(T_1 \oplus \cdots \oplus T_k)$ is projective and has the same annihilator as $E(_R R)$—namely, zero. Thus, Lemma 31.5 applies. The proof of the right-hand version is entirely similar.

If R is left QF-3 with minimal faithful module $Re = Re_1 \oplus \cdots \oplus Re_k$ as in (31.5), then from a basic set of primitive idempotents for R, we can choose $f_1, \ldots, f_k$ with $Rf_i/Jf_i \cong Soc\ Re_i$ $(i = 1, \ldots, k)$. If $f = f_1 + \cdots + f_k$, then $fR = f_1 R \oplus \cdots \oplus f_k R$ is injective and $Soc\ f_i R \cong e_i R/e_i J$ $(i = 1, \ldots, k)$ by (31.3); and no two of these are isomorphic. Since, by (31.5), each minimal left ideal is isomorphic to one of the $Soc(Re_i) \cong Rf_i/Jf_i$, we see that fR is faithful because $r_R(fR)$ contains no minimal left ideals. Now by (31.5) again, fR is a minimal faithful right ideal. This proves the concluding statement and that the left-hand version of (a) implies the right-hand version. Similarly, we see that the right-hand version of (a) implies the left-hand one. ☐

QF-2 Rings

A left or right artinian ring is a *QF-2 ring* in case each of its indecomposable projective left and right modules has a simple socle. Of course, QF rings are QF-2 (31.4). Thrall proved that finite-dimensional QF-2 algebras are QF-3, and (31.3) allows us to extend this to the artinian case.

31.7. Theorem. *Every left or right artinian QF-2 ring is QF-3.*

Proof. Let $J = J(R)$, and, for each $_R M$, define $L(M) = l$ in case $J^l M = 0$ and $J^{l-1} M \neq 0$. Let T be a minimal left ideal of R. Then there exists a primitive left ideal Re with $L(Re)$ maximal among those satisfying $Soc\ Re \cong T$. If e' is any primitive idempotent in R, then we see that $(Soc\ Re) \cdot Je' = 0$ because otherwise, right multiplication by some element of Je' gives a monomorphism of Re into Je', contrary to the maximality of $L(Re)$. Thus,

$$Soc\ Re \subseteq l_R(J) = Soc(R_R).$$

Now let f be a primitive idempotent in R such that $Soc\ Re \cong Rf/Jf$. Then

$$0 \neq f(Soc\ Re) \subseteq f(Soc(R_R))e = (Soc\ fR)e,$$

so that, since it is simple, $Soc\ fR \cong eR/eJ$. Thus, Re is injective by (31.3); and (31.6) applies. ☐

From (31.7), (31.5), and (31.4) one can easily obtain

31.8. Corollary. *The following statements about a left or right artinian ring R are equivalent*:
  (a) *R is QF*;
  (b) *R is QF-2 and $Soc(_R R) = Soc(R_R)$*;
  (c) *R is QF-2 and every simple left R-module embeds in R.*

Proof. Exercise (3.14).   □

The Faith-Walker Characterization of QF Rings

We conclude this section with a theorem that characterizes QF rings strictly in terms of projective and injective modules.

31.9. Theorem. *The following statements about a ring R are equivalent:*
  (a) *R is QF*;
  (b) *Every projective left R-module is injective*;
  (c) *Every injective left R-module is projective.*

Proof. (a) $\Rightarrow$ (b) and (a) $\Rightarrow$ (c) by (31.1).

(b) $\Rightarrow$ (a). Suppose that $_R R^{(A)}$ is injective for some infinite set A. Then by (25.1) and the discussion preceding it, R has the ascending (descending) chain condition on annihilator left (right) ideals. By (30.9), every principal right ideal is an annihilator right ideal, so R is left perfect (28.4(e)). Thus, as in the proof of (30.10), R is semiprimary. Now if $e_1, \ldots, e_n$ is a basic set of idempotents for R, then $Re_1, \ldots, Re_n$ must be the injective envelopes of the n distinct simple left R-modules, and we see that $_R R$ is a cogenerator (18.15). But then by (25.2) every left ideal is an annihilator left ideal, so the left self-injective ring R is left noetherian and hence QF by (30.10).

(c) $\Rightarrow$ (a). Suppose all injective left R-modules are projective. Then they all must be isomorphic to direct summands of free modules, and it follows that every left R-module embeds in a direct sum of copies of R. Thus, by (26.3), R is left noetherian, and moreover, by considering projections on the terms in $R^{(A)}$, we see that each simple left R-module is isomorphic to a minimal left ideal. Since any collection of pairwise non-isomorphic simple submodules of a module is independent, and since R is left noetherian, it follows that R has only finitely many isomorphism classes of simple modules. Let $T_1, \ldots, T_n$ denote one representative from each of these classes. Then, by hypothesis, the pairwise non-isomorphic indecomposable injective modules $E(T_1), \ldots, E(T_n)$ are projective. Their endomorphism rings are local (25.4), so they are projective covers of (pairwise non-isomorphic, by (17.18)) simple modules by (17.20). Thus, letting $E = E(T_1) \oplus \cdots \oplus E(T_n)$, we see that the projective module E maps onto each of $T_1, \ldots, T_n$ and, hence, must be a generator (17.10). Therefore, we have $R \oplus R' = E^{(m)}$ (17.6) and R is QF.   □

31. Exercises

1. Prove that if R is a left artinian QF-3 ring, then $Soc(R_R)$ is a finite direct sum of simples.

2. Prove that if C is a division subring of a division ring D such that $_CD$ has finite dimension, but D_C does not, then the ring of matrices

$$\begin{bmatrix} D & 0 & 0 \\ D & C & 0 \\ D & D & D \end{bmatrix}$$

is a left artinian QF-3 ring that is not right artinian.

3. Show that the ring of matrices, over any division ring, of the form

$$\begin{bmatrix} a & 0 & 0 & 0 \\ u & b & 0 & 0 \\ v & 0 & b & 0 \\ w & x & y & c \end{bmatrix}$$

is a QF-3 ring that is not QF-2.

4. (1) Prove that if R is a left or right artinian QF-3 ring with e a primitive idempotent in R, then Re is injective iff $Soc\ Re \subseteq Soc(R_R)$.

   (2) Prove Corollary 31.8.

5. A QF ring R is called a *Frobenius ring* in case $Soc(_RR) \cong_R R/J$. Prove:

   (1) An artinian ring R is Frobenius iff $Soc(_RR) \cong_R R/J$ and $Soc(R_R) \cong R/J_R$.

   (2) The basic ring of every QF ring is Frobenius.

   (3) If R is a finite-dimensional algebra over a field K, the following are equivalent:

   (a) R is Frobenius;

   (b) There is an isomorphism $\varphi :_R R \to Hom_K(R_R, K)$;

   (c) There is a non-degenerate R-balanced K-bilinear mapping $\theta : R \times R \to K$. [Hint: Try $\theta(r, s) = (\varphi(s))(r).$];

   (d) $R_R \cong Hom_K(_RR, K)$.

6. A QF ring R is called *weakly symmetric* in the case $Soc\ Re \cong Re/Je$ for all (primitive) idempotents e in R. Prove that:

   (1) If R is artinian, then the following are equivalent:

   (a) R is a weakly symmetric ring;

   (b) $Soc\ Re \cong Re/Je$ and $Soc\ eR \cong eR/eJ$ for all primitive idempotents e in R;

   (c) $Hom_R(Re/Je, R) \cong eR/eJ$ and $Hom_R(eR/eJ, R) \cong Re/Je$ for all primitive idempotents e in R.

   (2) The algebra of matricies, over a field K, of the form

$$\begin{bmatrix} a & y & 0 & 0 \\ 0 & b & 0 & 0 \\ 0 & 0 & b & x \\ 0 & 0 & 0 & a \end{bmatrix}$$

is Frobenius but not weakly symmetric.

7. A finite-dimensional algebra R over a field K is a *symmetric algebra* in case there is a bimodule isomorphism $\varphi : {}_R R_R \to Hom_K(R, K)$. Prove that:
   (1) The following are equivalent:
      (a) R is a symmetric algebra;
      (b) The functors $Hom_R(-, R)$ and $Hom_K(-, K): {}_R FM \to FM_R$ are isomorphic;
      (c) There is a non-degenerate R-balanced symmetric (i.e., $\theta(r, s) = \theta(s, r)(r, s \in R)$) K-bilinear mapping $\theta : R \times R \to K$.
   (2) If G is a finite group, then the group algebra KG is symmetric via $\theta : (\sum a_g g, \sum b_g g) \to \sum a_g b_{g^{-1}}$.
8. Let R be a two-sided QF-3 ring (no chain conditions) with minimal faithful modules $Re = Re_1 \oplus \cdots \oplus Re_m$ and $fR = f_1 R \oplus \cdots \oplus f_k R$ as in (31.5). Prove that:
   (1) $k = m$ and $f_1, \ldots, f_m$ can be renumbered so that $Soc\ Re_i \cong Rf_i/Jf_i$ and $Soc\ f_i R \cong e_i R/e_i J$. [Hint: Je_i is maximal by (25.4) and (17.20).]
   (2) ${}_{fRf} fRe_{eRe}$ defines a Morita duality such that Re and fR are reflexive.
9. Suppose that R is a left artinian ring and $0 \neq f = f^2$ in R. Let $E = E(Rf/J(R))$. Prove:
   (1) $l_{fR}(E) = 0$ and $r_E(fR) = 0$.
   (2) ${}_{fRf} fE$ is a cogenerator.
   (3) Let D(E) denote the full subcategory of ${}_R M$ whose objects are modules M such that there are sets X and Y and an exact sequence

$$0 \to M \to E^{(X)} \to E^{(Y)}.$$

Prove that if $H = Hom_{fRf}(fR, -)$ and $T = (fR \otimes {}_R -)$, then

$$T : \text{D}(E) \to {}_{fRf} M \quad \text{and} \quad H : {}_{fRf} M \to \text{D}(E)$$

and these functors define an equivalence of categories, i.e., $T \circ H \cong 1_{{}_{fRf} M}$ and $H \circ T \cong 1_{\text{D}(E)}$ (cf. Exercise 20.18). [Hint: The first isomorphism of functors follows from (20.11); the second uses (31.2) and the Five Lemma, among other things.]

§32. Serial Rings

With the exception of semisimple rings, serial rings provide the best illustration of the relationship between the structure of a ring and its categories of modules. They were one of the earliest examples of rings of finite module (or representation) type; their introduction by Nakayama some 50 years ago was fundamental to what has come to be known as the representation theory of artinian rings and finite-dimensional algebras.

Loewy Series and Uniserial Modules

For each module ${}_R M \neq 0$ over a semiprimary ring R, there is a smallest positive integer ℓ such that $J^\ell M = 0$. This number ℓ is called the *Loewy length* of M and we write $L(M) = \ell$. (If $M = 0$, then, of course, $L(M) = 0$.) The *upper Loewy series*, or *radical series*, for M is

$$M > JM > \cdots > J^\ell M = 0.$$

The *lower Loewy series* for M is

$$0 < r_M(J) < \cdots < r_M(J^\ell) = M.$$

Letting $J^0 = R$, we say that for each $k = 1, \ldots, \ell$

$$J^{k-1}M/J^k M$$

is the k-th *upper Loewy factor of M* and

$$r_M(J^k)/r_M(J^{k-1})$$

is the k-th *lower Loewy factor of M*. Each of these factors is semisimple and none are zero (unless M is). All of these concepts have obvious analogues for right modules. Moreover, we note that

$$r_M(J^k)/r_M(J^{k-1}) = Soc(M/r_M(J^{k-1})) \quad (k = 1, \ldots, \ell);$$

so we often write

$$Soc^k M = r_M(J^k)$$

and call the lower Loewy series the *socle series*.

A module is called *uniserial* in case its lattice of submodules is a finite chain, i.e., any two submodules are comparable. Thus, simple modules and $\mathbb{Z}_{p^n}$ (p a prime) are uniserial modules.

32.1. Lemma. *The following statement about a module $M \neq 0$ over a semi-primary ring R are equivalent:*

(a) *M is uniserial;*

(b) *M has a unique composition series;*

(c) *The upper Loewy series*

$$M > JM > \cdots > J^\ell M = 0$$

is a composition series for M;

(d) *The lower Loewy series*

$$0 < Soc\, M < Soc^2 M < \cdots < Soc^\ell M = M$$

is a composition series for M.

Proof. (a) $\Leftrightarrow$ (b). This is obvious.

(a) $\Rightarrow$ (c) and (d). Any non-simple Loewy factor would yield incomparable submodules.

(d) $\Rightarrow$ (a). Let $L \leq M$ and choose k maximal with respect to $Soc^k M \leq L$. Then $L \cap Soc^{k+1} M < Soc^{k+1} M$, so, by hypothesis, $L \cap Soc^{k+1} M = Soc^k M$, i.e.,

$$L/Soc^k M \cap Soc(M/Soc^k M) = 0.$$

Thus, since socles of R-modules are essential, $L = Soc^k M$ and (a) follows.

(c) $\Rightarrow$ (a). This is dual to (d) $\Rightarrow$ (a). $\square$

Serial Rings Characterized

An artinian ring is a *left* (*right*) *serial ring* in case each of its left (right) indecomposable projective modules is uniserial. Thus, R is (two-sided) *serial* iff R is semiprimary and $_RR$ and R_R are direct sums of uniserial modules.

32.2. Theorem. *The following statements about a left artinian ring R with $J = J(R)$ are equivalent:*

(a) *R is a serial ring;*
(b) *Every factor ring of R is QF-2;*
(c) *Every factor ring of R is QF-3;*
(d) *R/J^2 is serial.*

Proof. (a) $\Rightarrow$ (b). If $I \leq_R R_R$ and $e_1, \ldots, e_n$ is a complete orthogonal set of idempotents in R, then

$$R/I \cong Re_1/Ie_1 \oplus \cdots \oplus Re_n/Ie_n.$$

Thus, factor rings of serial rings are serial, and hence QF-2.

(b) $\Rightarrow$ (c). This is by (31.7).

(c) $\Rightarrow$ (d). Suppose $J^2 = 0$ and R is QF-3. If f is a primitive idempotent in R such that $fJ \neq 0$, then Rf/Jf is isomorphic to a minimal left ideal since $J \leq \mathrm{Soc}(_RR)$. But then fR is injective (31.6), so $fJ = \mathrm{Soc}\, fR$ is simple. Thus, R is right, and similarly left, serial.

(d) $\Rightarrow$ (a). Assume inductively, for $k \geq 1$, that $J^k e_i/J^{k+1} e_i$ is simple, so there is a projective cover

$$Re_j \to J^k e_i \to 0.$$

Then $Je_j/J^2 e_j \cong J^{k+1} e_i/J^{k+2} e_i$ unless the latter is 0. $\qquad \square$

The next result characterizes serial rings in terms of left modules and also serves to describe their left and their right modules.

32.3. Theorem. *If R is a left artinian ring, then the following are equivalent:*

(a) *R is a serial ring;*
(b) *Every left R-module is a direct sum of uniserial modules;*
(c) *Every finitely generated indecomposable left R-module is uniserial;*
(d) *The projective cover and the injective envelope of every simple left R-module are uniserial.*

Proof. (a) $\Rightarrow$ (b). Let $J = J(R)$ and $\ell = L(R)$, so $J^\ell = 0$ and $J^{\ell-1} \neq 0$. If e is a primitive idempotent in R such that $J^{\ell-1}e \neq 0$, then since R is QF-3, Re is injective by (31.6). Suppose now that M is a left R-module, and let $\mathscr{E}$ denote the set of Rex such that e is a primitive idempotent in R, $J^{\ell-1}ex \neq 0$ and $x \in M$; and let E be the (necessarily direct) sum of a maximal independent subset of $\mathscr{E}$. Then since each $Rex \cong Re$, via multiplication by x, E is injective and we have $M = E \oplus M'$, where E is a direct sum of serial modules. But $J^{\ell-1}M' = 0$ since, otherwise, some $Rex \in \mathscr{E}$ would be contained in M' contrary to maximality. Thus, M' is an $R/J^{\ell-1}$-module and this implication follows by induction on $L(R)$.

(b) $\Rightarrow$ (c). Obviously.

(c) $\Rightarrow$ (d). If $Soc^k E/Soc^{k-1} E$ is not simple, then E contains a finitely generated submodule that is not uniserial; and every submodule of the injective envelope of a simple module is indecomposable.

(d) $\Rightarrow$ (a). It follows from Exercise (30.3) that if S is a simple left R-module with $E = E(S)$, then since $JS = 0$, the injective envelope of S over R/J^2 is $r_E(J^2) = Soc^2 E$; and, of course, the projectives over R/J^2 are factors of those over R. Thus, by (32.2(d)), we may assume that $J^2 = 0$. Let f be a primitive idempotent in R. If $fJ = 0$, then fR is certainly uniserial. If $fJ \neq 0$, then J contains a copy of Rf/Jf, so there is a non-zero map from J to $E = E(Rf/Jf)$ which must be multiplication by an element of E (18.3). Thus, in this case $JE \neq 0$, so $c(E) = 2$, and if $E/JE \cong Re/Je$, then $E \cong Re$. But then, by (31.3), fR is injective with $Soc \, fR = fJ \cong eR/eJ$ and fR is uniserial. $\square$

The Kupisch Series

Let R be a serial ring with basic set of primitive idempotents $e_1, \ldots, e_n$ and $J = J(R)$. For each $i = 1, \ldots, n$, set

$$S_i = Re_i/Je_i \quad \text{and} \quad T_i = e_iR/e_iJ.$$

Thus, $S_1, \ldots, S_n$ and $T_1, \ldots, T_n$ are complete irredundant sets of simple left and right R-modules, respectively. The (right) *quiver* of R is the directed graph $\mathcal{Q}(R)$ with vertex set $\{e_1, e_2, \ldots, e_n\}$ and with an arrow $e_i \to e_j$ if and only if $e_iJe_j \nsubseteq J^2$. (See Exercise (32.14) for the quivers of an artinian ring.) Equivalently,

$$e_i \to e_j \text{ in } \mathcal{Q}(R) \quad \text{iff } Je_j/J^2e_j \cong S_i$$
$$\text{iff } e_iJ/e_iJ^2 \cong T_j$$
$$\text{iff } Re_i \to Je_j \to 0 \text{ is a projective cover}$$
$$\text{iff } e_jR \to e_iJ \to 0 \text{ is a projective cover.}$$

It follows from (27.18) that R is indecomposable iff $\mathcal{Q}(R)$ is connected as an undirected graph (i.e., ignoring directions $\mathcal{Q}(R)$ is topologically connected). Now since R is serial, for all $i = 1, \ldots, n$, the modules Je_i and e_iJ have unique projective covers of the form Re_j and e_jR, so in $\mathcal{Q}(R)$ the in and out valence of any vertex is at most 1; i.e., neither configuration

$$e_i \to e_j \leftarrow e_k \quad \text{nor} \quad e_i \leftarrow e_j \to e_k$$

can occur in $\mathcal{Q}(R)$. Thus, if R is also indecomposable, so that $\mathcal{Q}(R)$ is connected, then $\mathcal{Q}(R)$ is either a single directed path of length n or a single cycle of length n. That is,

32.4. Theorem. *If R is an indecomposable serial ring with $J = J(R)$, then a basic set of idempotents can be numbered so that the quiver $\mathcal{Q}(R)$ of R is either*

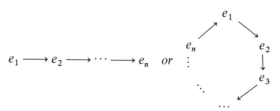

so that there are projective covers

$$Re_{i-1} \to Je_i \to 0 \quad (i = 2, \ldots, n)$$

and

$$Re_n \to Je_1 \to 0$$

if $Je_1 \neq 0$.   □

The list of indecomposable projective modules $Re_1, \ldots, Re_n$ of Theorem 32.4 is called the *Kupisch series* for R. We note that it is unique, except for cyclic permutation of $\{1, \ldots, n\}$ when $Je_1 \neq 0$. Also, if $c_i = c(Re_i)$, then

$$Je_i \cong Re_{i-1}/J^{c_i-1}e_{i-1} \quad (i = 2, \ldots, n)$$

and

$$Je_1 \cong Re_n/J^{c_1-1}e_n,$$

if $Je_1 \neq 0$. Thus, we must have

$$2 \le c_i \le c_{i-1} + 1 \text{ for } i = 2, \ldots, n \quad \text{and} \quad c_1 \le c_n + 1.$$

The numbers $c_1, \ldots, c_n$ form what is called the *admissible sequence for R*; and any sequence $c_1, \ldots, c_n$ satisfying these inequalities is simply called *an admissible sequence*. We shall presently see that each of these is the admissible sequence for some serial ring.

In what follows, if $k \in \mathbb{Z}$, we let $[k]$ denote the least positive residue of k modulo n. The next lemma, a consequence of the Kupisch series, is the key to many of the remaining results in this section.

32.5. Lemma. *Let R be an indecomposable serial ring with Kupisch series* $Re_1, \ldots, Re_n$ *and* $J = J(R)$. *Then there exist*

$$a_{i-1} \in e_{i-1}Je_i \quad \text{for } i = 2, \ldots, n \quad \text{and} \quad a_n \in e_nJe_1$$

such that, for any $k > 0$ *and any* $i \in \{1, \ldots, n\}$,

$$J^ke_i = Ra_{[i-k]} \ldots a_{[i-2]}a_{[i-1]} \quad \text{and} \quad e_iJ^k = a_ia_{[i+1]} \ldots a_{[i+k-1]}R.$$

Proof. By (32.4), there are projective covers $Re_{i-1} \to Je_i \to 0$ for $i = 2, \ldots, n$ and $Re_n \to Je_1 \to 0$ if $Je_1 \neq 0$. The images of $e_1, \ldots, e_n$ under these epimorphisms are elements $a_{i-1} \in e_{i-1}Je_i \setminus J^2$, $i = 2, \ldots, n$, and $a_n \in e_nJe_1 \setminus J^2$ if $Je_1 \neq 0$, and $a_n = 0$ if $Je_1 = 0$. Then the conclusion surely holds for $k = 1$. If $k > 1$, then, assuming the condition for $k - 1$, we have

$$J^ke_i = J^{k-1}Ra_{[i-1]}$$

$$= \left(\sum_{j=1}^{n} J^{k-1}e_j \right) a_{[i-1]}$$

$$= J^{k-1}e_{[i-1]}a_{[i-1]}$$

$$= Ra_{[i-k]} \ldots a_{[i-2]}a_{[i-1]}.$$

The proof for e_iJ^k is entirely similar.   □

From Lemma 32.5 we see at once that if R is serial with Kupisch series $Re_1, \ldots, Re_n$ and $c_i = c(Re_i)$ $(i = 1, \ldots, n)$, then

$$J^k e_i \cong Re_{[i-k]}/J^{c_i - k} e_{[i-k]}$$

if $J^k e_i \neq 0$, and the composition factors of Re_i are, starting from the top,

$$S_i, S_{[i-1]}, \ldots, S_1, S_n, S_{n-1}, \ldots, S_1, \ldots, S_{[i-c_i+1]}.$$

Similarly, if $d_i = c(e_i R)$ $(i = 1, \ldots, n)$, then

$$e_i J^k = e_{[i+k]} R/e_{[i+k]} J^{d_i - k}$$

if $e_i J^k \neq 0$ and the composition factors of $e_i R$ are, from the top,

$$T_i, T_{[i+1]}, \ldots, T_n, T_1, \ldots, T_n, \ldots, T_{[i+d_i-1]}.$$

In specific cases, this information can be nicely represented by diagram. For example, if the admissible sequence for R is $c_1 = 4$, $c_2 = 5$, $c_3 = 5$ structures of Re_1, Re_2, Re_3 and $e_1 R, e_2 R, e_3 R$ are indicated by

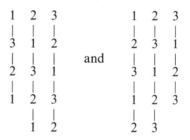

respectively.

Injective Modules

It is now relatively easy to identify the indecomposable injective modules, and hence all injective modules, over a serial ring R. Every serial ring is QF-3; so in terms of this characterization of injective modules, we can also characterize their minimal faithful modules.

Let R be a serial ring with Kupisch series $Re_1, \ldots, Re_n$ and admissible sequence $c_1, \ldots, c_n$ and again for each i let

$$S_i = Re_i/Je_i.$$

Let $_R M$ be indecomposable. By (32.3), M is uniserial and so has a projective cover $Re_i \to M \to 0$ for some unique i. Thus, if $c(M) = m$, then $m \leq c_i$ and

$$M \cong Re_i/J^m e_i.$$

It follows that the composition factors for M are, from the top,

$$M/JM \cong S_i, S_{[i-1]}, \ldots, S_{[i-m+1]} \cong Soc\, M.$$

Reading this in the other direction, if $Soc\, M \cong S_k$, then the composition factors for M are, from the bottom,

$$Soc\, M \cong S_k, S_{[k+1]}, \ldots, S_{[k+m-1]} \cong M/JM,$$

so that

$$M \cong Re_{[k+m-1]}/J^m e_{[k+m-1]}.$$

In other words, every indecomposable module is characterized to within isomorphism by its length and its socle. Now since $Re_i \to Je_{[i+1]} \to 0$ is a projective cover, $c_{[i+1]} > m$ if and only if

$$M \cong Re_i/J^m e_i \cong Je_{[i+1]}/J^{m+1} e_{[i+1]}$$

and there is a proper embedding of M into $Re_{[i+1]}/J^{m+1} e_{[i+1]}$.

Now let $_R E$ be an indecomposable injective with $Soc\, E \cong S_j$, so

$$E = E(Re_j/Je_j).$$

Then every left R-module M with $Soc\, M = S_j$ can be embedded in E. Therefore, E is the unique uniserial module of maximal length with socle S_j. With these last two observations we have established

32.6. Theorem. *Let R be an indecomposable serial ring with Kupisch series $Re_1, \ldots, Re_n$ and admissible sequence $c_1, \ldots, c_n$. For $1 \leq i \leq n$ and $m \leq c_i$, the indecomposable module $Re_i/J^m e_i$ is injective iff $c_{[i+1]} \leq m \leq c_i$. In particular, Re_i is injective iff $c_{[i+1]} \leq c_i$.* $\square$

This last statement means that the modules Re_i with $c_{[i+1]} \leq c_i$ are the indecomposable projective injective modules. As we have seen, serial rings are QF-3 rings. Thus, their minimal faithful modules are identified in

32.7. Corollary. *Let R be an indecomposable serial ring with Kupisch series $Re_1, \ldots, Re_n$ and admissible sequence $c_1, \ldots, c_n$. Then*

$$\bigoplus_{c_{[i+1]} \leq c_i} Re_i$$

is the minimal faithful left R-module. $\square$

The case $c_1 = 1$

The particular case of an indecomposable serial ring with $c_1 = 1$ is quite special. This is the case in which the quiver $Q(R)$ is a directed path

$$e_1 \to e_2 \to \cdots \to e_n.$$

As we shall see, these turn out to be factor rings of rings of upper triangular matrices. Indeed, let D be a division ring and let $R = \mathbb{UTM}_n(D)$ be the upper triangular matrix ring over D. This is an indecomposable serial ring with Kupisch series $Re_1, \ldots, Re_n$ where e_i is the matrix whose only non-zero entry is 1 in the ii^{th} position. The admissible sequence is $c_1 = 1, c_2 = 2, \ldots, c_n = n$. Note also, in this case, Re_1 is simple and $Soc\, Re_2 \cong \cdots \cong Soc\, Re_n \cong Re_1$, so

$$Soc(_R R) \cong \bigoplus_{i=1}^{n} Soc(Re_i) \cong (Re_1)^n$$

is projective. As a converse we have

32.8. Theorem. *Let R be a basic indecomposable serial ring with Kupisch series $Re_1, \ldots, Re_n$. If Re_1 is simple, then R is isomorphic to a factor ring of the $n \times n$ upper triangular matrix ring $\mathsf{UTM}_n(D)$ over a division ring D. Moreover, if $\mathrm{Soc}(_R R)$ is projective, then $R \cong \mathsf{UTM}_n(D)$.*

Proof. Suppose $c_1 = 1$. Then $c_k \leq k$ for $k = 1, 2, \ldots, n$, and, by Lemma 32.5, we see that the only composition factors of Re_j are the first c_j of S_j, $S_{j-1}, \ldots, S_1$ ($S_i = Re_i/Je_i$). It follows that for all i and j either $j - c_j < i \leq j$ and

$$e_i Re_j \cong e_i S_i$$

is simple over $e_i Re_i$ or $e_i Re_j = 0$. Similarly, $e_i Re_j$ is simple or zero over $e_j Re_j$. Thus, with $a_1, \ldots, a_n$ as promised in (32.5), if we set

$$e_{ij} = \begin{cases} e_i & \text{if } i = j \\ a_i \ldots a_{j-1} = e_i a_i \ldots a_{j-1} e_j & \text{if } 1 \leq i < j \leq n, \end{cases}$$

then we see that the $e_i Re_i$ are division rings and

$$e_i Re_j = e_i Re_i e_{ij} = e_{ij} e_j Re_j$$

whenever $1 \leq i \leq j \leq n$. It follows that there are division ring isomorphisms

$$\sigma_i : e_{i-1} Re_{i-1} \to e_i Re_i \quad (i = 2, \ldots, n)$$

such that

$$d_{i-1} a_{i-1} = a_{i-1} \sigma_i(d_{i-1}) \quad (d_{i-1} \in e_{i-1} Re_{i-1}),$$

and that

$$Re_j = \sum_{i=1}^{j} e_i Re_j = \sum_{i=1}^{j} e_i Re_i e_{ij}.$$

Now let $\sigma_1 = 1_{e_1 Re_1}$ and

$$\delta_i = \sigma_i \circ \cdots \circ \sigma_1 : e_1 Re_1 \to e_i Re_i \quad (i = 1, \ldots, n),$$

and let

$$D = \{\delta_1(x) + \delta_2(x) + \cdots + \delta_n(x) \mid x \in e_1 Re_1\}.$$

Then D is a division ring,

$$R = \sum_{i \leq j} De_{ij},$$

and if $d = \sum_{i=1}^{n} \delta_i(x)$, then, for $i = 2, \ldots, n$,

$$da_{i-1} = \delta_{i-1}(x) a_{i-1} = a_{i-1} \delta_i(x) = a_{i-1} d.$$

Thus, we see that whenever $1 \leq i \leq j \leq n$,

$$de_{ij} = e_{ij} d$$

for all $d \in D$, and it follows that the surjective mapping

$$[d_{ij}] \mapsto \sum_{i \leq j} d_{ij} e_{ij} \quad ([d_{ij}] \in \mathsf{UTM}_n(D))$$

is a ring homomorphism. For the last statement, we note that this is a D-vector space map and observe that if $Soc(_R R)$ is projective, then $e_{ij} \neq 0$ for all $1 \le i \le j \le n$. ☐

If R is an indecomposable serial ring with Kupisch series $Re_1, \ldots, Re_n$, then an argument similar to one in the proof of (32.8) shows that the division rings $e_i Re_i / e_i Je_i$ are all isomorphic, i.e., if R is basic, then R/J is a direct sum of n copies of the same division ring. One might hope that this division ring and the admissible sequence would determine the ring R, but this is dashed by the rings $\mathbb{Z}_4$ and $\mathbb{Z}_2[x]/(x^2)$. We shall, however, show that the hoped for result does hold for certain finite-dimensional algebras.

Split Serial Algebras

A finite-dimensional algebra R over a field $K \subseteq Cen\ R$ is a *split algebra* in case the endomorphism ring of every simple R-module consists entirely of scalar multiplications by element of K. Equivalently, in case $R/J(R)$ is isomorphic as a K-algebra to a direct sum of matrix rings over K; and if R is basic, then R is split if and only if $_K R = Ke_1 \oplus \cdots \oplus Ke_n \oplus J(R)$ when $e_1, \ldots, e_n$ is a basic set of idempotents. If K is algebraically closed, then R is automatically split (Exercise (32.7)).

If $\mathscr{R}$ is a finite semigroup with 0 (i.e., $0x = 0 = x0$ for all $x \in \mathscr{R}$), the *semigroup algebra* $\mathscr{K}\mathscr{R}$ is an algebra (maybe without identity) with K-basis $\mathscr{R}\setminus\{0\}$, and (with sums taken over $\mathscr{R}\setminus\{0\}$) multiplication satisfying

$$\left(\sum_x h_x x\right)\left(\sum_y k_y y\right) = \sum_z \left(\sum_{xy=z} h_x k_y\right) z$$

and $0x = 0 \in \mathscr{R}$. Similarly to Exercise (1.15), one can define this object as $K\mathscr{R} = \{f : \mathscr{R} \to K \mid f(0) = 0\}$.

We employ this notion to prove

32.9. Theorem. *If K is a field, then any two split basic indecomposable serial K-algebras with identical admissible sequences are isomorphic.*

Proof. Let R be such a K-algebra with Kupisch series $Re_1, \ldots, Re_n$. Then since R is split and basic, $c(_R R) = \dim(_K R) = \sum_{i=1}^n c_i$. Thus, it easily follows from (32.5) that if $a_{i-1} \in e_{i-1} Je_i \setminus J^2$ $(i = 2, \ldots, n)$ and $a_n \in e_n Je_1 \setminus J$ (if $Je_1 \neq 0$), then $\{e_1, \ldots, e_n\} \cup \{a_1, \ldots, a_{n-1}, a_n\}$ generates a subsemigroup $(\mathscr{R}, \cdot)$ of $(R, \cdot)$ whose non-zero elements form a K-basis for R. Therefore, $R \cong K\mathscr{R}$. So since the multiplication table of $\mathscr{R}$ is determined, via (32.5), by the admissible sequence $c_1, \ldots, c_n$, the theorem follows. ☐

This last theorem suggests a method of constructing a serial ring with a given admissible sequence $c_1, \ldots, c_n$. Let $e_1, \ldots, e_n$, $a_1, \ldots, a_n$, and 0 be distinct symbols (except possibly $a_n = 0$ if $c_1 = 1$). Then denote a k-tuple $(a_{[i-k]}, \ldots, a_{[i-2]}, a_{[i-1]})$ by $a_{[i-k]} \ldots a_{[i-1]}$, let

$$\mathscr{I} = \{a_{[i-k]} \ldots a_{[i-1]} \mid 1 \le k < c_i, i = 1, \ldots, n\} \cup \{0\},$$

and let

$$\mathcal{R} = \{e_1, \ldots, e_n\} \cup \mathcal{J}.$$

Define multiplication on R by

$$e_i e_j = e_i \quad \text{if } i = j,$$

$$e_j a_{[i-k]} \cdots a_{[i-1]} e_l = a_{[i-k]} \cdots a_{[i-1]} \quad \text{if } j = [i-k] \text{ and } l = i,$$

$$(a_{[j-t]} \cdots a_{[j-1]})(a_{[i-s]} \cdots a_{[i-1]}) = a_{[i-s-t]} \cdots a_{[i-1]} \quad \text{if } j = [i-s] \text{ and } s + t < c_i,$$

and define all other products to be 0. If any of x, y, or z is 0 or e_i, then, clearly, $(xy)z = x(yz)$. In fact, it is nearly obvious that associativity holds in every case except possibly when

$$x = a_{[k-u]} \cdots a_{[k-1]}, \qquad y = a_{[j-t]} \cdots a_{[j-1]}, \qquad z = a_{[i-s]} \cdots a_{[i-1]},$$

where $k = [j-t]$, $j = [i-s]$, and $t + u \geq c_j$. But even here, we have

$$(xy)z = 0z = 0$$

and, from $c_j = c_{[i-s]} \geq c_i - s$, we obtain

$$s + t + u \geq s + c_j \geq c_i$$

so that

$$x(yz) = a_{[i-u-t-s]} \cdots a_{[i-1]} = 0$$

also. Thus, $\mathcal{R}$ is a semigroup.

32.10. Theorem. *If $c_1, \ldots, c_n$ is an admissible sequence and K is a field, then there is a split serial K-algebra R with Kupisch series $Re_1, \ldots, Re_n$ and $c(Re_i) = c_i$, $i = 1, \ldots, n$.*

Proof. Of course, we set $R = K\mathcal{R}$; and we leave the rest of the proof as Exercise (32.8). ☐

The Transpose and Nakayama's Characterization

One of the most effective contemporary tools in the representation (or module) theory of artinian rings and artin algebras is the transpose of Auslander and Bridger [69], which we present here to obtain Nakayama's characterization of serial rings. But first we require some more information on projective modules.

An exact sequence of R-modules

$$P_1 \xrightarrow{f} P_0 \longrightarrow M \longrightarrow 0$$

is called a *minimal projective presentation* of M in case P_1 and P_0 are finitely generated projective and $\operatorname{Ker} f \ll P_1$ and $\operatorname{Im} f \ll P_0$. When R is semiperfect, every finitely presented R-module has a minimal presentation (see Exercise (20.17)), and then letting $J = J(R)$, minimality just means $\operatorname{Ker} f \leq JP_1$ and $\operatorname{Im} f \leq JP_0$. We begin by showing that these presentations are essentially unique.

32.11. Lemma. *If M and N have minimal projective presentations*

$$P_1 \xrightarrow{f} P_0 \longrightarrow M \longrightarrow 0 \quad and \quad Q_1 \xrightarrow{g} Q_0 \longrightarrow N \longrightarrow 0,$$

then $M \cong N$ if and only if there are isomorphisms φ_1 and φ_0 making the diagram

$$
\begin{array}{ccc}
P_1 & \xrightarrow{f} & P_0 \\
\varphi_1 \downarrow & & \downarrow \varphi_0 \\
Q_1 & \xrightarrow{g} & Q_0
\end{array}
$$

commute.

Proof. The condition is clearly sufficient. Conversely, given an isomorphism $\varphi : M \to N$, we use projectively of P_0 to get φ_0 making the right-hand square commute in the diagram

$$
\begin{array}{ccccccc}
P_1 & \xrightarrow{f} & P_0 & \xrightarrow{f_0} & M & \longrightarrow & 0 \\
\varphi_1 \downarrow & & \varphi_0 \downarrow & & \varphi \downarrow & & \\
Q_1 & \xrightarrow{g} & Q_0 & \xrightarrow{g_0} & N & \longrightarrow & 0
\end{array}
$$

Then $g_0 \varphi_0$ is epic; so since g_0 is a superfluous epimorphism, φ_0 is an epimorphism; and φ_0 splits since Q_0 is projective. But $Ker\ \varphi_0 \leq Ker\ \varphi f_0 = Ker\ f_0 \ll P_0$, so φ_0 is an isomorphism. Now $\varphi_0(Ker\ f_0) = Ker\ g_0$; so, by exactness, we can obtain φ_1 in the same manner as we found φ_0. $\square$

The $_RR_R$-dual ( )* and its properties presented in §20 are fundamental components of the transpose. We also need

3.12. Lemma. *If P is a finitely generated projective left R-module and $I \leq_R R$ then, regarding $Hom_R(P, I)$ as a subset of P^*,*

$$P^*I = Hom_R(P, I).$$

Proof. Since P^*, being (finitely generated) projective (20.17), is flat, we have isomorphisms

$$P^*I \cong Hom_R(P, R) \otimes_R I$$

$$\cong Hom_R(P, (R \otimes_R I))$$

$$\cong Hom_R(P, I)$$

by (19, 17), (20.10), and (19.6). The composite of these isomorphisms makes the diagram

$$
\begin{array}{ccc}
P^*I & \xrightarrow{\subseteq} & P^* \\
\downarrow & & \| \\
Hom_R(P, I) & \xrightarrow{i_*} & Hom_R(P, R)
\end{array}
$$

commute. $\square$

If an R-module M has a minimal projective presentation

$$P_1 \xrightarrow{f} P_0 \longrightarrow M \longrightarrow 0,$$

then the *transpose* of M is defined to be $TM = \text{Coker } f^*$ where $f^* = \text{Hom}_R(f, R)$.

32.13. Theorem. *Let R be a semiperfect ring. If M is a left (right) R-module with no non-zero projective direct summands and minimal projective presentation*

$$P_1 \xrightarrow{f} P_0 \longrightarrow M \longrightarrow 0,$$

then the exact sequence

$$P_0^* \xrightarrow{f^*} P_1^* \longrightarrow TM \to 0$$

is a minimal projective presentation of the right (left) R-module TM. Moreover, TM has no non-zero projective direct summands.

Proof. Let $J = J(R)$, suppose that R, M, and the exact sequence

$$P_1 \xrightarrow{f} P_0 \xrightarrow{f_0} M \longrightarrow 0$$

satisfy the hypothesis, and consider the exact sequence

$$P_0 \xrightarrow{f^*} P_1^* \xrightarrow{n} TM \longrightarrow 0,$$

where $TM = P_1^*/\text{Im } f^*$ and n is the natural epimorphism. Then P_0^* and P_1^* are finitely generated and projective by (20.17). If $\gamma \in P_0^*$, then

$$[f^*(\gamma)](P_1) = \gamma(f(P_1)) \le \gamma(JP_0) \le J;$$

so by (32.12)

$$\text{Im } f^* \subseteq P_1^* J.$$

If $\delta \in \text{Ker } f^*$, then $\text{Ker } f_0 = \text{Im } f \subseteq \text{Ker } \delta$, so there is a commutative diagram of R-maps

Thus, since M, and hence $\text{Im } \varphi$, has no non-zero projective direct summands, $\text{Im } \delta = \text{Im } \varphi \subseteq J$ (Exercise (32.11.2)). Now applying (32.12) again we have

$$\text{Ker } f^* \subseteq P_1^* J,$$

so f^* yields a minimal projective presentation of TM.

For the last statement, suppose that $TM = P_1^*/\text{Im } f^*$ has a non-zero projective direct summand. Then $P_1^* = Q \oplus Q'$, where $Q \neq 0$ and $\text{Im } f^* \subseteq Q'$ (Exercise (32.11.1)); it follows that $P_1^{**} = P \oplus P'$, where $P = \{\gamma \in P_1^{**} | \gamma(Q') = 0\} \cong Q^*$ (see (16.3)). But then $0 \neq P^* \subseteq \text{Ker } f^{**}$ since $\gamma \in P$ implies $f^{**}(\gamma) = \gamma \circ f^* = 0$. In view of the reflexivity of P_0 and P_1

(20.17),

$$
\begin{array}{ccc}
P_1^{**} & \xrightarrow{\,f^{**}\,} & P_0^{**} \\[4pt]
\cong \Big\uparrow & & \Big\uparrow \cong \\[4pt]
P_1 & \xrightarrow{\ \ f\ \ } & P_0
\end{array}
$$

this is contrary to $Ker\, f \ll P_1$.   □

If R is semiperfect with basic set of primitive idempotents $e_1, \ldots, e_n$, then $Re_i \mapsto e_i R \cong (Re_i)^*$ is a one-to-one correspondence between the isomorphism types of indecomposable projective left and right R-modules (27.10). The transpose provides a one-to-one correspondence between the remaining isomorphism types of finitely presented indecomposable left and right R-modules. Indeed, if we choose a fixed minimal projective presentation of each finitely presented R-module, then the transpose can be viewed as a pair of mappings described in

32.14. Corollary. *If R is a semiperfect ring, the transpose mappings T, between classes of the finitely presented left and right R-modules without non-zero projective direct summands, satisfy*
   (1) $T0 = 0$;
   (2) $TM \cong TN$ iff $M \cong N$;
   (3) $T(M \oplus N) \cong TM \oplus TN$;
   (4) $TTM \cong M$.

Proof. (1) is clear; (2) follows from Lemma 32.11; (3) holds because the direct sum of pair of minimal projective presentations is one too; (4) is a consequence of (32.11) and the reflexivity of P_1 and P_0 (20.17.1).   □

A left module M over R is called *local* in case $Rad(M)$ is a superfluous maximal submodule of M; equivalently, M is finitely generated and has a unique maximal submodule. Thus, every local module is indecomposable, and if R is semiperfect, then M is local if and only if it is an epimorphic image of an indecomposable projective module—namely, its projective cover.

When he introduced serial rings, Nakayama proved that all of their modules are direct sums of local (hence uniserial) modules (32.3(b)). Conversely, he also proved.

32.15. Theorem. *An artinian ring R is serial if each of its finitely generated indecomposable modules is local.*

Proof. Let $J = J(R)$ and let e and f_i be primitive idempotents in R such that

$$
\bigoplus_{i=1}^{k} Rf_i \to Re \to Re/Je \to 0
$$

is a minimal projective presentation of Re/Je. Then, by (32.13) and (4.7), we have a minimal projective presentation

$$eR \to \bigoplus_{i=1}^{k} f_i R \to T(Re/Je) \to 0,$$

and, by (32.14), $T(Re/Je)$ is indecomposable. Thus, by hypothesis $k = 1$, so $Je/J^2e \cong Rf_1/Jf_1$ is simple. It follows that R is left serial. (Note that we have only used localness of right indecomposable modules so far.) Similarly, R is right serial.   □

In conclusion we note the serial rings also have self-duality as has been shown by Dischinger and Müller [84] and Waschbusch [86]; the proofs are of such a technical nature that we choose not to include them here. Also Warfield [75] has provided an interesting account of non-artinian serial rings.

32. Exercises

1. Prove that if R is a serial ring with Kupisch series $Re_1, \ldots, Re_n$ with $c_i = c(Re_i)$ and $d_i = c(e_i R)$ for $i = 1, \ldots, n$, then:
   (1) $d_1, \ldots, d_n$ is a permutation of $c_1, \ldots, c_n$. [Hint: Try induction on $L(R)$.]
   (2) If $Soc(e_i R) \cong e_j R/e_j J$, then $E(Re_i/Je_i) \cong Re_j/J^{d_i}e_j$.
2. If R is a left artinian ring with basic set of idempotents $e_1, \ldots, e_n$, the *Cartan matrix* of R is $C(R) = [\![c_{ij}]\!]$, where c_{ij} is the number of composition factors in a composition series of Re_j that are isomorphic to Re_i/Je_i. Prove:
   (1) $c_{ij} = c(_{e_i Re_i} e_i Re_j)$.
   (2) $c(Re_j) = \sum_{i=1}^{n} c_{ij}$.
   (3) If R is serial, then $c(e_i R) = \sum_{j=1}^{n} c_{ij}$.
3. Let R be a basic serial ring with Kupisch series $Re_1, \ldots, Re_n$ and $c_i = c(Re_i)$. Prove that a subset $I \subseteq R$ is an ideal iff there are integers $0 \leq b_i \leq c_i$ with $b_i \leq b_{i-1} + 1$ $(i = 2, \ldots, n)$ and $b_1 \leq b_n + 1$ such that $I = \sum_{i=1}^{n} J^{b_i} e_i$.
4. Prove that if R is a serial ring the following are equivalent:
   (a) $Soc(_R R)$ is projective;
   (b) R is left hereditary;
   (c) R is right hereditary.
   [Note: (b) is equivalent to (c) over any perfect or noetherian ring R (see Rotman [79] for example).]
5. Let S be a semiperfect ring, and suppose that $_R I_R \leq J(S)$. Prove that if $S/I \approx R$ (Morita equivalent), then there is a ring $S' \approx S$ with an ideal $I' \leq J(S')$ such that $S'/I' \cong R$. [Hint: Exercise (17.16) and the proof of (17.12) may help.]
6. The $(m_1, \ldots, m_n)$ *block upper triangular matrix ring* over a ring D is the ring of matrices of the form $[\![A_{ij}]\!]$ where A_{ij} is an $m_i \times m_j$ matrix over D and $A_{ij} = 0$ if $i > j$. Prove:

(1) $S \approx \mathbb{UTM}_n(D)$ (Morita equivalent) with D a division ring iff there exist $m_1, \ldots, m_n$ such that S is isomorphic to the $(m_1, \ldots, m_n)$ block upper triangular matrix ring over D.

(2) If R is an indecomposable serial ring with $c_1 = 1$ (respectively, $c_1 = i$ for $i = 1, \ldots, n$), then R is a homomorphic image of (isomorphic to) a block upper triangular ring over D.

7. Let R be a finite-dimensional algebra over a field K. Prove:

   (1) If R is basic, then R is split iff $_K R = Ke_1 \oplus \cdots \oplus Ke_n \oplus J(R)$.

   (2) If K is algebraically closed, then R is split. [Hint: First show that K is the only finite-dimensional division algebra over K.]

8. A finite semigroup $\mathscr{R}$ with 0 such that $\mathscr{R} = \{e_1, \ldots, e_n\} \cup \mathscr{J}$ satisfying $\mathscr{R}\mathscr{J} \subseteq \mathscr{J}, \mathscr{J}\mathscr{R} \subseteq \mathscr{J}, \mathscr{J}^m = \{0\}$, and $e_i e_j = \delta_{ij} e_i$ (i.e., $\mathscr{J}$ is a nilpotent ideal and $e_1, \ldots, e_n$ are orthogonal idempotents $\notin \mathscr{J}$); and $\mathscr{R} = \bigcup_{ij} e_i \mathscr{R} e_j$ is called an *algebra semigroup*.

   (1) Prove that if $\mathscr{R}$ if an algebra semigroup with $R = K\mathscr{R}$, then $J(R) = K\mathscr{J}$ and $Re_i = K\mathscr{R}e_i$ $(i = 1, \ldots, n)$.

   (2) Complete the proof of Theorem 32.10.

9. A serial ring each of whose indecomposable projective modules has only one isomorphism type of composition factors is called a *uniserial ring*. Prove that the following statements about an artinian ring R are equivalent:

   (a) R is uniserial;

   (b) R is a direct sum of serial rings whose Kupisch series each have only one term;

   (c) R is isomorphic to a direct sum matrix rings over local serial rings;

   (d) Every factor ring of R is QF;

   (e) Every left and every right ideal of R is principal.

10. Prove that the lattice of submodules of a left R-module M is a (possibly infinite) chain if and only if for all $x, y \in M$, $Rx \le Ry$ or $Ry \le Rx$.

11. For a projective module $_R P$, prove:

   (1) If $I \le_R M$ and $M/I \cong P \oplus N$, then $M = P' \oplus N'$ with $P' \cong P$ and $I \le N'$.

   (2) If every epimorphic image of P has a projective cover, if $_R M$ has no projective direct summands; and if $\phi : M \to P$, then $\text{Im } \phi \ll P$. [Hint: (17.17) and (1) above.]

12. Use the transpose to show that if R is a left artinian ring with only finitely many isomorphism types of indecomposable finitely generated left modules, then R is right artinian and has the same number of isomorphism types of left and right finitely generated indecomposable modules.

13. Let $e_1, \ldots, e_n$ and $f_1, \ldots, f_m$ be primitive idempotents in a semiperfect ring R and suppose that M has a minimal projective presentation

$$Rf_1 \oplus \cdots \oplus Rf_m \overset{\alpha}{\to} Re_1 \oplus \cdots \oplus Re_n \to M \to 0.$$

Show that, writing direct sums of left (right) modules as row (column) vectors, one can regard α as right multiplication by a matrix $A = [\![a_{ij}]\!]$

with $a_{ij} \in f_i Re_j$; and then

$$
TM = \begin{bmatrix} f_1 R \\ \oplus \\ \vdots \\ \oplus \\ f_m R \end{bmatrix} \Bigg/ A \cdot \begin{bmatrix} e_1 R \\ \oplus \\ \vdots \\ \oplus \\ e_n R \end{bmatrix}
$$

14. Let R be a left artinan ring with $J = J(R)$, basic set of primitive intempotents $e_1, \ldots, e_n$, and simple modules $S_i = Re_i/Je_i$ $(i = 1, \ldots, n)$. Let h_{ij} denote the composition length of the S_i homogeneous component of $Je_j/J^2 e_j$. The *left quiver* of R is a directed graph $Q(_R R)$ with vertex set $\{e_1, \ldots, e_n\}$ and h_{ij} arrows

$$
e_j \xrightarrow{\alpha_k} e_i \quad (k = 1, \ldots, h_{ij}, \; 1 \le i, j \le n).
$$

If R is right artinian then $Q(R_R)$ is defined analogously.

(1) Show that R is indecomposable iff $Q(_R R)$ is connected.

(2) Prove that R is left serial iff at most one arrow exits each vertex.

(3) Let K be a field and let R be the algebra of matrices

$$
\begin{bmatrix}
e & a & x & y & z \\
0 & f & b & c & d \\
0 & 0 & e & 0 & 0 \\
0 & 0 & 0 & g & 0 \\
0 & 0 & 0 & 0 & h
\end{bmatrix}
$$

with entries in K. Calculate $Q(_R R)$ and $Q(R_R)$ to show that R is left but not right serial.

(4) Sketch the quiver of a "typical" indecomposable left serial ring.

15. Associated with a *quiver* (i.e., a finite directed graph), $\mathcal{2}$ is the *path semigroup* $P(\mathcal{2})$, a free semigroup with 0 on the vertices $e_1, \ldots, e_n$ and arrows $e_i \xrightarrow{\alpha} e_j$ of $\mathcal{2}$ subject to the relations

$$
e_i e_j = \delta_{ij} e_i, \qquad e_i \alpha e_j = \begin{cases} \alpha & \text{if } e_j \xrightarrow{\alpha} e_i \\ 0 & \text{otherwise}, \end{cases}
$$

Thus, if we designate $e_1, \ldots, e_n$ directed paths of length 0, we may identify $P(\mathcal{2}) \setminus \{0\}$ with the directed paths in $\mathcal{2}$. (For example, if $e_1 \xrightarrow{\alpha_k} e_2 \xrightarrow{\beta} e_3$ is $\mathcal{2}$, then the $\beta \alpha_k$, represent the only paths of length ≥ 2 in $\mathcal{2}$). If K is a field, we write

$$
K[\mathcal{2}] = KP(\mathcal{2})
$$

and this semigroup algebra is called the *K-path algebra* of $\mathcal{2}$. Let N denote the ideal of $K[\mathcal{2}]$ generated by the paths of length ≥ 2 in $\mathcal{2}$, and prove:

(1) If $I \leq N$ is an ideal and $R = K[\mathcal{Q}]/I$ is finite-dimensional, then $\mathcal{Q}(_R R) \cong \mathcal{Q}$, and R is a split basic algebra with basic set of idempotents $\{e_1 + I, \ldots, e_n + I\}$.

(2) If R is a split basic algebra with $\mathcal{Q} = \mathcal{Q}(_R R)$, then there is an ideal $I \leq N$ such that $K[\mathcal{Q}]/I \cong R$. [Hint: The number of arrows in $\mathcal{Q}$ is the K-dimension of J/J^2 if $J = J(R)$. Now see Exercise (1.15(4)).]

(3) If $\mathcal{Q}$ consists of one arrow and one vertex, then $K[\mathcal{Q}] \cong K[X]$.

Bibliography

Anderson, F. W. [69]: Endomorphism rings of projective modules. *Math Z.* **111**, 322–332 (1969).

Anderson, F. W. and Fuller, K. R. [72]: Modules with decompositions that complement direct summands, *J. Algebra* **22**, 241–253 (1972).

Artin, E., Nesbitt, C. J., and Thrall, R. M. [44]: Rings with minimum condition. Ann Arbor: University of Michigan, 1944.

Auslander, M. [55]: On the dimension of modules and algebras (III), global dimensions. *Nagoya Math. J.* **9**, 66–77 (1955).

Auslander M. and Bridger, M. [69]: *Stable Module Theory.* Memoirs Amer. Math. Soc., Vol 94, Providence, RI: American Mathematical Society, 1969.

Azumaya, G. [50]: Corrections and supplementaries to my paper concerning Remak-Krull-Schmidt's theorem. *Nagoya Math. J.* **1**, 117–124 (1950).

—— [59]: A duality theory for injective modules. *Amer. J. Math.* **81**, 249–278 (1959).

—— [66]: Completely faithful modules and self injective rings. *Nagoya Math. J.* **27**, 697–708 (1966).

Bass, H. [60]: Finitistic dimension and a homological generalization of semi-primary rings. *Trans. Amer. Math. Soc.* **95**, 466–488 (1960).

—— [62]: The Morita theorems. *Lecture Notes.* University of Oregon, Eugene, 1962.

—— [68]: Algebraic *K*-theory. New York: Benjamin, 1968.

Beachy, J. A. [71a]: On quasi-artinian rings. *J. London Math. Soc.* (2) **3**, 499–452 (1971).

—— [71b]: Generating and cogenerating structures. *Trans. Amer. Math. Soc.* **158**, 75–82 (1971).

Behrens, E. A. [72]: *Ring theory*. New York: Academic Press, 1972.

Bergman, G. M. [64]: A ring primitive on the right but not on the left. *Proc. Amer. Math. Soc.* **15**, 473–475 (1964).

Birckhoff, G. [66]: Lattice theory. *Amer. Math. Soc. Colloq. Publ.*, Vol. 25, 3rd ed. Providence, RI: American Mathematical Society, 1966.

Bourbaki, N. [58]: *Algèbre*. Chapter 8 (Fasc. 23). Paris: Hermann & Cie, 1958.

—— [61]: *Algèbre commutative*. Chapters 1 and 2 (Fasc. 27). Paris: Hermann & Cie, 1961.

Burgess, W. D., Fuller, K. R., Voss, E., and Zimmermann-Huisgen, B. [85]: The Cartan matrix as an indicator of finite global dimension for artinian rings, *Proc. Amer. Math. Soc.* **95**, 157–165 (1985).

Cartan, H. and Eilenberg, S. [56]: *Homological Algebra*. Princeton, NJ: Princeton University Press, 1956.

Chase, S. U. [60]: Direct products of modules. *Trans. Amer. Math. Soc.* **97**, 457–473 (1960).

Cohn, P. M. [61]: Quadratic extensions of skew fields. *Proc. London Math. Soc.* (3) **11**, 531–556 (1961).

—— [66a]: Some remarks on the invariant basis property. *Topology*, Vol. 5, pp. 215–228. Oxford-New York-London-Paris: Permagon Press, 1966.

—— [66b]: Morita equivalence and duality. *Mathematical Notes*. Queen Mary College, University of London, 1966.

Colby, R. R. and Rutter, E. A., Jr. [68]: Semiprimary QF-3 rings. *Nagoya Math. J.* **32**, 253–258 (1968).

Cozzens, J. H. [70]: Homological properties of differential polynomials. *Bull. Amer. Math. Soc.* **76**, 75–79 (1970).

Crawley P. and Jønsson, B. [64]: Refinements for infinite direct decompositions of algebraic systems. *Pacific J. Math.* **14**, 797–855 (1964).

Curtis, C. W. and Reiner, I. [62]: *Representation theory of finite groups and associative algebras*. New York-London: Interscience, 1962.

Dickson, S. E. [69]: On algebras of finite representation type. *Trans. Amer. Math. Soc.* **135**, 127–141 (1969).

Dischinger, F. and Müller, W. [84]: Einreihig zerlegbare artinsch Ringe sind selbsdual. *Arch. Math.* **43**, 132–136 (1984).

Dlab, V. [70]: Structure of prefect rings. *Bull. Austral. Math. Soc.* **2**, 117–124 (1970).

Dlab, V. and Ringel, C. M. [76]: *Indecomposable Representations of Graphs and Algebras*, Memoirs Amer. Math. Soc., Vol. 173. Providence, RI: American Mathematical Society 1976.

Eisenbud, D. and Griffith, P. A. [71]: Serial rings. *J. Algebra* **17**, 389–400 (1971).

Erkes, G. L. [75]: Rings of equivalent dominant and codominant dimension, *Proc. Amer. Math. Soc.* **48**, 297–306 (1975).

Faith, C. [66]: Rings with ascending condition on annihilators. *Nagoya Math. J.* **27**, 179–191 (1966).

—— [67a]: Lectures on injective modules and quotient rings. *Lecture Notes in Mathematics*, Vol. 49, Berlin-Heidelberg-New York: Springer-Verlag, 1967.

—— [67b]: A general Wedderburn theorem. *Bull. Amer. Math. Soc.* **73**, 65–67 (1967).

—— [73]: Modules finite over endomorphism rings. *Lecture Notes in Mathematics*, vol. 246, pp. 145–189. Berlin-New York: Springer-Verlag, 1972.

—— [76]: *Algebra II. Ring Theory*. Berlin-New York: Springer-Verlag, 1976.

Faith, C. and Walker, E. A. [67]: Direct-Sum representations of injective modules. *J. Algebra* **5**, 203–221 (1967).

Faith, C. and Utumi, Y. [64]: Quasi-injective modules and their endomorphism rings. *Arch. Math.* **15**, 166–174 (1964).

Fisher, J. W. [72]: Nil subrings of endomorphism rings of modules. *Proc. Amer. Math. Soc.* **34**, 75–78 (1972).

Fuller, K. R. [68a]: Generalized uniserial rings and their Kupisch series. *Math. Z.* **106**, 248–260 (1968).

—— [68b]: The structure of QF-3 rings. *Trans. Amer. Math. Soc.* **134**, 343–354 (1968).

—— [69]: On indecomposable injectives over artinian rings. *Pacific J. Math.* **29**, 248–260 (1969).

—— [70]: Double centralizers of injectives and projectives over artinian rings. *Illinois J. Math.* **14**, 658–664 (1970).

—— [72]: Relative projectivity and injectivity classes determined by simple modules. *London J. Math.* **5**, 423–431 (1972).

—— [87]: Algebras from diagrams. *J. Pure Appl. Alg.* **48**, 23–37 (1987).

—— [89]: *Artinian Rings*, Notas de Matematica 2, Universidad de Murcia, Murcia 1989.

Fuller, K. R. and Hill, D. A. [70]: On quasi-projective modules via relative projectivity. *Arch. Math.* **21**, 369–373 (1970).

Gabriel, P. [80]:, Auslander-Reiten sequences and representation-finite algebras. *Lecture Notes in Mathematics*, Vol. 831, pp. 1–71. Berlin-Heidelberg-New York: Springer-Verlag, 1980.

Goldie, A. W. and Small, L. W. [73]: A note on rings of endomorphisms. *J. Algebra* **24**, 392–395 (1973).

Gordon, R. [69]: Rings in which minimal left ideals are projective. *Pacific J. Math.* **31**, 679–692 (1969).

—— [71]: Rings defined by R-sets and a characterization of a class of semi-perfect rings. *Trans. Amer. Math. Soc.* **155**, 1–17 (1971).

Jacobson, N. [43]: The theory of rings. *Amer. Math. Soc. Surveys*, Vol. 2. Providence, RI: American Mathematical Society, 1943.

—— [64]: Structure of rings. *Amer. Math. Soc. Colloq. Publ.*, Vol. 37 (rev. ed). Providence, RI: American Mathematical Society, 1964.

Jans, J. P. [59]: Projective injective modules. *Pacific J. Math.* **9**, 1103–1108 (1959).

—— [64]: Rings and homology. New York-Chicago-San Francisco-Toronto-London: Holt, Rinehart & Winston, 1964.

—— [69]: On co-noetherian rings. *J. London Math. Soc.* (2) **1**, 588–590 (1969).

Kaplansky, I. [58]: Projective modules. *Ann. Math.* **68**, 372–377 (1958).

—— [69]: *Infinite abelian groups* (rev. ed.). Ann Arbor MI: University of Michigan Press, 1969.

Kupisch, H. [59]: Beiträge zur Theorie nichthalbeinfacher Ringe mit Minimalbedingung. *J. Reine Angew. Math.* **201**, 100–112 (1959).

Lambek, J. [66]: *Lectures on rings and modules.* Waltham-Toronto-London: Blaisdell, 1966.

MacLane, S. [63]: *Homology.* Berlin-Göttingen-Heidelberg: Springer-Verlag, 1963.

—— [71]: *Categories for the Working Mathematician.* New York-Heidelberg-Berlin: Springer-Verlag, 1971.

Matlis, E. [58]: Injective modules over noetherian rings. *Pacific J. Math.* **8**, 511–528 (1958).

Morita, K. [58a]: Duality of modules and its applications to the theory of rings with minimum condition. *Sci. Rep. Tokyo Kyoiku Daigaku, Sect. A* **6**, 85–142 (1958).

—— [58b]: On algebras for which every faithful representation is its own second commutator. *Math. Z.* **69**, 429–434 (1958).

—— [65]: Adjoint pairs of functors and Frobenius extensions. *Sci. Rep. Tokyo Kyoiku Daigaku, Sect A* **9**, 40–71 (1965).

Müller, B. J. [69]: On Morita duality. *Can. J. Math.* **21**, 1338–1347 (1969).

—— [70]: On semi-perfect rings. *Illinois J. Math.* **14**, 464–467 (1970).

Murase, I. [63]: On the structure of generalized uniserial rings, I. *Sci. Papers College Gen. Educ. Tokyo* **13**, 1–22 (1963); II. **13**, 131–158 (1963); III. **14**, 11–25 (1964).

Nakayama, T. [40]: Note on uniserial and generalized uniserial rings. *Proc Imperial Acad. Japan* **16**, 285–289 (1940).

—— [41]: On Frobeniusean algebras, II. *Ann. Math.* **42**, 1–21 (1941).

Osofsky, B. L. [66]: A generalization of quasi-Frobenius rings. *J. Algebra* **4**, 272–287 (1966).

Onodera, T. [68]: Uber Kogeneratoren, *Arch. Math.* **16**, 402–410 (1968).

Robert, E. de [69]: Projectifs et injectifs. *C. R. Acad. Sci. Paris Sér. A–B* **286**, 361–364 (1969).

Rosenberg, A. and Zelinsky, D. [59]: Finiteness of the injective hull. *Math. Z.* 70, 372–380 (1959).

—— [61]: Annihilators. *Portugaliae Math.* **20**, 53–65 (1961).

Rotman, J. J. [79]: *An introduction to homological algebra.* New York-San Francisco-London: Academic Press, 1979.

Sandomierski, F. L. [64]: Relative injectivity and projectivity. Ph.D. Thesis, Penn State University, 1964.

—— [69]: On semiperfect and perfect rings. *Proc. Amer. Math. Soc.* **21**, 205–207 (1969).

Schofield, A. H. [85]: Representations of rings over skewfields. *London Math. Soc. Lect. Note Ser.*, Vol. 92. Cambridge; Cambridge University Press, 1985.

Stoll, R. R. [63]: *Set theory and logic.* San Francisco-London: W. H. Freeman & Co., 1963.

Tachikawa, H. [73]: Quasi-Frobenius rings and generalizations. QF-3 and QF-1 rings. *Lecture Notes in* Mathematics, Vol. 351. Berlin-New York: Springer-Verlag, 1973.

—— [74]: QF-3 rings and categories of projective modules. *J. Algebra* **28**, 408–413 (1974).

Vámos, P. [68]: The dual notion of "finitely generated". *J. London Math. Soc.* **43**, 643–647 (1968).

Walker, C. P. [66]: Relative homological algebra and abelian groups. *Illinois J. Math.* **10**, 186–209 (1966).

Ware, R. [71]: Endomorphism rings of projective modules. *Trans. Amer. Math. Soc.* **155**, 233–256 (1971).

Warfield, R. B., Jr. [69a]: Decompositions of injective modules. *Pacific J. Math.* **31**, 263–276 (1969).

—— [69b]: A Krull-Schmidt theorem for infinite sums of modules. *Proc. Amer. Math. Soc.* **22**, 460–465 (1969).

—— [75]: Serial rings and finitely presented modules. *J. Algebra* 37, 187–222 (1975).

Waschbüsch, J. [86]: Self-duality of serial rings. *Comm. Algebra* 14, 581–589 (1986).

Index

Graduate Texts in Mathematics

continued from page ii